ATMOSPHERIC ACIDITY
Sources, Consequences and Abatement

Environmental Management Series

Edited by

J. CAIRNS JNR
University Center for Environmental and Hazardous Materials Studies, Virginia Polytechnic Institute and State University, Blacksburg, Virginia 24061-0415, USA

R. M. HARRISON
Institute of Public and Environmental Health, School of Biological Sciences, University of Birmingham, Edgbaston, Birmingham, B15 2TT, UK

Titles published in the *Environmental Management Series*

Instrumental Analysis of Pollutants
C. N. Hewitt

Pollution of the Arctic Atmosphere
W. T. Sturges

Organic Contaminants in the Environment
K. C. Jones

Methods of Environmental Data Analysis
C. N. Hewitt

Atmospheric Acidity: Sources, Consequences and Abatement
M. Radojević and R. M. Harrison

Titles published in the *Pollution Monitoring Series*

Quantitative Aquatic Biological Indicators
D. J. H. Phillips

Biomonitoring Air Pollutants with Plants
W. J. Manning and W. A. Feder

Effect of Heavy Metal Pollution on Plants 1 & 2
N. W. Lepp

Biological Monitoring of Heavy Metal Pollution
M. H. Martin and P. J. Coughtrey

Biological Indicators of Freshwater Pollution and Environmental Management
J. M. Hellawell

Pollution Threat of Heavy Metals in Aquatic Environments
G. Mance

Ecophysiology of Metals in Terrestrial Invertebrates
S. P. Hopkin

Ecology Effects of Thermal Discharges
T. E. L. Langford

ATMOSPHERIC ACIDITY
Sources, Consequences and Abatement

Edited by

M. RADOJEVIĆ
Department of Chemistry, University of Brunei Darussalam,
Bandar Seri Begawan 3186, Brunei Darussalam

and

R. M. HARRISON
Institute of Public and Environmental Health,
School of Biological Sciences, University of Birmingham,
Edgbaston, Birmingham, B15 2TT, UK

ELSEVIER APPLIED SCIENCE
LONDON and NEW YORK

ELSEVIER SCIENCE PUBLISHERS LTD
Crown House, Linton Road, Barking, Essex IG11 8JU, England

WITH 60 TABLES AND 91 ILLUSTRATIONS

British Library Cataloguing in Publication Data

Atmospheric Acidity: Sources, Consequences and Abatement.—(Environmental Management Series)
I. Radojević, M. II. Harrison, Roy M.
III. Series
363.73

ISBN 1-85166-777-6

Library of Congress Cataloging-in-Publication Data

Atmospheric acidity: sources, consequences, and abatement/edited by M. Radojević and R. M. Harrison.
p. cm.—(Environmental management series)
Includes bibliographical references and index.
ISBN 1-85166-777-6
1. Acid deposition—Environmental aspects. 2. Acid rain—Environmental aspects. 3. Air quality management. I. Radojević, M. (Miroslav). II. Harrison, Roy M., 1948— . III. Series.
TD195.42.A76 1992
363.73'86—dc20 92–6701
CIP

Photoset by Enset (Photosetting) Ltd, Bath.
Printed in Great Britain at the University Press, Cambridge.

Foreword

Environmental Management Series

The current expansion of both public and scientific interest in environmental issues has not been accompanied by a commensurate production of adequate books, and those which are available are widely variable in approach and depth.

The *Environmental Management Series* has been established with a view to co-ordinating a series of volumes dealing with each topic within the field in some depth. It is hoped that this Series will provide a uniform and quality coverage and that, over a period of years, it will build up to form a library of reference books covering most of the major topics within this diverse field. It is envisaged that the books will be of single, or dual authorship, or edited volumes as appropriate for respective topics.

The level of presentation will be advanced, the books being aimed primarily at a research/consultancy readership. The coverage will include all aspects of environmental science and engineering pertinent to management and monitoring of the natural and man-modified environment, as well as topics dealing with the political, economic, legal and social considerations pertaining to environmental management.

J. CAIRNS and R. M. HARRISON

Preface

Environmental problems associated with the atmosphere were recognised first in the form of local pollution, most notably urban smogs (smoke/SO_2 or photochemical type) and emissions from specific industrial works (e.g. alkali works or metal smelters). Only many years later were regional and then global problems recognised and given the research priority which they deserved. Probably the first of the regional pollution phenomena to gain widespread recognition was that of 'acid rain' but although the problem was becoming clear in the 1960s and early 1970s, it was not until the 1980s that its magnitude was acknowledged and a substantial research effort commenced.

Many distinct phenomena and problems became associated with the term 'acid rain', some having no relation to rainwater of low pH, some of which were processes not involving acids at all. The earliest problems to be recognised related to acidification of freshwaters and decay of historic buildings. Later, the acidification of soils and most prominently the problem of forest decline came to the fore. The latter is a highly complex problem, still by no means fully understood and involving a large range of stress factors, both pollutant and natural. Nonetheless, in the public mind all of these problems were associated with 'acid rain'.

In this book we deal primarily with acidic pollutants and their precursors, the consequences of their presence in, and removal from, the atmosphere and the means of their abatement. We do not include, other than in passing, pollutants such as tropospheric ozone which have been associated with 'acid rain' problems but are not themselves acids or direct acid precursors. We cover the basic chemistry and physics of the acids, the reactions responsible for their production, the processes involved in their atmospheric transport and deposition, measurement techniques, field studies, health and ecological effects and climatic impacts of acidity and strategies for control in terms of engineering technologies, legislation, economics and policy.

The book covers a breadth which no single author could cover authoritatively. We have therefore produced an edited volume. We have been especially fortunate in engaging as chapter authors internationally recognised experts in the various disciplines involved in understanding this complex multidisciplinary subject. The outcome is, we believe, a unique and authoritative treatment of this major environmental problem which takes the reader all the way from a definition of atmospheric acidity and the sources of airborne acids through the atmospheric processes, the human and environmental consequences to the means of control. It will be many decades before atmospheric acidity is understood in all aspects, and many more until it is controlled; this book aims to express the current state of knowledge.

We are grateful to the many authors for their most excellent contributions and their willingness to respond to editorial quibbles. We feel that they have provided a uniquely valuable compilation of knowledge on atmospheric acidity and hope that it will prove as enlightening to the reader as it has been to us during its preparation.

MIROSLAV RADOJEVIĆ
Leeds, UK
and
ROY M. HARRISON
University of Birmingham, UK

Contents

List of Contributors

P. BRIMBLECOMBE
School of Environmental Sciences, University of East Anglia, Norwich, NR4 7TJ, UK

R. N. BUTLIN
Building Research Establishment, Garston, Watford, Hertfordshire, WD2 7JR, UK

T. W. CHOULARTON
Department of Pure and Applied Physics, UMIST, PO Box 88, Manchester, M60 1QD, UK

A. G. CLARKE
Department of Fuel and Energy, University of Leeds, Leeds, LS2 9JT, UK

T. D. CROCKER
College of Business, University of Wyoming, Laramie, WY 82071, USA

R. M. DOWNER
Department of Chemistry, UMIST, PO Box 88, Manchester, M60 1QD, UK

B. E. A. FISHER
National Power, Senator House, 85 Queen Victoria Street, London, EC4V 4DP, UK

B. A. FORSTER
Department of Economics and Finance, University of Wyoming, Laramie, WY 82071, USA

D. FOWLER
Institute of Terrestrial Ecology, Bush Estate, Penicuik, Midlothian, EH26 0QB, UK

B. B. HICKS
Atmospheric Turbulence and Diffusion Division, NOAA/ARL, PO Box 2456, Oak Ridge, TN 37831, USA

H. HORVATH
Institute of Experimental Physics, University of Vienna, Dr Karl Lueger-Ring 1, 1010 Vienna, Austria

D. J. HUGHES
Faculty of Law, University of Leicester, Leicester, LE1 7RH, UK

A. IRWIN
Department of Human Science, Brunel University, Uxbridge, UB8 3PH, UK

D. J. JACOB
Department of Earth and Planetary Sciences, Division of Applied Science, Harvard University, 29 Oxford Street, Cambridge, MA 02138, USA

M. LIPPMANN
New York University Medical Center, Nelson Institute of Environmental Medicine, Long Meadow Road, Tuxedo, NY 10987, USA

G. O. LLOYD
Division of Materials Metrology, National Physical Laboratory, Teddington, Middlesex, TW11 0LW, UK

E. MÉSZÁROS
Institute for Atmospheric Physics, PO Box 39, H-1675 Budapest, Hungary

J. W. MUNGER
Department of Earth and Planetary Sciences, Division of Applied Science, Harvard University, 29 Oxford Street, Cambridge, MA 02138, USA

S. J. ORMEROD
Catchment Research Group, University of Wales College of Cardiff, PO Box 915, Cardiff, CF1 3TL, UK

M. RADOJEVIĆ
Department of Chemistry, University of Brunei, Darussalam, Bandar Seri Begawan 3186, Brunei Darussalam

J. M. WALDMAN
Department of Environmental and Community Medicine, UMDNJ—Robert Wood Johnson Medical School, 675 Hoes Lane, Piscataway, NJ 08854, USA

Chapter 1

Occurrence of Atmospheric Acidity

E. MÉSZÁROS

Institute for Atmospheric Physics, PO Box 39, H-1675 Budapest, Hungary

1.1 DEFINITION

According to the definition of Brönsted an *acid* (HA) is a substance which produces hydrogen ions (H^+, i.e. protons). Thus, if an acid is dissolved in water the following dissociation takes place:[1]

$$HA \rightleftarrows H^+ + A^- \tag{1.1}$$

where A^- is a negative ion. The process can be characterized by the dissociation constant (K_d) which is given by the following equation:

$$K_d = \frac{[H^+][A^-]}{[HA]} \tag{1.2}$$

where square brackets denote molar concentrations.[2] On the basis of eqn (1.2) strong acids can be defined as those with dissociation constants higher than 10^{-3} mol liter^{-1} (e.g. sulfuric (H_2SO_4), nitric (HNO_3) and hydrochloric acid (HCl)) and weak acids with low dissociation capability (such as formic (HCOOH) or acetic acid (CH_3COOH)).

On the other hand, any species, e.g. A^- in eqn (1.1), able to take up protons is called a *base*. If, in water, a base is dissolved, hydroxyl ions (OH^-) are formed because either protons are taken from water

Present address: University of Veszprém, H-8201 Veszprém, Egyetem, u.10.POB, 158.

molecules (e.g. ammonia, carbonate ions (CO_3^{2-})) or the base contains OH^- ions (e.g. different hydroxides). Since OH^- and H^+ combine easily, a solution is neutral and the molar concentration of H^+ is equal to $[OH^-]$. Consequently, for $[H^+] > [OH^-]$ the solution is called acidic, otherwise it is basic or alkaline. Since $[H^+]$ is low and can vary within several orders of magnitude, it is generally described by the negative logarithm of the hydrogen ion concentration (pH):

$$pH = -\log_{10}[H^+] \tag{1.3}$$

It should be noted that pure water also dissociates slightly to form H^+ and OH^- ions. At a temperature of 25°C

$$[H^+] = [OH^-] \sim 10^{-7} \text{ mol liter}^{-1}$$

which is equivalent to a pH value of 7, termed the *neutral point*. It follows from eqn (1.3) that one pH unit change is equivalent to a 10-fold change in H^+ ion concentration.

The above discussion implies that pH does not represent the *total acidity* of a solution; it characterizes only the concentration of acids in dissociated form. This part of acids is called the *free acidity*.

These relatively simple terms become more complicated in the case of atmospheric waters (clouds, fogs and precipitation). For example, if asked what does 'acid rain' exactly mean, the reply is not self-evident.

The first complication is due to the presence of carbon dioxide (CO_2) in the air. Hydrolysis of CO_2 molecules, absorbed by cloud-drops, fogdrops and raindrops yields bicarbonate (HCO_3^-) and subsequently CO_3^{2-} ions. Since during both dissociations H^+ ions are generated; at equilibrium:

$$[H^+] = [OH^-]+[HCO_3^-]+2[CO_3^{2-}]$$

By using appropriate physico-chemical data and atmospheric CO_2 concentrations, the equilibrium $[H^+]$ can be calculated. The result is a concentration equivalent to a pH value of 5·6 at a temperature of 10°C. This pH value is considered as the *atmospheric neutral point*. This means that atmospheric waters are acidic if their pH is lower than the above value.

However, the situation is further complicated by the fact that even in clean air, free from human influences, there are acid aerosol particles and gases which modify the pH of atmospheric waters. It can be demonstrated[3] that atmospheric sulfur compounds of biogenic origin can lower the pH to 4·5. This result is supported by direct atmospheric observations indicating that in remote areas of the world the pH of precipitation waters is generally lower than 5·6 (see Fig. 1.1). Thus, it can be concluded that the pH value of 5·6 cannot be considered as a reference for unpolluted air and *acidity* may be of *natural origin*. This acidity of natural origin is now modified by anthropogenic emissions of acid-forming gases which increase the H^+ ion concentration above the natural level and

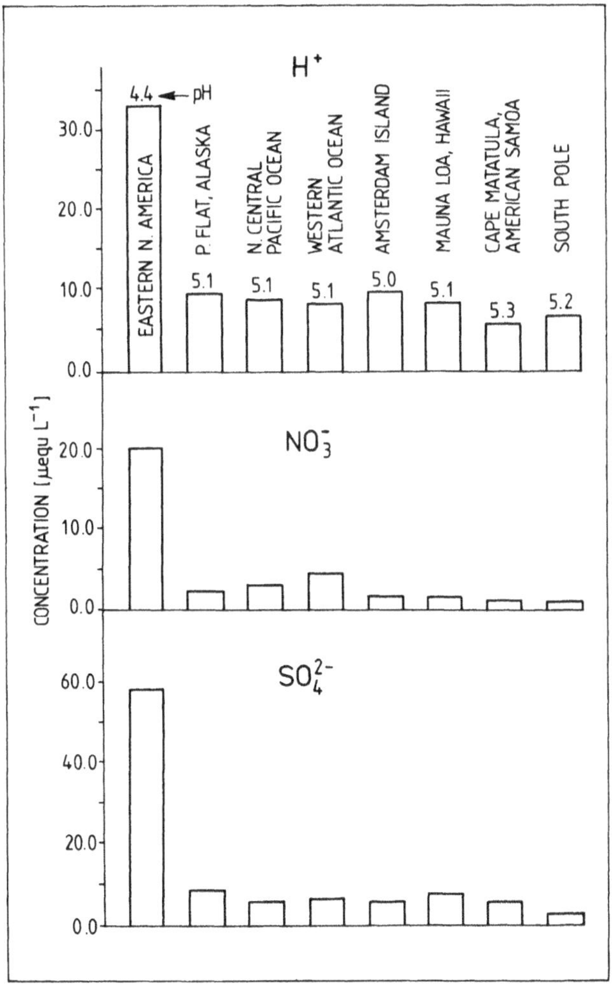

Figure 1.1. Acidity and composition of precipitation water under (or at) different conditions (or sites). (Reproduced with permission from Ref. 4.)

consequently lead to the formation of *acid rains* (more properly, acid depositions (see Section 1.3.3)) *of anthropogenic origin*. Unfortunately the distinction between natural (biogenic) and anthropogenic acidities is not always simple. The upper part of Fig. 1.1 gives an idea about free acidities observed under different conditions.[4]

It can be seen that in remote areas the pH is around 5·0. Thus, one can propose with caution that this value represents at first approximation the natural average background free acidity level. The figure also shows that this pH is lowered in polluted air (north-eastern America) by nearly one order of magnitude.

1.2 NATURE OF ATMOSPHERIC ACIDITY

Atmospheric observations carried out in different areas of the world show that SO_4^{2-}, nitrate (NO_3^-) and chloride (Cl^-) are the most important anions in precipitation. As an illustration, typical data obtained in southern Sweden in the 1970s[5] are presented in Table 1.1. For evaluating the anion concentrations one must take into account that on a larger scale the major part of Cl^- ions comes from sea salt containing sodium chloride. If it is supposed that the sodium is entirely of marine origin, from the tabulated data a continental (anthropogenic?) Cl^- concentration of 3 μeq liter^{-1} is calculated. Since this concentration can be neglected relative to SO_4^{2-} and NO_3^- levels, it can be concluded that under continental conditions acidity is caused mostly by H_2SO_4 and HNO_3. However, near the sources, e.g. power plants, the role of Cl^- from HCl may be locally important.[6]

Table 1.1
Average composition of precipitation water collected in southern Sweden

Cations	*Concentration (μeq liter^{-1})*	*Anions*	*Concentration (μeq liter^{-1})*
H^+	52	SO_4^{2-}	70
Ca^{2+}	14	NO_3^-	31
Mg^{2+}	8	Cl^-	18
K^+	3		
Na^+	15		
NH_4^+	31		
Total	123		119

Reproduced with permission from Ref. 5.

It is well documented that H_2SO_4 and HNO_3 form in the air and/or in atmospheric waters from precursor gases emitted by natural and anthropogenic sources. The most essential precursor gases are sulfur dioxide (SO_2) and nitrogen oxides, while in oceanic areas dimethyl sulfide (DMS) plays an important role in the control of acidity. It can further be supposed that over the continents hydrogen sulfide (H_2S) has a minor influence on atmospheric acidification. In the air these precursor gases are converted chemically into acid vapors which can condense to form acidic aerosol particles.

The precursor gases as well as the acids in vapor and aerosol phases are incorporated into atmospheric waters and determine, together with basic materials (see later), the pH of fog, cloud and precipitation elements. In the north-eastern US (see Fig. 1.1) the contribution of SO_4^{2-} ions to the acidity of precipitation is three times more important than that of NO_3^- ions. This tendency can also be observed under less-polluted conditions. The situation in Europe is very similar, as illustrated by Fig. 1.2. This figure represents the equivalent ratios of nitrate to non-marine (excess) SO_4.[7] It can be seen that while in western Europe the ratio is around 0·5 (close to the value for Sweden, see Table 1.1), it is equal to 0·2–0·3 over the eastern part of the continent. Thus, it can be concluded that under

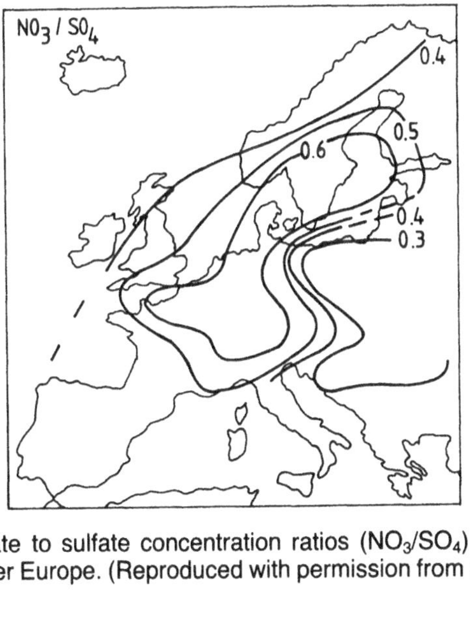

Figure 1.2. Nitrate to sulfate concentration ratios (NO_3/SO_4) in precipitation water over Europe. (Reproduced with permission from Ref. 7.)

continental conditions *the free acidity of atmospheric precipitation is mainly due to H_2SO_4* formed from SO_2.

The only gaseous neutralizing substance in the atmosphere is NH_3 originating from animal urea and from the decomposition of nitrogenous organic matters in different ecosystems. Since NH_3 can neutralize H_2SO_4 and HNO_3 either in aerosol particles or in cloud, fog and precipitation waters,[8] NH_4^+ is an important cation in both aerosol and aqueous phase[9] (see also Table 1.1). Under continental conditions wind-blown dust particles also provide alkaline materials. It has been speculated that, besides SO_4^{2-}, NO_3^- and NH_3, calcium-containing soil particles play an important role in the control of the acidity of European precipitation.[10] On the other hand, results of the analyses of precipitation samples collected under maritime conditions[11] suggest that neutralization of acids can also be caused by marine-derived bases (sea salt).

Finally, it was found that in remote areas anion and cation concentrations in precipitation may not be balanced if only inorganic ions are analyzed.[12] It has been proposed that this is due to organic acids present in the water.[13] Further studies revealed that organic acids play a minor role in the control of precipitation acidity under polluted conditions (e.g. in the north-eastern US[13]). However, it is reported that in clean areas, mostly *in the tropics, organic acids* contribute an important part of free and total acidities of rainwater.[14]

According to measurements carried out in the Amazon boundary layer, most of the precipitation-free acidity is caused by organic acids, mainly by HCOOH and CH_3COOH.[15] On the other hand, H_2SO_4 is responsible only for 10–20% of the H^+ ion concentration. Although a part of the organic acid can form through the aqueous phase oxidation of aldehydes,[16] it is more probable that organic acid content of precipitation is caused by the scavenging of gaseous HCOOH and CH_3COOH from the air. These gaseous species can be formed by the photo-oxidation of isoprene and terpenes.[17] It cannot be ruled out, however, that they are directly emitted by the vegetation.[18]

1.3 CONCENTRATIONS OF ACIDIC COMPOUNDS AND THEIR PRECURSORS IN THE ATMOSPHERE

The formation of acidic substances in the atmosphere is a complex process. One can find in the air not only acidic and alkaline compounds, but also a lot of precursor gases and other species which take part in the chemical reactions leading to acid formation. Table 1.2 gives a survey of the

Table 1.2
Major substances involved in the control of atmospheric acidity

Species	*Notes*
Precursors	
SO_2	Most important anthropogenic sulfur gas
$(CH_3)_2S$, H_2S	Most important acid-forming sulfur gases of biogenic origin
NO	Primary nitrogen oxide emitted by natural and anthropogenic sources
NO_2	Formed in the air from NO by chemical transformation
Isoprene, terpenes	Important for organic acid formation
Acids	
H_2SO_4	Identified in aerosol phase, substantially neutralized by NH_3
HNO_3	Identified mostly in gas phase, can be converted to NH_4NO_3 in aerosol phase
HCl	Directly emitted by human activities. Also formed from oceanic methyl chloride
Organic acids	Important in acidity control in clean areas
Basic substrates	
NH_3	Biogenic and anthropogenic (agriculture) product
Alkaline particles	Formed by mechanical disintegration of soils and rocks
Sea salt	A part of sea salt is converted in the air into gaseous HCl
Species involved in acid formation	
Volatile organic compounds	Important for NO–NO_2 transformation
O_3	Formed by oxygen atoms produced by the photo-dissociation of NO_2. Important for free-radical formation and SO_2 oxidation in aqueous phase ($pH > 5$)
OH	Free radical, responsible for SO_2, $(CH_3)_2S$, H_2S and NO_2 oxidation in gas phase
HO_2	Free radical, important for H_2O_2 formation
H_2O_2	Responsible for SO_2 oxidation in aqueous phase ($pH < 5$)
Metal ions (Fe, Mg, etc.)	Important for catalytic oxidation of SO_2 in aqucous phasc

major substances involved in the control of atmospheric acidity. It can be seen that their number is rather large. However, for the sake of simplicity only the concentration and distribution of the precursor gases and acidic substances as well as the atmospheric level of NH_3 will be discussed in this section. While NH_3 is an important alkaline substance in the air, NH_4^+ ions can raise considerably the acidity of soils (see above). Concerning the role and concentration of other substances in the table the interested readers are referred to special textbooks on atmospheric chemistry.[19,20]

1.3.1 Gas-phase compounds

It has been seen above that under continental conditions *sulfur and nitrogen oxides* play an important role in the control of the acidity of the air and precipitation. These gases are emitted into the air by burning sulfur- and nitrogen-containing fossil fuels, while nitrogen oxide (NO) can also be formed from atmospheric oxygen and nitrogen at high temperatures during combustion. Several observations have been carried out to determine the atmospheric concentrations of sulfur and nitrogen oxides mostly in Europe and North America. As an example, Fig. 1.3 illustrates the spatial distribution of SO_2-S and NO_2-N in Europe on the basis of the measurements carried out in different monitoring programs between 1978 and 1982.[21] It can be seen that maximum concentrations were detected in the air over the central, more populated parts of the continent. However, there is some difference between the distributions of SO_2 and nitrogen dioxide (NO_2) concentrations. While the SO_2 level is rather high even in eastern areas, NO_2 concentration maxima can be found over western regions of Europe. It is expected that this is explained by the differences in energy production and traffic density. It should be noted that in North America similar concentrations were measured. Thus, according to Ref. 22, the average concentration obtained at non-urban sites of the eastern US is equal to 150 nmol m^{-3}, which is very close to the European average (130 nmol m^{-3}). Another review shows that the NO and NO_2 non-urban concentrations in the US east of the Rocky Mountains range from 1 to 10 ppb (10^{-9} by vol.),[23] which are equivalent to 45 and 450 nmol m^{-3}, respectively; values which are close to those given in Fig. 1.3.

It goes without saying that SO_2 and NO_2 concentrations in locally polluted urban areas, where sources are concentrated, are much higher than the values obtained under regional (non-urban, countryside air) and remote continental conditions. Thus, the median SO_2 level measured in different cities in the network of the World Health Organization is 700 nmol m^{-3}.[24] In the center of Budapest, Hungary, the concentration of SO_2 is 10

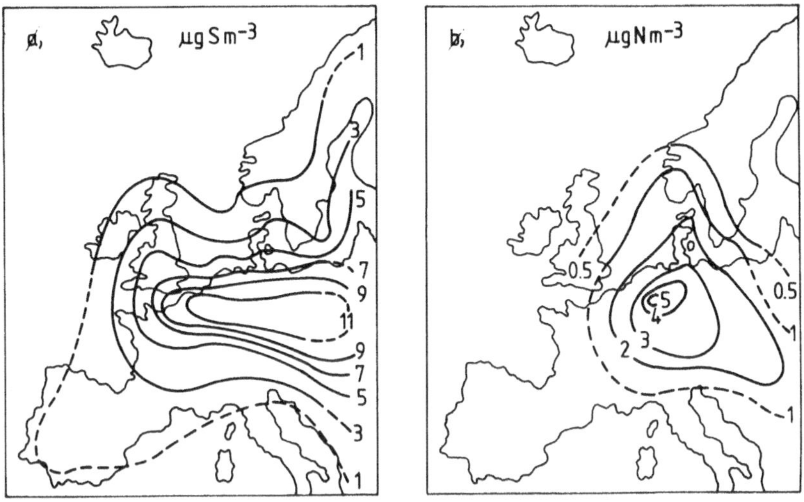

Figure 1.3. Spatial distribution of the concentration of (a) SO_2-S and (b) NO_2-N in the surface air over Europe.

times higher on average than at a regional site,[25] while the corresponding ratio for NO_2 is five. This difference between SO_2 and NO_2 is probably caused by the more uniform source distribution of nitrogen oxides (e.g. traffic). Observations also demonstrate that, owing to the annual variations of source intensities and meteorological conditions, SO_2 and NO_2 concentrations have a maximum during the winter and a minimum in summer.

Over the oceans and at higher altitudes SO_2 and NO_2 concentrations are smaller than the values measured in the surface air over the continents. Thus, above the sea surface SO_2 concentrations around 1·5 and 3·0 nmol m^{-3} were observed.[26,27] It is assumed that the presence of SO_2 above the ocean is caused by the chemical transformation of reduced organic sulfur compounds, like DMS, of biological origin. The SO_2 concentration is also low at higher tropospheric levels: it decreases with increasing altitude. According to Ref. 28, the values measured around 5 km are similar to those obtained in the air near the ocean surface. However, individual vertical profiles vary as a function of meteorological conditions determining the vertical exchange of different constituents. The average concentrations of NO and NO_2 measured on the Irish west coast were found to be 0·12 ppb (5·4 nmol m^{-3}) and 0·34 ppb (15 nmol m^{-3}), respectively,[29] while further studies under the same conditions gave slightly lower values.[30] Relatively few aircraft flights were carried out to determine the vertical profile of NO and NO_2 concentrations. The data obtained indicate that NO_x ($NO+NO_2$) levels decrease with height more slowly than the concentration of SO_2.[31,32]

DMS molecules in oceanic air are emitted by marine ecosystems. On the basis of available information, it is reported that the average DMS concentration in the marine atmosphere is equal to 4·7 nmol m^{-3}.[33] Results of recent observations of the vertical profile of DMS concentration in the marine troposphere show that under stable, cloudless conditions the surface air concentration decreases by two orders of magnitude during the first 3 km above the surface, while in the same layer the decrease is about a factor of 10 in the presence of large cumulus congestus clouds.[34]

Another possible acid-forming sulfur gas of biologic origin is H_2S. It is released into the air by anaerobic swamps, muds and eutrophic waters.[35] Over continental regions (Rhine-Main area, Germany) in the surface air the concentration of this gas is about 30 nmol m^{-3}. This concentration decreases rapidly with increasing height to become about 3 nmol m^{-3} at an altitude of 3 km.[28] Results of aircraft observations[28] also show that over continental polluted areas the possibility of anthropogenic H_2S emissions cannot be ruled out. In the marine atmosphere over the Atlantic Ocean an average H_2S concentration of 1·22 nmol m^{-3} was gained,[27] in agreement with the results of earlier observations.[36]

Very little information is available on the atmospheric level of *hydrogen chloride*. Thus, it is reported that in downtown Darmstadt (Germany) hydrogen chloride concentrations are between 30 and 60 nmol m^{-3},[37] while in a forest and in the meadow, some kilometers from the city, average values of 6·0 and 18 nmol m^{-3} were measured. Under oceanic conditions in the surface air the hydrogen chloride concentration is around 1 ppb (~ 40 nmol m^{-3}), while it varies between 2·0 and 19 nmol m^{-3} in an air layer of 0–7 km over Norway, Spain and the western Mediterranean.[38] It is very probable that the relatively high concentrations above the oceans are due to the release of HCl from sea salt particles (see later).

Under continental conditions in North Carolina, USA,[39] Sweden[40] and Hungary[25] the HNO_3 level is around 50 nmol m^{-3}. On the other hand, in the marine boundary layer the median value was found to be 0·06 ppb (~ 2·6 nmol m^{-3}). Interestingly, in the free troposphere a somewhat higher median value was obtained.[41] One can assume with caution that this situation is caused by the fact that HNO_3 is not emitted at the surface. Rather, it forms in the atmosphere by chemical reactions.

The precursors of *organic acids* in the air and precipitation are gases like isoprene and terpenes. The concentration of isoprene was found to be relatively high in tropical forests. Thus, in the Amazon boundary layer (Brazil) an average isoprene concentration of 2 ppb (89 nmol m^{-3}) was obtained,[42] which is very close to the value (1·6 ppb (72 nmol m^{-3})) gained in tropical Australia.[43] In contrast, the total concentration of terpenes (α- and β-pinene constituted 45% of the total quantity) in Amazonia was only 0·2 ppb (8.9 nmol m^{-3}). However, in a coniferous forest at the mid-latitudes (south-western France) the sum of the α- and β-pinene concentrations varied between 0·6 ppb (27 nmol m^{-3}) and 6 ppb (270 nmol m^{-3})[44] owing to the higher terpene emission of this forest type. Also in the Amazonia region, during the dry season gas phase HCOOH and CH_3COOH concentrations of 1·6 ppb (71 nmol m^{-3}) and 2·2 ppb (98 nmol m^{-3}) respectively, were measured.[15] For the south-western USA, HCOOH and CH_3COOH concentrations between 0·5 ppb (22 nmol m^{-3}) and 3 ppb (134 nmol m^{-3}) were reported;[45] values confirmed by recent observations carried out in central Europe (Austria).[46]

NH_3 concentrations in the surface air over the continents are typically between 10 and 100 nmol m^{-3},[25,47–49] except regions where the emission rate is very high. Thus, the average NH_3 concentration over the land area of the Netherlands was found to be 350 nmol m^{-3}.[50] The gaseous NH_3 level has a summer maximum which is related to the higher release during this season and/or to the lower solubility of NH_3 in water at higher temperatures.[25,47–49] Owing to sources on the surface the NH_3 concentration decreases with increasing altitude, which is more pronounced in summer.[28] Under maritime conditions much lower concentrations can be detected than over the continents. Thus, at Cape Grim (Tasmania, Australia) in oceanic air an average concentration of 3·3 nmol m^{-3} was measured.[48] Even lower concentrations (1 nmol m^{-3} on average) were observed recently over the north-eastern Pacific Ocean.[51] In data obtained at Cape Grim no seasonal trend was apparent.

Among other things, the significance of concentration measurements of acidic compounds in the study of atmospheric acidity is due to the fact that the *dry deposition* (deposition without precipitation of each species onto various surfaces) depends directly on its concentration in the air. The flux of material toward any surface is determined by the turbulent transfer in the air, by the molecular diffusion in a thin layer covering the surface, and by the uptake of the material by the surface. The overall process can be characterized by the following expression:

$$D = vC \qquad (1.4)$$

where D is the dry deposition rate, C is the concentration, and v is the so-called dry deposition velocity: the inverse of the resistance to material transfer. Table 1.3 gives the deposition velocity of some acidifying species.[52] In the table the average concentrations in regional air are also represented on the basis of the above discussion. The last column contains the dry deposition rates calculated by eqn (1.4). It can be seen that the dry deposition velocities are significant in particular for HNO_3 and HCl. This is due to the fact that different surfaces absorb these acids strongly and the downward flux is determined only by the turbulent transfer. However, the dry deposition of SO_2-S and NO_2-N is also very important as compared to the wet deposition of these elements (see later).

Table 1.3
Dry deposition velocity[52] (v); concentration (C) and dry deposition rate (D) of different species under regional (non-urban) conditions

Species	*Range* v $(cm\,s^{-1})$	*Mean* v $(cm\,s^{-1})$	C $(nmol\,m^{-3})$	D $(nmol\,m^{-2}\,year^{-1})$
SO_2	0·1–2·0	0·8	150	38
NO_2	0·01–1·0	0·4	140	18
NO	±0·1	0·01	70	< 1
HNO_3	1–20	< 5(~ 2)	70	~ 44
HCl	1–20	< 5(~ 2)	5·0	~ 3
NH_3	±2·0	1·0[a]	70	22

Reproduced with permission from Ref, 52.
[a] NH_3 flux can be bidirectional; surfaces may also act as sources. The value tabulated is valid only for periods when deposition occurs. If all cases are considered the average value of the deposition velocity is very low.[53]

1.3.2 Atmospheric aerosols

If they are not removed from the atmosphere by dry and wet deposition, sulfur- and nitrogen-containing precursor gases are converted in the air into aerosol particles consisting of SO_4^{2-}, NO_3^- and NH_4^+ ions. This is due to the fact that H_2SO_4 and HNO_3 molecules, formed from precursors by oxidation, condense together with water vapor to create minute acidic solution droplets. While H_2SO_4 condenses easily (for this reason it cannot be detected in gaseous form), a part of HNO_3 remains in the air in vapor form. The primary acidic droplets grow rapidly in the air by coagulation, since their size is small and their concentration is high. In the air acidic droplets are partly or totally neutralized as a function of the availability of NH_3 molecules. For this reason *ammonium sulfate* ($(NH_4)_2SO_4$) *and nitrate constitute the major part of the water-soluble fraction of the atmospheric aerosol.* SO_4^{2-} and NO_3^- particles grown by coagulation and by further condensation have generally a log-normal size distribution (the frequency of the sulfate or nitrate mass in different particle size intervals) with a geometric mean radius of 0·1–0·5 μm (for further details, see Ref. 19).

Figure 1.4 represents the spatial distribution of the mass concentration of SO_4^{2-} particles (expressed in sulfur) in Europe according to the measurements carried out in the period of 1978–1982,[21] while Fig. 1.5 gives the distribution of the same parameter in the US on the basis of observations made between 1970 and 1972.[54] One can see that in Europe two maxima can be found: in the western and eastern part of the territory considered, in agreement with the distribution of SO_2 (see Fig. 1.3). In the US two maxima are also detected. However, the maximum in California is insignificant. The main maximum is in the north-eastern part of the country similar to the distribution of sulfur emissions.[55] It also follows from the two figures that mean maximum concentrations over both continents are around 4 μg S m^{-3} (125 nmol m^{-3}), while the continental background in both cases is about 1 μg S m^{-3} (30 nmol m^{-3}). Results of US monitoring

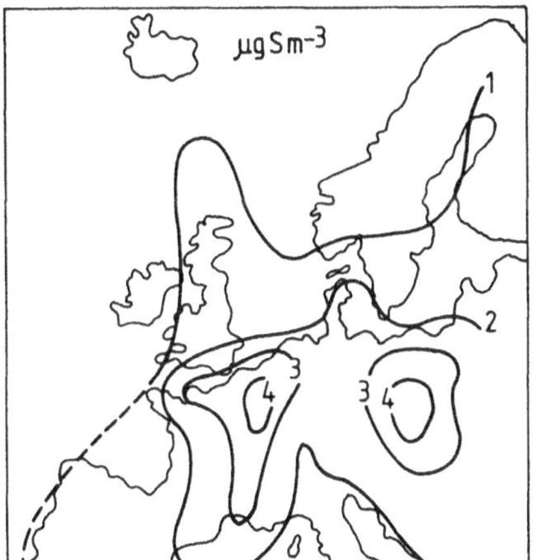

Figure 1.4. Spatial distribution of the concentration of SO_4^{2-}-S in the surface air over Europe.

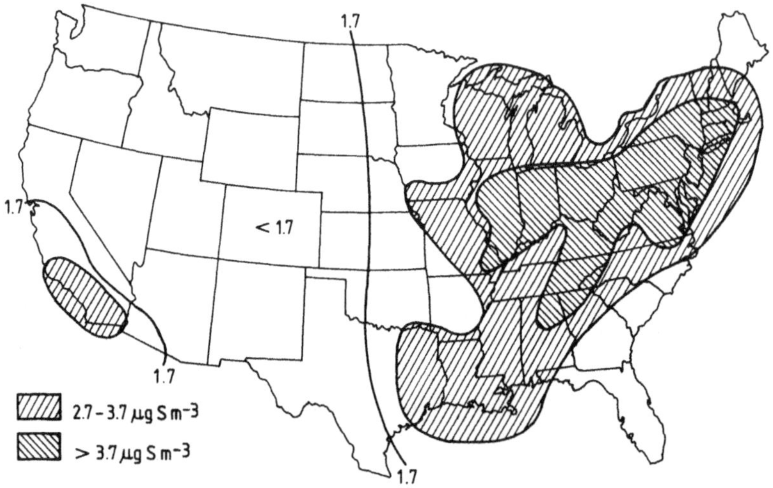

Figure 1.5. Spatial distribution of the concentration of SO_4^{2-}-S in the surface air over the US. (Reproduced with permission from Ref. 54.)

network[56] show that, while urban SO_4^{2-} concentrations measured during the same period are higher than non-urban levels, the differences are not as important as in the case of SO_2. Thus, the average ratio of urban to non-urban concentrations is equal to about two. This is due to the fact that SO_4^{2-} is not emitted directly by pollution sources and some time is necessary for its formation from SO_2. American studies also indicate that, in contrast to SO_2 concentration and source strength, SO_4^{2-} levels in the atmosphere have a summer maximum;[57] a fact caused by the higher rate of $SO_2 \rightarrow SO_4^{2-}$ conversion in summer.

Above the Atlantic, Pacific and Indian oceans, excess (non-sea-salt) SO_4^{2-} concentrations around 10 nmol m^{-3} were observed.[58] On the other hand, at Cape Grim (Tasmania, Australia) in maritime air masses an average concentration of 3 nmol m^{-3} was obtained.[59]

Unfortunately, the atmospheric concentration of NO_3^- ions is not well known. Generally speaking, in non-urban air over Europe[25,60] and North America,[61,62] the NO_3^- concentration is lower than the SO_4^{2-} level. For example, in Hungary in rural air the NO_3^- concentration is 70 nmol m^{-3} on average. The annual average NO_3^- concentrations measured in England at urban and rural sites were found to be very similar (~ 35 nmol m^{-3}).[60] On the other hand, within the marine boundary layer over the

Pacific the particulate NO_3^- concentration is very low, less than about 3·0 nmol m^{-3},[63] similar to the concentrations obtained in the surface air at Ny-Alesund, Spitzbergen.[64]

The concentrations of *NH_4^+* ions in Sweden in urban and rural air are very similar,[49] they are between 50 and 100 nmol m^{-3} proving that urban emissions are insignificant in this case. In England and Hungary, in the aerosol the NH_4^+ level is around 200 nmol m^{-3},[25,60] which indicates that there is sufficient NH_3 in the air to neutralize H_2SO_4, but not sufficient to neutralize the sum of H_2SO_4 and HNO_3. This means that the aerosol in non-urban air over Europe is slightly acidic. Similarly to Europe, it was reported that in the eastern third of the US SO_4^{2-} particles consisted of compounds ranging from H_2SO_4 to $(NH_4)_2SO_4$.[65] In the south-western US, a good correlation was found between the concentrations of NH_4^+ and SO_4^{2-} ions.[61] However, the SO_4^{2-} was not fully neutralized $(NH_4)_2SO_4$. For this reason the authors speculated that $(NH_4)HSO_4$ was also present, mainly if the SO_4^{2-} concentration was high. Further, in the North Polar region and over the Atlantic, aerosol particles with radii smaller than 0·1 μm were found to be acidic.[66] In the surface air over the Pacific Ocean much lower NH_4^+ concentrations were measured than over continental areas. Thus, they were found to be around 5–6 nmol m^{-3}.[51,63] On the basis of the analyses of aerosol samples collected in remote Everest highlands at altitudes between 5400 and 6500 m Japanese workers showed that SO_4^{2-} particles were in the form of a hydrated H_2SO_4 solution and/or in the form of supersaturated acidic droplets.[67] In other words at these heights the NH_3 concentration is not sufficient to neutralize H_2SO_4.

Many observational programs have been carried out to study the concentration and size distribution of *chloride* particles in the atmosphere.[19] However, since atmospheric chloride is mostly of maritime origin, the question of whether small chloride particles sampled in continental air are at least partly acidic, is rather poorly understood. Recent observations gained in England show that Cl^- ions in these small particles are neutralized by NH_3, but further measurements are needed in this field.[68]

Little information is available concerning the amount of *HCOOH and CH_3COOH in atmospheric aerosol*. However, according to our present knowledge it seems that this amount is insignificant relative to the fraction in gaseous form. Thus, it is reported that in the mixed layer over the Amazon Basin, Brazil, the concentration of HCOOH and CH_3COOH in aerosol particles is only 1·5 and 0·9% of the gas phase concentration, respectively.[15]

The dry deposition velocity of aerosol particles in the size range where the majority of the mass of SO_4^{2-}, NO_3^- and NH_4^+ ions is found (0·1–0·5 μm in radius) is very small:[69] it has a magnitude of ~ 0·01 cm s^{-1}. Consequently, their dry deposition can probably be neglected. However, these water-soluble particles control the visibility in the air as well as constituting the class of nuclei for water vapor condensation. In this way they initiate the precipitation formation and are removed from the air by wet deposition, the composition of which depends, among other factors, on the chemical nature of the particles.

1.3.3 Aqueous compounds

On condensation, cloud and fog droplets form on condensation nuclei consisting of water-soluble aerosol particles. After cloud and fog formation the sorption of soluble gases begins immediately. A part of the gases incorporated into the aqueous phase is converted irreversibly in water (e.g. SO_4^{2-} can be formed from SO_2 absorbed). On the other hand, small aerosol particles, remaining airborne after cloud/fog formation, are captured by water droplets or ice crystals due to their Brownian motion. These removal processes taking part in the clouds are termed *in-cloud scavenging*, or *rainout*.

There are two possibilities concerning the future fate of a cloud or fog. One possibility is that cloud and fog elements evaporate and the substances scavenged again become airborne. The other possibility is that the fog and cloud droplets and crystals grow into drizzle or precipitation elements which are deposited onto the earth surface. Below the cloud base raindrops and snowflakes incorporate further materials. This so-called *below-cloud scavenging* or *washout* can be particularly significant in the polluted atmosphere.

The above process, removing different species from the air and carrying them to the surface; is termed *wet deposition*. A special form of wet deposition is when cloud and fog droplets are captured from the air by trees or by any object on the surface (interception). It is also possible that water vapor condenses directly on the surface of the vegetation (or on other surfaces) to form liquid water in the form of dew.

Due to wet deposition an important part of acidifying substances and acid species is removed from the air. While wet deposition is a sink of atmospheric gases and particles, it creates an important material flux from the air to other media of our environment. Consequently, the study of the composition of atmospheric liquid waters is of crucial importance for estimating the acidification of soils and natural waters.

Owing to its significance in acidification processes, several programs have been performed during the last years to investigate the chemical composition of *fogwater*. The results of these programs demonstrate that fogwater is generally rather polluted and its H^+ ion concentration is high. Thus, in the Po Valley, Italy, pH values in fogwater between 3·49 and 4·32 and SO_4^{2-}, NO_3^- and NH_4^+ concentrations in the range of 100–1000 μeq liter^{-1} were observed.[70] In southern California a median pH value of 3·32 was measured,[71] while the lowest figure was as low as 2·7. Even a pH of 2·2 was detected in a dissipating fog under urban conditions in California.[72] In the Po Valley the SO_4 concentration was generally higher than that of NO_3^- ions, but in California the inverse was observed. The mean NO_3^- concentration in fog is equal to 1110 μmol liter^{-1},[71] while the corresponding figure for SO_4^{2-} is 292 μmol liter^{-1}. The high NO_3^- content in California is obviously due to the large traffic density and the intense photochemical activity of the area.

In California and in the Po Valley the organic acid content of fog droplets was also investigated. In California the HCOOH and CH_3COOH concentration varied between 45–142 μeq liter^{-1} and 1–134 μeq liter^{-1}, respectively.[73] Similar to these findings the results of the study carried out in Italy show that the HCOOH concentrations in fogwater range from 11 to 175 μmol liter^{-1},[74] while the corresponding interval for CH_3COOH is 10–269 μmo! liter^{-1}. In agreement with previous discussions these levels are much lower than the concentrations of inorganic acidic compounds.

Some studies have also been made to determine the composition of *dew* as summarized by Ref. 75. Owing to the effects of biologically derived alkaline ions rather high pH values are generally reported in dewwater. Thus, at a site in Warren (Michigan, USA) a median pH value of 6·5 was measured in natural dew.[75] This was due to the very high calcium concentration (690 μeq liter^{-1} on average) and to the relatively 'low' SO_4^{2-} concentration (242 μeq liter^{-1}). Dew was found to be much more concentrated than rainwater, except for H^+ ions. The study also showed that dew enhances considerably the dry deposition of different materials from the air as compared to the flux over dry surfaces.[75]

The composition and acidity of *cloudwater* have been measured by several research workers.[76] The significance of such measurements for our subject is the fact that in mountainous areas the acid input to the ecosystems by the interception of cloud droplets may be more important than the deposition due to falling precipitation. This concept was formulated on the basis of the results obtained in the Adirondack Mountains of New York State, USA, according to which the pH of cloudwater was found to

Table 1.4
The chemical composition of cloudwater in Sweden[79] according to wind trajectories at 1000 hPa

Sector	*No. of samples*	*Composition (μeq liter⁻¹)*								
		H^+	NH_4^+	Na^+	K^+	Mg^+	Ca^{2+}	Cl^-	NO_3^-	SO_4^{2-}
NE	47	33	17	7	< 0·8	1	0·6	6	9	34
NW	41	13	0·5	< 0·4	< 0·8	< 0·2	0·3	0·8	2	6
W	33	38	1	5	< 0·8	0·6	0·8	4	10	30
S	4	370	—	33	6	12	22	19	68	700

Reproduced with permission from Ref. 79.

be rather low; it was in the range of 2·66–4·66 in 90% of the cases.[77] These results are supported by more recent observations made at Whiteface Mountain (New York, USA).[78]

The chemical composition of cloudwater formed in air masses of different origin can be estimated on the basis of recent Swedish observations carried out at Areskutan (63°26′ N, 13°6′ E) at an altitude of 1250 m.[79] The authors divided their results according to the air trajectories arriving at the sampling site from different directions as shown in Table 1.4. It can be estimated that the composition of the NW sector refers to the hemispheric background, while the results gained on the occasion of S polluted continental air (compare these figures with precipitation chemistry data tabulated in Table 1.1). In the other two sectors transitional composition was measured.[80]

Owing to the importance of cloud chemistry a program was started in the US to regularly monitor cloud composition on six mountain summits. The results obtained in 1986 at one of these stations (Mt. Mitchell, State Park, North Carolina, USA)[81] show that the average pH in precipitation is about 4·4 while it ranges from 2·2 to 5·4 in cloudwater. On the basis of their observations, the authors estimate that the deposition caused by cloud interception is two to five times the deposition in form of precipitation.

It goes without saying that this conclusion is valid only for mountainous regions. In other areas the wet deposition is controlled by the composition and quantity of *precipitation water* (the deposition is the product of the concentration and precipitation amount), except during relatively infrequent foggy conditions. For this reason many national and international research and monitoring programs were initiated to measure the chemical composition of precipitation, including the European Atmospheric Chemistry Network (EACN) organized by Swedish scientists,[82] the National Acid Deposition Program in the US (NADP), the Canadian Network for Sampling Precipitation (CANSAP), the European Monitoring and Evaluation Programme (EMEP) of the Economic Commission for Europe and the Background Air Pollution Monitoring Network (BAPMoN) of the World Meterological Organization. The results of NADP and CANSAP are summarized by Refs 83 and 84, EMEP data are evaluated by Refs 7, 21 and 52, while precipitation chemistry data gained in BAPMoN are discussed by Refs 85 and 86. As an example Figs 1.6–1.9 illustrate the spatial distribution of pH, SO_4^{2-}, NO_3^- and NH_4^+ concentrations according to the data obtained in 1980 by NADP and CANSAP and by some smaller networks,[83] while Table 1.5 gives the range of the

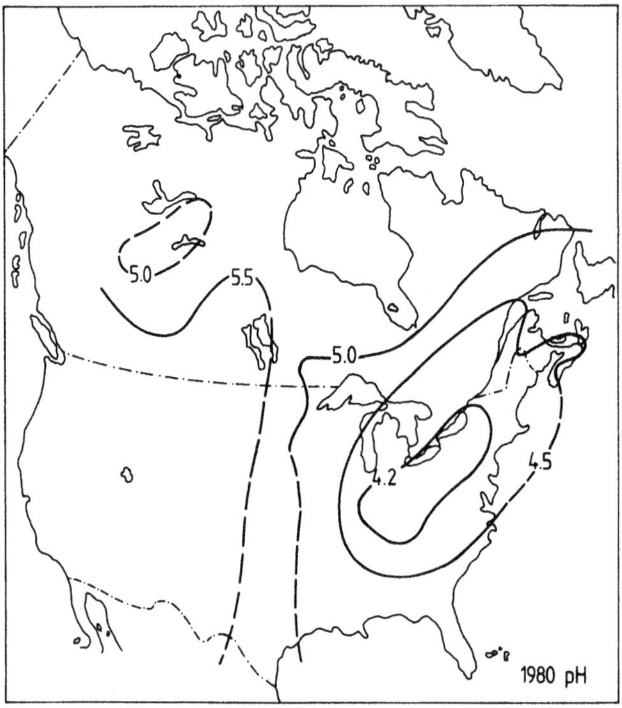

Figure 1.6. Spatial distribution of pH of precipitation water collected in North America. (Reproduced with permission from Ref. 83.)

wet deposition of these ions for eastern (north-eastern US and south-eastern Canada) and western parts of the territory where the minimum and maximum pH values occur. First of all, it should be mentioned that these values are very similar to those reported for Europe.[52] Secondly, maximum H^+, SO_4^{2-} and NO_3^- ion concentrations are found in areas with high SO_2 and NO emissions.[55] Thirdly, the spatial distribution of the concentration of NH_4^+ ions in precipitation is different from that of SO_4^{2-} and NO_3^+ ions. This is not surprising since NH_3 molecules are released from different sources than sulfur and nitrogen oxides. Thus, maximum NH_4 concentrations were found in the north central US where livestock feedlots are concentrated. On the other hand, the comparison of data in Tables 1.3 and 1.5 indicates that dry deposition of different acidifying species is comparable to their flux due to wet deposition.

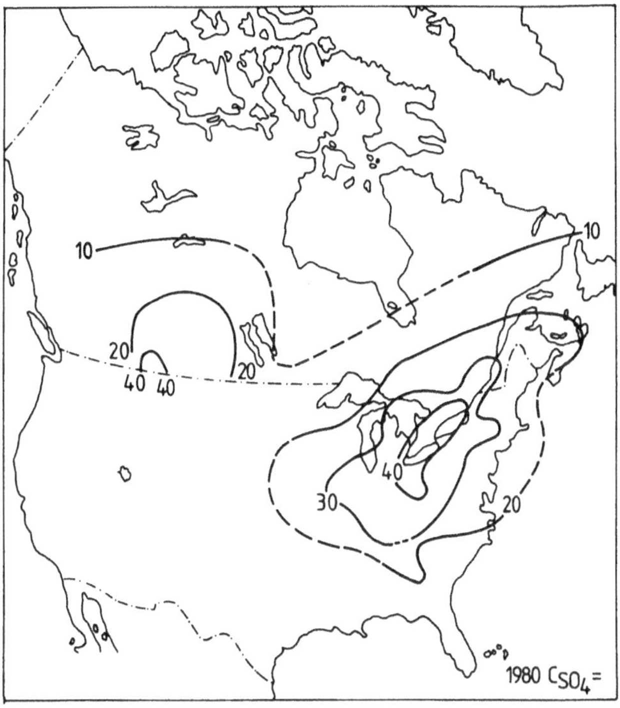

Figure 1.7. Spatial distribution of the concentration of SO_4^{2-} ions in precipitation water (expressed in μmol liter^{-1}) collected in North America. (Reproduced with permission from Ref. 83.)

Since the time period of precipitation chemistry measurements is longer than the interval of gas and aerosol phase observations, several studies have been made to estimate the possible trend in precipitation chemistry data. Thus, it was found that data obtained at Rothamsted (UK) show an increase of NO_3^- concentration from the end of the last century until now, mainly in the spring months.[87] A similar trend was reported on the basis of the results of Hungarian observations.[88] Hungarian data also reveal that since the turn of the century the NH_4^+ level of precipitation has remained practically constant. Precipitation chemistry data measured at Hubbard Brook (New Hampshire, USA) since 1964 were studied by several authors. The results indicate some increase in NO_3^- concentration and a relatively constant pH level.[84] At the same time the

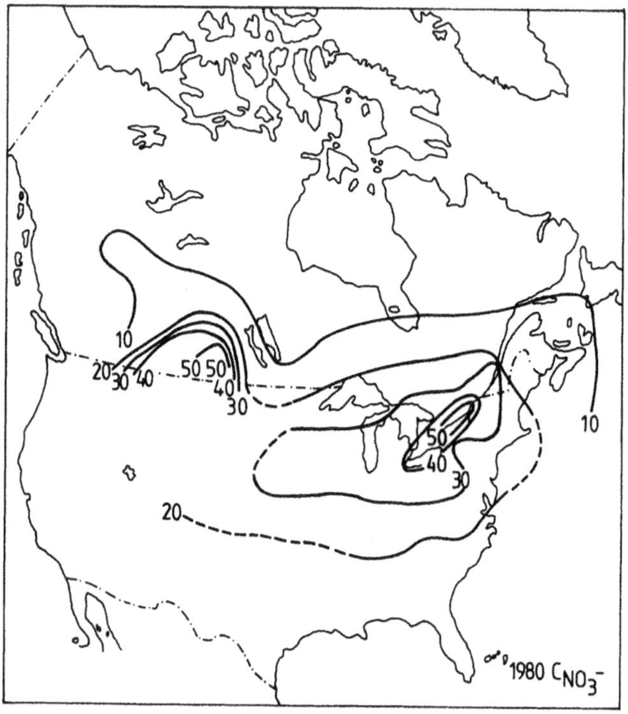

Figure 1.8. Spatial distribution of the concentration of NO_3^- ions in precipitation water (expressed in μmol liter^{-1}) collected in North America. (Reproduced with permission from Ref. 83.)

concentration of SO_4^{2-} ions has decreased significantly (2% per year) in accordance with the trend of SO_2 emissions in the north-eastern part of the US.[55] According to the analysis of EACN data obtained between 1955 and 1982 in Scandinavia, the SO_4^{2-} concentration in precipitation increased by 50% during the late-1950s and 1960s, while it declined by 20% since the early-1970s in agreement with changes in SO_2 emissions.[82]

Measurements of HCOOH and CH_3COOH in precipitation water were reviewed in several papers.[18,89] They demonstrate that, with the exception of one site in Virginia (USA) where samples were influenced by anthropogenic emissions, the concentration of these acids is high in particular in tropical and subtropical regions (ranging from 20 to 30 μmol liter^{-1}). At other continental and marine locations they vary between 5

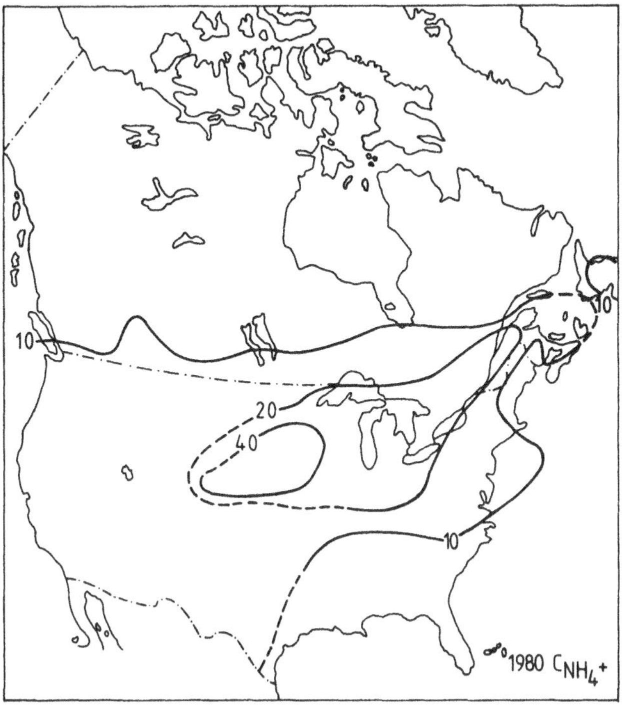

Figure 1.9. Spatial distribution of the concentration of NH_4^+ ions in precipitation water (expressed in μmol liter^{-1}) collected in North America. (Reproduced with permission from Ref. 83.)

and 20 μmol liter^{-1}. Under continental condititons the concentrations exhibit a maximum during the growing season supporting the existence of natural biogenic sources.

Table 1.5
Wet deposition of different ions in eastern (Max.) and western (Min.) North America

Species	*Max. (mmol m^{-2})*	*Min. (mmol m^{-2})*
H^+	40–80	1–10
NH_4^+	20–30	1–15[a]
SO_4^-	30–50	2–20
NO_3^-	20–40	1–10

[a]Some high data which are inconsistent with other results are not considered. (Data taken from Ref. 83.)

1.4 ATMOSPHERIC CYCLES AND RESIDENCE TIMES

As has been seen, the atmospheric cycles of water and acidifying substances are interrelated. Cloud droplets form on condensation nuclei consisting of SO_4^{2-}, NO_3^- and NH_4^+ ions, and a significant fraction of acidifying species is removed from the air by precipitation.

The source of atmospheric water is the evaporation of water molecules from water surfaces and, to a lesser extent, the release of water by the vegetation (transpiration). The major part of water vapor (454×10^{15} liter year^{-1}, or 25×10^6 Tmol year^{-1}) is released into the air by the oceans. This quantity is 86% of the total source strength (526×10^{15} liter year^{-1}, or 29×10^6 Tmol year^{-1}).[90]

The cycle of any compound (element) in the atmosphere can be characterized by the *residence time*, which is the average time spent by its molecules in the air. The residence time (t) can be given by the following simple equation:

$$t = \frac{M}{F} = \frac{M}{R} \tag{1.5}$$

where M is the quantity of the species in the atmosphere (or in a part of it), and F and R are the formation (source strength) and removal (sinks) rate, respectively. Since the mass of water in the atmosphere is 13×10^{15} liters ($0{\cdot}72\times10^6$ Tmol), its atmospheric residence time is equal to about 10 days. Owing to the aforementioned interrelationship, the atmospheric residence time of aerosol particles and water-soluble gases is very close to the value calculated for water. This means that the water cycle not only redistributes water on the earth surface, but it also plays an important role in the control of atmospheric pathways of other substances, including acidifying compounds.

The aim of this section is to discuss briefly the atmospheric cycle of those species which determine under continental (sulfur and nitrogen oxides and NH_3), and global (DMS and organic acids) conditions the acidity of dry and wet deposition. For the sake of completeness an approximative survey of atmospheric HCl emissions is also presented.

1.4.1 Sulfur

Table 1.6 gives, on the basis of present knowledge, the source and sink terms of the global atmospheric sulfur cycle. It is noted that atmospheric sulfur compounds with long residence times such as carbonyl sulfide and

Table 1.6
Terms of the global atmospheric cycle of sulfur[a]

Emissions[b]	*(Tmol year^{-1})*	*Depositions*	*(Tmol year^{-1})*
Anthropogenic SO_2	2·47	SO_2	1·50
Sea salt	1·37	SO_4^{2-} (sea salt)	1·37
Volcanoes	0·06	SO_4^{2-} (excess)	1·93
DMS (continent)	0·15		
DMS (oceans)	1·10		
Biomass burning	0·20		
Total	5·35		4·80

[a]Emissions are taken from Ref. 33, except the value for volcanoes,[19] while deposition data are from Ref. 19.
[b]The emission of H_2S and the sulfur coming into the air by wind-blown dust is not considered since this term is very uncertain.[33]
(Reproduced with permission from Ref. 91.)

carbon disulfide, which play an important role in stratospheric chemistry, are not considered. Considering the uncertainties involved in the estimation of the different terms (around 50%) the agreement between values calculated for total emission and deposition is good. It can be seen from the table that the strength of anthropogenic emissions is about half of the magnitude of the total sulfur release. The anthropogenic fraction is even higher if only gaseous emissions are considered, that is if the emission of sea salt particles is disregarded. This means that even on a global scale human activities considerably modify the atmospheric sulfur cycle and consequently the acidity of deposition. Data also show that natural gaseous emissions are higher above the oceans than above the continents. This implies that before the industrial era there was probably a net sulfur transport from the marine to the continental atmosphere. Anthropogenic effects have reversed the direction of this flux and at present the emission of man-made sulfur over the continents exceeds the magnitude of deposition.

The conclusion is valid in particular if an industrialized continent or an industrialized continental area is considered separately. This is illustrated by data in Tables 1.7 and 1.8 representing the results of sulfur budget calculations for Europe, and for the eastern part of North America. In Europe 21% of the total sulfur input or 23% of anthropogenic emissions leaves the air over the continent. The outflow from the eastern US and eastern Canada (Table 1.8) to the air above the Atlantic Ocean is equal

Table 1.7
Sulfur budget over the whole of Europe ($10{\cdot}5 \times 10^6$ km^2)

Term	*Tmol year^{-1}*
Inputs	
Man-made emissions	0·78
Natural emissions	0·03
Inflow from oceans	0·06
Total	0·87
Outputs	
Wet deposition	0·38
Dry deposition	0·31
Outflow[a]	0·18
Total	0·87

[a]Outflow is calculated by balancing the cycle.

to 24% of the total input and 26% of man-made emissions. Consequently, SO_4^{2-} concentrations over the western Atlantic Ocean were found to be higher (~ 40 nmol m^{-3}) than those reported for remote marine air.[93]

Table 1.8
Sulfur budget in the atmosphere over the eastern part of North America ($6{\cdot}0 \times 10^6$ km^2)

Term[a]	*Tmol year^{-1}*
Inputs	
Man-made emissions	0·50
Natural emissions	0·02
Inflow from west	0·02
Total	0·54
Outputs	
Wet deposition	0·17
Dry deposition	0·14
Outflow to oceans	0·13
Total	0·44

Reproduced with permission from Ref. 92.
[a]Budget terms are calculated independently. For this reason the budget is not balanced.

For estimating the residence times, knowledge of the atmospheric burden of SO_2 is 0·014 Tmol, while it is 0·025 Tmol in the case of SO_4^{2-} particles. Taking into account the corresponding deposition rates (see Table 1.6) and the quantity of SO_2 converted yearly into SO_4^{2-} (1·3 Tmol), residence times of 1·8 and 3·8 days can be obtained for SO_2 and SO_4^{2-}, respectively. Reference 33 discusses in detail the sulfur budget in oceanic air. It reports that the residence time of DMS in remote marine atmospheres is 1·5 days, while the excess SO_4^{2-} formation rate is 0·06 nmol m^{-3} $year^{-1}$. Combining this latter figure with an average excess SO_4^{2-} concentration of 4·7 nmol m^{-3}, a residence time of 3·3 days can be calculated for excess SO_4.[33]

Finally, it is noted that the residence time of H_2S, not discussed in this subsection, is very short; it is only about 12 h due to the rapid oxidation of H_2S molecules in the atmosphere.[35]

1.4.2 Nitrogen

Table 1.9 reproduces the results of global NO_x budget calculations of Ref. 31. According to this work the accuracy of source terms in the table is about a factor of two. It can be seen that the agreement between sources and sinks is not as good as for sulfur compounds: the total emission is close to the upper limit of the deposition. It is to be noted that Refs 19 and 94 estimate a total nitrogen oxides deposition of 2·85 Tmol $year^{-1}$ and 1·7 Tmol $year^{-1}$, respectively; values which practically confirm deposition data in Table 1.9. This means that the source terms tabulated should be considered with some caution.

Table 1.9
Global atmospheric budget of NO_x

Sources (Tmol year^{-1})		*Sinks (Tmol year^{-1})*	
Fossil-fuel combustion	1·5	Wet deposition	0·86–3·0
Biomass burning	0·86	Dry deposition	0·86–1·6
Lightning	0·57		
Microbial activity in soils	0·57		
Oxidation of NH_3	~ 0·36		
Total	3·86		1·72–4·6

Reproduced with permission from Ref. 31.

However, the data given clearly show that the emission from fossil-fuel combustion exceeds the value of other source terms. Other fluxes (e.g. biomass burning) can also be modified by man, if these are taken into account then it can be concluded that human activities play an important part in the control of the atmospheric cycle of nitrogen oxides.

Global NH_3 emission[19] is equal to 5·0 Tmol year^{-1}. This rather large amount of NH_3 is released mostly by biogenic sources which may also be modified by human activities (e.g. animal husbandry). Thus, there is some indication (see Ref. 8) that NH_3 emission in Europe has increased by 25% from 1960 to the early 1980s. While a small fraction of NH_3 is converted into nitrogen oxides (see Table 1.9), NH_3 is removed from the air mostly by dry and wet deposition (4·1 Tmol year^{-1}). In the case of wet deposition it leaves the atmosphere in the form of NH_4^+ ions.

The total nitrogen oxide mass in the atmosphere[19] is 0·11 Tmol, while the corresponding figure for NH_3 is equal to 0·07 Tmol. It follows from these values and from the emission terms discussed above that oxidized nitrogen compounds have an atmospheric residence time of about 10 days. For NH_3 the calculation results in a value of 5 days. However, it seems that over the ocean the residence times of NH_3 and NH_4^+ are much shorter. Thus, for these species in oceanic air residence times of 3·6 and 22 h were estimated.[51] This short NH_3 residence time is caused by the

Table 1.10
The European balance sheet for oxidized and reduced nitrogen compounds

Term	*NO_y-N*[a] *(Tmol year^{-1})*	*NH_x-N*[b] *(Tmol year^{-1})*
Emission		
Natural	0·02–0·03	0·02–0·11
Anthropogenic	0·25–0·56	0·27–0·50
NH_3 oxidation	< 0·01–0·02	–(< 0·01–0·02)
Total	0·27–0·61	0·29–0·63
Deposition		
Dry deposition	0·10–0·43	0·11–0·23
Wet deposition	0·21	0·19
Total	0·31–0·64	0·30–0·42

Reproduced with permission from Ref. 96.
[a]For NO_y, y = 1, 2 and 3.
[b]For NH_x, x = 3 and 4.

rapid absorption of NH_3 molecules by clouds and by their irreversible chemical transformation in aqueous solution.

The atmospheric mass balance of NO_x for north-western Europe has been investigated,[95] while for Europe as a whole the NH_3 and NO_x budgets have been estimated.[96] Table 1.10 summarizes the results of the latter authors. Considering the uncertainties of the calculations (see the original paper) they concluded that the European sources of nitrogen compounds are practically balanced by wet and dry deposition. However, if the budget of nitrogen oxides over NW Europe is considered separately, a net outflow is determined.[95] In agreement with this finding an important outflow from the North American continent (0·2 Tmol year^{-1}) was estimated.[31] This conclusion is confirmed by the relatively high NO_3 concentrations (14 nmol m^{-3} on average) found recently in the air over the Bermuda region, Pacific Ocean.[93]

1.4.3 Chlorine

The atmospheric cycle of gaseous chlorine is not well known. It is certain that the burning of fossil fuels containing chloride as well as waste incineration produce HCl. It is estimated that the western European countries[6] (for further details see the original paper) emit $17{\cdot}5\times10^{-3}$ Tmol yearly of HCl. The SO_2 emission of the same 13 countries, calculated according to Ref. 98, is equal to 0·24 Tmol year^{-1}, which is about 10% of the global anthropogenic emission (see Table 1.6). By accepting this ratio for HCl a total man-made emission of 0·18 Tmol year^{-1} can be calculated.

The most important natural HCl source is the reaction of sea salt with H_2SO_4 and HNO_3. The emission of sea-salt chloride[99] is estimated to be 4·22 Tmol year^{-1}.[90] There is some indication[97] that about 25% of this quantity is released as HCl which yields a yearly flux of 1·01 Tmol. Owing to the aquatic biota methyl chloride (CH_3Cl) is also released from the oceans into the air. The global strength of CH_3Cl emission[100] is 0·06 Tmol year^{-1}. It should be noted, however, that an important part of CH_3Cl molecules diffuse into the stratosphere;[101] consequently they do not take part in tropospheric chemistry.

Another natural source is provided by volcanic eruption. For this term a global value of 0·02 Tmol year^{-1} is proposed.[102] It is further believed that during biomass burning some gaseous CH_3Cl is formed. A tentative figure of 5×10^{-3} Tmol year^{-1} for the overall intensity of this source has been given.[103]

The above brief discussion is summarized in Table 1.11. The accuracy of values tabulated is not better than a factor of three. Thus, they must be

Table 1.11
Global atmospheric sources of gaseous chlorine

Source	*Tmol year^{-1}*
Released from sea salt	1·01
Anthropogenic	0·18[a]
Volcanoes	0·02
Formation from CH_3Cl	0·06
Biomass burning	0·005
Total	1·28

[a]Very uncertain value (see the text).

considered with caution. In spite of this fact it seems probable that the production of HCl by the reactions of sea salt dominate the gaseous chlorine cycle. However, this chemical flux does not produce *new* acids, since existing H_2SO_4 and HNO_3 molecules are transformed into HCl. In this way, it can be stated that human activities provide the major source of new acid molecules. The strength of anthropogenic sources, however, is negligible compared to the intensity of the emissions of sulfur- and nitrogen-containing acidifying gases (see Tables 1.6 and 1.9). This statement is valid for global as well as for European acid production.[6]

1.4.4 Organic acids

Very little quantitative information is available on the global atmospheric cycle of organic acids. Thus, on the basis of model calculations[17] it is estimated that the total HCOOH flux from atmospheric chemical processes is equal to 0·47 Tmol year^{-1}.[104] It is speculated that this flux is considerably higher than the strength of other source terms (e.g. emissions by formicine ants or by combustion).[104] The tentative value of HCOOH concentration in precipitation over oceanic and continental regions is equal to 2 and 8 μmol m^{-3}, respectively.[18] The volume of precipitation[90] falling on the surface of the oceans is 416×10^{15} liters year^{-1}, while the corresponding continental figure is 108×10^{15} liters year^{-1}. The combination of these values yields an oceanic and continental HCOOH deposition of 0·83 and 0·86 Tmol year^{-1}, respectively. In spite of the fact that these figures are very approximate, it seems that other source terms are necessary to explain the atmospheric cycles of HCOOH especially if the fact that dry deposition was not taken into account in this calculation is considered. This conclusion is in agreement with the work of Keene and Galloway[89] in which substantial HCOOH concentrations are reported for continental regions (e.g. Alaska) without isoprene-emitting vegetation.

According to theoretical studies[17], after 5 days of a model run, a HCOOH concentration of 22 nmol m^{-3} can be obtained for the boundary layer in Amazonia. At the same time the HCOOH quantity removed by deposition and by ventilation through the top of the layer is 6·7 nmol m^{-3} day^{-1} and 2·2 nmol m^{-3} day^{-1}, respectively. This means that the HCOOH residence time is equal to 2·5 days.

For the time being the quantitative estimation of the CH_3COOH cycle is not possible since the sources of this material are not exactly known. The correlation between the concentrations of HCOOH and CH_3COOH seems to indicate similar sources, but some anthropogenic release cannot be excluded. Available information makes it probable that the atmospheric residence time of acetic acid is short; it has an order of several hours to a few days.[89]

It may thus be concluded that, athough a qualitative picture is beginning to emerge, further research is necessary for understanding the atmospheric cycle of HCOOH and CH_3COOH. Such studies are obviously needed to obtain a more detailed knowledge of the occurrence of atmospheric acidity on a global scale.

REFERENCES AND NOTES

1. More exactly, in water, hydroxonium ions (H_3O^+) are formed, since H^+ ions interact with water molecules. However, this formal complication does not alter the concept outlined.
2. In this chapter, molar concentrations and concentrations expressed in equivalents (eq) are used alternately. For univalent ions the two concentrations are numerically equal, while for bivalent ions (e.g. sulfate (SO_4^{2-}) the number of moles must be multiplied by two to obtain the number of equivalents.
3. Charlson, R. J. & Rodhe, H., Factors controlling the acidity of natural rainwater. *Nature,* **295** (1982) 683–5.
4. Miller, J. M., Acid rain. *Weatherwise,* **37** (1984) 233–9.
5. Swedish Ministry of Agriculture, *Acidification Today and Tomorrow*. Risbergs Tryckeri A.B., Uddevalla, Sweden, 1982.
6. Lightowlers, P. J. & Cape, J. N., Sources and fate of atmospheric HCl in the UK and Western Europe. *Atmos. Environ.,* **22** (1988) 7–15.
7. Nodop, K., Nitrate and sulphate wet deposition in Europe. In *Physico-chemical Behaviour of Atmospheric Pollutants*, ed. G. Angelotti & G. Restelli. Reidel, Dordrecht, The Netherlands, 1986, pp. 520–8.

8. However, after being deposited into the soil, NH_3 (actually NH_4^+ ions) produce H^+ ions by nitrification processes. (Buijsman, E. & Erisman, J.-W., Wet deposition of ammonium in Europe. *J. Atmos. Chem.*, **6** (1988) 265–80.)
9. Quinn, P. K., Charlson, R. J. & Zoller, W. H., Ammonia, the dominant base in the remote marine troposphere: a review. *Tellus,* **39B** (1987) 413–25.
10. Granat, L., On the relation between pH and the chemical composition in atmospheric precipitation. *Tellus,* **24** (1972) 550–60.
11. Brezonik, P. L., Edgerton, E. S. & Hendry, C. D., Acid precipitation and sulfate deposition in Florida. *Science,* **208** (1980) 1027–9.
12. Galloway, J. N., Likens, G. E., Keene, W. C. & Miller, J. M., The composition of precipitation in remote areas of the world. *J. Geophys. Res.*, **87** (1982) 8771–86.
13. Likens, G. E., Edgerton, E. S. & Galloway, J. N., The composition and deposition of organic carbon in precipitation. *Tellus,* **35B** (1983) 16–24.
14. Keene, W. C., Galloway, J. N. & Holden Jr, J. D., Measurement of weak organic acidity in precipitation from remote areas of the world. *J. Geophys. Res.,* **88** (1983) 5122–30.
15. Andreae, M. O., Talbot, R. W., Andreae, T. W. & Harriss, R. C., Formic and acetic acid over the central Amazon region, Brazil. 1. Dry Season. *J. Geophys. Res.,* **93** (1988) 1616–24.
16. Jacob, D. J., The chemistry of OH in remote clouds and its role in the production of formic acid and peroxymonosulfate. *J. Geophys. Res.,* **901** (1986) 9807–26.
17. Jacob, D. J. & Wofsy, S.C., Photochemistry of biogenic emissions over the Amazon forest. *J. Geophys. Res.,* **93** (1988) 1477–86.
18. Keene, W. C. & Galloway, J. N., Considerations regarding sources for formic and acetic acids in the troposphere. *J. Geophys. Res.,* **91** (1986) 14466–74.
19. Mészáros, E., *Atmospheric Chemistry*. Elsevier, Amsterdam, The Netherlands, 1981.
20. Finlayson-Pitts, B. J. & Pitts Jr, J. N., *Atmospheric Chemistry*. J. Wiley & Sons, New York, 1986.
21. Popovics, M., Szepesi, D. & Horváth, L., Territorial and temporal variations of air and precipitation quality for Europe (in Hungarian). *Időjárás,* **90** (1986) 131–42.
22. Altshuller, A. P., Atmospheric sulfur dioxide and sulfate distribution of concentration at urban and non-urban sites in United States. *Environ. Sci. Technol.,* **7** (1973) 709–12.
23. Altshuller, A. P., Review paper: the role of nitrogen oxides in nonurban ozone formation in the planetary boundary layer over N. America and W. Europe and adjacent areas of ocean. *Atmos. Environ.,* **20** (1986) 245–68.
24. Global Atmospheric Monitoring System, *Urban Air Pollution, 1973–1980*. World Health Organization, Geneva, Switzerland, 1984.
25. Mészáros, E. & Horváth, L., Concentration and dry deposition of atmospheric sulfur and nitrogen compounds in Hungary. *Atmos. Environ.,* **18** (1984) 1725–30.
26. Nguyen, B. C., Bonsang, B. & Gaudry, A., The role of the ocean in the global sulfur cycle. *J. Geophys. Res.,* **88** (1983) 10903–14.

27. Herrmann, J. & Jaeschke, W., Measurements of H_2S and SO_2 over the Atlantic Ocean. *J. Atmos. Chem.*, **1** (1984) 111–23.
28. Georgii, H.-W., Large scale spatial and temporal distribution of sulfur compounds. *Atmos. Environ.*, **12** (1978) 681–90.
29. Cox, R. A., Some measurements of ground level NO, NO_2 and O_3 concentrations at an unpolluted maritime site. *Tellus*, **29** (1977) 356–62.
30. Helas, G. & Warneck, P., Background NO_x mixing ratios in air masses over the North Atlantic Ocean. *J. Geophys. Res.*, **86** (1981) 7283–90.
31. Logan, J. A., Nitrogen oxides in the troposphere: global and regional budgets. *J. Geophys. Res.*, **88** (1983) 10785–807.
32. Elshout, A., Aircraft measurements of acid substances and precursors. In *Evaluation of Atmospheric Processes Leading to Acid Deposition in Europe.* Task Force on Acid Deposition, COST 611, Report EUR 11411 (Comm. European Communities), 1987, pp. 136–54.
33. Andreae, M. O., The ocean as a source of atmospheric sulfur compounds. In *The Role of Air–Sea Exchange in Geochemical Cycling*, ed. P. Buat-Ménard. Reidel, Dordrecht, 1986, The Netherlands, pp. 331–62.
34. Ferek, R. J., Chatfield, R. B. & Andreae, M. O., Vertical distribution of dimethylsulfide in the marine atmosphere. *Nature*, **320** (1986) 514–16.
35. Jaeschke, W., Georgii, H.-W., Claude, H. & Malewski, H., Contributions of H_2S to the atmospheric sulfur-cycle. *Pure Appl. Geophys.*, **116** (1978) 465–77.
36. Slatt, B. J., Natusch, D. F. S., Prospero, J. M. & Savoie, D. L., Hydrogen sulfide in the atmosphere in the northern equatorial Atlantic Ocean and its relation to the global sulphur cycle. *Atmos. Environ.*, **12** (1978) 981–91.
37. Matusca, P., Schwarz, B. & Bächmann, K., Measurements of diurnal concentration variations of gaseous HCl in air in the subnanogram range. *Atmos. Environ.*, **18** (1984) 1667–75.
38. Singh, H. B. & Kasting, J. F., Chlorine–hydrocarbon photochemistry in the marine troposphere and lower stratosphere. *J. Atmos. Chem.*, **7** (1988) 261–85.
39. Shaw Jr, R. W., Stevens, R. K. & Bowermaster, J., Measurement of atmospheric nitrate and nitric acid; the denuder difference experiment. *Atmos. Environ.*, **16** (1982) 845–53.
40. Grennfelt, P., Investigation of gaseous nitrate in an urban and rural area. *Atmos. Environ.*, **14** (1980) 311–16.
41. Huebert, B. J. & Lazrus, A. L., Bulk composition of aerosols in the remote troposphere. *J. Geophys. Res.*, **85** (1980) 7337–44.
42. Zimmermann, P. R., Greenberg, J. P. & Westberg, C. E., Measurements of atmospheric hydrocarbons and biogenic emission fluxes in the Amazon boundary layer. *J. Geophys. Res.*, **93** (1988) 1407–16.
43. Ayers, G. P. & Gillett, R. W., Isoprene emissions from vegetation and hydrocarbon emissions from bushfires in tropical Australia. *J. Atmos. Chem.*, **7** (1988) 177–90.
44. Lopez, A., Fontan, J., Barthomeuf, M. O. & Minga, A., Présentation de l'expérience ATILA (Action des terpénes et isoprene dans l'atmosphere) effectuée en forêt tempéré au sudouest de la France (Forêt des Landes). *Atmos. Environ.*, **22** (1988) 1881–94.

45. Dawson, G. A., Farmer, J. C. & Moyers, J. L., Formic and acetic acids in the atmosphere of the southwest USA. *Geophys. Res. Lett.*, 7 (1980) 725–8.
46. Puxbaum, H., Rosenberg, C., Gregori, M., Lanzerstorfer, C., Ober, E. & Winiwarter, W., Atmospheric concentrations of formic and acetic acid and related compounds in eastern and northern Austria. *Atmos. Environ.*, **22** (1988) 2841–50.
47. Tjepkema, J. D., Cartica, R. J. & Hemond, H. F., Atmospheric concentration of ammonia in Massachusetts and deposition on vegetation. *Nature*, **294** (1981) 445–6.
48. Ayers, G. P. & Gras, J. L., The concentration of ammonia in southern ocean air. *J. Geophys. Res.*, **88** (1983) 10655–9.
49. Ferm, M., *Concentration Measurements and Equilibrium Studies of Ammonium, Nitrate and Sulfur Species in Air and Precipitation*. University of Göteborg, Göteborg, Sweden, 1986.
50. Erisman, J. W., Vermetten, A. W. M., Pintesterboer, E. F., Asman, W. A. H., Waijers-Ypelaan, A. & Slanina, J., Atmospheric ammonia: distribution, equilibrium with aerosols and conversion rate to ammonium. *Ammonia and Acidification* (EURASAP Symp.), National Institute of Public Health and Environmental Hygiene, Bilthoven, The Netherlands, 1987, pp. 59–72.
51. Quinn, P. K., Charlson, R. J. & Bates, T. S., Simultaneous observations of ammonia in the atmosphere and ocean. *Nature*, **335** (1988) 336–8.
52. Beilke, S. & Gravenhorst, G., Deposition. In *Evaluation of Atmospheric Processes Leading to Acid Deposition in Europe*. Report EUR 11411 (Comm. European Communities), Task Force on Acid Deposition, COST 611, 1987, pp. 113–35.
53. Horváth, L., On the vertical flux of gaseous ammonia above water and soil surfaces. In *Deposition of Atmospheric Pollutants*, ed. H.-W. Georgii & J. Pankrath. Reidel, Dordrecht, The Netherlands, 1982, pp. 17–22.
54. Trijonis, J., Patterns and trends in data for atmospheric sulfates and visibility. In *Acid Deposition Long-Term Trends*. National Academy Press, Washington, DC, USA, 1986, pp. 109–27.
55. Husar, R. B., Emissions of sulfur dioxide and nitrogen oxides and trends for eastern North America. In *Acid Deposition Long-Term Trends*. National Academy Press, Washington, DC, USA, 1986, pp. 48–92.
56. Environmental Quality, *The Sixth Annual Report*. US Government Printing Office, Washington, DC, USA, 1975.
57. Hidy, G. M., Mueller, P. K. & Tong, E. Y., Spatial and temporal distributions of airborne sulfate in parts of the United States. *Atmos. Environ.*, **12** (1978) 735–52.
58. Mészáros, E., On the atmospheric input of sulfur into the ocean. *Tellus*, **34** (1982) 277–82.
59. Ayers, G. P., Ivey, J. P. & Goodman, H. S., Sulfate and methanesulfonate in the maritime aerosol at Cape Grim, Tasmania. *J. Atmos. Chem.*, **4** (1986) 173–85.
60. Willison, M. J., Clarke, A. G. & Zeki, E. M., Seasonal variation in atmospheric aerosol concentration and composition at urban and rural sites in northern England. *Atmos. Environ.*, **19** (1985) 1081–9.

61. Macias, E. S., Zwicker, J., Ouimette, J. R., Hering, S. V., Friedlander, S. K., Cahill, T. A., Kuhlmey, G. A. & Richards, L. W., Regional haze case studies in the southwestern US. I. Aerosol chemical composition. *Atmos. Environ.,* **15** (1981) 1971–86.
62. Anlauf, K. G., Bottenheim, J. W., Brice, K. A., Fellin, P., Wiebe, H. A., Schiff, H. I., Mackay, G. I., Braman, R. S. & Gilbert, R., Measurement of atmospheric aerosols and photochemical products at a rural site in SW Ontario. *Atmos. Environ.,* **19** (1985) 1859–70.
63. Parungo, F. P., Nagamoto, C. T., Rosinsky, J. & Haagenson, P. L., A study of marine aerosols over the Pacific Ocean. *J. Atmos. Chem.,* **4** (1986) 199–226.
64. Heitzenberg, H., Hansson, H. C. & Lannefors, H., The chemical composition of arctic haze at Ny-Alesund, Spitzbergen. *Tellus,* **33** (1981) 162–71.
65. Charlson, R. J., Covert, D. S., Larson, T. V. & Waggoner, A. P., Chemical properties of tropospheric sulfur aerosols. *Atmos. Environ.,* **12** (1978) 39–53.
66. Winkler, P., Acidity of aerosol particles and of precipitation in the North Polar region and over the Atlantic. *Tellus,* **35B** (1983) 25–30.
67. Ono, A., Yamato, M. & Yoshida, M., Molecular state of sulfate aerosols in the remote Everest highlands. *Tellus,* **35B** (1983) 197–205.
68. Willison, M. J., Clarke, A. G. & Zeki, E. M., Chloride aerosols in central northern England. *Atmos. Environ.,* **23** (1989) 2231–9.
69. Garland, J. A., Field measurements of the dry deposition of small particles to grass. In *Deposition of Atmospheric Pollutants*, ed. H.-W. Georgii & J. Pankrath. Reidel, Dordrecht, The Netherlands, 1982, pp. 9–16.
70. Fuzzi, S., Orsi, G. & Mariotti, M., Wet deposition due to fog in the Po Valley, Italy. *J. Atmos. Chem.,* **3** (1985) 289–96.
71. Brewer, R. L., Gordon, R. J. & Shepard, L. S., Chemistry of mist and fog from Los Angeles urban area. *Atmos. Environ.,* **17** (1983) 2267–70.
72. Munger, J. W., Jacob, D. J., Waldman, J. M. & Hoffmann, M. R., Fogwater chemistry in an urban atmosphere. *J. Geophys. Res.,* **88** (1983) 5109–21.
73. Jacob, D. J., Munger, J. W., Waldman, J. D. & Hoffmann, M. R., The H_2SO_4–HNO_3–NH_3 system at high humidities and in fog. *J. Geophys. Res.,* **91** (1986) 1073–88.
74. Winiwarter, W., Puxbaum, H., Fuzzi, S., Facchini, M. C., Orsi, G., Beltz, N., Enderle, K. & Jaeschke, W., Organic acid gas and liquid-phase measurements in Po Valley fall–winter conditions in the presence of fog. *Tellus,* **40B** (1988) 348–57.
75. Mulawa, P. A., Cadle, S. H., Lipari, F., Ang, C. C. & Vandervennet, R. T., Urban dew: its composition and influence on dry deposition rates. *Atmos. Environ.,* **20** (1986) 1389–96.
76. Daum, P. H., Kelly, T. J., Schwartz, S. E. & Newman, L., Measurements of the chemical composition of stratiform clouds. *Atmos. Environ.,* **18,** 2671–84.
77. Falconer, R. F. & Falconer, P. D., Determination of cloud water acidity at a mountain observatory in the Adirondack Mountains of New York State. *J. Geophys. Res.,* **85** (1980) 7465–70.
78. Mohnen, V. A. & Kadlecek, J. A., Cloud chemistry research at Whiteface Mountain. *Tellus,* **41B** (1989) 79–91.

79. Ogren, J. & Rodhe, H., Measurements of the chemical composition of cloudwater at a clean site in central Scandinavia. *Tellus,* **38B** (1986) 190–6.
80. It is not so easy to explain the relatively high NH_4^+ concentrations obtained in the case of NE trajectories (continental emissions?).
81. Saxena, V. K., Stagner, R. E., Hendler, A. H., DeFelice, T. P., Yeh, R. J.-Y. & Lin. N.-H., Monitoring the chemical climate of the Mt. Mitchell State Park for evaluation of its impact on forest decline. *Tellus,* **41B** (1989) 92–109.
82. Rodhe, H. & Granat, L., An evaluation of sulfate in European precipitation 1955–1982. *Atmos. Environ.,* **18** (1984) 2627–39.
83. Barrie, L. A. & Hales, J. M., The spatial distributions of precipitation acidity and major ion wet deposition in North America during 1980. *Tellus,* **36B** (1984) 333–55.
84. Stensland, G. J., Whelpdale, D. M. & Oehlert, G., Precipitation chemistry. In *Acid Deposition Long-Term Trends.* National Academy Press, Washington, DC, USA, 1986, pp. 128–99.
85. Georgii, H.-W., Review of the acidity of precipitation according to the WMO network. *Időjárás,* **85** (1981) 1–9.
86. Wallén, C. C., *Sulphur and Nitrogen in Precipitation*: An attempt to use BAPMoN and other data to show regional and global distribution. Technical Document WHO/TD-N° 103, 1986, Geneva, Switzerland.
87. Brimblecombe, P. & Pitman, J., Long-term deposit at Rothamsted, southern England. *Tellus,* **32** (1980) 261–7.
88. Horváth, L., Trend of nitrate and ammonium content of precipitation water in Hungary for the last 80 years. *Tellus,* **35B** (1983) 304–8.
89. Keene, W. C. & Galloway, J. N., The biogeochemical cycling of formic and acetic acids through the troposphere: an overview of current understanding. *Tellus,* **40B** (1988) 322–34.
90. SMIC, *Inadvertent Climate Modification.* Massachusetts Institute of Technology Press, Boston, MA, USA, 1971.
91. Mészáros, E. & Várhelyi G., An evaluation of the possible effects of anthropogenic sulfate particles on the precipitation ability of clouds. *Időjárás,* **86** (1982) 76–81.
92. Galloway, J. N. & Whelpdale, D. M., An atmospheric sulfur budget for eastern North America. *Atmos. Environ.,* **14** (1980) 409–17.
93. Hastie, D. R., Schiff, H. J., Whelpdale, D. M., Peterson, R. E., Zoller, W. H. & Anderson, D. L., Nitrogen and sulfur over the Western Atlantic Ocean. *Atmos. Environ.,* **22** (1988) 2393–9.
94. Ehhalt, D. H. & Drummond, J. W., The tropospheric cycle of NO_x. In *Chemistry of Unpolluted and Polluted Troposphere*, ed. H.-W. Georgii & W. Jaeschke. Reidel, Dordrecht, The Netherlands, 1982, pp. 219–51.
95. Söderlund, R., NO_x pollutants and ammonia emissions—a mass balance for the atmosphere over NW Europe. *Ambio,* **6** (1977) 118–22.
96. Bónis, K., Mészáros, E. & Putsay, M., On the atmospheric budget of nitrogen compounds over Europe. *Időjárás,* **84** (1980) 57–68.
97. Horváth, L., Mészáros, E., Antal, E. & Simon, A., On the sulfate, chloride and sodium concentration in maritime air around the Asian continent. *Tellus,* **33** (1981) 382–6.

98. Eliassen, A., Hov, O., Iversen, T., Saltbones, J. & Simpson, D., *Estimates of Airborne Transboundary Transport of Sulphur and Nitrogen over Europe*. EMEP/MSC-W Report 1/88, Oslo, Norway, 1988.
99. This estimate does not contain the mass of very large particles which return very quickly from the air to the ocean.
100. Singh, H. B., Salas, L., Shigeishi, H. & Scribner, E., Atmospheric halocarbons, hydrocarbons and sulfur hexafluoride: global distributions, sources and sinks. *Science,* **203** (1979) 899–903.
101. Penkett, S. A., Derwent, R. G., Fabian, P., Borchers, R. & Schmidt, U., Methyl chloride in the stratosphere. *Nature,* **283** (1980) 58–60.
102. Cadle, R. D., Volcanic emissions of halides and sulphur compounds to the troposphere and stratosphere. *J. Geophys. Res.,* **80** (1975) 1650–2.
103. Crutzen, P. J., Heidt, L. E., Krasnec, J. P., Pollock, W. H. & Seiler, W., Biomass burning as a source of atmospheric gases CO, H_2, N_2O, NO, CH_3Cl and COS, *Nature,* **282** (1979) 253–6.
104. Graedel, T. E. & Eisner, T., Atmospheric formic acid from formicine ants: a preliminary assessment. *Tellus,* **40B** (1988) 335–9.

Chapter 2

Sources of Atmospheric Acidity

A. G. CLARKE
Department of Fuel and Energy, University of Leeds, Leeds, LS2 9JT, UK

2.1 NATURAL SOURCES

2.1.1 Introduction

The emissions of acid gases from anthropogenic sources and their impact on the environment are the main concern of this book. However, that impact can only be assessed if all the naturally occurring sources of these gases are also known and can be quantified. Given the widely dispersed nature of the natural sources and the problems of measurement of trace species at low concentrations, often in remote regions, the quantification is a very difficult task. Nevertheless, considerable progress has been made over the last decade. Many of the conclusions relating to the global cycles of sulphur and nitrogen have been brought together in the volumes arising out of the workshops arranged by the Scientific Committee on Problems of the Environment (SCOPE),[1-3] including the most recent volume on acidification in tropical countries.[3]

In this chapter both man-made and natural sources of atmospheric acidity will be reviewed, but the emphasis will be placed not so much on the global balances as on the scale of the natural sources in relation to the man-made sources. This requires that the very uneven geographical distribution of emissions be taken into account. Figure 2.1 illustrates this

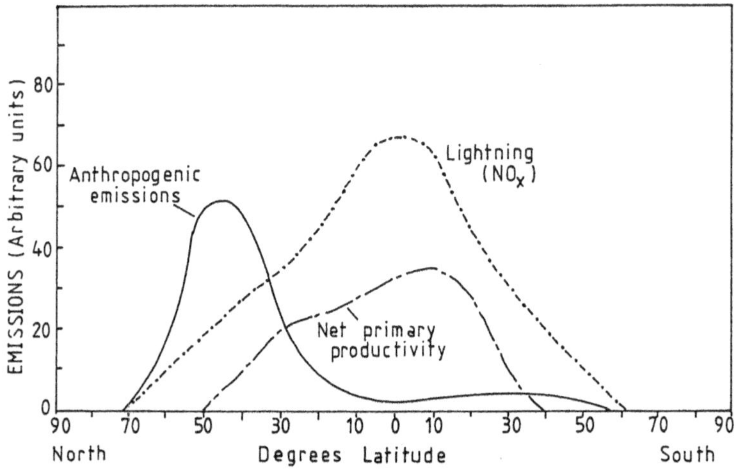

Figure 2.1. Relative distribution with latitude of anthropogenic emissions, lightning and net primary productivity. (Reproduced with permission from Ref. 4.)

point.[4] In contrast to emissions from industrial countries which are located predominantly at fairly high latitudes in the Northern Hemisphere, emissions from biomass conversion peak in equatorial areas since this is where the regions of large primary productivity occur. Climatological reasons also locate the peak of lightning activity (a source of NO_x) at low latitude. Thus, the fact that on a global scale natural emissions of an acid gas or its precursors may be comparable to anthropogenic emissions does not mean that those natural emissions are necessarily important for an 'acid rain' problem in any particular location.

Another factor which must be taken into account along with the geographical distribution is the lifetime of the individual chemical species. A long lifetime or residence time before removal from the atmosphere implies the possibility of an impact a long distance from the source area. A short lifetime implies only local significance. It is in this connection that global emissions figures can be quite misleading and should be interpreted with caution.

2.1.2 Sulphur compounds

2.1.2.1 Volcanoes

There are about 600 active volcanoes in the world although less than 10% of these would be in eruption in any one year. Occasions of recent major eruptions such as Mt. St Helens in the north-western US (1980) and El Chichon in Mexico (1982) have been opportunities to study in detail the gaseous and particulate emissions using airborne sampling equipment or remote-sensing techniques. It is generally assumed that the predominant gaseous sulphur species in the plume is sulphur dioxide (SO_2). However, the relative amounts of SO_2 and hydrogen sulphide (H_2S) can be quite variable. At Mt. St Helens H_2S was found to be dominant during a period of major eruption but subsequently the major species was SO_2 and the mean SO_2/H_2S ratio was about 7.[5] Carbon disulphide (CS_2) and carbonyl sulphide (OCS) levels were below the limit of detection.[6] At El Chichon H_2S was found still to be the major species 7 months after the main eruption.[7] For both volcanoes the particulate sulphate (SO_4^{2-}) was only a small fraction of the total sulphur—of the order of 5%. Table 2.1 shows the estimated fluxes of various species during the relatively quiet, later phases of the eruptions. Chlorine emissions were predominantly as hydrogen chloride (HCl) with a small proportion of particulate chloride (Cl^-). Nitrogen oxide (NO) emissions are small relative to both the sulphur and chlorine species.

Table 2.1
Emission fluxes from recent volcanic eruptions (10^3 kg day^{-1})

	El Chichon[7] *Nov. 1982*	*Mt. St Helens*[5,6] *Sept. 1980*	*Etna*[8] *May 1977*	*Etna*[8] *June 1975*
SO_2	400[a]	450	550	1850
H_2S	—	61		
CS_2, OCS		< 20, < 10		
Particulate SO_4^{2-}	16	24		
Total Cl	53	—		
Particulate Cl^-	2			
NO		1.1		
NH_3, HNO_3	Not detected			

[a]Figure for total sulphur: H_2S dominant.

In the atmosphere H_2S is converted to SO_2 and hence eventually to sulphuric acid (H_2SO_4) with potential acidifying effects on rainfall. However, rain affected by volcanic plumes in the near-field will more effectively remove the HCl than the SO_2. The acidity can then be dominated by HCl as has been shown by Johnson and Parnell near the Masaya volcano in Nicaragua.[9] The higher the measured acidity of the precipitation the higher the fraction contributed by Cl^- rather than SO_4^{2-}. This fraction could be as great as 80%.

Calculation of total annual sulphur emissions can be attempted on the basis of the relationship

$$SO_2 \text{ emission rate} = (\text{lava emission rate}) \times (\text{typical gas/solid ratio}) \times (SO_2 \text{ conc. in gases})$$

A review[10] on the basis of 20th century eruptions up to 1975 suggests that the annual global emission rate of lava is $\sim 3 \times 10^3$ Tg, the volatile component amounts to 3% and the SO_2 content is between 2 and 10% depending on the type of eruption.[10] The final figure is 1·45 Tg of sulphur (as SO_2) released per annum during eruptive behaviour. However, this is dwarfed by the estimate of 26·4 Tg S $year^{-1}$ emitted from fumaroles during the relatively quiescent state of the majority of volcanoes. The final total of 28 Tg S $year^{-1}$ is considerably larger than most previous estimates[11–14] (see Table 2.3).

2.1.2.2 *Biogenic emissions of reduced sulphur compounds*

In this section the emissions of H_2S, CS_2, OCS and organic sulphur compounds are considered. These include dimethyl sulphide (DMS—CH_3SCH_3), dimethyl disulphide (DMDS—CH_3SSCH_3) and methyl hydrogen sulphide (or methyl mercaptan; MS—CH_3SH). In the following section the fate of these species in the atmosphere will be considered and then the oxidation products such as dimethyl sulphoxide (DMSO—$(CH_3)_2SO$), dimethyl sulphone (DMSO$_2$—$(CH_3)_2SO_2$) and methane sulphonic acid (MSA—CH_3SO_3H) will be considered.

As well as its occurrence in volcanic gases, H_2S arises as a result of biological degradation of organic sulphur compounds and the reduction of SO_4^{2-} in anoxic conditions. Emissions have been studied from the oceans (where it is of minor importance compared to DMS), from coastal saline marshes, inland marshes and various soil types.[15] It is now clear that although the most intense emissions are from marshes where the concentration can exceed the detection limit of smell (mixing ratio $0{\cdot}5 \times 10^{-9}$ or 0·5 ppb) the total emissions from the large area of dry inland soils are comparable to if not larger than those from the smaller area of marsh land. In a study of various locations and soil types in North America,[15] it was estimated that 41% of total sulphur emissions came from coastal wetlands and inland marshes (7% of the study area) while 59% of the sulphur emissions came from the drier inland soils (93% of the area).[15] The relative amounts of H_2S, DMS and other reduced sulphur compounds were very variable but the highest emission rates (up to 600 g S m^{-2} $year^{-1}$) were in saline marshes where H_2S was dominant. There was a strong increase of emissions at lower latitudes. Extrapolating the results to the land area of the whole globe 64 Tg S was obtained as the annual emission rate of which 50–70% was H_2S; about 20% was DMS and the balance comprised smaller proportions of other compounds.

In contrast to emissions from land or coastal areas emissions of H_2S from the oceans are low probably due to the relatively highly oxygenated state of the surface layers of water. Sulphur is, however, emitted in the form of DMS which is produced by phytoplankton.[16–18] The emissions are not uniformly distributed throughout the oceans but are elevated in areas of high productivity and regions where upwelling occurs.[16] Highest concentrations in the marine atmosphere occur with high winds (fast sea–air transport), overcast conditions (slower removal by hydroxyl (OH) radicals) and non-continental air masses. Continental air streams have enhanced oxidant levels which promote more rapid removal. Typical concentrations range from 100 to 400 ng S m^{-3} with a mean of 150 ng S m^{-3} (mixing ratio 107×10^{-12} or 107 ppt).[17]

Reference 16 estimates that the global emissions are about 40 ± 20 Tg S $year^{-1}$.[16] Combining the data of Refs 15 and 16 leads to global DMS emissions of about 52 Tg S $year^{-1}$ and global biogenic emissions of sulphur of about 103 Tg S $year^{-1}$. In another review of the available data, the slightly lower total of 70 ± 30 Tg S $year^{-1}$ has been arrived at which is broken down according to the various compounds and their origins as shown in Fig. 2.2.[19]

The significance of the DMS emissions for acidification is in the production of a background concentration of SO_2 which has been estimated to be about 0·05 μg m^{-3} or 35 ppt as a result of oxidation processes.[18] Subsequently, SO_2 is converted to 'non-sea-salt sulphate' (0·04–0·35 μg S m^{-3}). This is further discussed below.

The concentration of OCS in the atmosphere is about 500 ± 50 ppt with very little variation vertically or latitudinally. This constancy suggests natural rather than anthropogenic sources together with a long atmospheric lifetime (2–2·5 years). The total atmospheric burden of OCS at about 4·6 Tg forms the largest single reservoir of sulphur compounds in the atmosphere and results in a significant input of sulphur to the stratosphere. The concentrations of CS_2 are much smaller than OCS and found to be highly variable with several hundred ppt in polluted air but typical values of only 10–20 ppt. The sources and sinks, both natural and an-

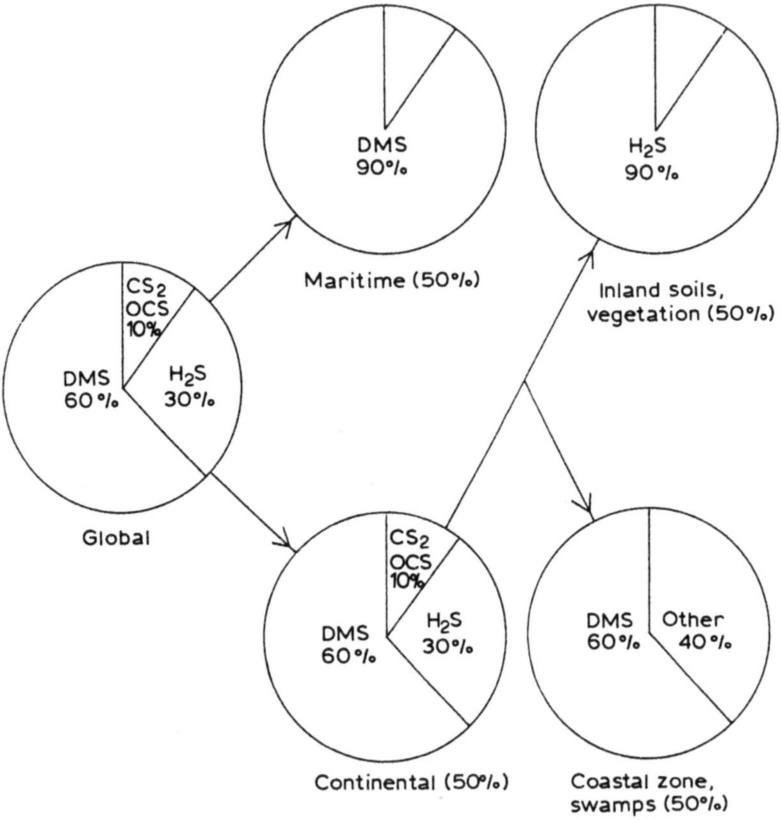

Figure 2.2. Contributions to global emissions of reduced sulphur species. (Adapted from Ref. 19.)

thropogenic, have been reviewed for these gases.[20] The oceans are an important source of OCS and CS_2 with coastal and high biomass productivity areas being as important, if not more so, than the open oceanic regions. This parallels the conclusion for DMS. Soil releases of sulphur compounds are dominated by H_2S with lesser amounts of DMS and the proportion of sulphur released as OCS or CS_2 is small (see above and Fig. 2.2). Photochemical conversion of CS_2 to OCS is fairly rapid so that the lifetime in the atmosphere is only about 13 days. Consequently, the total burden of CS_2 is estimated to be only 0·07 Tg although the sources and sinks of OCS and CS_2 are thought to be similar at about 2 Tg year^{-1}. This figure includes industrial releases of CS_2.[21] The industrial releases of OCS are negligible.

2.1.2.3 Oxidation of reduced sulphur compounds in the atmosphere

The dominant route for the removal of reduced sulphur compounds in the atmosphere is attack by OH radicals followed by reaction with oxygen. In addition to OH the possible importance of nitrogen trioxide (NO_3) and IO radicals as the initiating species in the marine atmosphere has been suggested.[17,22] A schematic diagram of the most important conversion routes is given in Fig. 2.3 and typical reaction rates with OH and corresponding atmospheric lifetimes are given in Table 2.2. The reaction kinetics and mechanisms of organic sulphur compounds with OH have been reviewed,[23] and some of the more recent reports of laboratory measurements are included in the references at the end of the chapter.[25–32] However, what most concerns us is the extent to which the sulphur in these compounds can be assumed to be converted in the atmosphere to SO_2 and hence to the sulphate ion (SO_4^{2-}). Very often this conversion has been

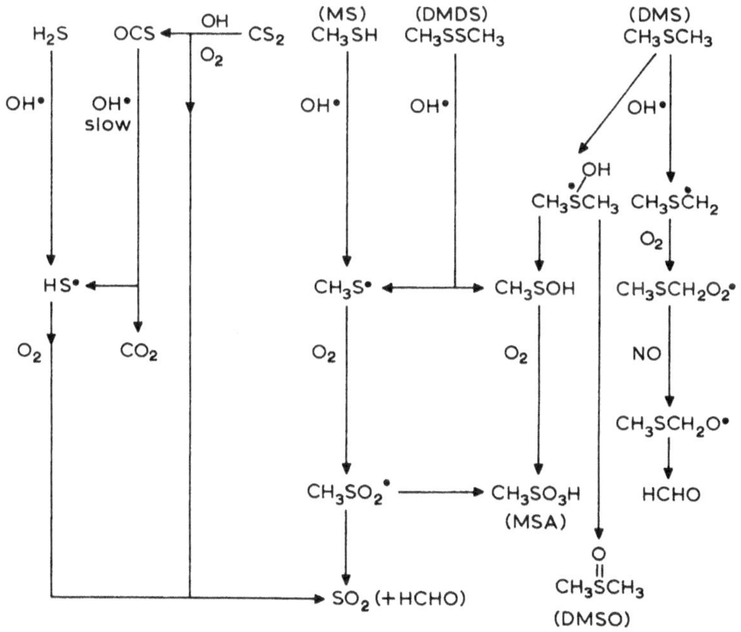

Figure 2.3. Routes for the interconversion and oxidation of reduced sulphur compounds in the atmosphere.

Table 2.2
Rate constants for the reaction of reduced S-compounds with OH; corresponding half-lives in the presence of [OH] = 10^6 cm^{-3} and other estimates of atmospheric residence times

Compound	*Rate constant*[23] *(cm^{-3} mol^{-1} s^{-1})*	*Half-life (days)*	*Residence time*	
			Ref. 24	*Ref. 20*
DMDS	$2{\cdot}33\times10^{-10}$	0·034 (0·8 h)		
MS	$9{\cdot}04\times10^{-11}$	0·089 (2·1 h)		
DMS	$9{\cdot}1\ \times10^{-12}$	0·88	0·75 days	
H_2S	$5{\cdot}0\ \times10^{-12}$	1·6	1·5 days	
CS_2	$0{\cdot}43\times10^{-12}$	18·6	45 days	13 days
OCS	$4\ \times10^{-14}$	200 (0·55 year)	160 days	2–2·5 years

assumed to be complete whereas, in fact, this is not the case for all the species involved.

Laboratory studies show that H_2S is quantitatively converted to SO_2 by gas phase photochemical reactions. CS_2 is converted to OCS and SO_2 but OCS is then only very slowly converted to carbon dioxide (CO_2) and SO_2. The chemistry of the organic sulphides is complex as can be seen from Fig. 2.3 and not all of the detailed reaction paths are clearly established. However, conversion to SO_2 is not quantitative. The major reaction products are SO_2, formaldehyde (HCHO) and MSA with lesser amounts of other species such as DMSO. From DMS the MSA product accounts for 40–60% of the original sulphur and the SO_2 only 20–30%. The majority of the reaction proceeds via the hydrogen abstraction route rather than by the OH· addition route. It follows that the total amount of SO_2 produced by conversion of biogenic sulphur compounds like DMS is significantly less than the emissions.[24]

MSA is quickly removed from the gas phase to the aerosol or droplet phase. Once it is absorbed into cloud or water droplets it is converted to sulphate on a time-scale of hours.[30] Various long-term measurements of MSA in remote marine aerosols have been reported.[29] The MSA amounted to 6·7% of the non-sea-salt sulphate values. Measurements of aerosol MSA, DMSO and DMSO2 in the vicinity of the British Isles have been obtained.[22] As indicated above, although the reaction mechanisms in the atmosphere of DMSO and DMSO2 are still unclear neither are thought to be major reaction products of DMS oxidation.

2.1.2.4 Sea spray

Seawater contains significant amounts of sulphate and wet deposition data for sulphate are normally corrected for the contribution from this source. This can be done on the basis of the sodium (Na^+), chloride (Cl^-) or magnesium (Mg^{2+}) content of the water assuming that there are no additional sources of these ions. Most commonly the SO_4^{2-}/Na^+ ratio is used since there can be anthropogenic Cl^- sources and terrestrial Mg^{2+} sources. This approach presumes that any 'excess' sulphate or 'non-sea-salt' sulphate must have a non-marine or anthropogenic origin. In fact the excess sulphate is non-zero even in remote marine environments. Two reasons have been suggested for this phenomenon. One is that in the process of spray formation of SO_4^{2-} is preferentially concentrated in the droplets; there is limited evidence for an effect of this sort.[31] Alternatively, it is proposed that the excess arises from atmospheric oxidation of reduced sulphur compounds (such as DMS). This suggestion is supported by the discovery that in addition to the normal sea-salt sulphate which is concentrated in large particles of 1–10 μm diameter there is fine particle sulphate of 0·1–1 μm diameter which might be expected to have arisen from a gas-to-particle conversion process. The only other alternative is that the precursor SO_2 came from long-range transport from continental regions over the oceans. In a study of the transatlantic transport of sulphur, the background marine excess sulphate in precipitation is given by Ref. 32 as 6–8 μeq S litre^{-1}. This compares with ~ 30 μeq S litre^{-1} for the western coastal regions of Europe and an approximate figure of 4 μeq S litre^{-1} arising from transatlantic transport from North America. The excess sulphate deposition in remote regions is 200 μg S m^{-2} day^{-1} as given by Ref. 16. This is about ⅔ of their estimate of the DMS daily input to the atmosphere, the remainder being presumed to be converted to other compounds as discussed above.

For many years the global ocean-to-land transfer of sulphur given by Ref. 33 (44 Tg S year^{-1}) was generally adopted as the best available estimate.[33] More recently, this estimate has been revised considerably upwards. Duce considers that it may be too low by a factor of 2–8.[34] As shown in Table 2.3 two recent evaluations include figures of 140–170 Tg S year^{-1}.

2.1.2.5 Global sulphur emissions

Although the man-made sources of sulphur in the atmosphere have not yet been considered it may be helpful at this stage to summarise the natural emission sources. This is done in Table 2.3. As can be seen, on a

Table 2.3
Global sulphur emissions according to different authors (Tg S $year^{-1}$)

	Granat et al. (1976)[35]	*Ryabo-shapko (1983)*[36]	*Cullis & Hirschler (1980)*[14]	*Möller (1984)*[19,37]	*Other (1981–5)*
Natural sources					
Volcanoes	3	28	5	< 2	
Aeolian emission[a]	—	20	—	—	
Biogenic gases from land	5	18	48	35	64[15]
Biogenic gases from water	27	23	50	35	39[16]
Sea spray	44	140	44	171	
Total natural emissions	79	229	147	243	
Anthropogenic emissions	65	113	104	75	79[38]

[a]Wind-blown dusts containing sulphate; mainly relevant to dry continental areas.

global scale most authors are agreed that the natural sulphur emissions are of the same order of magnitude but slightly greater than the man-made emissions although there is still considerable uncertainty about the figures.

2.1.3 Nitrogen compounds

2.1.3.1 Nitrogenous species in the atmosphere

The atmospheric nitrogen cycle is highly complex and involves a large number of species. For details, see the reviews in Refs 39–41. This section will be concerned only with NO_x sources but brief comments on two other species are appropriate here.

Nitrous oxide (N_2O) is released from soil through the action of bacteria. It is the major combined nitrogen species in the atmosphere at a level of about 300 ppb. However, its lifetime in the troposphere is very long (100–200 years) and so it is not a significant source of the acidic nitrogen oxides NO or nitrogen dioxide (NO_2). In the stratosphere it can be converted to NO_x species and there may then be slow stratosphere to troposphere transfer. But this is thought to be of little significance for acid rain.

Ammonia (NH_3) is the major alkaline species in the atmosphere and its role in that capacity is covered in other chapters. It is removed relatively quickly from the gas to the aerosol or droplet phase and its lifetime in the atmosphere is only a few days. However, a small proportion can be converted to NO by photochemical reactions commencing with OH· attack. Estimates of this source are given later.

2.1.3.2 *Lightning*

A typical flash of lightning between clouds and the ground produces a cylinder of air heated to about 30 000 K, roughly 5 cm diameter and as long as 8 km. The energy dissipated may be 50–1000 MJ. At these high temperatures nitrogen gas (N_2) and oxygen gas (O_2) equilibrate with nitrogen and oxygen atoms and with NO_x but as the gases cool the chemical reactions leading to NO_x formation cease below about 1800 K. The amount of NO_x formed can be investigated by atmospheric measurements of NO_x concentrations in the vicinity of electrically active storms (either on the ground or from aircraft) or by laboratory experiments on electrical discharges. Calculation of the total NO_x production rate is by one of the following relationships:

NO production rate = (NO production per unit energy input)×
(total energy deposited)

or = (NO production per lightning flash)×
(lightning flash frequency)

There is considerable uncertainty about each of these factors as discussed in the review in Ref. 42. The global flash frequencies have been derived from satellite data,[43,44] but this is not straightforward. There is a problem of detecting the flashes during bright daylight and as a consequence most of the data relate to night-time or dawn/dusk conditions. Even then the detection rates are very low—only a few per cent. Estimates of the global flash rate are about 40–120 s^{-1}.[43] Two features are clearly revealed by the satellite measurements. One is that only a small fraction of the total global lightning activity occurs over the oceans (37% at dawn, 15% at dusk[43]). The other is that the lightning activity peaks at low latitudes and is less significant at high latitudes (see Fig. 2.1).

Putting all the data together leads to a wide range of estimates of global NO_x production. Levine *et al.*[42] lists authors[45–50] estimating from 4 to above 30 Tg N $year^{-1}$. A figure as high as 70 Tg N $year^{-1}$ has recently been given.[51] Authors who have surveyed all the sources of NO_x in the troposphere have tended to adopt figures below 10 Tg N $year^{-1}$ for lightning (see Table 2.4 below). However, if some of the higher estimates are confirmed these global balances will have to be revised.

It is possible that NO_x from lightning could contribute to *locally* enhanced acidity of precipitation. In the USA the highest flash densities occur in Florida with about 1/400 of the total global flash rate. But even here the estimate of the annual production of fixed nitrogen is only $4{\cdot}7\times10^{-3}$ Tg N year^{-1} less than 1% of the estimated 0·5 Tg N year^{-1} from anthropogenic sources in Florida.[52]

2.1.3.3 *Biogenic production of NO_x from soils*

Numerous laboratory and field studies have been made of the emissions of NO and N_2O from soils. The gases arise from the action of nitrifying bacteria in aerobic conditions and denitrifying bacteria in anaerobic conditions. Nitrifying bacteria, which oxidise NH_4^+ to nitrite and nitrate are the most important. The emissions vary considerably with soil characteristics, O_2 level, water content, history of soil use, vegetation cover, wind speed and NO_x levels in the air above the soil. Field measurements can show emissions varying over several orders of magnitude, fertilised fields giving much higher figures than unfertilised fields. Whereas uncultivated soils in temperate regions can yield emissions of 0·4–2·2 ng N m^{-2} s^{-1},[53] emissions from arable land can be from 17 (unfertilised) up to 62 ng

Table 2.4
Global sources of NO_x according to different authors (Tg N year^{-1})

	Enhalt & Drummond (1982)[61]	*Logan (1983)*[59]	*Stedman & Shetter (1983)*[39]
Natural sources			
Lightning	5 (2–8)	8 (2–20)	3 (2)[a]
NH_3 oxidation	3·1 (1·2–4·9)	1–10	1 (2)
From stratosphere	0·6 (0·3–0·9)	0·5	1 (2)
Biogenic production	5·5 (1–10)	8 (4–16)	10 (2)
Biomass burning	11·2 (5·6–16·4)	12 (4–24)	5 (3)
Total	25	33	20
Anthropogenic sources			
Fuel combustion	13·5 (8·2–18·5)	21 (14–28)	20 (1·1)

[a]The single figures in parentheses are factors of uncertainty.

N m^{-2} s^{-1} (fertilised area).[54] Even higher figures have been given by Ref. 55, with figures up to 250 ng N m^{-2} s^{-1} on unfertilised soil and over 2000 ng N m^{-2} s^{-1} on fertilised soils.

Given the variability just indicated, estimates of global or regional source strengths are very uncertain. They range from the low value of 1 (range 0·2–3·2) Tg N $year^{-1}$ of Ref. 56 through 10 or 11 to 15 Tg N $year^{-1}$.[55,57,58]

2.1.3.4 Biomass burning

By this phrase is meant the burning of vegetation associated with land clearance in the tropics, the burning of savannahs, the use of wood as a fuel and forest wild fires. The two latter sources are relatively small on a global scale. Estimation of the NO_x emissions requires a knowledge of the total amount of biomass burnt and the yield of NO_x per unit mass. The NO_x yield can be determined from direct measurement of the NO_x concentrations in plumes, relating the NO_x/CO_2 concentration ratios to the mass of fuel burnt and the carbon content. Typical figures are about 10^{-3} (Ref. 59), although Ref. 60 suggests a higher figure of $4{\cdot}7\times10^{-3}$. An alternative method is to take account explicitly of the varying nitrogen content of different types of vegetation together with experimental determination of the conversion efficiencies for fuel-N to NO_x (which are about 25%). Reference 59 gives a range of possible values for the global total of biomass emissions of 4–24 Tg N $year^{-1}$, which is in agreement with other estimates (see Table 2.4).

2.1.3.5 Summary of global tropospheric sources of NO_x

The natural sources of NO or NO_2 discussed in the preceding sections are drawn together in Table 2.4 and compared with estimates of the NO_x from fossil-fuel combustion. The agreement of the three surveys included in the table is fairly good but the uncertainties are large as indicated in the table and as pointed out several times above. For example, none of the surveys has adopted one of the higher estimates of the lightning source which could double the total of the natural sources. On a global basis, the natural sources are comparable to the anthropogenic sources.

2.1.4 Chlorine compounds

Emissions of HCl from volcanoes have been mentioned above. Estimates of the global annual source strength range from 0·75 to 7·6 Tg $year^{-1}$.[13,62] Biomass burning may contribute 0·2–0·4 Tg Cl $year^{-1}$ and is therefore a relatively minor contribution.[60] The major natural source is the marine emission of methyl chloride (CH_3Cl) which is converted in the atmosphere to HCl following OH attack. The global emissions are estimated to be 6·69 Tg $year^{-1}$ or 4·7 Tg Cl $year^{-1}$.[63] Apart from anthropogenic releases of various chlorine compounds HCl is generated in the atmosphere by reactions of other acid gases with sea salt. These reactions do not add to the total atmospheric acidity and will not be considered further.

2.1.5 Organic acids

Analysis of precipitation collected at sites throughout the world has revealed the ubiquitous presence of organic acids.[64,65] Formic (HCOOH) and acetic acids (CH_3COOH) are generally found with HCOOH usually being the major species. Special sampling methods are required because these acids are rapidly consumed by bacteria and disappear even from refrigerated samples within a few weeks.[66]

HCOOH and CH_3COOH have little environmental significance since they are rapidly degraded after deposition to the ground. Their lifetime in the atmosphere is quite short—of the order of hours up to a few days—and there is no evidence of long-range transport. However, they can contribute significantly to the measured acidity of precipitation samples and hence affect the interpretation of acid deposition data. With pK_a values of 3·75 for HCOOH and 4·76 for CH_3COOH (at 25°C) the degree of dissociation into the anions and hydrogen ions (H^+) will depend on the pH of the aqueous phase and the proportion of the H^+ ion concentration which is derived from other acidic compounds. In typical 'acid' rainwater with pH 3·8–4·5 HCOOH will be more fully dissociated than CH_3COOH.

Figure 2.4 shows the results of Ref. 65 for the concentration of the two acids relative to the total H^+ ion concentration. The average combined contribution to the free acidity was 16% at a continental site in Virginia, USA, but rather higher fractions ranging from 25 to 59% were found at remote sites. Clearly there are natural sources which are globally distributed. The mass ratio of HCOOH and CH_3COOH in precipitation was about 2. There was a strong seasonal variation of the concentrations in precipitation with a peak in the late summer followed by a rapid fall-off. This is to be expected if the acids were arising from vegetation which was lost in the autumn. Measurements at another site in Virginia covering gases, particles and rainwater gave rather different results with larger amounts of CH_3COOH.[67]

It is clear that there are a multiplicity of sources for these acids but the following possibilities are generally accepted.

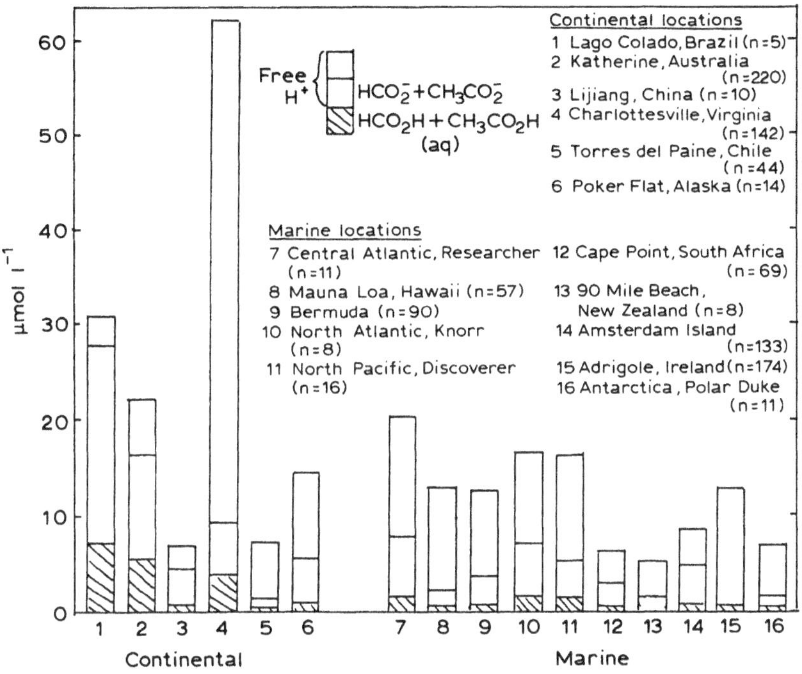

Figure 2.4. Comparison of the concentrations of H^+ from dissociated carboxylic acids with that from other acids together with the concentrations of the undissociated fraction. Data for land-based sites correspond to volume-weighted averages and those for ship-board collections (numbers 7, 10, 11 and 16) represent arithmetic averages. (Reproduced with permission from Ref. 65.)

(1) Aqueous phase oxidation of formaldehyde (HCHO). HCHO is commonly present in the atmosphere being one of the products of photochemical oxidation of hydrocarbons (including methane) as well as being emitted by anthropogenic sources. It is readily absorbed into water droplets and is subsequently oxidised rapidly to HCOOH by OH attack and O_2 addition.[68] Although a similar chemical oxidation can occur with other aldehydes it is not clear whether this is a major route for CH_3COOH production.

(2) Gas phase photochemical oxidation. HCHO can be oxidised in the gas phase by the route

$$HCHO + HO_2 \rightarrow O_2CH_2OH \xrightarrow{NO_2/NO} HCOOH$$

However, the relative importance of this route is uncertain. It depends on the concentration of other gases such as NO and Jacob and Wofsy considered it unimportant in the context of their study of the Amazon Forest.[69] There, the release of isoprene and related hydrocarbons from vegetation is thought to lead to HCOOH via ozonolysis.

(3) Direct release from vegetation. In the case of CH_3COOH it is thought that the direct-release route must be occurring because of the lack of clear evidence for a major chemical formation route.

(4) Anthropogenic sources. There are known to be small emissions from automobiles[70] and from biomass burning which may be significant in urban areas, especially in winter when the biogenic sources are relatively small.[67]

(5) Formicine ants. Graedel and Eisner have made approximate estimates of the amount of HCOOH that could be released into the atmosphere from ants.[71] Their calculated global flux is small relative to the atmospheric chemistry routes.

2.2 ANTHROPOGENIC SOURCES

2.2.1 Combustion of fossil fuels

2.2.1.1 Formation of SO_2 during combustion

Sulphur is present as a significant impurity in most industrial fuels except natural gas. In petroleum products sulphur levels may range from below 0·05 wt % for gasoline, below 0·3 wt % for diesel or gas oil to over 3 wt % for some residual fuel oils. The world-wide average for residual fuel oils is, however, rather lower at 1·8%.[14] British coals generally contain 1·0–2·5% S with domestic bituminous coals and anthracite being on the low side of the range and coals for power stations having a weighted mean of 1·6%.[72] The sulphur content of US coal varies considerably with the state of origin. Western coals are low in sulphur (0·5–1·0%), the Appalachian districts rather higher (1–3%) and the mid-western states very high (over 3%). The average sulphur content of coals for the whole US has dropped considerably over the last 100 years and especially since 1970 and is now about 1·6%.[73] Appropriate emission factors for the calculation of regional and global source strengths based on fuel sulphur contents have been given by several authors.[14,37,38,74]

The chemistry of combustion of sulphur compounds need not be of great concern. Details may be found in reviews such as that by Cullis and Mulcahy.[75] The organic sulphur compounds are degraded at high temperatures to elementary species such as H_2S, HS, S or S_2. These are rapidly converted to SO_2 via the SO· radical. For combustion of petroleum products the conversion may be assumed to be 100%. Only under extremely fuel-rich conditions more typical of fuel processing or gasification conditions are significant amounts of H_2S or OCS formed. The sulphur present in coal is partly in the form of organic compounds and partly in minerals such as iron pyrites (FeS_2). The mineral matter decomposes at flame temperatures to give SO_2 and metal oxides.

A small proportion of the SO_2 formed from the fuel sulphur may be retained by the ash. Traditionally in the UK the round figure of 10% retention has been used when calculating SO_2 emissions. However, this has now been reduced to 5% for the combustion of pulverised coal.[76] The precise figures will depend on the ash content of the coal, the degree of alkalinity of the ash[72] and the mode of combustion. For domestic coal burning as much as 20% of the sulphur may be retained.[74] Emission factors for solid fuels will therefore vary from 0·8 to 0·95 times the theoretical value obtained from the fuel sulphur content and may be rather uncertain.[37]

A small proportion of the SO_2 may be further oxidised to sulphur trioxide (SO_3) by atomic oxygen at high temperatures and hence lead to H_2SO_4 in the flue gas. This may be deposited on internal surfaces and lead to in-plant corrosion problems but a small amount may be emitted from the chimney. Allowance for this 'primary sulphate' may need to be made in models of plume oxidation of SO_2. For coal firing, the SO_4^{2-} to SO_2 ratio is generally of the order of 1%. For oil firing, values 5–10 times higher than those of a coal with similar sulphur content have been measured at some power stations.[77]

2.2.1.2 Formation of NO_x during combustion

If there is no nitrogen in the fuel NO_x can only be formed by the thermal route (Zeldovich mechanism) or the so-called 'prompt NO_x' route from atmospheric nitrogen. During coal or heavy fuel oil combustion the total NO_x is generally dominated by the nitrogen originally present in the fuel at levels of 1–2%. The fate of this nitrogen depends on its distribution along the various paths that lead to NO, N_2 and possibly N_2O. Unlike the situation with SO_2, modification of the combustion conditions (such as air or fuel staging) can alter the relative amounts of end-products and hence offer potential for the reduction of the total NO_x emissions.

Thermal NO_x. The extended Zeldovich mechanism for thermal NO_x formation, which includes the N+OH reaction, is as follows:

$$N+O_2 = NO+O$$

$$O+N_2 = NO+N$$

$$N+OH = NO+H$$

This process is very temperature dependent. The peak flame temperature needs to be over 1800 K before significant amounts of NO are formed and the rate of the reaction roughly doubles for every 40 K temperature rise. The rate increases with the oxygen atom concentration in the flame. Given equilibrium conditions this would suggest a dependence of the rate on $(O_2 \text{ concentration})^{1/2}$, however, super-equilibrium concentrations of oxygen atoms occur in the flame zone and the rate may then be higher than the simplest theory would predict.

The low temperatures during fluidised bed combustion (< 1000°C) mean that thermal NO_x can generally be ignored in comparison with fuel NO_x. During pulverised coal combustion the contribution of thermal NO_x is in the range 20–40% out of a total concentration of up to 700 ppmv. The exact values depend on the temperature–time history of the gases, which in turn depends on the burner design, the furnace geometry and the arrangement of heat transfer surfaces. In general, for large multi-burner units such as power station boilers, tangential firing produces less NO_x than opposed-fired systems which in turn produce slightly less than front wall-fired systems. The high temperatures present in cyclone furnaces mean that they are particularly poor performers in the NO_x control league.[78]

Prompt NO_x. Reaction of atmospheric nitrogen with radical species such as CH or CH_2 can result in NO production via intermediate species such as hydrogen cyanide (HCN). During combustion of nitrogen-free fuels this can occur earlier in the flame zone than NO from the Zeldovich mechanism—hence the term 'prompt NO'. For coal combustion this is generally small compared to NO from fuel nitrogen (< 5%). Hayhurst and Vince estimate that the contribution is also small for diesel and gasoline engines (< 10%) but that for natural gas it may be more significant (50–60%) for household and commercial appliances.[79]

Vehicle emissions. Since neither gasoline, kerosine nor diesel fuel contain significant amounts of nitrogenous hydrocarbons the vehicular and aircraft emissions of NO_x are derived from atmospheric nitrogen. The temperatures in an internal-combustion engine are high and the thermal NO_x emissions are considerably higher than during combustion in conventional boilers and furnaces despite the short residence time. Gas phase concentrations of NO_x in an uncontrolled spark-ignition engine may be several thousand ppmv. They are highest under high speed and high load conditions. Conventional engines operate at a mixture of air to fuel which is slightly leaner (more air) than the stoichiometric mixture. This unfortunately implies high temperatures and high NO_x. Operating with considerably more air (lean-burn technology) results in lower temperatures and lower NO_x and carbon monoxide (CO) emissions but may enhance the hydrocarbon levels. Recirculating the exhaust gas is another way of lowering the temperature and hence reducing NO_x. However, the most effective method of NO_x control currently available is catalytic reduction of the exhaust gases using a 'three-way' catalyst. Diesel engines operate with higher pressures and a higher air-to-fuel ratio than spark-ignition engines and the ratio varies with load. At full load a typical engine may generate 1–2000 ppmv of NO_x whereas at low load, with much greater air dilution, the concentration reduces to only a few hundred ppmv.

The dependence of vehicular emissions on speed and load means that emission factors must be based on an appropriate distribution of operating conditions. Performance in test drive cycles used for compliance with emissions legislation may not in fact give an adequate account of the on-the-road emissions. Accurate emissions estimates therefore require the following:

- on-the-road emissions per kilometre as a function of speed for various vehicle classes,
- vehicle–speed distributions for each class, combined with
- total vehicle–kilometre figures for each class.

Such data have been used for the latest UK estimates of NO_x emissions.[74,76]

Fuel nitrogen chemistry. Rapid pyrolysis of coal particles as they enter a flame results in a distribution of the fuel nitrogen between the char, the volatile tars and light gases. The tars undergo further cracking reactions and the nitrogen is released as simple molecules such as HCN or NH_3 (or radicals such as CN or NH_2 derived from these parent molecules) (Fig. 2.5). A similar process occurs with heavy fuel oils although the char component is of relatively minor importance.

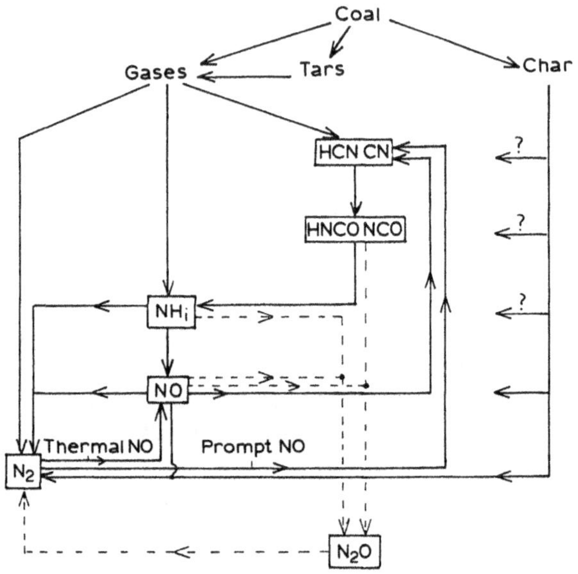

Figure 2.5. Conversion routes for nitrogen species during coal combustion.

The reactions which lead from HCN and NH_3 to the final products are indicated in the simplified schematic representation in Fig. 2.5. Full kinetic models may involve over 50 reactions just to represent the nitrogen chemistry.[80] In the presence of excess air and hence high O_2, O and OH concentrations, the NH_i species (i = 0–3) are converted to NO. Under fuel-rich conditions the NO reactions with NH or NH_2 leading to N_2 are important. This is the basic objective of the technique of staged combustion—to achieve maximum reduction of NO in the fuel-rich stage. NO can also be converted back to cyano species (and hence potentially to N_2) by reaction with hydrocarbon radicals. This is the basis of the reburn technique for NO_x reduction.

Note that with 'low-NO_x' burners it is the volatile nitrogen conversion to NO which is susceptible to control by changes in the pattern of near-field burner aerodynamics. The production of NO from char burn-out is on a longer time-scale and is therefore relatively unaffected.

The efficiency of conversion of char nitrogen into NO is much lower than that of volatile nitrogen—up to 20% compared to perhaps 40–50%.

Reasons for this are not entirely clear but may relate to NO reducing reactions at or near the particle surface:

$$CO+NO = CO_2+½N_2\text{(surface reaction)}$$

$$\text{C-surface}+NO = \text{CO-surface}+½N_2$$

These compete with oxidation of surface N to NO and with oxidation of other volatile N species released into the gas.

The general conclusions relating to the conversions of fuel nitrogen into NO are as follows.

(1) The amount of NO produced increases with fuel nitrogen content but the conversion efficiency decreases. Doubling the fuel nitrogen does not produce double the NO emissions.
(2) Fuel nitrogen conversion is insensitive to temperature, unlike thermal NO.
(3) Decreasing the air/fuel ratio decreases the NO emissions.
(4) The conversion of char nitrogen into NO is less efficient than the conversion of volatile nitrogen.

Given the considerable difficulty of predicting NO_x emissions when fuel nitrogen is present, estimates of emissions are based on factors derived from experimental surveys. Such factors have been given by the US EPA and other authors.[59,74,81,82]

NO_2 formation. In general NO is the primary pollutant and NO_2 is formed in the post-flame gases during flame quenching and by interaction with cold air. The steps involved are

$$H+O_2 = HO_2$$

$$HO_2+NO = NO_2+OH$$

The amount of NO_2 can vary depending on the system but is generally about 5% of the total NO_x in stationary combustion sources. Higher proportions are found in gas turbine combustion. Once in the atmosphere further oxidation takes place and the NO_2 can then equal or exceed the NO concentration.

N_2O formation. N_2O must be included in this assessment although it has frequently been overlooked in the past. Although it is stable in the atmosphere (lifetime ~ 100 years) and therefore has no relevance to the

current debate about acid rain, it is of current interest because of contributions to the atmospheric 'greenhouse' effect and to the problem of stratospheric ozone depletion. It has been shown that the atmospheric concentration of N_2O is increasing at a rate of approximately 0·25% per year and coal combustion could be a major anthropogenic origin of this increase. Possible formation mechanisms of N_2O are indicated in Fig. 2.5. Studies over the last few years seemed to suggest that up to 1 mole of N_2O might be formed for every 3 moles of NO and it would therefore be a significant contributor to the total nitrogen species in the flue gas. More recently it has been established that many of the early measurements were in error due to the formation of N_2O from SO_2 and NO_x in the presence of moisture on the walls of the sample flasks.[83] It now seems likely that in most situations N_2O emissions are relatively insignificant.

2.2.1.3 *Formation of HCl during combustion*

The largest anthropogenic source of HCl is the combustion of coal.[84] Chlorine occurs in coal in the form of soluble metal chlorides, predominantly sodium chloride but also other salts such as calcium chloride. Its origin is discussed by Raask.[72] In the UK typical coals contain 0·2–0·3% chloride by weight; high chlorine coals up to 0·7% are known. During combustion most of the chlorine is converted to HCl while the sodium is converted to the sulphate.[85] Iapalucci *et al.* estimate that 94% of the chlorine is emitted as HCl.[86] Flue gas desulphurisation will remove over 95% of the emissions.

Oils contain negligible amounts of chlorine but small quantities of ethylene dichloride are added to gasoline in association with lead anti-knock additives. Only a small fraction of this chlorine is thought to be emitted as HCl and the estimate of Lightowlers and Cape of 0·4 kt $year^{-1}$ for the UK[84] in 1984 must represent a considerable overestimate of the current situation as the amount of leaded gasoline sold is progressively being reduced.

The other major source is the incineration of waste. Chlorides occur in waste as salt, in vegetable matter and as polyvinyl chloride (PVC). Chlorine emissions from incinerators are very variable and difficult to predict due to the heterogeneous and inconsistent nature of the waste. Lightowlers and Cape assume HCl emissions to be equivalent to 38% of the weight of PVC and 50% of non-PVC chlorides.[84] Data on the total mass of waste and the PVC content can be used to obtain the emissions levels and these are given for 12 European countries.[84] New EC directives on incinerator emissions set limitations on acid-gas emissions and the levels will be gradually reduced.[87,88]

Other sources of HCl include glass-making, the steel industry (steel pickling and acid regeneration plant) and the release of chlorinated hydrocarbons into the atmosphere. In the latter case the HCl is slowly formed as a secondary pollutant, often a long way from the source.

2.2.2 Industrial process losses

The largest single source of SO_2 apart from combustion is the smelting of non-ferrous sulphide ores. Copper is the main contributor to the emissions with lead and zinc being less important ones. The scale of the problem may be realised when it is appreciated that for every 1 t of copper+nickel in the ore there may be 4·5 t of sulphur. The Canadian smelter of Inco Ltd in Sudbury alone had emissions of 1 Tg S $year^{-1}$ in 1988 although this had been reduced to 0·342 Tg S $year^{-1}$ by 1989 and will be further reduced to below 0·132 Tg S $year^{-1}$ by 1994. This represents 90% containment of the mined sulphur. For 1979 Varhelyi estimated global smelter emissions of 9·8 Tg S $year^{-1}$ of which copper represented 8·1 Tg S $year^{-1}$.[38] Other sources of sulphur include oil refineries (4·9 Tg S $year^{-1}$) and H_2SO_4 production (0·65 Tg S $year^{-1}$). Conversion of pulp into paper also produces localised emissions (0·25 Tg S $year^{-1}$)[14] of which the largest fraction is H_2S derived from the Kraft process. Combining the various industrial sources Varhelyi estimates that about 19% of the global sulphur emissions arise from industrial sources. However, given the progressively improving control of process losses these emission estimates should be treated with caution.

For nitrogen oxides the main sources of emissions are nitric acid manufacture and high-temperature processes such as cement and glass manufacture (although in emission estimates the latter are generally included with the combustion sources according to the fuel burnt). The process losses for NO_x are relatively unimportant compared to the situation for SO_2. They probably amount to no more than about 5% of the global anthropogenic emissions[59] (see Table 2.6).

2.2.3 Emissions data

Table 2.5 shows the distribution of SO_2, NO_x and HCl emissions among various source categories within the UK. As can be seen, power stations contribute nearly 75% of the total national emissions of SO_2 with refineries and other industry a further 18% and the domestic and commercial categories only 7%. In terms of the fuels burnt coal together with a small amount of solid smokeless fuel provides 81% of the SO_2 emissions and fuel oils 12% (1987 data[76]). This picture will gradually change over the next decade with changing fuel-use patterns including a possible slight

Table 2.5
Annual UK emissions (10^3 t) of SO_2, NO_x and HCl

Source category	*SO_2[a]*		*NO_x[a,c]*		*HCl[b]*	
	Emissions	%	*Emissions*	%	*Emissions*	%
Domestic	171	4	74	3	22	9
Commercial/public service	106	3	61	3	2	1
Power stations	2830	73	809	35	178	69
Refineries	162	4	43	2	—	—
Agriculture	8	—	4	—	—	—
Other industry	539	14	236	10	39	15
Rail transport	4	—	35	2	—	—
Road transport	46	1	1031	45	0·4	—
Incineration/agricultural burning	—	—	12	1	16	6
Total	3867	100	2303	100	257	100
Per cent of total potential acidity		68		28		4

[a]SO_2 and NO_x data are for 1987.[76]
[b]HCl based on coal data for 1983, adapted from Ref. 84.
[c]NO_x expressed as NO_2 equivalent.

increase in the nuclear power capacity together with the introduction of flue gas desulphurisation at some of the existing coal-fired power stations and at projected new power stations. There is also likely to be the introduction of electricity-generating capacity based on natural gas. These changes should ensure that the UK meets its commitments to recently agreed European legislation. The 'large combustion plants' directive sets targets for reduction of SO_2 emissions from all the member states of the EC.[88] The target for the UK is 60% reduction relative to the 1980 level by the year 2003. Although different states have different targets the overall EC emissions from large combustion plants (greater than 50 MW) should be decreased from 14·43 Tg in 1980 to 6·14 Tg in 2003, a decrease of 58%.

The UK NO_x emissions shown in Table 2.5 illustrate the dominance of road transport (45%) and power generation (35%) as the main sources. The corresponding figures for the whole of Europe are 43% and 28%.[82] Implementation of the EC directive[89] will lead to a 30% reduction in the contribution from large combustion plant between 1980 and 1998 when the emissions should be below 0.711 Tg $year^{-1}$. Introduction of stricter controls on vehicle emissions should ensure that this contribution will also gradually be reduced although progress may be slow given the likely growth in the number of vehicles.

The HCl emissions arise predominantly from power generation since this is where most of the coal in the UK is burnt. Table 2.5 also shows the per cent of total potential acidity contributed by the three pollutants. SO_2 accounts for two-thirds, NO_x for just under one-third and HCl the remainder. These figures are, of course, for the UK as a whole. Locally, the

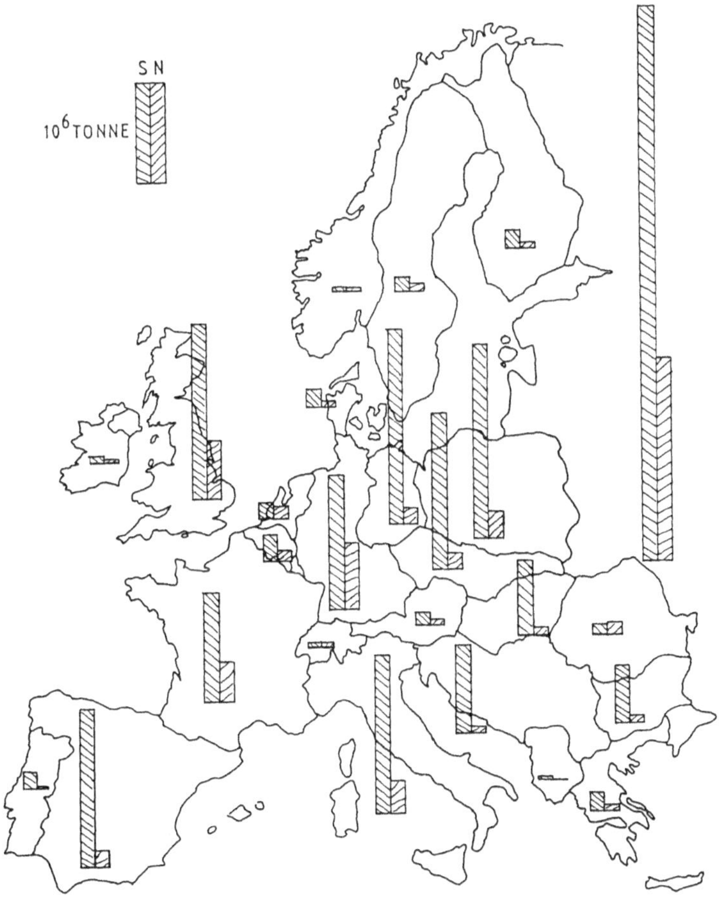

Figure 2.6. European national emissions of SO_2 and NO_x (10^6 t S or N year^{-1}). SO_2 data are for 1983,[90] NO_x data are for the late 1970s.[92] Former USSR emissions relate only to that part of the country within the EMEP grid.

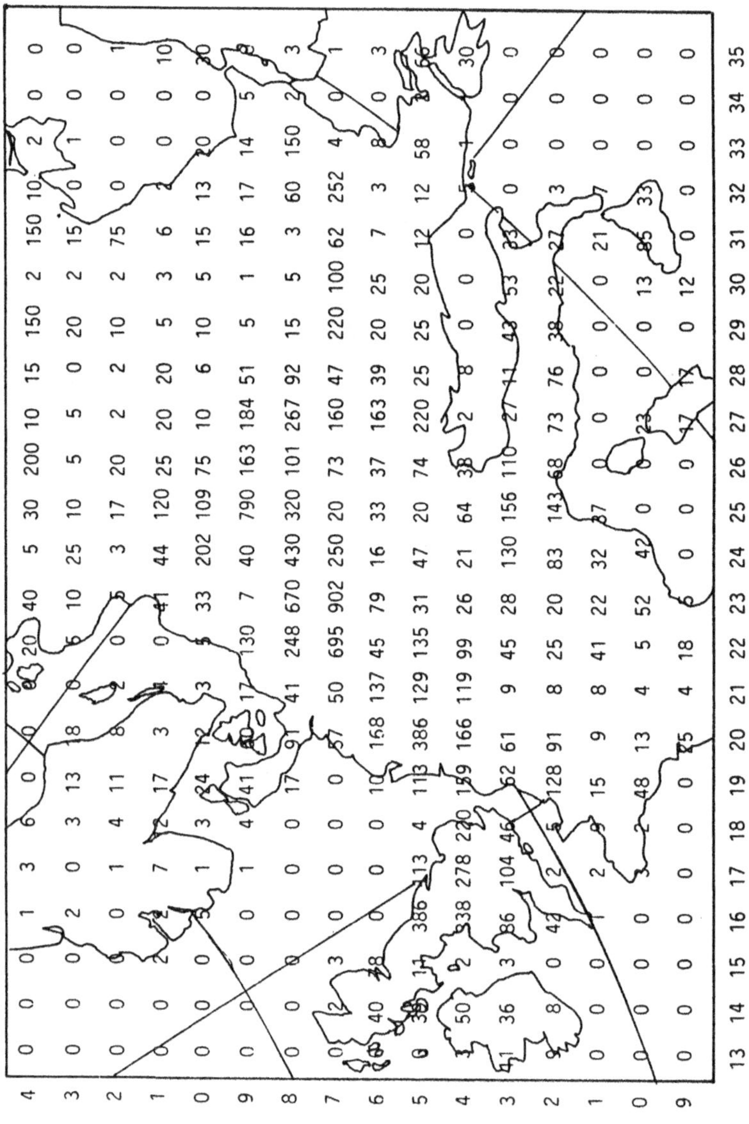

Figure 2.7. Distribution of SO_2 emissions (kt S $year^{-1}$) within a part of the EMEP grid for 1983. Grid size 150 km. (Adapted from Ref. 90.)

proportions may be significantly different. For example, in the vicinity of coal-fired power stations the HCl proportion is enhanced due to the relatively rapid rainout and washout of this gas (*cf.* Section 2.1.2.1 on volcanoes).

The spatial distribution of European SO_2 emissions is shown in Figs 2.6 and 2.7. These portray the inventories produced as part of the European Evaluation and Monitoring Programme (EMEP). Figure 2.6 indicates the national emissions of SO_2 and NO_x. Figure 2.7 shows the detailed distribution of SO_2 emissions in central Europe and is based on a 150 km square grid.[90] The grid is aligned to longitude 32°W and hence the grid numbering system shown in the figure does not commence at zero. The distribution of NO_x parallels that for SO_2 fairly closely, as might be expected.

Inventories with a greater spatial resolution are available for some countries. The (OECD) Organisation for Economic Cooperation and Development produced inventories of SO_2, NO_x and volatile organic carbons (VOC) for 12 European countries on a 50×50 km grid for 1980.[91] The countries included the UK for which data down to 10×10 km grid size for SO_2, NO_x and VOC are now available.[74] The most recent comprehensive inventory is that produced by the Commission of the European Com-

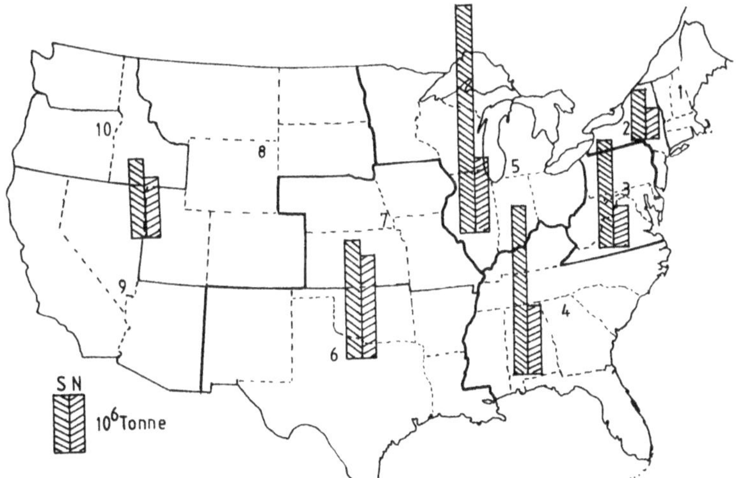

Figure 2.8. Distribution of SO_2 and NO_x emissions across the USA in 1980 expressed in terms of $\times 10^6$ t S or N year^{-1}. The full lines and numbers relate to EPA administrative regions which are in some cases grouped together for the purposes of this figure. Dashed lines indicate interstate boundaries. (Adapted from Ref. 92.)

munity and OECD CORINAIR Project. This presents the 1985 data for SO_2, NO_x and VOC on a spatial resolution corresponding to a French Departement or UK county.

The distribution of emissions of SO_2 and NO_x for the US is illustrated in Fig. 2.8 using 1980 data.[92,93] The dominance of the eastern states in the overall picture is evident. Emissions of SO_2 are on a downward trend due to the tightening of controls on power plants which contribute approximately two-thirds of the emissions. There was a 26% reduction between 1970 and 1987. NO_x emissions grew between 1970 and 1980 but have since fallen slightly. This relates to the more stringent emissions controls counterbalancing the increased vehicle numbers. The proportion of NO_x emissions from vehicles at 43% is very similar to that in Europe.

Various emissions inventories have been produced for the US. Earlier ones are listed in Ref. 94 (see also Refs 73, 92, 93). The importance of such inventories is that they form the basis of modelling studies of the transport and deposition of pollutants on a continental scale. These models can then be used to predict the likely effects of possible control strategies.

Table 2.6
Global distribution of anthropogenic SO_2 and NO_x emissions (Tg S or N year^{-1}) for 1979

Continent	SO_2[a]	%	NO_x[b]	%
North America	19·1	24	6·0	30·5
Europe	34·8	44	7·3	37
Africa	2·3	3		
Asia	18·2	23		
Oceania	0·7	1	6·4	32·5
Central and South America	4·1	5		
World	79·2	100	19·7	100

[a]SO_2 Ref. 38; Europe includes 75% of fuel use in the former USSR and Asia 25%.
[b]NO_x Ref. 59; Europe includes 80% of fuel use in the former USSR.
NO_x total excludes industrial sources estimated to be about 1·2 Tg N year^{-1} and air transport estimated to be 0·2 Tg N year^{-1}. Overall total thus 21·1 Tg N year^{-1}.

The distribution of emissions on a global scale is indicated in Table 2.6. Europe and North America account for over half of the SO_2 emissions and two-thirds of the NO_x emissions. Logan did not attempt a detailed breakdown of the continental NO_x emissions,[59] but the data of Dignon and Hameed suggest that South America, Africa and Oceania only contribute 7% of the global total.[95] Comparison of the anthropogenic emissions with the natural emissions was made in Tables 2.3 and 2.4. Historical trends in the emissions have been given by Möller[37] and Dignon and Hameed[95] (and other references cited therein). This aspect is covered in a later chapter.

REFERENCES

1. Ivanov, M. V. & Freney, J. R., eds, *The Global Biogeochemical Sulphur Cycle,* SCOPE 19. John Wiley & Sons, Chichester, 1983.
2. Bolin, B. & Cook, R. B., eds, *The Major Biogeochemical Cycles and their Interactions,* SCOPE 21. John Wiley & Sons, Chichester, 1983.
3. Rhode, H. & Herrera, R., *Acidification in Tropical Countries,* SCOPE 36. John Wiley & Sons, Chichester, 1988.
4. Isaksen, I. S. A. & Hov, Ø., Calculation of trends in the tropospheric concentration of O_3, OH, CO, CH_4 and NO_x. *Tellus,* **39B** (1987) 271–85.
5. Hobbs, P. V., Radke, L. F., Elgroth, M. W. & Hegg, D. A., Airborne studies of the emissions from the volcanic eruption of Mt. St Helens. *Science,* **211** (1981) 816–18.
6. Bandy, A. R., Maroulis, P. J., Wilner, L. A. & Torres, A. L., Estimates of the fluxes of NO, SO_2, H_2S, CS_2 and OCS from Mt. St Helens deduced from in situ plume concentration measurements. *Geophys. Res. Lett.,* **9** (1982) 1097–100.
7. Kotra, J. P., Finnegan, D. L., Zoller, W. H., Hart, M. A. & Moyers, J. L., El Chichon: composition of plume gases and particles. *Science,* **222** (1983) 1018–21.
8. Zetwoog, P. & Haulet, R., Experimental results on the SO_2 transfer in the Mediterranean obtained with remote sensing devices. *Atmos. Environ.,* **12** (1978) 795–6.
9. Johnson, N. & Parnell, R. A., Composition, distribution and neutralisation of 'acid rain' derived from the Masaya volcano, Nicaragua. *Tellus,* **38B** (1986) 106–17.
10. Yu Lein, A., Sulphur cycle in the lithosphere. In *The Global Biogeochemical Sulphur Cycle,* SCOPE 19, ed. M. V. Ivanov & J. R. Freney. John Wiley & Sons, Chichester, 1983, pp. 95–104.
11. Kellogg, W. W., Cadle, R. D., Allen, E. R., Lazrus, A. L. & Martell, E. A., The sulphur cycle. *Science,* **175** (1972) 587–96.
12. Stoiker, R. E. & Jepson, A., Sulphur dioxide contributions to the atmosphere by volcanoes. *Science,* **182** (1973) 577–8.
13. Cadle, R. D., Volcanic emissions of halides and sulphur compounds to the troposphere and stratosphere. *J. Geophys. Res.,* **80** (1975) 1650–2.
14. Cullis, C. F. & Hirschler, M. M., Atmospheric sulphur: natural and man-made sources. *Atmos. Environ.,* **14** (1980) 1263–78.

15. Adams, D. F., Farwell, S. O., Robinson, E., Pack, M. R. & Bamesburger, W. L., Biogenic sulfur source strengths. *Environ. Sci. Technol.*, **15** (1981) 1493–8.
16. Andrae, M. O. & Raemdonck, H., Dimethyl sulphide in the surface ocean and the marine atmosphere: a global view. *Science*, **221** (1983) 744–7.
17. Andrae, M. O., *et al.*, Dimethyl sulphide in the marine atmosphere. *J. Geophys. Res.*, **90** (1985) 12891–900.
18. Nguyen, B. C., Bonsang, B. & Gaudry, A., The role of the ocean in the global atmospheric sulphur cycle. *J. Geophys. Res.*, **88** (1983) 10903–14.
19. Möller, D., On the global natural sulphur emission. *Atmos. Environ.*, **18** (1984) 29–39.
20. Khalil, M. A. K. & Rasmussen, R. A., Global sources, lifetimes and mass balances of carbonyl sulphide (OCS) and carbon disulphide (CS_2) in the earth's atmosphere, *Atmos. Environ.*, **18** (1984) 1805–13.
21. Peyton, T. O., Steele, R. V. & Mabey, W. R., *Carbon Disulphide, Carbonyl Sulphide: Literature Review and Environmental Assessment.* SRI, Menlo Park, CA, USA, 1976.
22. Watts, S. F., Watson, A. & Brimblecombe, P., Measurements of the aerosol concentrations of methanesulphonic acid, dimethyl sulphoxide and dimethyl sulphone in the marine atmosphere of the British Isles. *Atmos. Environ.*, **21** (1987) 2667–72.
23. Atkinson, R., Kinetics and mechanisms of the gas phase reactions of the hydroxyl radical with organic compounds under atmospheric conditions. *Chem. Rev.*, **86** (1986) 69–201.
24. Sze, N. D. & Ko, M. K. W., Photochemistry of COS, CS_2, CH_3SCH_3 and H_2S: implications for the atmospheric sulfur cycle. *Atmos. Environ.*, **14** (1980) 1223–39.
25. Cox, R. A. & Sheppard, D., Reactions of OH radicals with gaseous sulphur compounds. *Nature*, **284** (1980) 330–1.
26. Hatakeyama, S. & Akimoto, H., Reactions of OH radicals with methanethiol, dimethyl sulphide and dimethyl disulphide in air. *J. Phys. Chem.*, **87** (1983) 2387–95.
27. Hatakeyama, S., Izumi, K. & Akimoto, H., Yield of SO_2 and formation of aerosol in the photooxidation of DMS under atmospheric conditions. *Atmos. Environ.*, **19** (1985) 135–41.
28. Grosjean, D. & Lewis, R., Atmospheric photooxidation of methyl sulphide, *Geophys. Res. Lett.*, **9** (1982) 1203–6.
29. Saltzman, E. S., Savoie, D. L., Zika, R. G. & Prospero, J. M., Methane sulphonic acid in the marine atmosphere. *J. Geophys. Res.*, **88** (1983) 10897–902.
30. Panter, R. & Penzhorn, R. D., Alkyl sulphonic acids in the atmosphere. *Atmos. Environ.*, **14** (1980) 149–51.
31. Garland, J. A., Enrichment of sulphate in maritime aerosols. *Atmos. Environ.*, **15** (1981) 787–91.
32. Whelpdale, D. M., Eliassen, A., Galloway, J. N., Dovland, H. & Miller, J. M., The transatlantic transport of sulphur. *Tellus*, **40B** (1988) 1–15.
33. Eriksson, E., The yearly circulation of chloride and sulphur in nature. *Tellus*, **12** (1960) 63–109.

34. Duce, R. A., Biogeochemical cycles and the air–sea exchange of aerosols. In *The Major Biogeochemical Cycles and their Interactions,* SCOPE 21, ed. B. Bolin & R. B. Cook. John Wiley & Sons, Chichester, 1983, pp. 427–56.
35. Granat, L., Rhode, H. & Hallberg, R. O., The global sulphur cycle. In *Nitrogen, Phosphorus and Sulphur—Global Cycles,* SCOPE 7, ed. B. H. Svensson & R. Soderland. *Ecol. Bull.* (Stockholm), **22** (1976) 89–134.
36. Ryaboshapko, A. G., The atmospheric sulphur cycle. In *The Global Biogeochemical Sulphur Cycle,* SCOPE 19, ed. M. V. Ivenov & J. R. Freney. John Wiley & Sons, Chichester, 1983, pp. 203–78.
37. Möller, D., On the global man-made sulphur emission. *Atmos. Environ.,* **18** (1984) 19–27.
38. Varhelyi, G. Continental and global sulfur budgets—anthropogenic SO_2 emissions. *Atmos. Environ.,* **19** (1985) 1029–40.
39. Stedman, D. H. & Shetter, R. E., The global budget of atmospheric nitrogen species. In *Trace Atmospheric Constituents,* ed. S. E. Schwartz. John Wiley, New York, 1983, pp. 411–54.
40. Rosswall, T., The nitrogen cycle. In *The Major Biogeochemical Cycles and their Interactions,* SCOPE 21, ed. B. Bolin & R. B. Cook. John Wiley & Sons, Chichester, 1983, pp. 46–50.
41. Crutzen, P. J. Atmospheric interactions—homogeneous gas reactions of C, N, and S containing compounds. In *The Major Biogeochemical Cycles and their Interactions,* SCOPE 21, ed. B. Bolin & R. B. Cook. John Wiley & Sons, Chichester, 1983, pp. 67–112.
42. Levine, J. S., Augustsson, T. R., Anderson, I. C. & Hoell, J. M., Tropospheric sources of NO_x: lightning and biology. *Atmos. Environ.,* **18** (1984) 1797–804.
43. Turman, B. N. & Edgar, B. C., Global lightning distributions at dawn and dusk. *J. Geophys. Res.,* **87** (1982) 1191–206.
44. Orville, R. E. & Spencer, D. W., Global lightning flash frequency. *Monthly Weather Rev.,* **107** (1979) 934.
45. Tuck, A. F., Production of nitrogen oxides by lightning discharges. *Quart. J. Royal Meteorol. Soc.,* **102** (1976) 749–55.
46. Noxon, J. F., Atmospheric nitrogen fixation by lightning. *Geophys. Res. Lett.,* **3** (1976) 463–5.
47. Peyrous, R. & Lapeyre, R. M., Gaseous products created by electrical discharges in the atmosphere and condensation nuclei resulting from gas phase reactions. *Atmos. Environ.,* **16** (1982) 959–68.
48. Levine, J. S., Rogowski, R. S., Gregory, G. L., Howell, W. E. & Fishman, J., Simultaneous measurements of NO_x, NO and O_3 production in a laboratory discharge: atmospheric implications. *Geophys. Res. Lett.,* **8** (1981) 357–60.
49. Chameides, W. L., Effect of variable energy input on nitrogen fixation in instantaneous linear discharges. *Nature,* **277** (1979) 123–5.
50. Drapcho, D. L., Sisterson, D. & Kumar, R., Nitrogen fixation by lightning activity in a thunderstorm. *Atmos. Environ.,* **17** (1983) 729–34.
51. Popp, C. J., Franzblau, E. & Chavez, R., Sources of NO_x and SO_2 as precursors of acid deposition in the southwestern United States: New Mexico. In *Acid Rain: Scientific and Technical Advances,* ed. R. Perry, R. M. Harri-

son, J. N. B. Bell & J. N. Lester. Selper Ltd, London, UK, 1987, pp. 13–18.
52. Green, A. E. S. & Uman, M. A., Lightning in Florida as a source of acid rain. In *Acid Deposition Causes and Effects: A State Assessment Model,* ed. A. E. S. Green & W. H. Smith. Gov. Inst., Rockville, MD, USA, 1984, pp. 209–19.
53. Johansson, C. H., Rodhe, H. & Sanhueza, E., Emission of NO in a tropical savannah and cloud forest during the dry season. *J. Geophys. Res.,* **93** (1988) 7180–92.
54. Johansson, C. H. & Granat, L., Emissions of nitric oxide from arable land. *Tellus,* **36B** (1984) 25–37.
55. Slemr, F. & Seiler, W., Field measurements of NO and NO_2 from fertilized and unfertilized soils. *J. Atmos. Chem.,* **2** (1984) 1–24.
56. Johansson, C. H., Field measurements of emissions of nitric oxide from fertilized and unfertilized forest soils in Sweden. *J. Atmos. Chem.,* **1** (1983) 429–42.
57. Galbally, I. E. & Roy, C. R., Loss of fixed nitrogen from soils by nitric oxide exhalation. *Nature,* **275** (1978) 734–5.
58. Lipschultz, F., Zafiriou, O. C., Wofsy, S. C., McElroy, M. B., Valois, F. W. & Watson, S. W., Production of NO and N_2O by soil nitrifying bacteria. *Nature,* **294** (1981) 641–3.
59. Logan, J. A., Nitrogen oxides in the troposphere: global and regional budgets. *J. Geophys. Res.,* **88** (1983) 10785–807.
60. Crutzen, P. J., Heidt, L. E., Krasnec, J. P., Pollock, W. H. & Seiler, W. Biomass burning as a source of atmospheric gases CO, H_2, N_2O, NO, CH_3Cl, COS. *Nature,* **282** (1979) 253–6.
61. Enhalt, D. & Drummond, J. W., The tropospheric cycle of NO_x. In *Chemistry of the Unpolluted and Polluted Troposphere,* ed. H. W. Georgii & W. Jaeschke. D. Reidel, Dordrecht, The Netherlands, 1982, pp. 219–51.
62. Bartels, O. G., An estimate of volcanic contributions to the atmosphere and volcanic gases and sublimates as the source of the radioisotopes ^{10}Be, ^{32}S, ^{32}P and ^{22}Na, *Health Phys.,* **22** (1972) 387–92.
63. National Research Council Panel on Atmospheric Chemistry, *Halocarbons: Effects on Stratospheric Ozone.* National Academy of Sciences, Washington, DC, USA, 1976, p. 45.
64. Keene, W. C. & Galloway, J. N., Organic acidity in precipitation of North America. *Atmos. Environ.,* **18** (1984) 2491–7.
65. Keene, W. C. & Galloway, J. N., The biogeochemical cycling of formic and acetic acids through the troposphere: an overview of current understanding. *Tellus,* **40B** (1988) 322–34.
66. Keene, W. C., Galloway, J. N. & Holden, J. D., Measurement of weak organic acidity in precipitation from remote areas of the world. *J. Geophys. Res.,* **88** (1983) 5122–30.
67. Talbot, R. W., Beecher, K. M., Harris, R., C. & Cofer, W. R., Atmospheric geochemistry of formic and acetic acids at a mid-latitude temperate site. *J. Geophys. Res.,* **93** (1988) 1638–52.
68. Chameides, W. L. & Davis, D. D., Aqueous source of formic acid in clouds. *Nature,* **304** (1983) 427–9.

69. Jacob, D. J. & Wofsy, S. C., Photochemistry of biogenic emissions over the Amazon forest. *J. Geophys. Res.*, **93** (1988) 1477–86.
70. Kawamura, K., Ng, L.-I. & Kaplan, I. R., Determination of organic acids (C_1–C_{10}) in the atmosphere, motor exhausts and engine oils. *Environ. Sci. Technol.*, **19** (1985) 1082–6.
71. Graedel, T. E. & Eisner, T., Atmospheric formic acid from formicine ants: a preliminary assessment. *Tellus*, **40B** (1988) 335–9.
72. Raask, E., Sulphate capture in ash and boiler deposits in relation to SO_2 emission. *Prog. Energy Combust. Sci.*, **8** (1982) 261–76.
73. Husar, R. B., Emissions of sulphur dioxide and nitrogen oxides and trends for Eastern North America. In *Acid Deposition: Long Term Trends*. National Research Council, National Academy Press, Washington, DC, USA, 1986, pp. 48–92.
74. Eggleston, H. S. & McInnes, G. *Methods for the Compilation of UK Air Pollutant Emission Inventories,* Report LR 634 (AP) M. Warren Spring Laboratory, Stevenage, UK, 1987.
75. Cullis, C. F. & Mulcahy, M. F. R., Kinetics of combustion of gaseous sulphur compounds. *Combust. Flame,* **18** (1972) 225–2.
76. Department of the Environment, *Digest of Environmental Protection and Water Statistics No. 11*. HMSO, London, UK, 1988.
77. Homolya, J. B. & Cheney, J. L., A study of primary sulphate emissions from a coal-fired boiler with FGD. *J. Air Pollut. Control Assoc.*, **29** (1979) 1000–4.
78. Sarofim, A. F. & Flagan, R. C., NO_x control for stationary combustion sources. *Prog. Energy Combust. Sci.*, **2** (1976) 1–25.
79. Hayhurst, A. N. & Vince, I. M., Nitric oxide formation from N_2 in flames: the importance of 'prompt' NO. *Prog. Energy Combust. Sci.*, **6** (1980) 35–51.
80. Glarborg, P., Kee, R. J. & Miller, J. A., Kinetic modelling and sensitivity analysis of nitrogen oxide formation in well-stirred reactors. *Combust. Flame,* **65** (1986) 177–202.
81. US Environmental Protection Agency, *Compilation of Air Pollution Emission Factors,* Publ. AP-42, 3rd edn. Research Triangle Park, NC, USA, 1977.
82. Semb, A. & Amble, E. *Emission of nitrogen oxides from fossil fuel combustion in Europe,* EMEP Report 13/81. Norwegian Institute for Air Research, Lillestrom, Norway, 1981.
83. Muzio, L. J., Teague, M. E., Kramlich, J. C., Cole, J. A., McCarthy, J. M. & Lyon, R. K. Errors in grab sample measurements of N_2O from combustion sources. *J. Air Pollut. Control Assoc.*, **39** (1989) 287–93.
84. Lightowlers, P. J. & Cape, J. N., Sources and fate of atmospheric HCl in the UK and Western Europe. *Atmos. Environ.*, **22** (1988) 7–15.
85. Halstead, W. D. & Raask, E., The behaviour of sulphur and chlorine compounds in pulverised coal-fired boilers. *J. Inst. Fuel,* **42** (1969) 344–9.
86. Iapalucci, T. L., Demski, R. J. & Bienstock, D. *Chlorine in Coal Combustion,* Report 7260. US Bureau of Mines, Pittsburgh, PA, USA, 1969.
87. Directive 89/369/EEC, On the prevention of air pollution from new municipal waste incineration plants. *Off. J. Euro. Comm.*, **L163** (1989) 32.

88. Directive 89/429/EEC, On the reduction of air pollution from existing incineration plants. *Off. J. Euro. Comm.*, **L203** (1989) 50.
89. Directive 88/609/EEC, On the limitation of certain pollutants into the air from large combustion plants. *Off. J. Euro. Comm.*, **L336** (1988) 1.
90. Saltbones, J. & Dovland, H., *Emissions of Sulphur Dioxide in Europe in 1980 and 1983*, EMEP/CCC Report 1/86. Norwegian Institute for Air Research, Lillestrom, Norway, 1986.
91. Lübkert, B. & De Tilly, S., The OECD–MAP emission inventory for SO_2, NO_x and VOC in Western Europe. *Atmos. Environ.*, **23** (1989) 3–15.
92. Gschwandtner, G., Gschwandtner, K., Eldridge, K., Mann, C. & Mobley, D., Historic emissions of sulfur and nitrogen oxides in the United States from 1900 to 1980. *J. Air Pollut. Control Assoc.*, **36** (1986) 139–49.
93. US Environmental Protection Agency, *National Air Pollutants Emission Estimates 1940–1987*, EPA 450/4–88–022. Washington, DC, USA, 1989.
94. Bubenick, D. V., ed., *Acid Rain Information Book*, 2nd edn. Noyes, Park Ridge, NJ, USA, 1984, p. 87.
95. Dignon, J. & Hameed, S. Global emissions of nitrogen and sulfur oxides from 1860 to 1980. *J. Air Poll. Control Assoc.*, **39** (1989) 180–6.

Chapter 3

SO_2 and NO_x Oxidation Mechanisms in the Atmosphere

MIROSLAV RADOJEVIĆ*
Department of Fuel and Energy, University of Leeds, Leeds, LS2 9JT, UK

3.1 INTRODUCTION

Oxidation reactions of sulphur dioxide (SO_2) and oxides of nitrogen (NO_x) are an important source of acidity in atmospheric deposition and they have been the subject of intensive study over the years. The current understanding of these processes is reviewed here.

There have been a number of field studies of SO_2 oxidation in the atmosphere and some of these are summarised in Table 3.1. Many of these studies investigated SO_2 oxidation in power-plant plumes although some have determined the oxidation rate over larger areas. More recently, the oxidation rate of SO_2 has been measured in clouds and fogs. Oxidation rates of SO_2 determined in field studies vary considerably. The slowest rates were measured in the relatively pristine air of the Arctic ($< 0{\cdot}2\%\ h^{-1}$), whereas the fastest rates were determined in urban areas (5–702% h^{-1}) and in clouds (2–1900% h^{-1}). Moderate rates were observed in power-plant plumes (0·2–8% h^{-1}).

**Present address:* Department of Chemistry, University of Brunei Darussalam, Bandar Seri Begawan 3186, Brunei Darussalam.

Table 3.1
Oxidation rates of SO_2 measured in field studies

Reference	*SO_2 conversion rate ($\% h^{-1}$)*	*Comments*
1	3·5	Smelting area
2	702	Industrial area near Tokyo
3	6–25	Industrial city in France
4	1·2 –13	Los Angeles Basin
5	~ 0·2	Power-plant plume
6	0·5	Power-plant plume
7	0·41–4·9	Power-plant plume
8	1·4 noon < 0·5 night	Power-plant plume
9	< 3	Power-plant plume
10	2–8	Power-plant plume
11	1	Smelting area
12	0·9 –8·5	Power-plant plume in cloud
13	2–1900	Wave cloud
14	5·3(±1·5)–21(±8)	Clouds
15	0·04–0·2	Arctic air
16	5·7 day 5·8 night	Los Angeles Basin
17	3±2	Fog
18	< 0·2 –2·3	Power-plant plume
19	8·1–16·5	Urban area
20	30±4 in fog 3·1±0·8 outside fog	Power-plant plume

Various reactions have been proposed to account for the oxidation rates observed in field studies and these may be grouped into three classes:

(1) oxidation in the gas phase (homogeneous gas-phase reactions);
(2) oxidation on the surface of particles (heterogeneous gas–solid reactions); and
(3) oxidation in aqueous droplets (homogeneous liquid-phase reactions).

Pathway (3) in fact involves heterogeneous gas–liquid processes, however, the oxidation reactions take place inside the aqueous solution. Heterogeneous liquid–solid reactions may also occur on the surfaces of insoluble solid particles suspended inside aqueous droplets. Mechanisms (2) and (3) are therefore more aptly termed 'multiphase' processes.

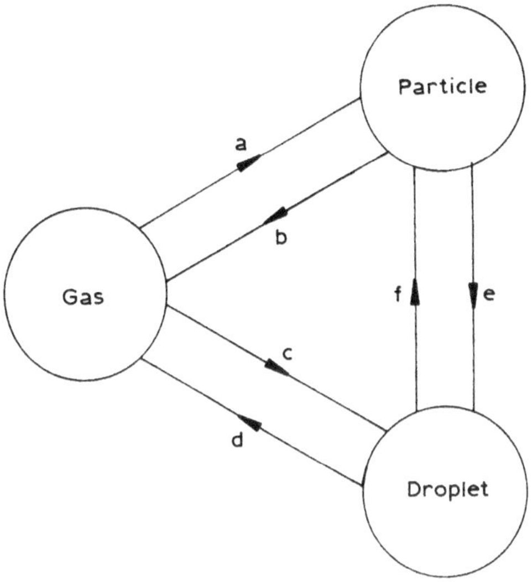

Fig. 3.1. Multiphase interactions in the atmosphere. a and b—adsorption/desorption equilibrium; gas to particle conversion. c and d—Henry's law equilibrium. e and f—condensation/evaporation; melting/freezing; deliquescence/efflorescence; scavenging of particles.

Multiphase interactions relevant to the physico-chemical transformation of atmospheric constituents are shown in Fig. 3.1 and chemical reactions may take place in each of the three phases. Before proceeding to discuss individual SO_2 and NO_x oxidation reactions a brief introduction to atmospheric chemistry shall be given. The general chemistry of the atmosphere has been reviewed in several publications.[21–26]

3.2 ATMOSPHERIC CHEMISTRY

The general chemistry of the atmosphere is dominated by photochemical gas-phase reactions initiated by solar radiation. Electrons in atoms and molecules may be excited to higher energy states by absorption of near-ultraviolet (UV) and visible radiation and these excited species may undergo numerous reactions including direct chemical reaction. Absorption of radiation by molecules may also result in their photo-dissociation (photolysis) to produce atoms and radicals.

The central role of hydroxyl radicals (OH) in tropospheric chemistry has been recognised for some time. Free radicals such as OH can act as strong oxidising agents for a variety of trace atmospheric gases including SO_2 and NO_x species. The production of OH radicals is initiated by the photolysis of ozone. UV light with wavelengths below 315 nm splits ozone into molecular oxygen and singlet atomic oxygen $O(^1D)$ which may either relax to the triplet state $O(^3P)$ or react with water vapour to form OH radicals.

$$O_3 + h\nu \xrightarrow{\lambda < 315\ \text{nm}} O(^1D) + O_2 \quad (3.1)$$

$$O(^1D) + M \longrightarrow O(^3P) + M \quad (3.2)$$

$$O(^1D) + H_2O \longrightarrow 2OH \quad (3.3)$$

This represents the most important route of OH· radical formation in the troposphere. Singlet oxygen $O(^1D)$ may also react with methane (CH_4) and molecular hydrogen (H_2) to produce OH· radicals.

$$O(^1D) + CH_4 \longrightarrow CH_3 + OH \quad (3.4)$$

$$O(^1D) + H_2 \longrightarrow H + OH \quad (3.5)$$

Other potential sources of OH· radicals include photolysis of nitrous acid (HNO_2) and hydrogen peroxide (H_2O_2):

$$HNO_2 + h\nu \xrightarrow{\lambda < 400\ \text{nm}} OH + NO \quad (3.6)$$

$$H_2O_2 + h\nu \xrightarrow{\lambda < 360\ \text{nm}} 2OH \quad (3.7)$$

Photolysis of HNO_2 may be important in polluted air, however, photolysis of H_2O_2 is unlikely to be a net source of OH· radicals. OH radicals may also be formed in the oxidation of NO by hydroperoxyl (HO_2) radicals:

$$NO + HO_2 \longrightarrow NO_2 + OH \quad (3.8)$$

Measurements of OH radicals in the trophosphere have been reviewed and these range from 0·5 to 5×10^6 molecules cm^{-3} as daytime means.[27] These measurements, together with laboratory determinations of OH· radical reactions,[28,29] imply that OH controls the rate at which many atmospheric species are oxidised and removed from the atmosphere.

The major sinks for OH radicals are reactions with carbon monoxide (CO) and CH_4:

$$CO + OH \longrightarrow CO_2 + H \quad (3.9)$$

$$CH_4 + OH \longrightarrow CH_3 + H_2O \quad (3.10)$$

CH_3 and H radicals combine with molecular oxygen to produce methylperoxyl (CH_3O_2) and HO_2 radicals, respectively:

$$CH_3+O_2+M \longrightarrow CH_3O_2+M \qquad (3.11)$$

$$H+O_2+M \longrightarrow HO_2+M \qquad (3.12)$$

The HO_2 radical can regenerate the OH radical via reaction with NO, reaction (3.8), and with ozone (O_3):

$$HO_2+O_3 \longrightarrow 2O_2+OH \qquad (3.13)$$

NO may also react with CH_3O_2 radicals:

$$CH_3O_2+NO \longrightarrow CH_3O+NO_2 \qquad (3.14)$$

HO_2 can then be produced by

$$CH_3O+O_2 \longrightarrow CH_2O+HO_2 \qquad (3.15)$$

Nitrogen dioxide (NO_2) may be photolysed to produce $O(^3P)$ which reacts with oxygen gas (O_2) to give O_3.

$$NO_2+h\nu \xrightarrow{\lambda < 435\ nm} NO+O(^3P) \qquad (3.16)$$

$$O(^3P)+O_2+M \longrightarrow O_3+M \qquad (3.17)$$

O_3 may react with NO to produce NO_2:

$$NO+O_3 \longrightarrow NO_2+O_2 \qquad (3.18)$$

These reactions are all very important in the formation of photochemical smog from CO, NO_x and hydrocarbon pollutants emitted by motor vehicle exhausts. Photochemical smog episodes are characterised by high O_3, H_2O_2 and peroxyacetylnitrate (PAN) levels, the latter being formed by reactions of acetyl (CH_3CO) radicals. The photochemistry of organic compounds in the troposphere is quite complex and it results in a multitude of compounds and free radicals.

Atmospheric gases may dissolve in cloud-drops, fog-drops and raindrops where they can undergo aqueous phase reactions. Chemical species may be leached into solution from atmospheric aerosol particles scavenged by these droplets. Gases may also be adsorbed onto the surfaces of dust particles where they may undergo further reaction.

3.3 SO_2 OXIDATION

3.3.1 Gas-phase reactions

Homogeneous gas-phase oxidation of SO_2 is a potentially important mechanism of SO_2 conversion in the atmosphere. Gas-phase reactions may be conveniently classified as

(i) direct photochemical oxidation;
(ii) oxidation by reactive molecular (O_2, O_3, NO_x), atomic ($O(^3P)$) and free radical (OH, HO_2, CH_3O_2) species; and
(iii) oxidation by products of the O_3+alkene reaction.

3.3.1.1 Photochemical oxidation

Within the troposphere SO_2 cannot undergo photo-dissociation but absorption of solar radiation produces electronically excited SO_2 states, and tropospheric SO_2 photochemistry involves reactions of these excited molecules. The primary photochemical reactions of SO_2 are

$$SO_2(^1A_1)+h\nu(340\text{–}400\ \text{nm}) \rightarrow SO_2(^3B_1) \quad (3.19)$$

$$SO_2(^1A_1)+h\nu(240\text{–}330\ \text{nm})\dagger \rightarrow SO_2(^1A_2) \quad (3.20a)$$

$$\rightarrow SO_2(^1B_1) \quad (3.20b)$$

The route in eqn (3.19) is inefficient as a source of excited SO_2 molecules in the atmosphere because of the change in spin multiplicity required for the transition. The predominant fate of the singlet states formed in eqn (3.20) is physical deactivation back to the ground state by collision with other molecules. However, a small fraction of collisions, ~ 9%, is effective in inducing intersystems crossing to $SO_2(^3B_1)$, the major photochemically active species of SO_2 in the atmosphere.[30] It is generally believed that $SO_2(^3B_1)$ reacts with O_2 to form sulphur trioxide (SO_3) and two possibly significant pathways exist for this reaction:

$$\text{(i) } SO_2(^3B_1)+O_2(^3\Sigma_g^-) \rightarrow SO_4(\text{cyclic}) \quad (3.21)$$

$$SO_4(\text{cyclic})+O_2 \rightarrow SO_3+O_3 \quad (3.22)$$

$$\text{(ii) } SO_2(^3B_1)+O_2(^3\Sigma_g^-) \rightarrow SO_3+O(^3P) \quad (3.23)$$

The oxidation is thought to involve the intermediate (cyclic) SO_4 species,

†The shortest wavelength available in the troposphere is 290 nm. One limit on the importance of this process is the low intensity of light in the 290–330 nm region in the lower atmosphere.

but it has not been experimentally identified. The detailed mechanism of this process is still unclear, however, it is too slow to be of major importance in the atmospheric oxidation of SO_2. A review of the tropospheric photochemistry of SO_2 has concluded that the maximum rates of the direct photochemical conversion of SO_2 are less than 0·04% h^{-1}.[30] Observed rates of SO_2 oxidation in air are much higher and they must therefore be due to reactions of ground state SO_2 with reactive species in the troposphere.

3.3.1.2 Oxidation by reactive species in the troposphere

These reactions involve oxidation by singlet molecular oxygen ($O_2(^1\Delta_g)$ and $O_2(^1\Sigma_g^+)$, O_3, NO_x, oxygen atom ($O(^3P)$) and free radicals (OH, HO_2, CH_3O_2, CH_3COO_2). Some reported rate constants for reactions of SO_2 with $O(^3P)$ and molecular species in the atmosphere are given in Table 3.2, together with evaluated SO_2 conversion rates in the polluted troposphere. These rates are well below the conversion rates measured in field studies and rule out any significant contribution to tropospheric

Table 3.2

Rate constants for SO_2 oxidation by $O(^3P)$, O_2, O_3 and NO_x together with estimated polluted atmosphere conversion rates (references are given in brackets)[a]

Reaction	*k* (*cm^3 molecule$^{-1}s^{-1}$*)	*SO_2 conversion rate* (*$\%h^{-1}$*)
$SO_2+O(^3P)(+M) \rightarrow SO_3(+M)$	2×10^{-14}(32)	$1{\cdot}8\times10^{-3}$
	$1{\cdot}3\times10^{-13}$(33)	$1{\cdot}2\times10^{-2}$
$SO_2+O_2(^1\Sigma_{g+}) \rightarrow SO_4$; $\rightarrow SO_3+O(^3P)$; $\rightarrow SO_2+O_2(^1\Delta g)$	$6{\cdot}6\times10^{-16}$(34)	$1{\cdot}4\times10^{-7}$
$SO_2+O_2(^1\Delta_g) \rightarrow SO_4$; $\rightarrow SO_3+O(^3P)$; $\rightarrow SO_2+O_2(^3\Sigma g^-)$	$3{\cdot}9\times10^{-20}$(35)	$1{\cdot}4\times10^{-6}$
$SO_2+O_3 \rightarrow SO_3+O_2$	$< 8\times10^{-24}$(36)	$< 1{\cdot}4\times10^{-5}$
	$< 2\times10^{-22}$(37)	$< 3{\cdot}6\times10^{-4}$
$SO_2+NO_2 \rightarrow SO_3+NO$	$< 2\times10^{-24}$(37)	$< 3{\cdot}7\times10^{-6}$
$SO_2+NO_3 \rightarrow SO_3+NO_2$	$< 7\times10^{-21}$(36)	$< 6{\cdot}3\times10^{-8}$
	$< 1\times10^{-21}$(37)	$< 9\times10^{-9}$
$SO_2+N_2O_5 \rightarrow SO_3+N_2O_4$	$< 4\times10^{-3}$(36)	$< 3{\cdot}6\times10^{-8}$

[a]Concentrations in molecules cm^{-3} chosen for the calculation: $O(^3P)$—$2{\cdot}5\times10^5$ (26); $O_2(^1\Sigma_{g+})$—6×10^2 (31); $O_2(^1\Delta_g)$—10^8 (31); O_3—5×10^{12} (30); NO_2—$5{\cdot}1\times10^{12}$ (26); NO_3—$2{\cdot}5\times10^7$ (31); N_2O_5—$2{\cdot}5\times10^9$ (31).

oxidation of SO_2, although reactions with $O(^3P)$ may be important at very high NO_2 concentrations in sunlight.

Oxidation of SO_2 by OH radicals in the gas phase is of major importance in the troposphere. This is a three body addition reaction usually represented as

$$OH+SO_2(+M) \rightarrow HOSO_2(+M) \qquad (3.24)$$

The reaction is third-order at low pressures and in the transition region between third-and-second order at pressures characteristic of the troposphere. It has been studied in kinetic experiments over the pressure range 10^{-3}–10^3 torr (0·133Pa–133kPa) and with helium, argon, N_2, water, carbon dioxide (CO_2), SO_2, sulphur hexafluoride (SF_6) and air as third bodies. There is some discrepancy between the reported values of the rate constant for this reaction at any given pressure[38,39] and the rate constant depends on the nature of the third body as well as its pressure. Values of the effective second-order rate constant for the reaction at atmospheric pressure are given in Table 3.3 together with estimated atmospheric conversion rates. Some of the experimental studies were carried out using the OH+CO reaction as a reference. It now appears that the authors used incorrect values of the rate constant for the OH+CO reaction when evaluating experimental results. This reference reaction has been found

Table 3.3

Experimental measurements of the effective second-order reaction rate constant for the reaction $SO_2+OH(+M) \rightarrow HOSO_2(+M)$ at pressures and at temperatures relevant to the atmosphere, and estimated SO_2 conversion rates in polluted atmosphere[a]

Method and reference	*T (°C)*	*M*	*k (cm^3 molecule$^{-1}s^{-1}$)*	*SO_2 conversion rate (% h^{-1})*
Flash photolysis–resonance fluorescence				
40	27	N_2	9×10^{-13}	1·6
41	25	Ar	$6{\cdot}7\times10^{-13}$	1·2
42	25	N_2	10^{-12}	1·8
Static photolysis competitive with OH+CO reaction				
43	25	N_2	6×10^{-13}	1·1
44	21	Air	$5{\cdot}6\times10^{-13}$	1·0
Static photolysis competitive with $OH+n-C_4H_{10}$ reaction				
45	30	Air	$1{\cdot}22\times10^{-12}$	2·2

[a] [OH]—5×10^6 molecules cm^{-3}.[27]

to exhibit an unexpected pressure dependence and the values of the rate constant for SO_2 oxidation by OH radicals reported in these studies have to be corrected. A value of $1{\cdot}1\ (\pm 0{\cdot}3)\times 10^{-12}$ cm^3 molecule^{-1} s^{-1} (1 atm (101 kPa) air, 25°C) for the effective second-order rate constant for SO_2 reaction with OH radicals in the atmosphere has been recommended on the basis of four detailed studies carried out at atmospheric pressures and correcting the rate constants from competitive studies.[30] In a recent competitive study using the $OH+n-C_4H_{10}$ reaction as reference, a value of $k = 1{\cdot}22\ (\pm 0{\cdot}13)\times 10^{-12}$ cm^3 molecule^{-1} s^{-1} has been determined.[45] This is in good agreement with the aforementioned value, as is the value of $k = 10^{-12}$ cm^3 molecule^{-1} s^{-1} recommended for 25°C and tropospheric pressures in a study of the reaction using flash photolysis–resonance fluorescence.[33]

The overall mechanism of SO_2 oxidation by OH radicals is complex and still unclear, but it is certain that the reaction ultimately leads to sulphuric acid (H_2SO_4) aerosol formation in the atmosphere containing water vapour.[40] An experimental study carried out specifically with the aim of elucidating the mechanism[46] suggests that the main fate of $HOSO_2$ radicals formed by the $OH+SO_2$ addition reaction is reaction with O_2 to give HO_2 and SO_3:

$$HOSO_2+O_2 \rightarrow HO_2+SO_3 \quad (3.25)$$

followed by

$$SO_3+H_2O \rightarrow H_2SO_4 \quad (3.26)$$

SO_3 formed via oxidation by OH or by other gas-phase reactions rapidly reacts with water in the atmosphere to form H_2SO_4. That the SO_2+OH reaction leads predominantly to the production of HO_2 radicals has been confirmed in smog chamber experiments.[47] The overall oxidation of SO_2 by OH radicals would hence appear to be best represented by

$$OH+SO_2(+O_2, H_2O) \rightarrow H_2SO_4+HO_2 \quad (3.27)$$

There are considerable differences between reported measurements of rate constants for the oxidation of SO_2 by HO_2, CH_3O_2 and methoxyl (CH_3O) radicals as shown in Table 3.4, which also gives the SO_2 conversion rates estimated for the polluted troposphere. An early study of the HO_2+SO_2 reaction suggests that it may be important in the atmosphere;[48] however, more recent measurements do not confirm this.[49,50] The reasons for the discrepancy between the values of the rate constant reported in

Table 3.4
Rate constants for the oxidation of SO_2 by HO_2, CH_3O_2 and CH_3O radicals and estimated SO_2 conversion rates in the polluted atmosphere[a]

Reaction and reference	*k* (cm^3 $molecule^{-1}s^{-1}$)	*SO_2 conversion rate* (% h^{-1})
$SO_2+HO_2 \rightarrow SO_3+OH$		
48	$8{\cdot}7\times10^{-16}$	1·9
49	$\leqslant 10^{-18}$	$\leqslant 2{\cdot}16\times10^{-3}$
50	$\leqslant 2\times10^{-17}$	$\leqslant$ 0·043
$SO_2+CH_3O_2 \rightarrow SO_3+CH_3O$		
51	$\leqslant 3{\cdot}32\times10^{-15}$	$\leqslant$ 1·2
52	$5{\cdot}3\times10^{-15}$	1·9
53	$8{\cdot}2\times10^{-15}$	2·9
54	5×10^{-17}	0·018
55	$1{\cdot}4\times10^{-14}$	5·0
$SO_2+CH_3O \rightarrow CH_3OSO_2$		
56	$\sim 6\times10^{-15}$	~ 0·01
55	$\sim 5{\cdot}5\times10^{-13}$	~1

[a]Concentrations (in molecules cm^{-3}): HO_2—6×10^9 (31); CH_3O_2—10^9 (31); CH_3O—5×10^6(30).

Table 3.4 are not yet clear and there is no basis at present for recommending preferred values. It would appear, however, that SO_2 reaction with CH_3O_2 is more complex than was previously thought and a number of possible complicating reactions may be involved in the mechanism. SO_2 oxidation by CH_3O_2 has been studied with the aim of resolving the discrepancy,[55] and it was concluded that the process can be rationalised in terms of a reversible addition of CH_3O_2 to SO_2:

$$SO_2+CH_3O_2 \rightleftharpoons CH_3O_2SO_2 \quad (3.28)$$

The net oxidation of SO_2 following this reaction proceeds via a complex mechanism which is still not fully understood and depends on the experimental technique and conditions (CH_3O_2 concentration, flash intensity, absence or presence of NO). A rate constant of $1{\cdot}4(\pm0{\cdot}2)\times10^{-14}$ cm^3 $molecule^{-1}$ s^{-1} was determined for eqn (3.28).[55] The study reported in Ref. 54, which measured values of the rate constant considerably lower than those reported by other workers, employed much lower flash intensities and radical concentrations than in other studies, and the experimental conditions appear to be more representative of the troposphere.

Hence, the earlier claims that reactions of SO_2 with HO_2 and CH_3O_2 radicals could be competitive with oxidation by OH radicals should be treated with caution until the discrepancy is fully resolved and the mechanism clarified.

There is little information on the kinetics of SO_2 oxidation by CH_3O and CH_3COO_2 radicals. It has been concluded that oxidation of SO_2 by CH_3O and peroxyacyl ($RCOO_2$) radicals is negligible for general atmospheric conditions.[30] For SO_2 oxidation by CH_3O radicals in the polluted troposphere a conversion rate of 0·01% h^{-1} was estimated based on an approximate estimate of 6×10^{-15} cm^3 $molecule^{-1}$ s^{-1} for the reaction rate constant and an atmospheric concentration of 5×10^6 molecules cm^{-3} for CH_3O.[30] A value of ~ $5{\cdot}5 \times 10^{-13}$ cm^3 $molecule^{-1}$ s^{-1} has been estimated for the rate constant for this reaction in a separate study,[55] which would lead to SO_2 conversion rates of ~ 1% h^{-1} if the same concentration of CH_3O radicals is used in the calculation. Such a rate is not insignificant and the rate constant for SO_2 oxidation by CH_3O needs to be firmly established. Rates of SO_2 oxidation by CH_3COO_2 and *trans*-$C_4H_9O_2$ in the urban troposphere have been estimated to be < $2{\cdot}5 \times 10^{-5}$% h^{-1},[57] which is insignificant in comparison with the rates of the other mechanisms shown in Table 3.4.

3.3.1.3 Oxidation by products of the O_3+ alkene reaction

The rate of gas-phase oxidation of SO_2 by O_3 is negligible in the troposphere as shown in Table 3.2; however, when an alkene is present in the SO_2+O_3+air mixture, a fairly rapid oxidation takes place even in the dark. It is postulated that the species responsible for this fast oxidation is a product of the O_3+ alkene reaction. This mechanism is very complex and several different reaction schemes can be proposed to account for the oxidation. These have been reviewed in Ref. 30, which estimates SO_2 oxidation rates by a reactive species formed in the O_3+ alkene reaction in the sunlight-irradiated, polluted, atmosphere (total alkene = 0·1 ppm (v/v), O_3 = 0·15 ppm (v/v), SO_2 = 0·05 ppm (v/v)). Upper limits of 0·23 and 0·12% h^{-1} were set for the SO_2 conversion rate at 50 and 100% relative humidity, respectively. It was assumed that SO_2 was oxidised by the Criegee intermediate, $R\dot{C}HOO\cdot$, via the following reaction:

$$R\dot{C}HOO\cdot + SO_2 \rightarrow RCHO + SO_3 \qquad (3.29)$$

It was concluded that this mechanism is not insignificant even though the rates are not large in comparison to those expected for the SO_2+OH reaction shown in Table 3.3.[30] However, in contrast to the oxidation of SO_2 by

OH and other radicals which is strongly solar modulated SO_2 oxidation by products of O_3+alkene reactions will continue to occur during the night when both reactants are present, but the rates are not expected to be large. Results from a recent smog chamber study indicate an SO_2 conversion rate of 0·48% h^{-1} for this mechanism,[58] which is greater than the values estimated by Ref. 30. It was also found that reaction of the Criegee intermediate with SO_2 may produce both SO_4^{2-} and organo-sulphur compounds and the relative amounts of these products were controlled by the humidity level.[58]

3.3.2 Particle reactions

Heterogeneous oxidation reactions of SO_2 on particles have been studied for a long time by workers interested in the industrial manufacture of H_2SO_4[59] and flue gas desulphurisation,[60] but these studies are of limited relevance to the atmospheric conversion of SO_2 as the experimental conditions were very different from those prevailing in the atmosphere. Over the past 20 years interest has been expressed in these reactions by atmospheric chemists and it has been suggested that SO_2 reactions on particle surfaces could be competitive with gas- and aqueous-phase reactions.[61,62] Sulphur compounds in the S(IV) state have been identified in aerosols,[63–65] and it is proposed that SO_2 is adsorbed on the surface of particles such as soot and metal oxides (e.g. Fe_2O_3, ZnO, etc.) which are present in polluted atmospheres[66,67] and oxidised to H_2SO_4.

The heterogeneous oxidation of SO_2 on particles is affected by temperature, the relative humidity and the presence of oxidants other than O_2 (e.g. NO_2, O_3) and these effects have been investigated in the laboratory. Experimental techniques employed in the study of heterogeneous SO_2 reactions vary considerably and the simplest experiments involve the contacting of SO_2 with solids and analysis of the quantity adsorbed. Analytical techniques involve chemical methods,[68] radio-tracing techniques[69–71] and electron spectroscopy for chemical analysis (ESCA).[72] Fixed-bed reactors,[73,74] tubular-flow reactors,[75] aerosols deposited on filter papers[76–78] and genuine reacting aerosol systems[69,70,79,80] have been employed in these experiments and artificial and atmospheric aerosols have been used. As the reaction may also be controlled by mass transfer it has been suggested that studies employing aerosol particles deposited on filters or the walls of reaction vessels may yield unrepresentatively low extents of reaction.[80]

The mechanism of SO_2 oxidation on particle surfaces is one of heterogeneous catalysis. SO_2 oxidation takes place on certain active sites either by adsorbed O_2 or NO_2. In either case oxidation results in poisoning of the active sites and the reaction is therefore limited by the availability of active sites, and many studies report that the relative humidity increases the capacity of aerosol for reaction with SO_2. Since this process involves reaction between species adsorbed on the surface the rate may be predicted by the Langmuir–Hinshelwood equation,[81] which has been used to model SO_2 oxidation on particle surfaces.[69,70,76,79]

3.3.2.1 Reactions with carbonaceous particles

In early studies of SO_2 oxidation on activated carbon it was found that formation of sulphate occurred only in the presence of O_2.[82–84] The reaction proceeded until a monolayer of H_2SO_4 was formed on the particle surface and the presence of water vapour increased the rate of oxidation. The catalysis was found to take place on only a limited number of active sites occupying about 10% of the surface. SO_2 was physically adsorbed on the remaining area but did not react to form sulphate and could easily be removed by evacuation. This early work demonstrated that only a limited quantity of sulphate could be produced via this mechanism and this has since been confirmed in other experimental studies. Many of the early studies have been concerned with typical flue gas conditions relevant to flue gas desulphurisation or in-plant gas–particle interactions at high temperature (> 150°C). Results from these studies are not directly applicable to the atmosphere and a review of the literature prior to 1971 is available.[60]

In recent years it has been proposed that heterogeneous oxidation on carbonaceous particles (soot) in the atmosphere may be important in SO_2 conversion,[61,85] and experimental studies with relevance to the atmosphere have been carried out. SO_2 oxidation on soot and graphite particles has been studied using electron spectroscopy for chemical analysis (ESCA).[61] Particles exposed to SO_2 at 400 ppm (v/v) were found to contain sulphite, S(IV), and sulphate, S(VI), and more sulphate was observed in the humidified than in dry air. Reference 61 also reports a strong correlation between the concentration of ambient carbon and sulphate in Los Angeles air and this tempted the authors to ascribe an important role to heterogeneous oxidation of SO_2 on soot particles in the atmosphere. However, the experimental study was carried out at SO_2 concentrations considerably higher than those encountered in the general atmosphere.

Recently, two detailed studies using different experimental systems have shown that the earlier work greatly exaggerated the role of this reaction in the atmosphere.[69,76] SO_2 oxidation on soot aerosols inside a reaction chamber has been studied using a radio-tracing technique.[69] Increasing the SO_2 concentration from 1 to 7 ppm (v/v) was found to result in an increase in both the initial reaction rate and the final amount of S(VI) formed. Most of the reactions were studied at low relative humidity (< 4%) and increasing the relative humidity to 80% was found to have little effect on the reaction. In an earlier study involving carbon black particles deposited on filter papers, it was found that the amount of sulphate formed increased by 80% when NO_2 was present at 50 ppm (v/v) in an air stream containing 2000 ppm (v/v) SO_2 at 50% relative humidity.[86] The effect of NO_2 was re-examined in more detail using the suspended aerosol system and it was observed that the initial rate of sulphate formation increased with increasing NO_2 concentration (0·03 to 17·7 ppm(v/v)) until a maximum of 10 times higher than in the absence of NO_2 was reached. The final amount of sulphate formed after 2 h was found to increase with NO_2 up to near $[SO_2] = [NO_2]$. At higher NO_2 concentrations the final amount of sulphate formed was depressed. These reactions were all studied in the dark and it was found that the presence of soot made a negligible difference to the oxidation of SO_2 by gas-phase reactions with photochemically produced radicals in SO_2/NO_x/air mixtures. The authors of Ref. 86 concluded that heterogeneous SO_2 oxidation on soot aerosols is of little consequence in the general atmosphere, but they did not discount the role of this mechanism in certain low temperature flue gases or plumes.

SO_2 oxidation on activated charcoal deposited on filter papers has been studied, and the effect of relative humidity and NO_2, ammonia (NH_3), and O_3 on the reaction was investigated.[76] The adsorption rates were monitored by measuring the effluent concentrations of gases passed through the reaction filter using continuous analysers. Furthermore, at

Table 3.5
Estimated atmospheric SO_2 oxidation rates based on extrapolation of results from a laboratory study using activated carbon[76]

Reaction[a]	*SO_2 conversion rate (% h^{-1})*	
	60% relative humidity	*80% relative humidity*
SO_2-soot	5×10^{-4}	—
SO_2+soot+NO_2(0·12 ppm (v/v))	—	9×10^{-4}
SO_2+soot+NH_3(0·01 ppm (v/v))	$0{\cdot}8\times10^{-4}$	—

[a]SO_2—0·1 ppm (v/v); soot—10 μg m^{-3}; soot specific surface—600 m^2 g^{-1}.

the end of some experiments the catalyst was washed with distilled water and analysed for SO_4^{2-}, NH_4^+ and $NO_3^-+NO_2^-$ by wet chemical techniques. Experiments were carried out at relative humidities ranging from 3 to 100% and also in soot suspensions in distilled water and it was concluded that to a good approximation the rate of SO_2 oxidation on the particles was proportional to the relative humidity. Addition of NO_2 was found to enhance the total mass of the sulphate produced. Atmospheric SO_2 conversion rates estimated for this mechanism are given in Table 3.5, showing that heterogeneous SO_2 oxidation on carbonaceous particles is irrelevant to the oxidation of SO_2 in the atmosphere even if NO_2 and NH_3 are present in the air.

The heterogeneous mechanism of SO_2 oxidation on carbon particles involves the adsorption of SO_2 and O_2 onto active sites on the catalyst surface where they interact to form SO_3 which remains attached to the active sites, poisoning the catalyst. This process can be expressed by the following reactions:

$$O_2(g) \rightleftharpoons 2O(ads) \tag{3.30}$$

$$SO_2(g) \rightleftharpoons SO_2(ads) \tag{3.31}$$

$$SO_2(ads)+O(ads) \rightarrow SO_3(ads) \tag{3.32}$$

Equations (3.30) and (3.31) are at equilibrium and eqn (3.32) is rate determining. If NO_2 is present in the air it can also be adsorbed onto the active sites where it can react with SO_2:

$$NO_2(g) \rightleftharpoons NO_2(ads) \tag{3.33}$$

$$NO_2(ads)+SO_2(ads) \rightarrow NO(g)+SO_3(ads) \tag{3.34}$$

NO_2 may also decompose with NO release to the gas phase:

$$NO_2(ads) \rightarrow NO(g)+O(ads)\dagger \tag{3.35}$$

or

$$2NO_2(ads) \rightarrow NO(g)+NO_3(ads) \tag{3.36}$$

NO_3 may also be formed via

$$NO_2(ads)+O(ads) \rightarrow NO_3(ads) \tag{3.37}$$

Both SO_3 and NO_3 formation result in site poisoning. Oxidation of SO_2 by NO_2, reaction (3.34), takes place at ambient temperatures while at higher

†O(ads) is assumed not to react with SO_2.

temperatures NO_2 is decomposed on carbon in the presence of SO_2 without affecting the amount of SO_3 formed.[69] A Langmuir–Hinshelwood model based on these reactions was found to agree well with experimental results.[69,76,87] A similar, but somewhat more detailed Langmuir–Hinshelwood model was used to explain experimental results of Ref. 76.

It has been shown by electron paramagnetic resonance that free electrons on carbon particles can serve as active centres for the catalytic oxidation of SO_2.[88] Since the concentration of free electrons is related to the origin and thermal history of carbon particles, the heterogeneous oxidation of SO_2 via this mechanism may be expected to vary considerably depending on the age and origin of carbonaceous particles. Therefore, no single equation may be recommended for predicting the average rates of SO_2 oxidation on atmospheric soot particles produced at different temperatures and conditions and hence different surface reactivities. Furthermore, the oxidation of SO_2 to SO_3 ceases when a certain portion of the surface is covered with the oxidation product. This saturation effect limits the role of soot catalysis in atmospheric SO_2 oxidation.

3.3.2.2 Reactions with metal oxides and other particles

Other particles which have been employed in experimental studies of the heterogeneous SO_2 oxidation include metal oxides such as magnetite (Fe_3O_4), ferric oxide (Fe_2O_3), lead oxide (PbO), lead dioxide (PbO_2), calcium oxide (CaO), chromium trioxide (Cr_2O_3), vanadium pentoxide (V_2O_5), aluminium oxide (Al_2O_3), manganese dioxide (MnO_2), zinc oxide (ZnO) and magnesium oxide (MgO).[70,71,74,75,77–90] Iron oxides (Fe_2O_3 and Fe_3O_4) appear to be the most effective catalysts of SO_2 oxidation. SO_2 oxidation on Fe_2O_3 particles has been studied, and the experimental results were explained in terms of a heterogeneous catalysis mechanism in which reaction between SO_2 and chemisorbed oxygen to form SO_3 is the limiting process.[77] Nearly all of the adsorbed SO_2 was observed to oxidise to H_2SO_4. The reaction was first-order relative to the fraction of active sites occupied by adsorbed oxygen and the amount of sulphate formed corresponded to the coverage of 35% of the particle surface with a monomolecular layer. The total capacity of the particles for oxidation of SO_2 was approximately constant over a range of SO_2 concentrations (5 to 19 ppm(v/v)) and relative humidities. In another study involving a variety of particles it was found that the adsorption reaction was first-order in SO_2 and zero-order in O_2 and humidity.[75] The amount of SO_2 adsorbed by ZnO fume was found to be independent of SO_2 concentrations (4 to 18 ppm(v/v)) but strongly dependent on the relative humidity.[78]

While studying a variety of metal oxides, it was found that the uptake of SO_2 increased with increasing SO_2 concentration and relative humidity,[70] and 2·5 ppm(v/v) NO_2 was found to increase the adsorption of SO_2 by MgO 10-fold at 50% relative humidity. Removal rates of SO_2 over the first hour of the experimental runs were extrapolated to the atmosphere assuming that the reaction is first-order in SO_2 and an atmospheric aerosol burden of 100 $\mu g\ m^{-3}$. These rates are given in Table 3.6 showing that oxidation of SO_2 on these particles is not of significance in the atmosphere, even at the high concentrations employed in the calculations. Most workers agree that this mechanism is not important in the general atmosphere, except near emission sources.[71,75,77,78]

Table 3.6
Estimated removal rates of SO_2 by metal oxide particles assuming an atmospheric burden of 100 $\mu g\ m^{-3}$ based on experimental results[70]

Aerosol matter	*Relative humidity (%)*	*SO_2 removal rate ($\%\ h^{-1}$)*
Fe_2O_3	0	0·002
Fe_3O_4	0	0·003
	85	0·005
MgO	0	0·002
	70	0·069
CuO	0	0·002
	50	0·009

Dry sodium chloride (NaCl) does not react significantly with SO_2, however at high humidities NaCl deliquesces into electrolyte droplets and these reactions are treated in the section on aqueous reactions (see Section 3.3.3.3). Likewise, reactions with other metal salts which exist as wet aerosols at high humidities (e.g. $MnCl_2$, $CuCl_2$, $FeCl_3$, $MnSO_4$) fall into the category of aqueous reactions.

3.3.2.3 Reactions with atmospheric particles

Anthropogenic activity releases aerosol particles with a variety of physico-chemical properties into the atmosphere and experimental studies have been carried out using genuine atmospheric samples. Particular emphasis has been placed in these studies on fly ash particles from power plants as a possible medium for SO_2 oxidation. Reaction capacities measured for atmospheric and industrial particles in some experimental studies are summarised in Table 3.7, together with the experimental conditions. It can be seen that there are considerable differences between the results of various studies.

Some authors suggest that SO_2 oxidation on atmospheric particles may contribute significantly to SO_2 oxidation and acidity formation.[79,80,92] The reaction capacities were measured which are a factor of 10^2–10^3 greater than those determined by other authors. Reasons for these discrepancies are still not clear.

SO_2 reactions with atmospheric particles have been studied using a controlled reaction chamber described by Ref. 91.[79,80,92] The rate of sulphate

Table 3.7
Reaction capacities for SO_2 observed with atmospheric particles in laboratory experiments

Reference	*SO_2 (ppm(v/v))*	*Aerosol matter*	*Relative humidity (%)*	*Final amount adsorbed ($mgSO_2 g^{-1}$ aerosol)*
74	14·4	Fly ash	20	0
			95	0·12
75	3–100	Fly ash	0	0·5
			50	0·2
			95	1·4
73	8·6	Atmospheric dust	55	0·075–0·152
		Dusts from industrial plants	55	0·096–0·345
		Fly ash	55	0·060–0·331
70	1–20	Fly ash	0	1·5
			50	2·8
			70	3·2
79	1–2	Fly ash	30	200
			60	300
			82	~ 380
		Cement dust	30	20
			60	30
			90	80
91	1–2	Fly ash	20	80
			90	200

formation was found to increase with the relative humidity and cement dust had a lower reaction capacity than fly ash. Results were analysed in terms of m_w/m_0 (where m_w is the mass of water and m_0 the mass of the dry aerosol). For aerosol experiments ($m_w/m_0 \leq 8$) the rate of SO_2 oxidation was found to be ½-order in SO_2 while in solutions containing suspensions of fly ash and cement dust the reaction was zero-order in SO_2. For relative humidities less than 99·5% ($m_w/m_0 \leq 5$) the amount of sulphate formed at the end of the reaction was found to be proportional to $\exp(m_w/m_0)$.

A saturation effect with fly ash has been observed.[75] Some of the samples showed an increase of 2 to > 300% in the capacity to adsorb SO_2 after being washed with distilled water, and treatment with gaseous NH_3 permitted the regeneration of 50% of the initial capacity of the reacted sample. It was suggested that most fly ash particles have their activity exhausted on, or immediately after, emission, and that part of the initial capacity may be regenerated by mixing with clean air containing NH_3. SO_2 has been found to chemically adsorb onto dusts from stack emissions of industrial plants and be converted to sulphate, while on urban particles no sulphate was observed and the only process taking place was physical adsorption of SO_2 which could easily be desorbed from the aerosol by evacuation.[73] It was found that alkaline particles are the most effective in the conversion of SO_2 to SO_3; this is in contradiction with the results of Ref. 92, which states that heterogeneous reactions proceed more readily on neutral or acidic surfaces than on basic surfaces. It has been concluded that these rates are too slow to be of importance in the atmosphere,[73] conclusions which also differ from those of Refs 79, 80 and 92.

The uptake of SO_2 by pulverised fuel ash inside a reaction chamber has been studied using a radiochemical technique, and it was concluded that these reactions are unlikely to be of importance in the atmosphere.[70] Extrapolating the laboratory results to the atmosphere by assuming a fly ash aerosol burden of 100 $\mu g\, m^{-3}$ SO_2 removal rates of 0·003 and 0·005% h^{-1} were calculated at 0 and 70% relative humidity, respectively. SO_2 oxidation on urban atmospheric carbonaceous particles resulting mainly from vehicle emissions has been investigated.[76] The technique involved collecting particles on a filter paper and then passing SO_2 in air through the filter in the laboratory and measuring the concentration in the effluent air stream. The rate of uptake of SO_2 and the capacity of the particles for SO_2 adsorption were found to be enhanced by the relative humidity (60–90%). Reaction capacities in the range of 2–8% of the initial mass of the catalyst were determined. The effect of NO_2 on the reaction was investigated and it was found that at atmospheric concentrations the direct oxidation of SO_2 by O_2 predominates over the reaction with NO_2. It was concluded that these reactions are irrelevant compared to photochemical or aqueous droplet reactions in the atmosphere as shown by the conversion rates of SO_2 based on these experiments which are given in Table 3.8.

Table 3.8
Estimated SO_2 conversion rates for heterogeneous oxidation on carbonaceous aerosols in the atmosphere[76]

Reaction[a]	*SO_2 conversion rate (% h^{-1})*	
	60% relative humidity	*93% relative humidity*
SO_2[a]+urban aerosol	$5{\cdot}6\times10^{-3}$	$7{\cdot}4\times10^{-3}$
SO_2[a]+urban aerosol +NO_2 (0·12 ppm (v/v))	$2{\cdot}4\times10^{-3}$	—

[a]*SO_2—0·1 ppm (v/v).*

SO_2 reactions with carbonaceous particles collected from the flue ducts of an oil-burning power station have been studied, and the main interest was reaction in flue gases.[93] The reaction rate was not determined, but it was found that the amount of adsorption increased when water and O_2 are present in the gas stream.

It has been suggested that mass transfer may control the reaction capacities and rates and that in the experiments utilising aerosols deposited on filter papers only the outer layer of the deposited particles may react with SO_2 thus yielding lower results than those obtained using genuine reacting aerosols systems.[79] This does not, however, resolve the discrepancy, as slow rates have been observed using unsupported reacting aerosols and a sensitive measurement technique.[70] These rates were comparable to those determined by Ref. 76, in which dust particles collected on filter papers were employed.

3.3.3 Aqueous reactions

In the atmosphere, aqueous-phase oxidation can proceed in cloudwater, fogwater and rainwater, deliquescent aerosol droplets at high relative humidities, and in surface liquid films on aerosol particles. Gaseous SO_2 is in equilibrium with dissolved S(IV) species in the aqueous phase as described by the following equilibrium reactions:

$$SO_2(g)+H_2O(l) \overset{H}{\rightleftharpoons} SO_2\,.\,H_2O \tag{3.38}$$

$$H = \frac{[SO_2\,.\,H_2O]}{P_{SO_2}} = 1{\cdot}242\ \text{mol litre}^{-1}\,\text{atm}^{-1}\,(101\ \text{kPa})\ \text{at}\ 25^\circ\text{C (Ref. 94)} \tag{3.39}$$

$$SO_2\,.\,H_2O \overset{K_1}{\rightleftharpoons} H^+ + HSO_3^- \tag{3.40}$$

$$K_1 = \frac{[HSO_3^-][H^+]}{[SO_2\,.\,H_2O]} = 0{\cdot}0132\ \text{mol litre}^{-1}\ \text{at}\ 25^\circ\text{C (Ref. 94)} \tag{3.41}$$

$$HSO_3^- \overset{K_2}{\rightleftharpoons} H^+ + SO_3^{2-} \tag{3.42}$$

$$K_2 = \frac{[SO_3^{2-}][H^+]}{[HSO_3^-]} = 6{\cdot}42\times10^{-8}\ \text{mol litre}^{-1}\ \text{at}\ 25^\circ\text{C (Ref. 94)} \tag{3.43}$$

$SO_2\,.\,H_2O$ is physically dissolved SO_2, HSO_3^- is the bisulphite ion and SO_3^{2-} is the sulphite ion. The sum total of these species in solution is designated by S(IV). HSO_3^- is the dominant species over much of the atmospheric pH range (pH 3–7). H is Henry's law constant for SO_2, and K_1 and K_2 are the first and second dissociation constants.

Ideally, these constants should be defined in terms of activities and not of concentrations, however, for dilute solutions such as cloud-drops, fog-drops and raindrops with ionic strengths less than 10^{-3} mol litre^{-1} the value of the activity coefficients is close to unity and it is possible to employ concentrations in equilibrium calculations. Exceptions are deliquescent aerosol and haze droplets with high ionic strengths, and for these full account must be taken of solution non-ideality.[95,96]

Once dissolved in the droplets S(IV) species may be oxidised to sulphate (S(VI)) via a variety of mechanisms. Factors which may affect SO_2 oxidation in atmospheric droplets are summarised in Table 3.9. As aqueous oxidation reactions of SO_2 play an important role in many industrial processes these mechanisms have been studied extensively over the last 100 years and it is only in the last 30 years that studies have appeared which focus specifically on the role of these reactions in the atmosphere.

Experimental studies of SO_2 oxidation have employed homogeneous bulk solution, multiphase gas–bulk solution and gas–droplet systems,

Table 3.9
Factors affecting SO_2 oxidation in atmospheric droplets

General	*Specific*
Oxidants	O_2, O_3, H_2O_2, NO_x
Free radicals	$OH, HO_2, RO_2, RCOO_2, Cl, Cl_2^-$
Catalysts	$Fe^{3+}, Mn^{2+}, Cl^-, Br^-$, solid carbon and synergistic effects between several catalysts
Neutralising agents	NH_3, alkaline metal oxides, HCl desorption from chloride solutions
Ionic strength	Effects on solution equilibria Salt effects on ion–ion reaction kinetics
Inhibitors	Most organic compounds, Sb, Se, As Organic surface films

Adapted from Ref. 95.

and a variety of techniques have been used to monitor the reaction. Some workers have employed spectrophotometric techniques involving the monitoring of optical absorbance in the UV region[97–102] and the use of this technique has been subjected to some criticism since it involves irradiation of the solution with intensities higher than those available in the troposphere.[103] Mechanisms of aqueous oxidation have been studied over a wide range of experimental conditions, many of them not of relevance to the atmosphere.

3.3.3.1 *Uncatalysed oxidation by O_2*

This mechanism has been studied extensively in laboratory studies and conflicting results have been reported with regard to the kinetics, mechanism and atmospheric implications. A critical review of the available literature has been published in an attempt to resolve some of the discrepancies.[104]

The overall reaction for SO_2 oxidation by O_2 may be expressed as

$$SO_3^{2-} + \tfrac{1}{2}O_2 \rightarrow SO_4^{2-} \qquad (3.44)$$

Most of the kinetic studies report that the reaction is first-order in sulphite and zero-order in oxygen, at experimental conditions relevant to the atmosphere. The rate may therefore be expressed as

$$\frac{d[SO_4^{2-}]}{dt} = k[SO_3^{2-}] \qquad (3.45)$$

The rate constant (k) is a function of pH, however, even at any given pH there is a discrepancy of several orders of magnitude in the reported value of k and the purity of experimental solutions appears to be largely responsible for the reported scatter in the data.[104] At pH values of atmospheric interest (pH < 7) the rate may be expressed by the following equation:

$$\frac{d[SO_4^{2-}]}{dt} = k'[SO_3^{2-}][H^+]^{1/2} \tag{3.46}$$

This rate expression agrees with the experimental results from a number of kinetic studies,[105–108] but not all.[109–111] A value of $k' = 0{\cdot}32$ (mol litre^{-1})$^{-1/2}$ s^{-1} at 25°C was found to best describe the process at conditions representative of the atmosphere.[104]

Aqueous oxidation of SO_2 by O_2 is thought to proceed via a free radical chain mechanism involving SO_3^- and SO_5^- radicals, which have been identified in experimental solutions.[112] A common form of the mechanism is shown in Fig. 3.2 together with a derived rate law which is in reasonable agreement with experimental results, except that the O_2

Initiation	$SO_3^{2-} + M^+ \rightarrow SO_3^- + M$	(a)
Propagation	$SO_3^- + O_2 \rightarrow SO_5^-$	(b)
	$SO_5^- + HSO_3^- \xrightarrow{pH<7} HSO_5^- + SO_3^-$	(c)
	$SO_5^- + SO_3^{2-} \xrightarrow{pH>7} SO_5^{2-} + SO_3^-$	(d)
Oxidation	$SO_3^{2-} + HSO_5^- \xrightarrow{pH<7} HSO_4^- + SO_4^{2-}$	(e)
	$SO_3^{2-} + SO_5^{2-} \xrightarrow{pH>7} 2SO_4^{2-}$	(f)
Termination	$SO_3^- + SO_5^- \rightarrow S_2O_6^{2-} + O_2$	(g)
	SO_5^- + inhibitor $\rightarrow$ non-reactive products	(h)

$$\frac{d[SO_4^{2-}]}{dt} = \left(\frac{k_a k_b k_c}{K_2 k_g}\right)^{1/2} [SO_3^{2-}][H^+]^{1/2}[M^+]^{1/2}[O_2]^{1/2}$$

for pH < 7

Fig. 3.2. The free radical mechanism and theoretical rate law.

dependence has been observed only at high pH values.[106] Some authors have preferred to include SO_4^- radical reactions in the mechanism.[112,113]

Further modifications of this mechanism[114] and alternative mechanisms[115,116] have been proposed. The free radical mechanism may also be photo-initiated, but this is unlikely to be of importance in the troposphere or in most laboratory studies.[104]

It is now generally accepted that the uncatalysed air oxidation of SO_2 by O_2 is of little importance in the atmospheric oxidation of SO_2.[98,104,111,117] In any case there are sufficient catalysts present in the atmosphere to allow catalytic mechanisms to dominate over the uncatalysed reaction.

3.3.3.2 Metal catalysed oxidation by O_2

It has long been known that traces of certain transition metal ions and their co-ordination complexes may catalyse oxidation of SO_2 in solution. The earliest work on metal catalysis of SO_2 oxidation in solution investigated the catalytic action of Cu(II) and it was claimed that concentrations as low as 10^{-13} mol litre^{-1} exerted a measurable effect.[118] Other studies also reported that copper was an effective catalyst of SO_2 oxidation;[105,119,120] however, many studies have failed to observe any catalytic effect of Cu^{2+} on SO_2 oxidation.[121–123] Studies which observed catalysis by Cu(II) were carried out in unbuffered solutions at pH > 8, while those which failed to observe any effect have been performed in acidic solutions or droplets. It has been reported that the effectiveness of metal catalysts is pH dependent[119,124] and that Cu(II) was most active as a catalyst of SO_2 oxidation between pH6 and 12·2;[125] this might explain the discrepancy in the literature. A number of other transition metals may catalyse SO_2 oxidation in solution (Co, Fe, Mn, V) and comparative studies of the catalytic effectiveness of various metal ions indicate that Fe and Mn are the most effective catalysts in acidic solutions.[123,126]

Rate laws determined for Fe- and Mn-catalysed oxidation by O_2 and the experimental conditions are summarised in Table 3.10.

Laboratory studies can be divided into those employing dilute solutions or droplets which are relevant to the oxidation in fog-drops, cloud-drops and rain-drops and those carried out in concentrated solutions and deliquescent aerosols of metal salts which are relevant to the oxidation in atmospheric aerosol droplets at high humidities. In the former experiments catalyst concentrations are low (10^{-8}–10^{-4} mol litre^{-1}) while in the latter they may be as high as 0·1 mol litre^{-1}. Studies using deliquescent aerosol droplets are difficult to interpret kinetically and many experiments have failed to produce quantitative data on the reaction rate constant.

For the Fe(III)-catalysed oxidation there is general agreement that the rate of oxidation is first-order in Fe(III) concentration. A second-order dependence has been observed in one study with respect to S(IV) concentration,[128] while some studies reported a variable, pH-dependent reaction order.[139,146] Most studies report a first-order dependence with respect to S(IV) concentration and these were carried out at lower catalyst concentrations ($\leq 10^{-4}$ mol litre^{-1}). At higher concentrations of Fe(III) the mechanism of the reaction may differ and direct oxidation of SO_2 by Fe(III) may be significant.[136] The overall rate of oxidation via this mechanism at conditions relevant to the atmosphere may be expressed by an equation of the form

$$\frac{d[SO_4^{2-}]}{dt} = k[\text{Fe(III)}][\text{S(IV)}] \quad (3.47)$$

The value of k appears to decrease with increasing acidity ($k \propto [H^+]^{-1}$) in the pH range 0–3.[143,145] Both Fe(II) and Fe(III) salts have been employed in experiments and these show that the Fe(II)/Fe(III) couple is involved in the catalysis with Fe(III) being the active catalytic species.[135,147] As Fe catalysis is inhibited by ionic strength and sulphate ions, it has been proposed that these effects may be responsible for the discrepancies between reported studies.[145]

There is considerably less agreement between the results of experimental studies of the Mn-catalysed oxidation of SO_2. First-order, second-order and non-integral order reactions, with respect of Mn(II) concentration, have been reported. Investigators have observed zero-order[121,128,130] and first-order rates[133,134,142] with respect of S(IV) concentration.

Two detailed studies of this reaction have been carried out in an attempt to resolve this discrepancy.[141,143] Rates were measured over a wide range of experimental conditions and complex behaviour, which was found to depend on the concentrations of both Mn(II) and S(IV), was noted. Both studies found rates consistent with the following rate law:

$$\frac{-d[\text{S(IV)}]}{dt} = k[\text{Mn(II)}][\text{S(IV)}][H^+]^{-1} \quad (3.48)$$

at

$$[\text{S(IV)}] \leq 10^{-6} \text{ mol litre}^{-1} \text{ and } [\text{Mn(II)}] < 10^{-3} \text{ mol litre}^{-1}$$

Table 3.10
Summary of rate studies of the Mn- and Fe-catalysed oxidation of SO_2 by O_2 in aqueous solution

Reference	*Experimental system*	*pH*	*T (°C)*	*[S(IV)] (mol litre⁻¹)*	*[Metal] (mol litre⁻¹)*	*Rate expression*	*Rate constant*	*Comments*
121	Bulk	3–4	30–35	10^{-4}–10^{-3}	$0{\cdot}3$–$8{\cdot}0\times10^{-5}$	$R = k[Mn^{2+}]^{1{\cdot}7}$	$k = 44\,(mol\,l^{-1})^{-0{\cdot}7}s^{-1}$	Dilatometric measurements Rate independent of aqueous SO_2
127	Bulk and droplet	$0{\cdot}7$–3	~25	$1{\cdot}89$–$2{\cdot}52\times10^{-3}$	$2{\cdot}65\times10^{-5}$–6×10^{-3}	$R = k[Mn^{2+}]^2$ for $[Mn^{2+}] < 10^{-4}$ mol litre^{-1}	$k = 5{\cdot}78\times10^2\,l\,mol^{-1}s^{-1}$	Iodometric titration in bulk solutions, also 700 μm diameter droplets exposed to SO_2 (120–298 ppm(v/v))
128	Bulk	$1{\cdot}4$–$2{\cdot}8$ $1{\cdot}2$–$2{\cdot}65$	10 $1{\cdot}5$–20	10^{-5}–10^{-3} $0{\cdot}72$–$20{\cdot}5\times10^{-4}$	$2{\cdot}7$–$8{\cdot}3\times10^{-4}$ $1{\cdot}8\times10^{-4}$–$0{\cdot}0215$	$R = k[Mn^{2+}][H^+]^{-1}$ $R = k[SO_2.aq]^2[Fe^{3+}][H^+]^{-1}$	$k = 1{\cdot}7\times10^{-5}mol\,l^{-1}s^{-1}$ $k = 40\,l\,mol^{-1}s^{-1}$	Iodometric titration Study in O_2 k is for 10°C
129	—	—	—	—	—	$R = \dfrac{k_1[Fe^{3+}]}{[H^+]}\left\{\dfrac{1-k_2[Fe^{2+}]}{[O_2]}\right\}[H_2SO_3]$	—	Based on Ref. 128
130	Bulk	~3–4	25	$1{\cdot}06$–$2{\cdot}3\times10^{-3}$	$3{\cdot}3\times10^{-5}$–$0{\cdot}066$	$R = k[Mn^{2+}]^2$ for $[Mn^{2+}) < 66\times10^{-5}$ mol litre^{-1}	$k = 4{\cdot}66\times10^{-2}l\,mol^{-1}s^{-1}$	Batch and flow systems Study in O_2 pH estimated
131	Bulk	—	25	1–$3{\cdot}4\times10^{-5}$	$0{\cdot}1$–$7{\cdot}5\times10^{-5}$	$R = \dfrac{[HSO_3^-]_i^{2{\cdot}2}[Mn^{2+}]^2}{k_1[Mn^{2+}]^{2+}k_2[HSO_3^-]_i^{2{\cdot}2}}$	$k_i = 1{\cdot}32\times10^{-3}s\,l\,mol^{-1}$ $k_2 = 3{\cdot}2\times10^{-4}s\,l\,mol^{-1}$	Conductivity measurements Study in pure air. pH not measured
132	Droplet	< 1	25	—	$0{\cdot}17$	$R = k[Mn^{2+}]^2$	$k = 4\times10^3\,l\,mol^{-1}s^{-1}$	Deliquescent aerosol study at 95% relative humidity $[SO_2]$gas = $5{\cdot}128$ ppm (v/v) Droplet diameter = $3{\cdot}7$ μm
133	Bulk	~7	20	~7×10^{-6}	$1{\cdot}8\times10^{-6}$	$R = k[S(IV)]$	$k_{Fe} = 8{\cdot}05\times10^{-5}s^{-1}$ $k_{Mn} = 3{\cdot}33\times10^{-5}s^{-1}$	West–Gaeke method Air oxidation
134	Droplet	~ 1	25	—	$0{\cdot}34$–$5{\cdot}2\times10^{-3}$	$R = k[Mn^{2+}][S(IV)]$	—	Deliquescent droplet study. Drops 15 μm in diameter suspended on Teflon beads and exposed to $4{\cdot}5$ ppm (v/v) SO_2 at 95% relative humidity
135	Bulk	$4{\cdot}9$	20±2	10^{-5}–10^{-3}	$0{\cdot}5$–5×10^{-7}	$R = k[Fe^{3+}][S(IV)]$	$k = 100\,l\,mol^{-1}s^{-1}$	Radiochemical method Air oxidation
136	—	—	—	—	—	$R = \{k_o+k_F[Fe^{3+}]\}K_1^2[Fe^{3+}]\cdot[SO_2.aq]^2[H^+]^{-3}$ for $[Fe^{3+}] \leq 1{\cdot}06\times10^{-2}$ mol litre^{-1}	$k_o = 1{\cdot}52\times10^2\,l\,mol^{-1}s^{-1}$ $k_F = 2{\cdot}22\times10^4\,l^2\,mol^{-2}s^{-1}$	Based on Ref. 128 K_1 is the dissociation constant of sulphurous acid
113	Droplet	2–$5{\cdot}6$	8,25	—	10^{-6}–10^{-4}	$R = k[SO_3^{2-}]$	$k_{Fe} = 5{\cdot}8\times10^2s^{-1}$ $k_{Mn} = 9{\cdot}4\times10^2s^{-1}$	2.1 mm diameter droplets exposed to $[SO_2]$ gas =

						$R = \frac{[HSO_3^-][H^+][Fe^{3+}]}{k_1+k_2[HSO_3^-]}$	$k_1 \gg k_2[HSO_3^-]$ for small $[HSO_3^-]$	Rate expression reported by Ref. 138
137	Bulk	< 4·5	—	—	—			
139	Bulk	3–8	20	10^{-5}–2×10^{-3}	10^{-5}–10^{-7}	$R = k[S(IV)]^n[Fe^{3+}]$	—	Electrode measurements Air oxidation Value of k not reported
140	Droplet	—	25	—	0·1	$R = k[S(IV)]$	$k = 0{\cdot}04 s^{-1}$	Deliquescent aerosols of $(NH_4)_2SO_4$ containing Mn Droplets 1–10 μm in diameter exposed to SO_2 (0·1–1 ppm (v/v)) at 95% relative humidity
141	Droplet	—	24±2	—	0·17	$R = k[H^+]^{-1}$	$k = 8{\cdot}3(\pm2{\cdot}5)\times10^{-5}$ $mol^2 l^{-2}$	Deliquescent aerosols of $MnSO_4$ at 91–97% relative humidity $[SO_2]g$ = 14–92 ppm (v/v)
126	Bulk	1–5	25	10^{-2}–0·5	10^{-5}–5×10^{-4}	$R = k[Fe]^{1{\cdot}16}[HSO_3^-]^{0{\cdot}6}[H^+]^{-0{\cdot}99}$	$k = 10^{[2{\cdot}32-1{\cdot}26.x]}$ $mol^{0{\cdot}23}l^{-0{\cdot}23}s^{-1}$	Iodometric titration
	Bulk	1–5	25	<10^{-3}–0·06	10^{-5}–3×10^{-4}	$R = k_1[Mn]^2+k_2[Mn][HSO_3^-]$	$k_1 = 10^{[2{\cdot}83-4{\cdot}07.x]}$ $l mol^{-1}s^{-1}$ $k_2 = 10^{[-0{\cdot}08-1{\cdot}01.x]} mol l^{-1}s^{-1}$	Rate constant depends on the ionic strength $x = I^{1/2}/(1+I^{1/2})$
142	Bulk	2·7–7·5	21–30	0·5–$1{\cdot}5\times10^{-5}$	1–5×10^{-6}	$R = k[Mn][S(IV)]$	$k = 5{\cdot}1\pm0{\cdot}5\times10^{-3}$ $l mol^{-1}s^{-1}$	Continuously flowing stirred reactor system Value of k is for 23·7°C
143	Bulk	0–3	25	10^{-3}–10^{-4} < 10^{-4} ~5×10^{-4} 10^{-5}–10^{-3}	10^{-5}–10^{-3} < 10^{-6} 10^{-6}–10^{-4} 10^{-6}–10^{-4}	$R_{Mn} = k[Mn^{2+}]^2[H^+]^{-1}$ $R_{Mn} = k[Mn^{2+}][S(IV)][H^+]^{-1}$ $R_{Fe} = k[Fe^{3+}][S(IV)][H^+]^{-1}$ $R_{Mn+Fe} = R_{Mn} + \left\{1+\frac{1{\cdot}70\times10^3[Mn^{2+}]^{1{\cdot}5}}{6{\cdot}31\times10^{-6}+[Fe^{3+}]}\right\}$	$k = 4{\cdot}7 s^{-1}$ $k = 2{\cdot}5 s^{-1}$ $k = 0{\cdot}82 s^{-1}$	Spectrophotometric measurements R_{Mn+Fe} is the rate for synergistic catalysis by Mn and Fe
144	Bulk	4·2	23·8	2×10^{-5}	10^{-7}–10^{-6}	$R_{Mn} = k[Mn^{2+}][S(IV)]$ $R_{Fe} = k[Fe^{3+}][S(IV)]$ $R_{Mn+Fe} = k[Mn^{2+}][Fe^{3+}]\cdot[S(IV)]$	$k = 1{\cdot}71(\pm0{\cdot}06)\times$ $10^3 l mol^{-1}s^{-1}$ $k = 1{\cdot}75(\pm0{\cdot}05)$ $10^3 l mol^{-1}s^{-1}$ $k = 3{\cdot}6(\pm0{\cdot}6)\times10^{10}$ $l^2mol^{-2}s^{-1}$	Continuously flowing stirred reactor R_{Mn+Fe} is the rate for synergistic catalysis by Mn and Fe
145	Bulk	0–3·6	25	10^{-6}–10^{-3}	3×10^{-7}–5×10^{-5}	$R = k[Fe^{3+}][S(IV)][H^+]^{-1}$	$k = 6{\cdot}0(\pm1)s^{-1}$	Optical measurements Inhibition by ionic strength was noted

At $[S(IV)] > 10^{-4}$ mol litre^{-1}, second-order rates in Mn(II) and zero-order rates in S(IV) were observed in these two studies and this may explain some of the discrepancies between reported studies. Mn-catalysed oxidation of S(IV) exhibits unusual kinetics; this has been noted in other studies.[121,122,126] Also, ionic strength was found to inhibit the reaction.[148]

As in atmospheric droplets a mixture of trace metals is present a number of studies have employed combinations of catalysts.[113,123,143,144,149] A 'mixed salt' catalytic effect, whereby mixtures of Fe(III) and Mn(II) result in much faster oxidation of S(IV) than when the catalysts are present individually at the same concentration, has been reported and it has been suggested that this synergistic effect may be important in atmospheric droplets. This effect has been studied in some detail in one study and a complex rate law determined.[143] Other metals were also investigated but only Mn(II) and Fe(III) exhibited synergism.[123,143]

One of the most detailed studies of trace metal catalysis of SO_2 oxidation in droplets is that of Ref. 113, and rate constants evaluated there have been used extensively in atmospheric models.[150,151] However, droplets 2·1 mm in diameter were employed and there is evidence to suggest that significant mass transfer limitation of the oxidation rate may have taken place in the experiments and that the reported rate constants may be low by 20% to as much as an order of magnitude.[152]

Orders of catalyst efficiency established for the various trace metals are shown in Table 3.11. An assessment of the abundance of transition metals

Table 3.11
Order of efficiency of transition metal catalysts in experimental studies

Reference	*Order of catalyst efficiency*
121	$Fe^{2+} \gg Mn^{2+} \gg Cu^{2+a}$
153	$Mn^{2+} > Cu^{2+} > Fe^{2+} > Co^{2+}$
127	$Mn^{2+} > Fe^{2+} > Ni^{2+}$
154	$Mn^{2+} > Fe^{2+}$
131	$Fe^{2+} > Fe^{3+} > Mn^{2+} \gg Co^{2+}, Ni^{2+}, Cu^{2+}, Zn^{2+}, Cr^{3+}$
133	$Fe^{3+} > Cu^{2+} > Mn^{2+} > Al^{2+}, Co^{2+}$
134	$MnSO_4 > MnCl_2 > CuSO_4$
123	$Mn^{2+} > Fe^{2+} > Fe^{3+}$
155	$Fe^{3+} > Cu^{2+} \gg Mn^{2+a}$
156	$Mn^{2+} > Fe^{3+} > Co^{2+}$
157[b]	Fe > Mn ~ Cu > Ti > V ~ Ni ~ Cr > Zn

[a]No catalytic effect observed.
[b]Not an experimental but a theoretical study.

Table 3.12
Summary of rate studies of SO_2 oxidation in rainwater samples

Reference	*pH*	*T* (°C)	*[S(IV)]* ($mol\,l^{-1}\,litre^{-1}$)	*Rate* ($mol\,s^{-1}$)	*Comments*
133	~7	20	~7×10^{-6}	$R = 1{\cdot}89\times10^{-5}[S(IV)]$	West–Gaeke method in non-polluted sample
114	3·5–5·6	25	~8×10^{-3}	$R = 4{\cdot}71\,(\pm1{\cdot}78)\times10^{-3}[S(IV)]$[a]	Iodometric titration in 22 polluted samples buffered to pH 5·5. Rates were about an order of magnitude slower in unpolluted samples
159	3·7–4·4	25	10^{-6}–3×10^{-4}	$R = 0{\cdot}142[Mn]-1{\cdot}02\times10^{-8}$ at $[S(IV)] > 10^{-4}$ mol litre^{-1}; $R = 1{\cdot}42\times10^{5}(Mn]\cdot[S(IV)]^{3/2}$ at $[S(IV)] < 5\times10^{-5}$ mol litre^{-1}	Radiochemical method in 9 samples stored for 1 year
160	3–5	25	~7×10^{6}	$R = 0{\cdot}19[SO_3^{2-}]$	West–Gaeke method in 8 urban samples
161	3·3–5·6	25	10^{-5}–10^{-4}	$R = 6{\cdot}65\times10^{6}\cdot[Fe]^{1{\cdot}87}[S(IV)]$	West–Gaeke method in 40 urban samples. Rate expression is based on 26 samples
162	3·7–4·8	25	$62{\cdot}5\times10^{-6}$	$R = 438[M^+][S(IV)]$ unfiltered[a]; $R = 386[M^+][S(IV)]$ filtered	Ion chromatography in 8 polluted samples $[M^+] = [Mn^{2+}]+[Fe^{3+}]$

[a]Evaluated from original data quoted in the relevant reference.

in the atmosphere, their chemical forms, available oxidation states and bonding tendency showed that Fe(III) and Mn(II) are the most important catalysts of SO_2 oxidation in atmospheric droplets;[157] however, from the experimental studies it is not clear which is more effective. There is some indication that the effectiveness of catalysts depends on the nature of the metal salts employed in the experiments.[122,123,158]

Because of the discrepancies between experimental studies and the problems involved in extrapolating the results of such studies in pure water systems to real atmospheric droplets containing many trace species which may affect the oxidation rate, kinetic studies of SO_2 oxidation have recently been reported which employed rainwater samples as the reaction medium. The results of these studies are summarised in Table 3.12 together with the experimental conditions. Most studies report rates first order in S(IV)[67,114,133,161,162] and values of the first order reaction rate constant in the range of 10^{-5}–10^{-2} s^{-1} have been measured.

In some of the studies regression analysis was employed to identify the catalyst or catalysts responsible. A high degree of correlation has been found between the S(IV) oxidation rate and the Mn concentration, and a lower but significant correlation was found with V but not with Cr, Co, Ni, Fe or Cu.[159] On the other hand, a strong correlation between the rate constant and the Fe concentration has been observed, but no significant correlation with Mn.[161] Strong correlations were found between the first-order rate constant and the concentrations of several trace metals, including both Mn and Fe, in rainwater.[67] In a recent study of S(IV) oxidation

Table 3.13
Minimum concentrations of Mn and Fe catalyst required for catalysis of S(IV) oxidation by O_2 and concentrations measured in rainwater, fogwater and cloudwater

Reference	*Concentration (mol litre^{-1})*	
	Mn	*Fe*
131	10^{-6}	—
113	10^{-6}	—
121	10^{-5}	—
135	—	10^{-8}
133	—	10^{-6}
164 (rainwater)	$2{\cdot}2\times10^{-8}$–$1{\cdot}4\times10^{-6}$	$1{\cdot}1\times10^{-6}$–$4{\cdot}8\times10^{-5}$
165 (fogwater)	$3{\cdot}3\times10^{-7}$–$1{\cdot}5\times10^{-5}$	$1{\cdot}6\times10^{-6}$–$4{\cdot}3\times10^{-4}$
166 (cloudwater)	—	$3{\cdot}6\times10^{-7}$–$1{\cdot}2\times10^{-4}$

in cloudwater samples collected at a rural site, the rate was also found to be highly correlated with Fe but not Mn concentrations.[163]

Comparisons of results from the rainwater studies with those from experimental studies in pure water are difficult because of

- differences between reported studies;
- differences between single metal and multi-metal systems where synergism is possible; and
- the complex nature of rainwater where inhibitors and other factors may be important.

Differences between the studies reported in rainwater samples could reflect differences in rainwater composition.

Minimum concentrations of metal catalysts required for catalysis, as established in experimental studies, are compared to levels measured in aqueous atmospheric samples in Table 3.13. The concentrations of Mn(II) in rainwater appear to be lower than those required for catalysis in clean water experiments and catalysis may be expected only at higher concentrations measured in some fogwater samples. Fe(III) is present in all types of aqueous samples at concentrations sufficient to induce catalysis. Table 3.13 shows that trace metal concentrations in fogwater and cloudwater can be higher than those measured in rainwater and metal-catalysed oxidation of SO_2 by O_2 may be particularly important in these systems. Synergistic catalysis may be anticipated at the concentrations of Fe(III) and Mn(II) determined in cloudwater and fogwater.

An experimental study of the simultaneous leaching of trace metals from fly ash into solution and catalysis of SO_2 oxidation by O_2 has been reported.[167] Experiments were carried out at pH 3·1 and it was concluded that Fe(III) plays a more important role than Mn(II) in the oxidation of SO_2.

There is still considerable uncertainty regarding the mechanism of the metal-catalysed oxidation of SO_2 by O_2 and a number of different sequences have been proposed. The proposed mechanisms can be classified according to those involving

- free radicals;
- metal–sulphite complexes; and
- both free radicals and metal–sulphite complexes.

The simplest free radical mechanism is the thermally initiated chain mechanism shown previously in Fig. 3.2. Initiation with the following reaction has been proposed:[147]

$$Fe(III)+SO_3^{2-} \rightarrow SO_3^- + Fe(II) \quad (3.49)$$

Fe(III) is regenerated by the reaction of Fe(II) with SO_5^- radical. A similar mechanism involving initiation by hydroxylated Fe(III) has been proposed.[135]

Most of the non-radical mechanisms are based on the sequence of reactions proposed by Ref. 158 for the Mn(II)-catalysed reaction. The mechanism is initiated by complex formation between Mn(II) and sulphite. The complex rapidly adds on an oxygen molecule by direct addition and then undergoes rearrangement to give sulphate. This mechanism may be described by the following reactions:

$$Mn^{2+}+2SO_2^{2-} \rightleftharpoons (Mn(SO_3)_2)^{2-} \quad (3.50)$$

$$(Mn(SO_3)_2)^{2-}+O_2 \rightleftharpoons O_2(Mn(SO_3)_2)^{2-} \quad (3.51)$$

$$O_2(Mn(SO_3)_2)^{2-} \rightleftharpoons (Mn(SO_4)_2)^{2-} \quad (3.52)$$

Similar mechanisms have been proposed by others for the metal-catalysed reactions.[132,136,168] A mechanism for Fe catalysis of S(IV) oxidation that is similar to that of Ref. 158 for Mn catalysis has been proposed, and it also accounts for the direct oxidation of S(IV) by Fe(III).[136] Some authors have suggested that photochemical reactions of Fe–sulphite complexes may be important in the mechanism as they observed a significant effect of light on the oxidation of SO_2 catalysed by trace metals.[169–173] However, no difference was observed between metal-catalysed oxidation rates measured in daylight and absolute darkness.[153]

A combined mechanism involving free radicals and metal–sulphite complexes has been proposed by Ref. 113 based on the earlier work of Ref. 125. The mechanism involves the following sequence of reactions:

$$[M^{x+}(S(IV))_3]^{-(6-x)}+O_2 \rightarrow [M^{x+}(S(IV))_3]^{-(5-x)}+O_2^- \quad (3.53)$$

$$O_2^-+H^+ \rightarrow HO_2 \quad (3.54)$$

$$[M^{x+}(S(IV))_3]^{-(5-x)}+SO_3^{2-} \rightarrow [M^{x+}(S(IV))_3]^{-(6-x)}+SO_3^- \quad (3.55)$$

I SO_3^- chain

$$SO_3^-+O_2 \rightarrow SO_5^- \quad (3.56)$$

$$SO_5^-+SO_3^{2-} \rightarrow SO_4^{2-}+SO_4^- \quad (3.57)$$

$$SO_4^- + SO_3^{2-} \rightarrow SO_3^- + SO_4^{2-} \quad (3.58)$$

II HO_2 chain

$$HO_2 + SO_3^{2-} \rightarrow SO_4^{2-} + OH \quad (3.59)$$

$$OH + SO_3^{2-} \rightarrow SO_4^{2-} + H \quad (3.60)$$

$$H + O_2 \rightarrow HO_2 \quad (3.61)$$

A metal–sulphite complex acts to initiate the process and this complex is regenerated by reaction with the sulphite ion and is therefore considered to be the active catalyst. The reaction then proceeds via two chains with HO_2 and SO_3^- as chain carriers and also involves other radicals. Other mechanisms involving both free radicals and metal–sulphite complexes have been proposed for Mn(II)- and Fe(III)-catalysed reactions.[169,174]

There are considerable difficulties in applying any of the proposed mechanisms to the atmosphere as the solution chemistry of trace metals is also important and different complexes of the same metal may have different effects. Recent modelling studies have addressed the problem of trace metal speciation in atmospheric droplets,[172,175] but there is a paucity of field data.

3.3.3.3 Cl^--catalysed oxidation by O_2

Cl^- catalysis of SO_2 oxidation by O_2 has been noted in two early studies.[176,177] The only bulk solution study of the Cl^--catalysed oxidation of SO_2 with relevance to the atmosphere is that of Ref. 108. SO_2 oxidation by O_2 was studied in a variety of chloride salt solutions and the following rate expression obtained:

$$\frac{-d[S(IV)]}{dt} = 32{\cdot}8[S(IV)]^2[Cl^-]^{1{\cdot}3}[H^+]^{-0{\cdot}5} \text{ mol l}^{-1}\text{ min}^{-1} \quad (3.62)$$

over the following range of experimental conditions: $[S(IV)] = 3\text{–}5\times10^{-5}$ mol litre^{-1}, $[Cl^-] = 10^{-3}\text{–}1$ mol litre^{-1} and pH $= 3\text{–}7$. At Cl^- concentrations > 1 mol litre^{-1} the reaction was too fast to study with the experimental technique of Ref. 108. A similar, but weaker, catalytic effect was observed for Br^- while F^- was observed to inhibit the reaction at 0·1 mol litre^{-1}. Sulphate, nitrate and bicarbonate salts of alkali and alkaline earth metals at 0·1 mol litre^{-1} did not affect SO_2 oxidation by O_2. Second order rates of S(IV) oxidation were also observed in seawater,[178] and the fast oxidation rates were consistent with catalysis by Cl^-, the concentration of which is 0·55 mol litre^{-1} in seawater. Rates were faster in artificial than in natural seawater and a possible explanation may be the presence of organic inhibitors in natural seawater.

Evidence for the possible importance of the Cl^--catalysed oxidation of SO_2 in the atmosphere comes mainly from droplet experiments employing deliquescent salt aerosols at high humidities. Rates of SO_2 oxidation in these experiments are summarised in Table 3.14 together with results from dilute droplet studies. A large effect is expected for Mn salts, however, rates measured in chloride salt aerosols are also considerably faster than those measured in dilute droplets and in some of the studies are comparable to the rates measured in deliquescent Mn salt aerosols. These experiments are relevant to the oxidation of SO_2 in deliquescent salt aerosols in the atmosphere and salt concentrations in these droplets may be several molar and the pH very low ($pH \leqslant 1$).[132,134,180] Rates and equilibria in these droplets may be influenced by the high ionic strengths and ion-pair formation and HCl desportion may be significant.[95,180] The laboratory studies indicate that the relative humidity exerts the strongest effect on the reaction rate in NaCl aerosols.[92,134,178–180] Above the deliquescence point of the salt, the rate of SO_2 oxidation increases with increasing relative humidity, however, even at relative humidities below the deliquescence point significant reaction was observed.[178] This was attributed to the presence of an aqueous phase at these conditions due to the hysteresis effect.[181] Recently, it has been suggested that Cl^--catalysed oxidation of SO_2 by O_2 may be important in fogs and clouds formed on sea salt condensation nuclei.[182,183] Oxidation rates of $> 30\%\ h^{-1}$ were observed in cloud chamber experiments using sea salt. SO_2 concentrations were in the range of 0·25–0·7 ppm(v/v) and liquid water contents in the range of 1–2·9 $g\ m^{-3}$. Salt concentrations were estimated to be between 8×10^{-5} and 5×10^{-3} mol $litre^{-1}$ in these experiments which was comparable to concentrations in marine fogs in the range of 10^{-4}–10^{-3} mol $litre^{-1}$.[165,184] These authors observed faster SO_2 oxidation when natural rather than artificial sea salt was used to generate fogs. This is contrary to the observations with deliquescent sea salt aerosols where faster rates were observed in artificial than in natural aerosols.[178] Another implication of this reaction is on the deposition of SO_2 to the sea surface. In rough weather the SO_2 deposition rate to the sea surface could be increased by the scrubbing effect of sea spray.[178]

The mechanism of the Cl^--catalysed reaction is not fully understood but it appears to involve Cl and Cl_2^- radicals. Cl^- may be involved in electron transfer to either SO_5^- or SO_4^- radicals although the latter seems more likely.[108]

Table 3.14
SO_2 reaction rates measured in deliquescent aerosols and dilute droplets

Reference	*Catalyst*	*SO_2 (ppm(v/v))*	*Relative humidity (%)*	*SO_2 reaction rate (10^{-3}mol(kgH_2O)$^{-1}$min^{-1})*
Deliquescent aerosols				
155	NaCl	250	95	355
132	$MnSO_4$	5–128	95	69
134	NaCl	18	81	5·7
179[a]	NaCl+$MnSO_4$	1	78	14 329
91	$MnSO_4$	1–3	95	1·5
140	$(NH_4)_2SO_4$+$MnSO_4$	0·7	90	3·3
92	NaCl	0·8	95	~ 0·9
180	$MgCl_2$	1·4–19·4	50	1·1
178	Sea salt	0·1–2	85	0·83
Dilute droplets				
113	10^{-4}M $MnCl_2$	0·9	—	0·05
110	None	1–4300	—	$7{\cdot}23–11{\cdot}7\times10^{-5}$

[a] *NH_3 present in the gas phase.*

$$Cl^- + SO_4^- \rightarrow SO_4^{2-} + Cl \quad (3.63)$$

Cl radicals produced in this reaction can equilibrate to form Cl_2^- and both may propagate the chain by reaction with sulphite or they may be involved in chain termination. Any ClO^- produced in the system will oxidise further sulphite to sulphate. Chlorine radicals can react with a variety of species in solution[185,186] and they may be particularly important in the free radical chemistry of sea salt aerosols at high humidities.

3.3.3.4 Carbon-catalysed oxidation by O_2

Recently, it has been shown that soot particles in aqueous suspension may catalyse the oxidation of SO_2 by O_2 and bulk and droplet studies of the reaction have been reported. Experimentally determined rate laws are given in Table 3.15 together with other details. It has been noted that carbon particles significantly catalysed the aqueous oxidation of SO_2 by air and/or NO_2 and that the rate is independent of pH.[190] In another study of SO_2 oxidation by air and NO_2 catalysed by carbon it was established that the sulphate production was due to catalysis by carbon particles and not by soluble species dissolved from the particles.[191]

The experimental results suggest that the mechanism of this process is one of surface catalysis in which the first step is adsorption of O_2 onto active sites.[187,192] The adsorbed oxygen complexes oxidise S(IV) to sulphate. By assuming that the reaction follows the conditions of Langmuir's adsorption equilibrium a theoretical rate law was derived which was readily reducible to the experimental rate expressions shown in Table 3.15. Different values of the constants have been measured with different types of carbon particles.[192] This dependence of the catalytic activity on the nature of soot particles has been noted in Section 3.3.2.1 and it is related to the number of active sites per unit mass of particles which depends on their origin and history.

That S(IV) oxidation catalysed by aqueous suspensions of carbon may be of significance in the atmospheric conversion of SO_2 has been suggested by a number of workers;[193,194] however, there may be considerable uncertainties in the extrapolation of laboratory studies to the atmosphere. It still remains to be demonstrated that laboratory soots used in the experimental studies correspond to those present in the atmosphere or that suspensions of soot at ambient levels of $\leqslant 10\ \mu g\ m^{-3}$ in air[66] are similar to those in laboratory systems. In one study of SO_2 oxidation in aqueous suspensions of genuine atmospheric carbonaceous aerosols it was concluded that this is not a significant process.[76]

Table 3.15
Rate studies of the carbon-catalysed oxidation of SO_2 in aqueous solution

Reference	*pH*	*T (°C)*	*[S(IV)] (mol litre⁻¹)*	*Rate expression*	*Rate constants*	*[Cₓ] (g litre⁻¹)*	*Comments*
187	1·45–7·5	20	7×10^{-8}–10^{-3}	$R+k[C_x][O_2]^{0\cdot69}\cdot\frac{\{\alpha[S(IV)]^2\}}{1+\beta[S(IV)]+\alpha[S(IV)]^2}$	$k = 1\cdot69\times10^{-5}\,mol^{0\cdot31}l^{-0\cdot31}g^{-1}s^{-1}$ $\alpha = 1\cdot5\times10^{12}\,mol^{-2}l^2$ $\beta = 3\cdot0\times10^6\,mol^{-1}l$	20×10^{-3}–3·2	Oxidation by air and O_2 in bulk soot suspensions Iodometric titration and ion chromatography No catalysis observed at pH > 7·6
188,189	2·8–5	Room temperature	10^{-5}–10^{-3}	$R = k[C_x]\cdot\frac{\{\alpha[S(IV)]^2\}}{1+\beta[S(IV)]+\alpha[S(IV)]^2}$	$k = 1\cdot036\times10^{-7}\,mol^1g^{-1}s^{-1}$ $\alpha = 4\cdot915\times10^8\,mol^{-2}l^2$ $\beta = 2\cdot956\times10^5\,mol^{-1}l$	—	Droplet study SO_2 = 0·007–22 ppm (v/v) pH estimated

3.3.3.5 Oxidation by O_3

Experimental rate laws for SO_2 oxidation by O_3 are summarised in Table 3.16 together with the experimental conditions. The results of bulk solution studies are supported by cloud/fog chamber experiments.[196,198]

The observed rates have been resolved in terms of the partial contributions of the $SO_3^{2-}+O_3$ and $HSO_3^-+O_3$ reactions.[195] In some studies catalysis by Mn and Fe of the oxidation of SO_2 by O_3 has been observed,[123,199] while other studies failed to observe any catalysis by these and other transition metals.[143,200] One of the former investigations employed the West–Gaeke method to monitor S(IV) oxidation,[199] whereas the latter studies used a spectrophotometric technique. It has been suggested that the discrepancy may be due to trace metal interference in the West–Gaeke method.[200] A small but measurable effect of ionic strength on the reaction rate was noted.[200] The rate constant increased with the ionic strength of solutions (0·001 to 0·94 mol litre^{-1}) in a manner suggesting that the reaction involved two univalent negatively charged species.[200]

There is good agreement between the reported studies of SO_2 oxidation by O_3, with the exception of the work of Ref. 106, where the rate expression predicts rates about an order of magnitude slower.[143,200,201] It employed chemical analysis, while the other studies used spectrophotometric techniques.[106] It has been suggested that the higher rates in studies employing spectrophotometric methods may be due to imparting of additional energy to the reacting system.[103] However, varying the monitoring UV intensity over two orders of magnitude had no effect on the oxidation rate of S(IV) by O_3,[200] demonstrating that the reaction was not influenced by photo-oxidation. Possible reasons for the slower rates observed in Ref. 106 may be consumption of O_3 in reactions with impurities in solution, lack of attainment of O_3 gas–liquid equilibrium and interference in the West–Gaeke method employed in the study.[200] It should also be noted that the study employed O_3 concentrations at least two orders of magnitude lower than those used in the other experimental studies.[106]

It has been suggested that the mechanism of SO_2 oxidation by O_3 involves free radicals on the basis of isotopic ^{18}O experiments,[202] and several free radical mechanisms have been proposed.[98,106] A non-radical mechanism involving the formation of intermediate O_3–S(IV) adducts has also been proposed.[200,201] A theoretical rate law was derived for this mechanism, which agreed with experimental results and explained the absence of a strong effect of light, trace metal catalysts and inhibitors

Table 3.16
Summary of rate studies of SO_2 oxidation by O_3 in aqueous solution

Reference	*Experimental system*	*pH*	*T (°C)*	*[S(IV)] (mol litre⁻¹)*	*[O₃] (mol litre⁻¹)*	*Rate expression*	*Rate constants*	*Comments*
97	Bulk	4·65	9·6	10^{-5}–10^{-4}	3–5×10^{-6}	$R = k[HSO_3^-][O_3]$	$k = 3{\cdot}32\times10^5 l\,mol^{-1}s^{-1}$	Spectrophotometric measurements
195	Bulk	−1·3–4·02	16,25	10^{-5}	0·5–5×10^{-4}	$R = k[S(IV)][O_3]$	$k = 1{\cdot}63\times10^5 l\,mol^{-1}s^{-1}$ pH 2·11 $k = 5{\cdot}97\times10^5 l\,mol^{-1}s^{-1}$ pH 3·12 $k = 1{\cdot}74\times10^6 l\,mol^{-1}s^{-1}$ pH 4·02	Spectrophotometric measurements Rate also resolved in terms of fractional contribution of the S(IV) species Values of k are for 25°C
106	Bulk	4–6·2	5–25	~10^{-5}	0·34–1·2×10^{-8}	$R = k[HSO_3^-][O_3][H^+]^{-0{\cdot}1}$	$k = 4{\cdot}4\times10^4 l^{0{\cdot}9} mol^{-0{\cdot}9}s^{-1}$	West–Gaeke method value of k is for 25°C
98	Bulk	1–5	5–25	0·5–5×10^{-4}	~2×10^{-5}	$R = k[S(IV)][O_3][H^+]^{-0{\cdot}4}$	$k = 1{\cdot}79\times10^4 l^{0{\cdot}6} mol^{-0{\cdot}6}s^{-1}$	Spectrophotometric measurements Value of k is for 10°C
103	—	—	—	—	—	$R = \{k_1[SO_2.aq]+k_2\cdot[HSO_3^-]+k_3[SO_3^{2-}]\}[O_3]$	$k_2 = 3{\cdot}73\times10^5 l\,mol^{-1}s^{-1}$ $k_3 = 3{\cdot}12\times10^8 l\,mol^{-1}s^{-1}$	Based on data of Ref. 98 First term in expression can be ignored Values of k are for 25°C
102	Bulk	3–6·2	10,25	10^{-5}–10^{-4}	0·5–8·5×10^{-5}	$R = (k_1+k_2 10^{pH})[S(IV)\cdot[O_3]$	$k_1 = 4{\cdot}21\times10^5 l\,mol^{-1}s^{-1}$ $k_2 = 100 l\,s^{-1}$	Spectrophotometric measurements Values of k are for 25°C
143	Bulk	0–3	25	5×10^{-4}	10^{-4}	$R = k[S(IV)][O_3][H^+]^{-0{\cdot}5}$	$k = 1{\cdot}9\times10^4 l^{0{\cdot}5} mol^{-0{\cdot}5}s^{-1}$	Spectrophotometric measurements
196	Droplet	4·48–5·53	23–25	—	—	$R = k[S(IV)][O_3]$	$k = 8{\cdot}4\times10^6 l\,mol^{-1}s^{-1}$	Cloud chamber study SO_2 = 0·023–0·31 ppm (v/v) O_3 = 0·12–0·15 ppm (v/v)
197	Droplet	4·7–6·3	20	~10^{-4}	—	$R = k[S(IV)][O_3][H^+]^{-0{\cdot}51}$	$k = 1{\cdot}23\times10^4 l^{0{\cdot}49} mol^{-0{\cdot}49}s^{-1}$	Cloud chamber study SO_2 = 15–450 ppb (v/v) O_3 = 15–120 ppb (v/v)

noted in the study.[200] The mechanism also explained the observed ionic strength dependence. Most authors agree that aqueous oxidation of SO_2 by O_3 may be important in the conversion of tropospheric SO_2 and the formation of acidic sulphate aerosols.

3.3.3.6 *Oxidation by H_2O_2*

Experimental studies of SO_2 oxidation by H_2O_2 are summarised in Table 3.17 together with the empirical rate laws, and these show that the reaction is catalysed by buffer salts and subject to general acid catalysis. Most studies show that the rate of S(IV) oxidation is second-order, being first-order in both H_2O_2 and S(IV) concentration. The value of the second-order reaction rate constant is a function of the nature and concentration of buffer, the ionic strength, and the solution pH. A rate expression which explains the pH dependence observed in various studies has been derived.[210] Of the different buffer salts which have been employed only catalysis by carbonate buffers is of interest in the atmosphere. Catalysis by carbonate buffers has been studied only at high pH and high buffer concentrations[203] and data are required for typical atmospheric conditions. The catalytic effect of buffers is attributed to the acidity of proton-donating species such as HCO_3^-. The reaction may also be affected by the ionic strength of solutions.[98,101] Experiments show that the reaction was not affected by potential catalysts (Fe, Mn) or inhibitors (EDTA, hydroquinone);[101] however, one preliminary report states that the presence of Fe and formaldehyde in reacting solutions markedly altered the reaction kinetics.[206]

The agreement between different experimental studies of SO_2 oxidation by H_2O_2 is considerably better than that between other aqueous SO_2 oxidation mechanisms and the rate constant measured in pure water studies has been confirmed in experiments employing rainwater samples as the reaction medium.[208] The rate of SO_2 oxidation by H_2O_2 varies much less with droplet pH than the rate via other reactions, and this is due to the increase in the rate constant with increasing acidity, which compensates to a large extent for the decrease in SO_2 solubility at low pH. This relative insensitivity of the mechanism to droplet pH, the high solubility of H_2O_2, and the established presence of H_2O_2 in the atmosphere, suggest that this may be the dominant route for SO_2 oxidation in atmospheric droplets.[98,150,151,172,175,211] The results of the bulk solution experiments have been confirmed in a recent droplet study.[212]

There is some dispute as to whether the oxidation of SO_2 by H_2O_2 in aqueous solution proceeds by a molecular or a free radical mechan-

Table 3.17
Summary of rate of SO_2 oxidation by H_2O_2 in aqueous solution

Reference	*Experimental system*	*pH*	*T (°C)*	*[S(IV)] (mol litre⁻¹)*	*[H₂O₂] (mol litre⁻¹)*	*Rate expression*	*Rate constants*	*Comments*
203	Bulk	7–12·75	0·4–35	$0{\cdot}55–3{\cdot}95\times10^{-2}$	$1{\cdot}12\times10^{-2}$	$R = k[S(IV)][H_2O_2]$	$k\alpha$[buffer]	Spectrophotometric measurements in buffered solution
204	Bulk	4–8	12	2×10^{-2}	5×10^{-3}	$R = k[HSO_3^-][H_2O_2][H^+]^{0{\cdot}7}$	$k = 1{\cdot}4\times10^6\,l^{1{\cdot}9}\,mol^{-1{\cdot}7}s^{-1}$	Spectrophotometric measurements in buffered solutions Rate expression and constant as evaluated by Ref. 103 for zero buffer concentration
98	Bulk	4·3–8·2	6·5–41	$0{\cdot}8–2{\cdot}5\times10^{-2}$	$2{\cdot}68–8{\cdot}05\times10^{-3}$	$R = k[S(IV)][H_2O_2]\cdot[H^+]^{0{\cdot}735}$	$k = 1{\cdot}94\times10^6\,l^{1{\cdot}735}\,mol^{-1{\cdot}735}s^{-1}$	Spectrophotometric measurements in buffered solutions Value of k is for 10°C
101	Bulk	0–2·8	25	$^{\sim}5\times10^{-4}$	$> 10^{-3}$	$R = \dfrac{k[SO_2.H_2O][H_2O_2]}{0{\cdot}1+[H^+]}$	$k = 8\times10^4\,l\,mol^{-1}s^{-1}$	Spectrophotometric measurements
205	Bulk	4·6	22	$3{\cdot}086\times10^{-6}$	$^{\sim}1{\cdot}5\times10^{-7}$	$R = k[HSO_3^-][H_2O_2]$	$k = 1{\cdot}05(\pm0{\cdot}41)\times10^3\,l\,mol^{-1}s^{-1}$	Chemiluminescent measurements Study carried out at ambient H_2O_2 concentration of $^{\sim}$1ppb(v/v) in the gas phase
206	Bulk	0–4·5	15	$5\times10^{-5}–2{\cdot}4\times10^{-4}$	$10^{-4}–10^{-3}$	$R = \dfrac{k_1[SO_2.H_2O][H_2O_2]}{1+k_2[H^+]}$	$k_1 = 7{\cdot}6\times10^5\,l\,mol^{-1}s^{-1}$ $k_2 = 16\,l\,mol^{-1}s^{-1}$	Spectrophotometric measurements in buffered and unbuffered solutions
207	Bulk	4–5·8	22	$10^{-6}–10^{-5}$	$^{\sim}10^{-7}$	$R = k[HSO_3^-][H_2O_2][H^+]$	$k = 8{\cdot}03(\pm0{\cdot}18)\times10^7\,l^2mol^{-2}s^{-1}$	Chemiluminescent measurements in acidified solutions
208	Bulk	3·3–5·3	22	4×10^{-6}	4×10^{-6}	$R = (S(IV)][H_2O_2][H^+]$	$k = 9{\cdot}6(\pm0{\cdot}5)\times10^7\,l^2mol^{-2}s^{-1}$ in pure water $k = 8{\cdot}1\times10^7\,l^2mol^{-2}s^{-1}$ and in rainwater	Fluorescent measurements
209	Bulk	4–5·1	18	$1–3\times10^{-3}$	$0{\cdot}1–3\times10^{-6}$	$R = k[S(IV)][H_2O_2][H^+]$	$k = 7{\cdot}2(\pm2{\cdot}0)\times10^7\,l^2mol^{-2}s^{-1}$	Fluorescent measurements in acidified solutions

ism.[103,213] Evidence for the molecular mechanism comes from isotopic ^{18}O studies,[214–216] which show that the oxygen transferred to the product sulphate comes from H_2O_2. The molecular mechanism is generally believed to proceed via the formation of peroxymonosulphurous acid as an intermediate in the oxidation of SO_2. This intermediate then undergoes acid-assisted rearrangement to give H_2SO_4. Such a mechanism consisting of nucleophilic displacement by H_2O_2 on HSO_3^- has been proposed:[204]

$$HSO_3^- + H_2O_2 \rightarrow (O)_2S\text{—}OOH + H_2O \tag{3.64}$$

$$HA + (O)_2S\text{—}OOH \rightarrow H_2SO_4 + A^- \tag{3.65}$$

where HA is H_3O^+ or a buffer molecule. This mechanism explains the effect of H^+ and buffer concentration on the rate and leads to a rate expression consistent with experimental results.[101,206,210] It also accounts for the absence of photo-initiation, catalysis by trace metals and inhibition. On the other hand, it has been proposed that a free radical mechanism may be responsible for the reaction,[103,213] since free radicals have been identified in experimental solutions, but this seems unlikely on the basis of most of the formerly mentioned kinetic experiments. However, it was noted in a laboratory study that free radicals may have an important effect on the reaction.[205] It found that O_3 promoted the formation of H_2O_2 in solution, which was ascribed to reactions of OH and HO_2 radicals and organic peroxide radicals originating from precursors such as O_3 and NO_x with water or secondary precursors such as unsaturated hydrocarbons. It may be possible that free radical reactions may be important in the formation and destruction of H_2O_2 in the atmosphere, and they may also oxidise S(IV) directly.[175,211,217]

3.3.3.7 *Oxidation by* NO_x

Studies of this mechanism carried out with relevance to the atmosphere are summarised in Table 3.18 together with the empirical rate laws.

NO_x species dissolve in droplets to produce mainly NO_3^- and NO_2^- species.[221] NO and NO_3^- do not appear to react with S(IV) but NO_2^- reacts rapidly with S(IV).[218,222]

Table 3.18
Summary of rate of SO_2 oxidation by oxides of nitrogen in bulk aqueous solution

Reference	*pH*	*[S(IV)] (mol litre^{-1})*	*[Oxidant] (mol litre^{-1})*	*Rate expression*	*Rate constants*	*Comments*
217	0·5–3·2	5×10^{-4}	$1{\cdot}5\times10^{-3}$	$R \leq k[NO_x][S(IV)]$	$k = 0{\cdot}011\,mol^{-1}s^{-1}$	Spectrophotometric measurements $[NO_x] = [NO_3^-]+[NO.aq]$
			$0{\cdot}3–10\times10^{-3}$	$R = k[S(IV)][N(III)][H^+]^{1/2}$	$k = 142\,l^{3/2}mol^{-3/2}s^{-1}$	Temperature not specified
194	4–7	0·01–0·1	$10^{-3}–10^{-2}$	$R = k_1[H^+]^2[NO_2^-]$ $+k_2[H^+][NO_2^-][HSO_3^-]$ $+k_3[NO_2^-][HSO_3^-]^2$	$k_1 = 8\times10^5l^2mol^{-2}s^{-1}$ $k_2 = 3{\cdot}8\times10^3l^2mol^{-2}s^{-1}$ $k_2 = 9\times10^4l^2mol^{-2}s^{-1}$	Experimental technique not reported k_2 is a function of the ionic strength Temperature not specified
219	4·3–7·0	0·02–0·45	$2–30\times10^{-3}$	$R = k[H^+][NO_2^-][HSO_3^-]$	$k = 4{\cdot}8\times10^3l^2mol^{-2}s^{-1}$	Spectrophotometric measurements in buffered solutions at 22°C
220	5·8,6·4	$10^{-7}–10^{-5}$	—	$R = k[NO_2][S(IV)]$	$k = 2\times10^{-6}l\,mol^{-1}s^{-1}$	Conductivity measurements in gas–liquid reactor at 22°C P_{NO_2} gas $= 8\times10^{-9}–1{\cdot}4\times10^{-7}$ atm

There is some evidence that NO_2 may affect the metal-catalysed oxidation of SO_2 by oxygen. In a droplet study, it was found that 130 ppb(v/v) NO_2 had no effect on SO_2 oxidation but it decreased by 60% the rate of oxidation in 10^{-4} mol l^{-1} Mn(II).[123] NO_2 also cancelled the positive effect of 140 ppb(v/v) O_3 on the oxidation of SO_2 in droplets containing Mn(II) catalyst. In deliquescent $MgCl_2$ aerosols at high humidities it was found that the initial rate of SO_2 absorption increased with increasing NO_2 concentration in the range 0·5–3·5 ppm(v/v), but it became zero-order in the gas phase NO_2 at higher concentrations.[70] In these droplets a synergistic effect between Cu(II) and NO_2 was noted. Cu(II) greatly increased the measured SO_2 absorption rate by $MgCl_2$ droplets in the presence of 2·5 ppm(v/v) NO_2 while in the absence of NO_2, Cu(II) acted as an inhibitor[70]. NO_2^- at concentrations observed in rainwater was found to have no significant effect on the oxidation of S(IV).[222]

A number of different theories have been proposed to account for the oxidation of SO_2 by NO_x in the lead chamber process for H_2SO_4 manufacture, and the early literature has been reviewed.[223] There is general agreement that the oxidation of SO_2 to H_2SO_4 is not directly affected by O_2 but that an intermediate compound involving NO_x is responsible for the reaction. From the results of studies mentioned above it appears that NO_2^-, resulting from the dissolution and reaction of NO_x in aqueous solution, is responsible for the observed oxidation of SO_2. It has been proposed that oxidation of SO_2 by NO_2^- proceeds via hydroxylamine disulphonate ($HON(SO_3)_2^{2-}$) and amine thiosulphate ($N(SO_3)_3^{3-}$) which hydrolyse in solution to give sulphate.[194,219,224] A complex mechanism for SO_2 oxidation by NO_2 has been proposed on the basis of a detailed experimental study.[224] A number of possible reactions and intermediates are specified in it but at pH < 7 the predominant reaction product is sulphate. A free radical mechanism involving OH radicals produced in the reaction of NO_2 with H_2O has also been proposed.[155] Most of the studies have concluded that oxidation of SO_2 by NO_x in aqueous droplets is of little consequence in the general atmosphere.[218,222,225] However, there is disagreement regarding the importance of this mechanism close to sources. On the basis of model calculations, it was concluded that this mechanism may be the dominant route for SO_2 oxidation close to sources of NO_x and in heavily polluted atmospheres.[194] On the other hand, others have concluded that NO_x will have relatively little effect on aqueous SO_2 oxidation even at conditions prevailing in flue gases.[225]

3.3.3.8 *Oxidation via other reactions*

A large number of substances are known to oxidise SO_2 in aqueous solution,[115] and recently interest has been expressed in the role of free radicals in aqueous SO_2 oxidation reactions.[176,186,217,226–228] It has been proposed that SO_2 oxidation by OH and HO_2 radicals dissolved in atmospheric droplets may be a significant process of sulphate formation:[227]

$$OH + HSO_3^- \rightarrow SO_4^- + H_2O$$
$$k = 9{\cdot}5 \times 10^9\ \mathrm{l\ mol^{-1}\ s^{-1}}\ \text{(Ref. 185)} \tag{3.66}$$

$$HO_2 + SO_3^{2-} \rightarrow SO_4^{2-} + OH$$
$$k = 1{\cdot}0 \times 10^7\ \mathrm{l\ mol^{-1}\ s^{-1}}\ \text{(Ref. 185)} \tag{3.67}$$

OH also reacts with SO_3^{2-} but this is not important in atmospheric droplets with pH < 7. Other radicals which have also aroused interest are Cl and Cl_2^-.[228] It has also been suggested that CH_3COO_2, HO_2NO_2, CH_3O_2H and other higher molecular weight organic peroxides (RO_2H) produced by gas-phase reactions may be sufficiently soluble in water droplets and strong enough oxidising agents in solution to contribute significantly to the aqueous oxidation of SO_2 in the atmosphere.[229,230] There are, however, little applicable experimental or field data with which to test this hypothesis.

3.3.3.9 *Inhibition of the aqueous oxidation of SO_2*

Although it has been known for a long time that many organic compounds cause retardation of aqueous SO_2 oxidation at concentrations as low as 10^{-7}–10^{-5} mol litre^{-1},[105,107,115] interest in inhibition of the atmospheric oxidation of SO_2 has been expressed more recently.[143] Inhibiting action has also been referred to as 'poisoning' and 'negative catalysis' in the literature. Alcohols, glycols, amines, amides, aldehydes, ketones and phenols have been identified as inhibitors in experimental studies of SO_2 oxidation by O_2. Inorganic arsenites, antimonite and selenium compounds have also been identified as potential inhibitors.[115]

The rate of S(IV) oxidation in the presence of inhibitors can be described by the following expression:

$$-\mathrm{d}[\mathrm{S[IV)}]/\mathrm{d}t = k[\mathrm{S(IV)}]A/(B+m) \tag{3.68}$$

where A and B are constants that are a function of the inhibitor, m is the concentration of the inhibitor, and k is the rate constant for the uninhibited reaction. Such expressions have been employed to account for experimental results.[105,107] Inhibitors of S(IV) oxidation can act in a number of different ways.

(1) As the air oxidation of sulphite is believed to proceed via a free radical chain reaction, inhibitors could slow down the rate of oxidation by removing radicals via the termination reaction of the free radical mechanism (Fig. 3.2). Alcohols are believed to act in this way.
(2) Inhibitors could form adducts with the sulphite. Formaldehyde appears to act in this manner and it has been suggested that complex formation between sulphite and transition metals may also stabilise S(IV) in the atmosphere.[231,232]
(3) Inhibitors could complex the trace metals thus making them unavailable for the initiation reaction in the free radical mechanism or for catalysis of the oxidation. EDTA may act in this manner.

Some compounds may not act as inhibitors in the strict sense of the word, but may hinder the gas–liquid exchange in atmospheric droplets, and organic surface films are expected to act in this way.[233] Inhibition of the oxidation of SO_2 by O_3 has also been reported.[200] Although inhibition has been demonstrated in many experimental studies there is a need for more experimental studies of this phenomenon with special relevance to the atmosphere.

3.3.4 Relative importance of SO_2 oxidation mechanisms

SO_2 conversion rates for various gas-phase reactions expected in a polluted atmosphere have been summarised in Tables 3.2–3.4 and only oxidation by free radicals appears to be important. The significance of these reactions will vary with the solar intensity. SO_2 conversion rates via the dominant SO_2+OH reaction of $1\pm0{\cdot}5$ and $3{\cdot}7\pm1{\cdot}9\%\ h^{-1}$ were evaluated for winter and summer daytime conditions.[57] At night these rates drop to nearly zero.

SO_2 conversion rates on aerosol particles at typical atmospheric levels were given in Tables 3.5, 3.6 and 3.8 and these are several orders of magnitude lower than those of the more important gas-phase reactions. It may, therefore, be concluded that reactions on particles are insignificant in the conversion of SO_2 in the general atmosphere.

Field studies of SO_2 oxidation in clouds have measured rates considerably faster than those in the gas phase, as shown in Table 3.1, confirming the importance of aqueous reactions. SO_2 conversion rates for the three

Table 3.19
Atmospheric SO_2 conversion rates in a cloud with a liquid water content of 0·5 g m^{-3} at 25°C for different mechanisms[a]

pH	*SO_2 conversion rate (% h^{-1})*		
	H_2O_2	O_3	$O_2 \cdot Fe$
3	278	0·016	0·23
4	261	0·48	2·15
5	256	14·7	21·1
6	227	414	187

[a] pH_2O_2 = 1 ppb (v/v), pO_3 = 30 ppb (v/v), [Fe] = 10×10^{-6} mol $litre^{-1}$. Calculations performed using the kinetic data of Ref. 143 for oxidation by O_3, Ref. 208 for oxidation by H_2O_2 in rainwater samples and Ref. 161 for the Fe-catalysed oxidation by O_2 in rainwater samples. Values of Henry's law constant for H_2O_2 and O_3 were from Ref. 217 while the SO_2/S(IV) equilibrium constants were from Ref. 94.

most significant aqueous phase mechanisms, oxidation by H_2O_2, by O_3, and by O_2 in the presence of Fe catalyst, are given in Table 3.19 in a form comparable to gas-phase conversion rates. It is apparent that oxidation by H_2O_2 will dominate the aqueous route over the pH range expected in cloud droplets (pH 3–6). However, H_2O_2 may be rapidly consumed by reaction with SO_2 as suggested by field studies.[234–236] In the absence of H_2O_2, reactions with O_3 and O_2 catalysed by Fe become significant. The ozone reaction may be especially important in marine and remote regions where the pH is high and the trace metal concentrations low while the Fe-catalysed oxidation may dominate at polluted sites in the more acidic droplets with high catalyst concentrations. It should be noted that Fe concentrations as high as 420×10^{-6} mol $litre^{-1}$ have been measured in urban fogwater[164] and at these levels the catalytic oxidation by O_2 would dominate over the oxidation by H_2O_2 for much of the pH range.

Since close to 50% of the troposphere may be filled by clouds on a global basis,[237] aqueous reactions could play an important role in the transformation of SO_2. Model calculations of SO_2 oxidation indicate that the aqueous sink is as large as the gas-phase sink in the troposphere, and possibly even larger.[238] Another potentially important sink for tropospheric SO_2 is oxidation in sea salt and other deliquescent aerosols which are present in droplet form over a range of ambient relative humidities. SO_2 conversion rates for sea salt aerosols have been evaluated and at high relative humidities these may be competitive with gas-phase reactions, especially at non-polluted sites.[178] SO_2 oxidation in deliquescent sea salt aerosols may provide a possible explanation for the non-sea-salt sulphate observed in marine aerosols.

It is also worth mentioning the role of several important species in the aqueous oxidation of SO_2. Dissolution of NH_3 raises the pH of reacting droplets, enabling further SO_2 to dissolve in the droplets, and this enhancement due to NH_3 has been noted in modelling studies.[240] Similarly, calcium and magnesium leached from alkaline dust particles may also raise the droplet pH and enhance SO_2 absorption and oxidation.[241] Finally, formaldehyde dissolves in droplets to form hydroxymethane sulphonate complexes with S(IV) and a significant fraction of dissolved SO_2 may be present in this form according to field studies.[231,232] The consequences of this effect on SO_2 oxidation kinetics have still not been fully elucidated.

3.4 OXIDATION OF NO_x

The important role of NO_x in the general chemistry of the troposphere has already been noted in Section 3.2. Reactions of NO and NO_2 to form HNO_2 and HNO_3 acids is of interest because of the potential contribution of these compounds to atmospheric acidity. A comparative field study into SO_2 and NO_2 oxidation to sulphate and nitrate respectively has concluded that, whereas droplet-phase reactions are important for SO_2 oxidation gas-phase reactions are predominant for the oxidation of NO_2.[242]

NO may be oxidised to NO_2 by a variety of molecular and free radical species including O_2, O_3, H_2O_2, HO_2, NO_3, RO_2 and $RCOO_2$. Reactions with O_2 and H_2O_2 are too slow to be of importance.[57,243]

Significant reactions are shown in Table 3.20, together with selected values of the rate constant and evaluated NO conversion rates. Rates of oxidation by O_3 and HO_2 are extremely rapid and these reactions therefore represent an important route of NO conversion to NO_2. OH radicals may also react with NO in a three-body reaction to produce HNO_2:

$$NO+OH(+M) \rightarrow HNO_2(+M) \qquad (3.69)$$

The major source of HNO_3 in the troposphere is oxidation by OH radicals

Table 3.20
Rate constants for the oxidation of NO to NO_2 and estimated NO conversion rates in the polluted troposphere (references are given in parentheses)[a]

Reaction	*k* *(cm³molecule⁻¹s⁻¹)*	*NO conversion rate (% s⁻¹)*
$NO+O_3 \rightarrow NO_2+O_2$	$1{\cdot}6\times10^{-14}$(244)	8
$NO+HO_2 \rightarrow NO_2+OH$	$8{\cdot}1\times10^{-12}$(245)	5
$NO+CH_3O_2 \rightarrow NO_2+CH_3O$	$6{\cdot}2\times10^{-13}$(31)	0·06

[a]Concentrations of oxidants are those given in Tables 3.2, 3.3 and 3.4.

in a three-body reaction:

$$NO_2+OH(+M) \rightarrow HNO_3(+M) \qquad (3.70)$$

In the troposphere one can expect an effective second-order rate constant for this reaction of $\sim 10^{-11}$ cm^3 molecule^{-1} s^{-1} according to Ref. 57, which gives a conversion rate of $\sim$ 18% h^{-1} for NO_2 via this reaction in the polluted troposphere containing 5×10^6 mol cm^{-3} of OH.

NO_2 may also react with O_3 to produce NO_3 radicals:[245]

$$NO_2+O_3 \rightarrow NO_3+O_2 \qquad (3.71)$$

Reaction of NO_2 with HO_2 gives rise to peroxynitric acid (HO_2NO_2):[246,247]

$$HO_2+NO_2 \rightleftharpoons HO_2NO_2 \qquad (3.72)$$

This compound is, however, unstable in the troposphere and it decomposes to re-form NO_2 and HO_2.[246]

Peroxyacyl radicals formed by reaction of NO_2 with $RCOO_2$ radicals are more stable:[57]

$$NO_2+RCOO_2 \rightleftharpoons RCOO_2NO_2 \qquad (3.73)$$

PAN ($CH_3COO_2NO_2$) is a characteristic product of photochemical smog. It has been suggested that HO_2NO_2 and $RCOO_2$ radicals may dissolve in water droplets where they may participate in SO_2 oxidation.[229,230]

HNO_3 readily dissolves in water droplets and dissociates to the nitrate (NO_3^-) ion. HNO_2 is considerably less soluble and its contribution to cloudwater acidification is not significant.[248] NO_2^- concentrations in atmospheric hydrometeors are therefore considerably lower than those of NO_3^-. In rainwater NO_2^-/NO_3^- ratios of about 0·01 were determined.[222]

NO_x species may also dissolve in water to give NO_2^- and NO_3^- according to the following reactions:

$$2NO_2(g)+H_2O(l) \rightleftharpoons NO_3^- + NO_2^- + 2H^+ \quad (3.74)$$

$$NO(g)+NO_2(g)+H_2O(l) \rightleftharpoons 2NO_2^- + 2H^+ \quad (3.75)$$

$$3NO_2(g)+H_2O(l) \rightleftharpoons 2NO_3^- + H^+ + NO(g) \quad (3.76)$$

These reactions have been investigated in detail, and it was concluded that they are not significant over the time-scales of relevance to clouds.[221]

It has been proposed that N_2O_5 formed in the gas phase from NO_2 and NO_3 radicals according to

$$NO_2 + NO_3(+M) \rightarrow N_2O_5(+M) \quad (3.77)$$

may dissolve in droplets to produce nitrate:[249,250]

$$N_2O_5(g)+H_2O(l) \rightleftharpoons 2NO_3^- + 2H^+ \quad (3.78)$$

However, there are uncertainties in model calculations employed to investigate the significance of this reaction.[249] It has also been claimed that NO_3 radicals may be scavenged by droplets where they may initiate a vigorous free radical chemistry and contribute to the oxidation of dissolved SO_2.[250]

NO_2^- may be oxidised to nitrate in solution by oxidants such as O_2, O_3 and H_2O_2, however, these reactions are an insignificant source of nitrate in atmospheric droplets.[239,251] Reactions of dissolved NO_x species with S(IV) have been discussed in Section 3.3.3.7 and these are also of little consequence.

Given the present state of knowledge regarding NO_x chemistry in the troposphere it may be concluded that the major acidification process involves oxidation to HNO_3 in the gas phase followed by rapid dissolution and dissociation in aqueous droplets to give NO_3^- and H^+.

REFERENCES

1. Katz, M., Photoelectric determination of atmospheric SO_2 employing dilute starch–iodine solution. *Anal. Chem.*, **22** (1950) 1040–7.
2. Shirai, T., Hamada, S., Takahashi, H., Ozawa, T., Ohmuro, T. & Kawakami, T., Photooxidation of sulphur dioxide in air. *Kogyo-Kagaku Zasshi*, **65** (1962) 1906–11.
3. Benarie, M., Menard, T. & Nonat, A., Etude de la transformation de l'anhydride sulfureux en acide sulfurique en relation avec les donnes

climatologiques, dans un ensemble urbain a caractere industriel, Rouen. *Atmos. Environ.*, **7** (1973) 403–21.
4. Roberts, P. T. & Friedlander, S. K., Conversion of SO_2 to sulphur particulate in the Los Angeles atmosphere. *Environ. Health Perspect.*, **10** (1975) 103–8.
5. Meagher, J. F., Stockburger, L., Bailey, E. M. & Huff, O., The oxidation of sulphur dioxide to sulphate aerosols in the plume of a coal fired power plant. *Atmos. Environ.*, **12** (1978) 2197–203.
6. Dittenhoefer, A. C. & Pena, R. G. de, A study of production and growth of sulphate particles in plumes from a coal-fired power plant. *Atmos. Environ.*, **12** (1978) 297–306.
7. Cantrell, B. K. & Whitby, K. T., Aerosol size distributions and aerosol volume formation for a coal-fired power plant plume. *Atmos. Environ.*, **12** (1978) 323–33.
8. Husar, R. B., Patterson, D. E., Husar, J. D., Gillani, N. V. & Wilson Jr, W. E., Sulphur budget of a power plant plume. *Atmos. Environ.*, **12** (1978) 549–98.
9. Gillani, N. V., Husar, J. D., Patterson, D. E. & Wilson Jr, W. E., Project MISTT: kinetics of particulate sulphur formation in a power plant plume out to 300 km. *Atmos. Environ.*, **12** (1978) 589–98.
10. Pueschel, R. F. & Van Valin, C. C., Cloud nucleus formation in a power plant plume. *Atmos. Environ.*, **12** (1978) 307–312.
11. Lusis, M. A. & Wiebe, H. A., The rate of oxidation of sulphur dioxide in the plume of a nickel smelter stack. *Atmos. Environ.*, **10** (1976) 793–8.
12. Gillani, N. V. & Wilson, W. E., Gas to particle conversion of sulphur in power plant plumes. II. Observations of liquid phase conversions. *Atmos. Environ.*, **17** (1983) 1739–52.
13. Hegg, D. A. & Hobbs, P. V., Measurements of sulphate production in natural clouds. *Atmos. Environ.*, **16** (1982) 2663–8.
14. Bamber, D. J., Clark, P. A., Glover, G. M., Healey, P. G. W., Kallend, A. S., Marsh, A. R. W., Tuck, A. F. & Vaughan, G., Air sampling flights round the British Isles at low altitudes: SO_2 oxidation and removal rates. *Atmos. Environ.*, **18** (1984) 1777–90.
15. Barrie, L. A. & Hoff, R. M., The oxidation rate and residence time of sulphur dioxide in the Arctic atmosphere. *Atmos. Environ.*, **18** (1984) 2711–22.
16. Cass, G. R. & Shair, F. H., Transport of sulphur oxides within the Los Angeles sea breeze/land breeze circulation system. In *Proceedings of the Second Joint Conference on Applications of Air Pollution Meteorology*. American Meteorological Society, 1980, pp. 320–7.
17. Jacob, D. J., Shair, F. H., Waldman, J. M., Munger, J. W. & Hoffmann, M. R., Transport and oxidation of SO_2 in a stagnant foggy valley. *Atmos. Environ.*, **21** (1987) 1305–14.
18. Liebesch, E. J. & Pena, R. G. de, Sulphate aerosol production in coal-fired power plant plumes. *Atmos. Environ.*, **16** (1982) 1323–31.
19. McMurry, P. H. & Wilson, J. C., Droplet phase (heterogeneous) and gas phase (homogeneous) contributions to secondary ambient aerosol formation as a function of relative humidity. *J. Geophys Res.*, **88** (1983) 5101–8.

20. Eatough, D. J., Arthur, R. J., Eatough, N. L., Hill, M. W., Mangelson, N. F., Richter, B. E., Hansen, L. D. & Cooper, J. S., SO_2(g) to sulphate conversion rate in an oil fired power plant plume in a fogbank. Paper 31.1 presented at the 76th Annual Meeting of the Air Pollution Control Association, Atlanta, GA, USA, 19–24 June 1983.
21. Finlayson, B. J. & Pitts Jr, J. N., Photochemistry of the polluted troposphere. *Science,* **192** (1976) 111–19.
22. Chameides, W. L. & Davis, D. D., Chemistry in the atmosphere. *Chem. Engng News,* **60** (1982) 38–52.
23. Harrison, R. M., Atmospheric chemical processes relevant to 'acid rain' problems. In *Acid Rain: Scientific and Technical Advances,* ed. R. Perry, R. M. Harrison, J. N. B. Bell & J. N. Lester. Selper, London, UK, 1987, pp. 3–12.
24. Seinfeld, J. H., *Atmospheric Physics and Chemistry of Air Pollution.* Wiley, New York, USA, 1986.
25. Cocks, A. & Kallend, T., The chemistry of atmospheric pollution. *Chem. In Brit.,* **24** (1988) 884–912.
26. Campbell, I. M., *Energy and the Atmosphere: A Physical–Chemical Approach.* John Wiley & Sons, London, UK, 1977.
27. Hewitt, C. N. & Harrison, R. M., Tropospheric concentrations of the hydroxyl radical—a review. *Atmos. Environ.,* **19** (1985) 545–54.
28. Atkinson, R., Darnall, K. R., Lloyd, A. C., Winer, A. M. & Pitts Jr, J. R., Kinetics and mechanisms of the reactions of the hydroxyl radical with organic compounds in the gas phase. *Adv. Photochem.,* **11** (1979) 375–488.
29. Darnall, K. R., Atkinson, R. & Pitts, J. N., Rate constants for the reaction of the OH radical with selected alkanes at 300K. *J. Phys. Chem.,* **83** (1978) 1581–4.
30. Calvert, J. G., Su, F., Bottenehim, J. W. & Straus, O. P., Mechanism of the homogeneous oxidation of sulphur dioxide in the troposphere. *Atmos. Environ.,* **12** (1978) 197–226.
31. Demerjian, K. L., Kerr, J. A. & Calvert, J. G., The mechanism of photochemical smog formation. *Adv. Environ. Sci. Technol.,* **4** (1974) 1–262.
32. Davis, D. D., A kinetic review of atmospheric reactions involving H_xO_y compounds. *Can. J. Chem.,* **52** (1974) 1405–14.
33. Schofield, K., Evaluated chemical rate constants for various gas phase reactions. *J. Phys. Chem. Ref. Data,* **2** (1973) 25–77.
34. Kear, K. & Abrahamson, E. W., Electronic energy transfer in the gas phase: the quenching of $O_2(^1\Sigma g^+)$. *J. Photochem.,* **3** (1974/75) 409–16.
35. Penzhorn, R. D., Gusten, H., Schurath, U. & Becker, K. H., Quenching of singlet molecular oxygen by some atmospheric pollutants. *Curr. Res.,* **10** (1974) 907–9.
36. Daubendied, R. L. & Calvert, J. G., A study of the N_2O_5–SO_2–O_3 reaction system. *Environ. Lett.,* **8** (1975) 103–16.
37. Davis, D. D., Prusazcyk, J., Dwyer, M. & Kim, P., A stop flow time-of-flight mass spectrometer kinetics study. Reaction of ozone with nitrogen dioxide and sulphur dioxide. *J. Phys. Chem.,* **78** (1974) 1775–9.
38. Leu, M., Rate constants for the reaction of OH with SO_2 at low pressure. *J. Phys. Chem.,* **86** (1982) 4558–62.

39. Moortgat, G. K. & Junge, C. E., The role of SO_2 oxidation for the background stratospheric sulphate layer in the light of new reaction rate data. *Pageoph.*, **115** (1977) 759–74.
40. Davis, D. D., Ravishankara, A. R. & Fischer, S., SO_2 oxidation via the hydroxyl radical: atmospheric fate of HSO_x radicals. *Geophys. Res. Lett.*, **6** (1979) 113–16.
41. Atkinson, R., Perry, R. A. & Pitts, J. N., Rate constants for the reactions of the OH radical with NO_2 (M = Ar and N_2) and SO_2 (M = Ar). *J. Chem. Phys.*, **65** (1976) 306–10.
42. Harris, G. W., Atkinson, R. & Pitts, J. N., Temperature dependence of the reaction $OH+SO_2+M \rightarrow HSO_3+M$ for M = Ar. *Chem. Phys., Lett.*, **69** (1980) 378–82.
43. Castleman, A. W. & Tang, I. N., Kinetics of the association reaction of SO_2 with the hydroxyl radical. *J. Photochem.*, **6** (1976/77) 349–54.
44. Cox, R. A., The photolysis of gaseous nitrous acid—a technique for obtaining kinetic data in atmospheric photooxidation reactions. *Int. J. Chem. Kinet. Symp.*, **1** (1975) 379–98.
45. Izumi, K., Mizuochi, M., Yoshioka, M., Murano, K. & Fukuyama, T., Redetermination of the rate constant for the reaction of HO radicals with SO_2. *Environ. Sci. Technol.*, **18** (1984) 116–18.
46. Stockwell, W. R. & Calvert, J. G., The mechanism of HO–SO_2 reaction. *Atmos. Environ.*, **17** (1983) 2231–5.
47. Meagher, J. F., Olszyna, K. J. & Luria, M., The effects of SO_2 gas phase oxidation on hydroxyl smog chemistry. *Atmos. Environ.*, **18** (1984) 2095–104.
48. Payne, W. A., Stief, L. J. & Davis, D. D., A kinetic study of the reaction of HO_2 with SO_2 and NO. *J. Am. Chem. Soc.*, **95** (1973) 7614–19.
49. Graham, R. A., Winer, A. M., Atkinson, R. & Pitts, J. N., Rate constants for the reaction of HO_2 with HO_2, SO_2 CO, N_2O, *trans*-2-butene, and 2,3-dimethyl-2-butene at 300K. *J. Phys. Chem.*, **83** (1979) 1563–7.
50. Burrows, J. P., Cliff, D. I., Harris, G. W., Thrush, B. A. & Wilkinson, J. P. T., Atmospheric reactions of the HO_2 radical studied by laser magnetic resonance spectroscopy. *Proc. Royal Soc. (Lond.)*, **A368** (1979) 463–81.
51. Whitbeck, M. R., Bottenheim, J. W., Levine, S. Z. & Calvert, J. G., A kinetic study of the CH_3O_2 and $(CH_3)_3CO_2$ radical reactions by kinetic flash spectroscopy. In *Abstracts of 12th Informal Conference on Photochemistry*. US National Bureau of Standards, Gaithersberg, MD, USA, July 1976, pp. K1-1–K1-5.
52. Kan, C. S., McQuigg, R. D., Whitbeck, M. R. & Calvert, J. G., Kinetic flash spectroscopic study of the CH_3O_2–CH_3O_2 and CH_3O_2–SO_2 reaction. *Int. J. Chem. Kinet.*, **11** (1979) 921–33.
53. Sanhueza, E., Simonaitis, R. and Heicklen, J., The reaction of CH_3O_2 with SO_2. *Int. J. Chem. Kinet.*, **11** (1979) 907–14.
54. Sander, S. P. & Watson, R. T., A kinetic study of the reaction of SO_2 with CH_3O_2. *Chem. Phys. Lett.*, **77** (1981) 473–5.
55. Kan, C. S., Calvert, J. G. & Shaw, J. H., Oxidation of sulphur dioxide by methylperoxy radicals. *J. Phys., Chem.*, **85** (1981) 1126–32.
56. Calvert, J. G. & McQuigg, R. D., The computer simulation of the rates and

mechanisms of photochemical smog formation. *Int. J. Chem. Kinet. Symp.*, **1** (1975) 113–54.
57. Calvert, J. G. & Stockwell, W. R., Mechanism and rates of the gas phase oxidations of sulphur dioxide and nitrogen oxides in the atmosphere. In *SO_2, NO and NO_2 Oxidation Mechanism: Atmospheric Considerations*, ed. J. G. Calvert, *Acid Precipitation Series*, Vol. 3. Butterworth, Boston, MA, USA, 1984, pp. 1–62.
58. Olszyna, K. J. & Meagher, J. F., Contribution of hydroxyl radical addition mechanisms to SO_4^{2-} formation—a smog chamber study. Paper 30.3 presented at the 76th Annual Meeting of the Air Pollution Control Association, Atlanta, GA, USA, 19–24 June 1983.
59. Duecker, W. W. & West, J. R. *Manufacture of Sulphuric Acid.* Reinhold, New York, 1959.
60. Hartman, M., Polek, J. R. & Coughlin, R. W., Removal of sulphur dioxide from flue gas by sorption and catalytic reaction on carbon, *A.I. Chem. Eng. Prog. Symp. Series 115*, **67** (1971) 7–22.
61. Novakov, T., Chang, S. G. & Harker, A. B., Sulphates as pollution particulates: catalytic formation on carbon (soot) particles. *Science,* **186** (1974) 259–61.
62. Judeikis, H. S. & Siegel, S., Particle-catalysed oxidations of atmospheric pollutants. *Atmos. Environ.*, **7** (1973) 619–31.
63. Novakov, T., Muller, P. K., Aleccer, A. E. & Otvos, J. W., Chemical composition of Pasadena aerosol by particle size and time of day. III. Chemical states of nitrogen and sulphur by Photoelectron Spectroscopy. *J. Colloid. Interface Sci.*, **39** (1972) 225–34.
64. Craig, N. L., Harker, A. B. & Novakov, T., Determination of the chemical states of sulphur in ambient pollution aerosols by X ray photoelectron spectroscopy. *Atmos. Environ.*, **8** (1974) 15–21.
65. Dasgupta, P. K., De Cesare, K. B. & Brummer, M., Determination of S(IV) in particulate matter. *Atmos. Environ.*, **16** (1982) 917–27.
66. Novakov, T., Microchemical characterisation of aerosols. In *Nature, Aim and Methods of Microchemistry,* ed. H. Malissa, M. Grasserbauwer & R. Belchers. Springer-Verlag, Wien, Austria, 1981, pp. 141–65.
67. Hara, H. & Okita, T., Oxidation of S(IV) in rainwater collected in Japan. *Atmos. Environ.*, **24A** (1990) 1427–32.
68. Devitofrancesco, G., Liberti, A., Petronio, B. A. & Conte, C., Interaction between sulphur dioxide and atmospheric particulate matter. *Ann. Chim.*, **64** (1974) 27–33.
69. Britton, L. G. & Clarke, A. G., Heterogeneous reactions of sulphur dioxide and SO_2/NO_2 mixtures with a carbon soot aerosol. *Atmos. Environ.*, **14** (1980) 829–39.
70. Williams, P. T., The reactions of sulphur dioxide with solid and liquid aerosols. PhD thesis, University of Leeds, Leeds, UK, 1980.
71. Urone, P., Lutsep, H., Noyes, C. M. & Parcher, J. R., Static studies of sulphur dioxide reactions in air. *Environ. Sci. Technol.*, **2** (1968) 611–18.
72. Barbaray, B., Contour, J. P. & Mouvier, G., Sulphur dioxide oxidation over atmospheric aerosol-X-ray photoelectron spectra of sulphur dioxide adsorbed on V_2O_5 and carbon. *Atmos. Environ.*, **11** (1977) 351–6.

73. Liberti, A., Brocco, D. & Possanzini, M., Adsorption and oxidation of sulphur dioxide on particles. *Atmos. Environ.*, **12** (1978) 255–61.
74. Corn, M. & Cheng, R. T., Interaction of sulphur dioxide with insoluble suspended particulate matter. *J. Air Pollut. Control Assoc.*, **22** (1972) 870–5.
75. Judeikis, H. S., Steward, T. B. & Wren, A. G., Laboratory studies of heterogeneous reactions of SO_2. *Atmos. Environ.*, **12** (1978) 1633–41.
76. Harrison, R. M. & Pio, C. A., Kinetics of SO_2 oxidation over carbonaceous particles in the presence of H_2O, NO_2, NH_3 and O_3. *Atmos. Environ.*, **17** (1983) 1261–75.
77. Chan, K. C. & Quon, J. E., Capacity of ferric oxide particles to oxidise sulphur dioxide in air. *Environ. Sci. Technol.*, **7** (1973) 532–8.
78. Dyson, W. L. & Quon, J. E., Reactivity of zinc oxide fume with sulphur dioxide in air. *Environ. Sci. Technol.*, **10** (1976) 476–81.
79. Dlugi, R., Jordan, S. & Lindemann, E., Influence of particle properties on heterogeneous SO_2 reactions. In *Physico-Chemical Behaviour of Atmospheric Pollutants,* ed. B. Versino & H. Ott. D. Reidel, Dordrecht, The Netherlands, 1982, pp. 308–19.
80. Dlugi, R. & Jordan, S., Versuche zur bildung von sulfaten auf partikeln. In *Physico-Chemical Behaviour of Atmospheric Pollutants,* ed. B. Versino & H. Ott. Commission of the European Communities, Luxembourg, 1980, pp. 293–7.
81. Atkins, R. W., *Physical Chemistry,* 2nd edn. OUP, Oxford, UK, 1982, p. 1032.
82. Davtyan, O. K. & Ovchinnikova, E. N., Chemisorption and the oxidation of sulphur dioxide on solid catalysts at room temperature. *Doklady Akad. Nauk. SSR,* **104** (1955) 857–60.
83. Davtyan, O. K. & Ovchinnikova, E. N., The mechanism of oxidation, hydrogenation and electrochemical oxidation on solid catalysts. I. Oxidation of sulphur dioxide on activated carbon at 20°C in the presence of water vapour. *Russ. J. Phys. Chem.*, **35** (1961) 347–50.
84. Davtyan, O. K. & Tkach, Y. A. The mechanism of oxidation, hydrogenation and electrochemical oxidation on solid catalysts. II. The catalytic activity of surface ‘oxides’ on carbon. *Russian J. Phys. Chem.*, **35** (1961) 486–9.
85. Chang, S. G. & Novakov, T., Soot Catalysed Oxidation of Sulphur Dioxide. In *Man's Impact on the Troposphere,* ed. J. D. Levine & D. R. Schryer. NASA RP-1022, 1978, pp. 349–69.
86. Clarke, A. G. & Britton, L. G., Heterogeneous reactions of sulphur dioxide and nitrogen oxides on carbon. In *Proceedings of the Fourth International Clean Air Congress.* Tokyo, 1977, pp. 469–71.
87. Otake, T., Tone, S., Yokoto, Y. & Yoshimura, K., Kinetics of sulphur dioxide oxidation over activated carbon. *J. Chem. Engng Japan,* **4** (1971) 43–7.
88. Siedlewski, J., The mechanism of catalytic oxidation on activated carbon. The influence of free carbon radicals on the adsorption of SO_2. *Int. Chem. Engng,* **5** (1965) 297–301.
89. Vadjić, V. & Gentilizza, M. The effect of MnO_2 and some manganese salts on the behaviour of sulphur dioxide in the air investigated on model systems. *Sci. Total Environ.*, **44** (1985) 245–51.

90. Smith, B. M., Wagman, J. & Fish, B. R., Interaction of airborne particles with gases. *Environ. Sci. Technol.*, **9** (1969) 558–632.
91. Haury, G., Jordan, S. & Hofmann, C., Experimental investigation of the aerosol catalysed oxidation of SO_2 under atmospheric conditions. *Atmos. Environ.*, **12** (1978) 281–7.
92. Dlugi, R., SO_2—oxidation in aerosol particles and droplets. *J. Aerosol Sci.*, **14** (1983) 292–7.
93. Tartarelli, R., Davini, P., Morelli, F. & Corsi, P., Interactions between SO_2 and carbonaceous particulates. *Atmos. Environ.*, **12** (1978) 289–93.
94. Maahs, H. G., Sulphur-dioxide/water equilibria between 0° and 50°C. An examination of data at low concentrations. In *Heterogeneous Atmospheric Chemistry*, ed. D. R. Schryer, *Geophysical Monograph 26.* American Geophysical Union, Washington, DC, USA, 1982, pp. 187–95.
95. Clarke, A. G., Williams, P. T. & Radojević, M., Oxidation rates of SO_2 in deliquescent salt aerosols. In *Proceedings of 2nd European Symposium on Physico-Chemical Behaviour of Atmospheric Pollutants*, ed. B. Versino & H. Ott. D. Reidel, Dordrecht, The Netherlands, 1982, pp. 203–11.
96. Clarke, A. G., Electrolyte solution theory and the oxidation rate of sulphur dioxide in water. *Atmos. Environ.*, **15** (1981) 1591–5.
97. Penkett, S. A., Oxidation of SO_2 and other atmospheric gases by ozone in aqueous solution. *Nature*, **240** (1972) 105–6.
98. Penkett, S. A., Jones, B. M. R., Brice, K. A. & Eggleton, A. E. J., The importance of atmospheric ozone and hydrogen peroxide in oxidising sulphur dioxide in cloud and rainwater. *Atmos. Environ.*, **13** (1979) 123–37.
99. Sutherland, J. W. & Glines, A., Oxidation of sulphur dioxide in acidic aqueous solution. Paper presented before the Division of Environmental Chemistry, New Orleans, LA, USA, March 1977, American Chemical Society, pp. 221–3.
100. Martin, L. R., Damschen, D. E. & Judeikis, H. S., The reaction of nitrogen oxides with SO_2 in aqueous aerosols. *Atmos. Environ.*, **15** (1981) 191–5.
101. Martin, L. R. & Damschen, D. E., Aqueous oxidation of sulphur dioxide by hydrogen peroxide at low pH. *Atmos. Environ.*, **15** (1981) 1615–21.
102. Maahs, H. G., Measurements of the oxidation rate of sulphur(IV) by ozone in aqueous solution and their relevance to SO_2 conversion in nonurban tropospheric clouds. *Atmos. Environ.*, **17** (1983) 341–5.
103. Dasgupta, P. K., The importance of atmospheric ozone and hydrogen peroxide in oxidising sulphur dioxide in cloud and rainwater. Discussion. *Atmos. Environ.*, **15** (1980) 272–5.
104. Radojević, M., On the discrepancy between reported studies of the uncatalysed aqueous oxidation of SO_2 by O_2. *Environ. Technol. Lett.*, **5** (1984) 549–66.
105. Fuller, E. C. & Crist, R. H., The rate of oxidation of sulphite ions by oxygen. *J. Am. Chem. Soc.*, **63** (1941) 1644–50.
106. Larson, T. V., Horike, N. R. & Harrison, H., Oxidation of sulphur dioxide by oxygen and ozone in aqueous solution: a kinetic study with significance to atmospheric rate processes. *Atmos. Environ.*, **12** (1978) 1597–611.
107. Schroeter, L. C., Kinetics of air oxidation of sulphurous acid salts. *J. Pharm. Sci.*, **52** (1963) 559–63.

108. Clarke, A. G. & Radojević, M., Chloride ion effects on the aqueous oxidation of SO_2. *Atmos. Environ.*, **17** (1983) 617–24.
109. Beilke, S., Lamb, D. & Muller, J., On the uncatalysed oxidation of atmospheric SO_2 by oxygen in aqueous systems. *Atmos. Environ.*, **9** (1975) 1083–90.
110. Miller, J. M. & Pena, R. G. de, Contribution of scavenged sulphur dioxide to the sulphate content of rain. *J. Geophys. Res.*, **77** (1972) 5906–16.
111. Lamb, D., Miller, D. F., Robinson, N. F. & Gertler, A. W., The importance of liquid water concentration in the atmospheric oxidation of SO_2. *Atmos. Environ.*, **21** (1987) 2333–44.
112. Hayon, E., Treinin, A. & Wilf, J., Electronic spectra, photochemistry and autoxidation mechanism of the sulphite-bisulphite-pyrosulphite systems. The SO_2^-, SO_3^-, SO_4^- and SO_5^- radicals. *J. Am. Chem. Soc.*, **94** (1972) 45–57.
113. Barrie, L. A. & Georgii, H. W., An experimental investigation of the absorption of sulphur dioxide by water drops containing heavy metal ions. *Atmos. Environ.*, **10** (1976) 743–9.
114. Larson, T. V., The kinetics of sulphur dioxide oxidation by oxygen and ozone in atmospheric hydrometeors. PhD thesis, University of Washington, WA, USA, 1976.
115. Schroeter, L. C., *Sulphur Dioxide—Applications in Foods, Beverages and Pharmaceuticals,* 1st edn. Pergamon Press, Oxford, 1966.
116. Schroeter, L. C., Sulphurous acid salts as pharmaceutical antioxidants—review article. *J. Pharm. Sci.*, **50** (1961) 891–901.
117. Clarke, A. G. & Radojević, M., Chloride ion effect on the aqueous oxidation of SO_2. Authors' reply. Discussion. *Atmos. Environ.*, **17** (1983) 1857–8.
118. Titoff, A., Beitrage zur Kenntnis der negativen Katalyse in homogenen System. *Z. Phys. Chem.*, **45** (1903) 641–83.
119. Reinders, W. & Vles, S. I., Reaction velocity of oxygen with solutions of some inorganic salts. III. The catalytic oxidation of sulphites. *Recueil des Travaux Chimiques des Pays-Bas,* **44** (1925) 249–68.
120. Barron, C. H. & O'Hern, H. A., Reaction kinetics of sodium sulphite oxidation by the rapid mixing method. *Chem. Engng Sci.*, **21** (1966) 397–404.
121. Hoather, R. C. & Goodeve, C. F., The oxidation of sulphurous acid. III. Catalysis by manganous sulphate. *Trans. Faraday Soc.*, **30** (1934) 1156–61.
122. Pritchett, P. W., The catalysed reaction of oxygen with sulphurous acid and its effect on the absorption of oxygen and sulphur dioxide into water. PhD thesis, University of Delaware, DE, USA, 1960.
123. Barrie, L. A., An experimental investigation of the absorption of sulphur dioxide by cloud and raindrops containing heavy metals. PhD thesis, Frankfurt am Main, Germany, 1975.
124. Conklin, M. H. & Hoffmann, M. R., Metal ion–sulphur(IV) chemistry. 2. Kinetic studies of the redox chemistry of copper(II)–sulphur(IV) complexes. *Environ. Sci. Technol.*, **22** (1988) 891–8.
125. Schmidkunz, H., Chemilumineszenz der Sulphitoxidation. PhD thesis, University of Frankfurt, Frankfurt, Germany, 1963.
126. Huss Jr, A., Lim, P. K. & Eckert, C. A., Oxidation of aqueous sulphur dioxide. 1. Homogeneous manganese(II) and iron(II) catalysis at low pH. *J. Phys. Chem.*, **86** (1982) 4224–8.

127. Johnstone, H. F. & Coughanowr, D. R., Absorption of sulphur dioxide from air–oxidation in drops containing dissolved catalysts. *Indust. Engng Chem.*, **50** (1958) 1169–72.
128. Neytzell-de Wilde, F. G. & Taverner, L., Experiments relating to the possible production of an oxidising acid leach liquor by auto-oxidation for the extraction of uranium. In *Proceedings of 2nd UN International Conference on Peaceful Uses of Atomic Energy*, Geneva, 1–13 Sept. 1958, United Nations, Geneva, Vol. 3, 1958, 301–17.
129. Karraker, D. G., The kinetics of the reaction between sulphurous acid and ferric ion. *J. Phys. Chem.*, **67** (1963) 871–4.
130. Coughanowr, D. R. & Krause, F. E., The reaction of SO_2 and O_2 in aqueous solutions of $MnSO_4$. *I. and E. C. Fund.*, **4** (1965) 61–6.
131. Bracewell, J. M. & Gall, D., The catalytic oxidation of sulphur dioxide in solution at concentration occurring in fog droplets. In *Proceedings of Symposium on Physico-chemical Transformation of Sulphur Dioxide Compounds in the Atmosphere and the Formation of Acid Smog*, No. 2. OECD, Mainz, Germany, 1967, pp. 17–26.
132. Matteson, M. J., Stöber, W. & Luther, H., Kinetics of the oxidation of sulphur dioxide by aerosols of manganese sulphate. *I. and E. C. Fund.*, **8** (1969) 677–87.
133. Tsunogai, S., Oxidation rate of sulphite in water and its bearing on the origin of sulphate in meteoric precipitation. *Geochem. J*, **5** (1971) 175–85.
134. Cheng, R. T., Corn, M. & Frohliger, J. O., Contribution to the reaction kinetics of water soluble aerosols and SO_2 in air at ppm concentrations. *Atmos. Environ.*, **5** (1971) 987–1008.
135. Brimblecombe, P. & Spedding, D. J., The catalytic oxidation of micromolar aqueous sulphur dioxide. I. Oxidation in dilute solutions containing iron(III). *Atmos. Environ.*, **8** (1974) 937–45.
136. Freiberg, J., The mechanism of iron catalysed oxidation of SO_2 in oxygenated solutions. *Atmos. Environ.*, **9** (1975) 661–72.
137. Sutherland, J., Private communication. (Quoted in Ref. 138.)
138. Hegg, D. A. & Hobbs, P. V., Oxidation of sulphur dioxide in aqueous systems with particular reference to the atmosphere. *Atmos. Environ.*, **12** (1978) 241–53.
139. Fuzzi, S., Study of iron(III) catalysed sulphur dioxide oxidation in aqueous solution over a wide range of pH. *Atmos. Environ.*, **12** (1978) 1439–42.
140. Kaplan, D. J., Himmelblau, D. M. & Kanaoka, C., Oxidation of sulphur dioxide in aqueous ammonium sulphate aerosols containing manganese as catalyst. *Atmos. Environ.*, **15** (1981) 743–73.
141. Crump, J. G., Flagan, R. C. & Seinfeld, J. H., An experimental study of the oxidation of sulphur dioxide in aqueous manganese sulphate aerosols. *Atmos. Environ.*, **17** (1983) 1277–89.
142. Ibusuki, T. & Barnes, H. M., Manganese(II) catalysed sulphur dioxide oxidation in aqueous solution at environmental concentrations. *Atmos. Environ.*, **18** (1984) 145–51.
143. Martin, L. R., Kinetic studies of sulphite oxidation in aqueous solution. In *SO_2, NO and NO_2 Oxidation Mechanisms: Atmospheric Considerations*, ed. J. G. Calvert, *Acid Precipitation Series*, Vol. 3. Butterworth, Boston, MA, USA, 1984, pp. 63–100.

144. Ibusuki, T. & Takeuchi, K., Sulphur dioxide oxidation by oxygen catalysed by mixtures of manganese(II) and iron(III) in aqueous solution at environmental reaction conditions. *Atmos. Environ.*, **21** (1987) 1555–60.
145. Martin, L. R. & Hill, M. W., The iron catalysed oxidation of sulphur: reconciliation of the literature rates. *Atmos. Environ.*, **21** (1987) 1487–90.
146. Brimblecombe, P. & Spedding, D. J., The reaction order of the metal ion catalysed oxidation of sulphur dioxide in aqueous solution. *Chemosphere*, **1** (1974) 29–31.
147. Bronikowski, T., Sulphur dioxide oxidation in aqueous solutions containing metal cations. *React. Kinet. Catal. Lett.*, **10** (1979) 139–42.
148. Martin, L. R. & Hill, M. W., The effect of ionic strength on the manganese catalysed oxidation of sulphur(IV). *Atmos. Environ.*, **21** (1987) 2267–70.
149. Altwicker, E. R. & Nass, K. K., Evidence for enhanced mass transfer and synergistic catalysis of aqueous phase sulphur dioxide oxidation by mixtures of manganese and iron. *Atmos. Environ.*, **17** (1983) 187–90.
150. Middleton, P., Kiang, C. S. & Mohnen, V. A., The relative importance of various urban sulphate aerosol production mechanisms—a theoretical comparison. In *Heterogeneous Atmospheric Chemistry*, ed. D. R. Schryer, *Geophysical Monograph 26*. American Geophysical Union, Washington, DC, USA, 1982, pp. 221–30.
151. Middleton, P., Kiang, C. S. & Mohnen, V. A., Theoretical estimates of the relative importance of various urban sulphate aerosol production mechanisms. *Atmos. Environ.*, **14** (1980) 463–72.
152. Freiberg, J. E. & Schwartz, S. A., Oxidation of SO_2 in aqueous droplets: mass transport limitation in laboratory studies and the ambient atmosphere. *Atmos. Environ.*, **15** (1981) 1145–54.
153. Junge, E. & Ryan, T. G., Study of the SO_2 oxidation and its role in atmospheric chemistry, *Quart. J. Royal Meteorol. Soc.*, **84** (1958) 46–55.
154. Johnstone, H. F. & Moll, A. J., Formation of sulphuric acid in fogs. *Indust. Engng Chem.*, **52** (1960) 861–3.
155. Nash, T., The effect of nitrogen dioxide and of some transition metals on the oxidation of dilute bisulphite solutions. *Atmos. Environ.*, **13** (1979) 1149–54.
156. Altwicker, E. R. & Kleinstreuer, C., Sulphur dioxide absorption, oxidation and oxidation inhibition in falling drops: an experimental/modelling approach. In *Heterogeneous Atmospheric Chemistry*, ed. D. R. Schryer, *Geophysical Monograph 26*. American Geophysical Union, Washington, DC, USA, 1982, pp. 178–86.
157. Weschler, C. J. & Graedel, T. E., Theoretical limitations on heterogeneous catalysis by transition metals in atmospheric aerosols. In *Heterogeneous Atmospheric Chemistry*, ed. D. R. Schryer, *Geophysical Monograph 26*. American Geophysical Union, Washington, DC, USA, 1982, pp. 196–202.
158. Bassett, H, & Parker, W., The oxidation of sulphurous acid. *J. Chem. Soc.*, **47** (1951) 1540–60.
159. Penkett, S. A., Jones, B. M. R., Brice, K. A. & Eggleton, E. A. J., A study of SO_2 oxidation in stored rainwater samples. *Atmos. Environ.*, **13** (1979) 139–47.
160. Gravenhorst, G., Beilke, S., Betz, M. & Georgii, H. W., Sulphur dioxide

absorbed in rainwater. In *Effects of Acid Precipitation on Terrestrial Ecosystems,* ed. T. C. Hutchinson & M. Havas. Plenum, New York, USA, 1980, pp. 41–55.

161. Clarke, A. G. & Radojević, M., Oxidation of SO_2 in rainwater and its role in acid rain chemistry. *Atmos. Environ.,* **21** (1987) 1115–23.
162. Tanaka, S., Yamanaka, K. & Hashimoto, Y., Measurement of concentration and oxidation rate of S(IV) in rainwater in Yokohama, Japan. In *The Chemistry of Acid Rain: Sources and Atmospheric Processes,* ed. R. W. Johnson, G. E. Gordon, W. Calkins & A. Z. Elzerman, *ACS Symposium Series 349.* American Chemical Society, Washington, DC, USA, 1987, pp. 158–69.
163. Radojević, M., Tyler, B. J., Penderghest, N. & Hall, S., Trace metals in cloudwater and their influence on SO_2 oxidation. In *Heavy Metals in the Environment,* Vol. 1, ed. J. G. Farmer. CEP Consultants, Edinburgh, 1991, pp. 330–3.
164. Cawse, P. A., *A Survey of Atmospheric Trace Elements in the UK (1972–73),* Report No. AERE-R 7669. UK Atomic Energy Authority, Harwell, HMSO, London, UK, 1974.
165. Munger, J. W., Jacob, D. J., Waldman, J. M. & Hoffmann, M. R., Fogwater chemistry in an urban atmosphere. *J. Geophys. Res.,* **88** (1983) 5109–21.
166. Waldman, J. M., Munger, J. W., Jacob, D.J. & Hoffmann, M. R., Chemical characterisation of stratus cloudwater and its role as a vector for pollutant deposition in a Los Angeles pine forest. *Tellus,* **37B** (1985) 91–108.
167. Cohen, S., Chang, S.-G., Markowitz, S. S. & Novakov, T., Role of fly ash in catalytic oxidation of S(IV) slurries. *Environ. Sci. Technol,.* **15** (1981) 1498–502.
168. Cains, P. W. & Carabine, M. D., Oxidation of sulphur dioxide in aerosol droplets catalysed by manganous sulphate. *J. Chem. Soc. Faraday Trans. I,* **74** (1978) 2689–702.
169. Hoffmann, M. R. & Boyce, S. D., Catalytic autoxidation of aqueous sulphur dioxide in relationship to atmospheric systems. *Adv. Environ. Sci. Technol.,* **12** (1983) 147–89.
170. Hoffmann, M. R. & Jacob, D. J., Kinetics and mechanism of dissolved sulphur dioxide in aqueous solution: an application to nighttime fog water chemistry. In *SO_2, NO and NO_x Oxidation Mechanisms: Atmospheric Considerations,* ed. J. G. Calvert, *Acid Precipitation Series,* Vol. 3, Butterworth, Boston, MA, USA, 1984, pp. 101–72.
171. Weschler, C. J., Mandich, M. L. & Graedel, T. E., Speciation, photosensitivity and reactions of transition metal ions in atmospheric droplets. *J. Geophys. Res.,* **91** (1986) 5189–204.
172. Graedel, T. E., Weschler, C. J. & Mandich, M. L. Influence of transition metal complexes on atmospheric droplet chemistry. *Nature,* **317** (1985) 240–2.
173. Faust, B. C. & Hoigné, J., Photolysis of Fe(III) hydroxy complexes as a source of OH radicals in cloud, fog and rain. *Atmos. Environ.,* **24A** (1990) 79–89.

174. Huss Jr, A., Lim, P. K. & Eckert, C. A., Oxidation of aqueous sulphur dioxide. 2. High-pressure studies and proposed reaction mechanisms. *J. Phys. Chem.*, **86** (1982) 4229–33.
175. Graedel, T. E., Mandich, M. L. & Weschler, C. J., Kinetic model studies of atmospheric droplet chemistry. 2. Homogeneous transition metal chemistry in raindrops. *J. Geophys. Res.*, **91** (1986) 5205–21.
176. Hemmeler, A., Su alcuni tentativi di applicazione delle azioni 'antiossigeno' e 'proossigeno' all'analisi chimica. *Ann. Chim. Appl.*, **37** (1947) 515–30.
177. Wickert, K., Nolten, K., Kroll, J. & Ipach, E., Entgasung von Wassern zur Verhinderung der Korrosion. *Arch. Metallkunde,* **3** (1949) 105–13.
178. Clarke, A. G. & Radojević, M., Oxidation rates of SO_2 in sea-water and sea-salt aerosols. *Atmos. Environ.*, **18** (1984) 2761–7.
179. Eggleton, A. E. J. & Atkins, D. H., *Results of the Tee Side Investigation,* Report AERE-R6983. UK Atomic Energy Authority Research Group, Health Physics and Medical Division, Atomic Energy Research Establishment, Harwell, HMSO, London, UK, 1982.
180. Clarke, A. G. & Williams, P. T., The oxidation of sulphur dioxide in electrolyte droplets. *Atmos. Environ.*, **17** (1983) 607–15.
181. Winkler, P. & Junge, C. E., Comments on 'anomalous deliquescence of sea spray aerosols'. *J. Appl. Meteorol.*, **10** (1971) 159–63.
182. Miller, D. F., Lamb, D. & Gerlet, A. W., SO_2 oxidation in cloud drops containing NaCl or sea salt as condensation nuclei. *Atmos. Environ.*, **21** (1987) 991–3.
183. Gertler, A. W., Miller, D. F., Lamb, D. & Katz, U., SO_2 and NO_2 reactions in cloud droplets. In *Chemistry of Particles, Fogs and Rain,* ed. J. L. Durham, *Acid Precipitation Series,* Vol. 2. Butterworth, Boston, MA, USA, 1984, pp. 131–60.
184. Houghton, H. G., On the chemical composition of fog and cloud water. *J. Meteor.*, **12** (1955) 355–7.
185. Graedel, T. E. & Goldberg, K. I., Kinetic studies of raindrop chemistry. 1. Inorganic and organic processes. *J. Geophys. Res.*, **88** (1983) 10865–82.
186. Hasegawa, K. & Neta, P., Rate constants and mechanisms of reaction of Cl_2^- radicals. *J. Phys. Chem.*, **82** (1978) 854–7.
187. Brodzinsky, R., Chang, S.-G., Markowitz, S. S. & Novakov, T., Kinetics and mechanism for the catalytic oxidation of sulphur dioxide on carbon in aqueous suspensions. *J. Phys. Chem.*, **84** (1980) 3354–8.
188. Benner, W. H., Brodzinsky, R. & Novakov, T., *Oxidation of SO_2 in Droplets which Contain Soot Particles,* Report No. LBL-12386. Lawrence Berkeley Laboratory, University of California, CA, USA, 1981.
189. Benner, W. H., Brodzinsky, R. & Novakov, T., Oxidation of SO_2 in droplets which contain soot particles. *Atmos. Environ.*, **16** (1982) 1333–9.
190. Rogowski, R. S., Schryer, D. R. Coffer III, W. R., Edahl Jr, R. A. & Munavalli, S. Oxidation of SO_2 by NO_2 and air in an aqueous suspension of carbon. In *Heterogeneous Atmospheric Chemistry,* ed. D. R. Schryer, *Geophysical Monograph 26.* American Geophysical Union, Washington, DC, USA, 1982, pp. 174–7.
191. Schryer, D. R., Schryer, J., Coffer III, W. R. & Vay, S. A., Carbon catalysis in the aqueous oxidation of SO_2 by NO_2 and air. *Atmos. Environ.*,

18 (1984) 2281–4.
192. Brodzinsky, R., Catalytic oxidation of S(IV) on activated carbon in aqueous solution: kinetics and mechanism. PhD thesis, Lawrence Berkeley Laboratory, University of California, CA, USA, Report No. LBL-12218, 1981.
193. Novakov, T., Soot-catalysed atmospheric reactions. In *Heterogeneous Atmospheric Chemistry,* ed. D. R. Schryer, *Geophysical Monograph 26.* American Geophysical Union, Washington, DC, USA, 1982, pp. 215–20.
194. Chang, S.-G., Toossi, R. & Novakov, T., The importance of soot particles and nitrous acid in oxidising SO_2 in atmospheric aqueous droplets. *Atmos. Environ.,* **15** (1981) 1287–92.
195. Erickson, R. E., Yates, L. M., Clark, R. L. & McEwen, D., The reaction of sulphur dioxide with ozone in water and its possible atmospheric significance. *Atmos. Environ.,* **11** (1977) 813–17.
196. Miller, D. F., Gertler, A. W., Whitbeck, M. R. & Lamb, D., Ozone induced oxidation of SO_2 in simulated clouds. *J. Geophys. Res.,* **91** (1986) 14439–44.
197. Nahir, T. M. & Dawson, G. A., Oxidation of sulphur dioxide by ozone in highly dispersed water droplets. *J. Atmos. Chem.,* **5** (1987) 373–83.
198. Penkett, S. A. & Garland, J. A., Oxidation of sulphur dioxide in artificial fogs by ozone. *Tellus,* **26** (1974) 284–9.
199. Harrison, H., Larson, T. V. & Monkman, C. S., Aqueous phase oxidation of sulphites by ozone in the presence of iron and manganese. *Atmos. Environ.,* **16** (1982) 1039–41.
200. Maahs, H. G., Kinetics and mechanisms of the oxidation of S(IV) by ozone in aqueous solution with particular reference to SO_2 conversion in non-urban tropospheric clouds. *J. Geophys. Res.,* **88** (1983) 10721–32.
201. Hoffmann, M. R., On the kinetics and mechanism of oxidation of aquated sulphur dioxide by ozone. *Atmos. Environ.,* **20** (1986) 1145–54.
202. Esphenson, J. E. & Taube, H., Tracer experiments with ozone as oxidising agent in aqueous solution. *Inorg. Chem.,* **4** (1965) 704–9.
203. Mader, P.M., Kinetics of the hydrogen peroxide–sulphite reaction in alkaline solution. *J. Am. Chem. Soc.,* **80** (1958) 2634–9.
204. Hoffmann, M. R. & Edwards, J. O., Kinetics of the oxidation of sulphite by hydrogen peroxide in acidic solution. *J. Phys. Chem.,* **79** (1975) 2096–8.
205. Heikes, B. G., Lazrus, A. L., Kok, G. L., Kunen, S. M., Gandrud, B. W., Gitlin, S. N. & Sperry, P. D., Evidence for aqueous phase hydrogen peroxide synthesis in the troposphere. *J. Geophys. Res.,* **87** (1982) 3045–51.
206. McArdle, J. V. & Hoffmann, M. R., Kinetics and mechanism of the oxidation of aquated sulphur dioxide by hydrogen peroxide at low pH. *J. Phys. Chem,* **87** (1983) 5425–9.
207. Kunen, S. M., Lazrus, A. L., Kok, G. L. & Heikes, B. G., Aqueous oxidation of SO_2 by hydrogen peroxide. *J. Geophys. Res.,* **88** (1983) 3671–4.
208. Lee, Y.-N., Shen, J., Klotz, P. J., Schwartz, S. E. & Newman, L., Kinetics of hydrogen peroxide–sulphur(IV) reaction in rainwater collected at a Northeastern US site. *J. Geophys. Res.,* **91** (1986) 13264–74.
209. Lind, J. A., Lazrus, A. L. & Kok, G. L., Aqueous phase oxidation of sulphur(IV) by hydrogen peroxide, methylperoxide, and peroxyacetic acid. *J. Geophys. Res.,* **92** (1987) 4171–7.

210. Overton Jr, J. H., Validation of the Hoffmann and Edwards' S(IV)–H_2O_2 mechanism. *Atmos. Environ.*, **19** (1985) 687–90.
211. Radojević, M., H_2O_2 in the atmosphere and its role in the aqueous oxidation of SO_2. *Environ. Technol. Lett.*, **4** (1983) 499–508.
212. Jaeschke, W. A. & Herrmann, G. J., SO_2 oxidation by hydrogen peroxide in suspended droplets. In *Chemistry of Acid Rain: Sources and Atmospheric Processes,* ed. R. W. Johnson, G. E. Gordon, W. Calkins & A. Z. Elzerman, *ACS Symposium Series 349.* American Chemical Society, Washington, DC, USA, 1987, pp. 142–57.
213. Dasgupta, P. K., Further discussion. *Atmos. Environ.*, **14** (1980) 620–1.
214. Halperin, J. & Taube, H., The transfer of oxygen atoms in oxidation–reduction reactions. IV. The reaction of hydrogen peroxide with sulphite and thiosulphate, and of oxygen, manganese dioxide and of permanganate with sulphite. *J. Am. Chem. Soc.*, **74** (1952) 380–2.
215. Winter, E. R. S. & Briscoe, H. V. A., Oxygen atom transfer during the oxidation of aqueous sodium sulphite. *J. Am. Chem. Soc.*, **73** (1951) 496–7.
216. Holt, B. D., Kumar, R. & Cunningham, P. T., Oxygen-18 study of the aqueous phase oxidation of sulphur dioxide. *Atmos. Environ.*, **15** (1981) 557–66.
217. Chameides, W. L., The photochemistry of a remote marine stratiform cloud. *J. Geophys. Res.*, **89** (1984) 4739–55.
218. Martin, L. R., Damschen, D. E. & Jedeikis, H. S., The reaction of nitrogen oxides with SO_2 in aqueous aerosols. *Atmos. Environ.*, **15** (1981) 191–5.
219. Oblath, S. B., Markowitz, S. S., Novakov, T. & Chang, S. G., Kinetics of the formation of hydroxylamine disulphonate by reactions of nitrites with sulphites. *J. Phys. Chem.*, **85** (1981) 1017–21.
220. Lee, Y. & Schwartz, S. E., Kinetics of oxidation of sulphur(IV) by nitrogen dioxide. In *Precipitation Scavenging, Dry Deposition and Resuspension,* ed. H. R. Pruppacher, R. G. Semonin & W. G. N. Slinn. Elsevier, New York, USA, 1983, pp. 453–66.
221. Schwartz, S. E., Gas aqueous reactions of sulphur and nitrogen oxides in liquid water clouds. In *SO_2, NO and NO_2 Oxidation Mechanisms: Atmospheric considerations,* ed. J. G. Calvert, *Acid Precipitation Series,* Vol. 3. Butterworth, Boston, MA, USA, 1984 pp. 173–208.
222. Radojević, M., Nitrite in rainwater. *Atmos. Environ.*, **20** (1986) 1309–10.
223. Ridler, E. S., Chamber process for the manufacture of H_2SO_4. In *Manufacture of Sulphuric Acid,* ed. W. W. Duecker & J. R. West. Reinhold, New York, 1959, pp. 103–34.
224. Sato, T., Matani, S. & Okabe, T., The oxidation of sodium sulphite with nitrogen dioxide with special reference to analytical methods for nitrogen sulphur compounds produced in the reaction system. *Nippon Kagaku Kaishi,* **7** (1979) 869–78.
225. Ellison, T. K. & Eckert, C. A., The oxidation of aqueous SO_2. 4. The influence of nitrogen dioxide at low pH. *J. Phys. Chem.*, **88** (1984) 2335–9.
226. McElroy, W. J., The aqueous oxidation of SO_2 by OH radicals. *Atmos. Environ.*, **20** (1986) 323–30.
227. Chameides, W. L. & Davis, D. D., The free radical chemistry of cloud droplets and its impact upon the composition of rain. *J. Geophys. Res.*, **87** (1982) 4863–77.

228. Huie, R. E. & Neta, P., Rate constants for some oxidations of S(IV) by radicals in aqueous solution. *Atmos. Environ.*, **21** (1987) 1743–7.
229. Calvert, J. G., Lazrus, A., Kok, G. L., Heikes, B. G., Walega, J. G., Lind, J. & Cantrell, C. A., Chemical mechanisms of acid generation in the troposphere. *Nature,* **317** (1985) 27–35.
230. Calvert, J. G. & Stockwell, W. R., Acid generation in the troposphere by gas phase chemistry. *Environ. Sci. Technol.*, **17** (1983) 428–43A.
231. Munger, J. W., Jacob, D. J. & Hoffmann, M. R., The occurrence of bisulphite-aldehyde addition products in fog and cloudwater. *J. Atmos. Chem.*, **1** (1984) 335–50.
232. Radojević, M., Tyler, B. J., Wicks, A. J., Gay, M. J. & Choularton, T. W., Field studies of the SO_2 aqueous S(IV) equilibrium in clouds. *Atmos. Environ.*, **24** (1990) 323–8.
233. Graedel, T. E., Gill, P. S. & Weschler, C. J., Effects of organic surface films on the scavenging of atmospheric gases by raindrops and aerosol particles. In *Precipitation Scavenging, Dry Deposition and Resuspension,* ed. H. R. Pruppacher, R. G. Semonin & W. G. N. Slinn. Elsevier, New York, USA, 1983, pp. 417–27.
234. Farmer, J. C. & Dawson, G. A., Condensation sampling of soluble atmospheric trace gases. *J. Geophys. Res.*, **87** (1982) 8931–42.
235. Daum, P. H., Kelly, T. J., Schwartz, S. E. & Newman, L., Measurements of the chemical composition of stratiform clouds. *Atmos. Environ.*, **18** (1984) 2671–84.
236. Kelly, T. J., Daum, P. H. & Schwartz, S. E., Measurements of peroxides in cloudwater and rain. *J. Geophys. Res.*, **90** (1985) 7861–71.
237. Brock, J. R. & Durham, J. L., Aqueous aerosols as a reactant in the atmosphere. In *SO_2, NO and NO_2 Oxidation Mechanisms: Atmospheric Considerations,* ed. J. G. Calvert, *Acid Precipitation Series,* Vol. 3. Butterworth, Boston, MA, USA, 1984, pp. 209–49.
238. Hegg, D. A., The importance of liquid phase oxidation of SO_2 in the troposphere. *J. Geophys. Res.*, **90** (1985) 3773–9.
239. Lee, Y.-N. & Lind, J. A., Kinetics of aqueous phase oxidation of nitrogen(III) by hydrogen peroxide. *J. Geophys. Res.*, **91** (1986) 2793–800.
240. McKay, H. A. C., The atmospheric oxidation of sulphur dioxide in water droplets in the presence of ammonia. *Atmos. Environ.*, **5** (1971) 7–14.
241. Williams, P. T., Radojević, M. & Clarke, A. G., Dissolution of trace metals from particles of industrial origin and its influence on the composition of rainwater. *Atmos. Environ.*, **22** (1988) 1433–42.
242. Kadowaki, S., On the nature of atmospheric oxidation processes of SO_2 to sulphate and of NO_2 to nitrate on the basis of diurnal variations of sulphate, nitrate and other pollutants in an urban area. *Environ. Sci. Technol.*, **20** (1986) 1249–53.
243. Gray, D., Lissi, E. & Heicklen, J., The reaction of hydrogen peroxide with nitrogen dioxide and nitric oxide. *J. Phys. Chem.*, **76** (1972) 1919–24.
244. Hampson Jr, R. F. & Garvin, D., *Chemical Kinetic and Photochemical Data for Modelling Atmospheric Chemistry,* NBS Technical Note 866. National Bureau of Standards, Washington, DC, USA, 1975.

245. Howard, C. J. & Evenson, K. M., Kinetics of the reaction of HO_2 with NO. *Geophys. Res. Lett.*, **4** (1977) 437–40.
246. Cox, R. A., Derwent, R. G. & Hutton, A. J. L., Significance of peroxynitric acid in atmospheric chemistry of nitrogen oxides. *Nature*, **270** (1977) 328–9.
247. Simonaitis, R. & Heicklen, J., Reactions of HO_2 with NO and NO_2 and of OH with NO. *J. Phys. Chem.*, **80** (1976) 1–7.
248. Park, J.-Y. & Lee, Y.-N., Aqueous solubility and reactivity of nitrous acid. In *Symposium on Acid Rain: I Sources and Atmospheric Processes,* The Division of Petroleum Chemistry Inc. New York Meeting, 13–18 April 1986. American Chemical Society. pp. 437–40.
249. Heikes, B. G. & Thompson, A. M., Effects of heterogeneous processes on NO_3, HONO and HNO_3 chemistry in the troposphere. *J. Geophys. Res.*, **88** (1983) 10883–95.
250. Chameides, W. L., Possible role of NO_3 in the nighttime chemistry of a cloud. *J. Geophys. Res.*, **91** (1986) 5331–7.
251. Damschen, D. E. & Martin, L. R., Aqueous aerosol oxidation of nitrous acid by O_2, O_3 and H_2O_2. *Atmos. Environ.*, **17** (1983) 2005–11.

Chapter 4

The Long-Range Transport of Atmospheric Acidity

B. E. A. FISHER
National Power, Senator House, 85 Queen Victoria Street, London, EC4U 4DP, UK

4.1 PROCESSES LEADING TO THE FORMATION OF ACID RAIN

The major pollutants of concern arising from the burning of oil and coal are sulphur dioxide (SO_2), nitrogen oxides (NO_x) and hydrocarbon gases. These are generally emitted directly to the atmosphere from chimneys and vehicle exhausts, and in the case of large sources, such as power stations or smelters, from very tall chimneys. This ensures that by mixing with the atmosphere the ground-level concentrations of such material are very low. The life cycle of these pollutants in the atmosphere is investigated in long-range transport studies. Usually they remain airborne for several days. Some industrial or natural emissions are more chemically inert and remain airborne for much longer periods becoming involved in very long-range transport or global dispersion, but these are not relevant to the present discussion.

If sulphur oxides (SO_x) or NO_x are carried by the air into a rain system they can be removed by the rain, making the rain more acid than it would be if only influenced by natural processes. This is what is generally meant by 'acid rain'. Removal in rain is only one of the atmospheric cleaning mechanisms. When the direct uptake of these gases onto land or sea surfaces is included, one may talk of 'acid deposition'. This term is somewhat ambiguous, since the acidity is actually generated after deposition by chemical reactions in plants, soil or water. Even referring to the acidity of rain on a given occasion can be somewhat misleading, since very acid rain or cloudwater can be produced by evaporation from rain or cloud droplets. These difficulties can be lessened if the origin and fate of the major components determining the acidity of precipitation are discussed individually, namely the SO_x and NO_x, rather than the acidity itself.

4.2 MAIN ATMOSPHERIC TRANSPORT PROCESSES

Most of the SO_x at the time of emission is in the form of SO_2, a so-called 'primary' pollutant. NO_x species are mainly emitted in the form of nitric oxide (NO), but because of its fairly rapid reaction with ozone to form nitrogen dioxide (NO_2), both NO and NO_2 are considered primary pollutants in the nitrogen oxide cycle.

Urban pollution and the near-field spread of emissions from large point sources, such as power stations, are governed by atmospheric dispersion. Only a small fraction of the emission is removed within 100 km of the source by the process of dry deposition. This is the process by which SO_x or NO_x are taken up by natural or artificial surfaces, such as grass, trees, sea or building materials. It occurs continuously though it may vary in its effectiveness depending on the gas involved, atmospheric conditions and the type of underlying surface. The rate of dry deposition of, say, SO_2 is equal to the deposition velocity multiplied by the ground-level concentration of SO_2, both referred to the same nominal height of a few metres above the surface. The deposition velocity of SO_2 has been measured more than for other species, under various types of conditions. The magnitude of the deposition velocity of SO_2 largely determines its average lifetime in the atmosphere in the absence of rain.

If SO_2 remains airborne on average for, say, 2 days, and is carried by the wind at a speed of, say, 7 ms^{-1} it will travel 1200 km on average, so 1000 km is the typical length scale of the transport process. The SO_2 concentration and hence its associated deposition decreases with increasing travel distance. Over the course of a year, variations in large-scale weather systems cause the emissions from any particular source to be carried in all wind directions to a greater or lesser extent. This ensures that annual average concentrations decrease with increasing travel distance, just because of variations in the path of air-mass trajectories. Atmospheric concentrations also decrease because of removal by dry deposition, removal by precipitation (wet deposition) and chemical transformation into other sulphur compounds. Concentrations much beyond 1000 km of travel will be small. On special occasions some material may travel much further than this, but generally a source of SO_2 will make a contribution to dry deposition at a receptor site if the receptor lies approximately within this distance of the source. If a receptor lies close to a number of large sources or is within an urban area, the long-range transport contribution is of less practical significance.

4.3 MATHEMATICAL MODELLING OF THE LONG-RANGE TRANSPORT OF SULPHUR

Assessing the magnitude of source–receptor relationships is a task for mathematical modelling. Since the relative contributions of nearby and distant sources can vary greatly between receptor sites modelling plays an important role. Early approaches to modelling sulphur oxides were based on describing mathematically the main transport processes in very idealised ways.[1] It was not expected that these models would be able to predict sulphur concentration or deposition accurately during an episode, that is on a day to day basis. However, using the best estimates of sulphur emissions available, these models were able to reproduce the broad features of the observed sulphur concentration and deposition patterns.[2,3] In these models parameter values were used to describe the main processes, which were sensible and consistent with studies of the individual detailed processes. This gives confidence that the present day sulphur cycle is reasonably well understood.

These simple models have the limitation that they do not take into account many of the complex features of atmospheric transport. They are based on the implicit assumption that the long-term average concentration or deposition of sulphur on a regional scale is largely determined by source distribution, the average wind behaviour and the average duration of dry weather. (Once an air mass encounters large-scale precipitation, it is assumed in the model that sulphur in the air is largely removed by wet deposition.) As an assessment tool these models have the limitation that they may not give accurate predictions of changes in concentrations and depositions as a result of major changes in the emission rates of the main pollutant gases, where these affect the sulphur cycle, since the models are based on parameter values appropriate to present day conditions.

For these reasons there has been a move away from simple sulphur models towards comprehensive models, in which all known chemical processes, as well as the variability in physical processes together with very detailed meteorology, are included.[4,5] Comprehensive models are therefore computationally much more elaborate than simple models and can with ease outstrip the most advanced computer facilities. In addition, considerable effort is required to prepare the necessary meteorological, physical and chemical data. Results from the comprehensive models have only begun to appear in recent years. In the following sections when the main processes affecting atmospheric transport are reviewed (Fig. 4.1), it should become clear what factors ought to be included in a comprehensive model.

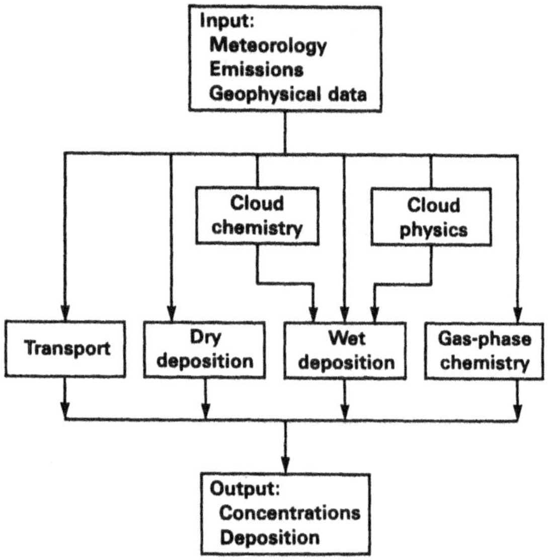

Fig. 4.1. Schematic diagram of the modular structure of comprehensive models.

4.4 THE LONG-RANGE TRANSPORT OF NITROGEN

Development of the understanding of the nitrogen cycle has not followed that of the sulphur cycle, although the acidity of rain is largely determined by a balance between sulphate (derived from airborne SO_x species), nitrates (derived from airborne NO_x species) and ammonium (derived from ammonia). The former two species tend to make rain more acid, while the latter makes it less acid. There may also be, on occasions, significant contributions from chloride of non-marine origin and soil-related components. Sources of NO_x are fairly well known in industrial regions, and behaviour of NO_x in specific conditions can be predicted in principle. However, there are more NO_x species of importance in the nitrogen cycle than corresponding species in the sulphur cycle. The physical and chemical behaviour differs markedly for each NO_x species, depending strongly on the prevailing meteorological conditions. This strong variability according to chemical species and atmospheric conditions means that it is not easy to define a simple model of NO_x containing well-defined bounds on parameter values. There is not therefore the nitrogen equivalent of the simple sulphur model.

The problem is compounded by the lack of an extensive database of observed nitrogen species. Although nitrate and ammonium in precipitation are monitored routinely, only NO_2 (and NO) of the gaseous species are regularly measured and then mainly in urban regions. The most easily deposited nitrogen species is thought to be nitric acid gas, which is a secondary species formed from NO_2 at a rate strongly dependent on time of day, time of year and latitude, because the oxidation process is associated with photochemical reactions in the atmosphere. There are other secondary airborne nitrate species which are not routinely measured, such as particulate nitrate (including ammonium nitrate) and peroxyacetyl nitrate (PAN), which are not thought to be so readily deposited onto ground surfaces. For this reason it is not possible to test whether a simple NO_x model, with presumably the correct emissions, is partitioning NO_x properly between the more rapidly and less rapidly deposited species and therefore whether the model can be used to estimate the relative contributions of nearby or more distant sources to the observed wet deposition at a monitoring site.

Comprehensive long-range transport models include detailed chemical schemes representing the main relevant chemical reactions in the atmosphere. They also take into account spatial and temporal variations in reaction rates and, as far as is known, variability in physical removal parameters, such as deposition velocity. The models can be used to predict NO_x as well as SO_x. The ability of comprehensive models to simulate the atmospheric sulphur and nitrogen cycles over distances up to several thousand kilometres has not as yet been fully evaluated.

4.5 THE AMMONIA CYCLE

The origin of ammonium in precipitation is ammonia derived mainly from certain agricultural activities, though the magnitude of ammonia emissions have not been established with great accuracy. Ammonia is expected to be rapidly absorbed into cloud droplets or taken up on moist surfaces. Since ammonia is emitted essentially at ground level, it is difficult to determine the representative large-scale emission from land surfaces, which may be the net consequence of localised emission and deposition processes. In the absence of a large body of observations of gaseous ammonia and airborne ammonium species (in the form of particles or droplets), it is not possible to test the likely accuracy of ammonia emissions inventories.[6] Attempts at modelling have tended to underestimate the long-range transport of ammonia species, leading to suspicions that emissions have been underestimated.[7] The rate of uptake of SO_2 in cloud droplets can be regulated by the amount of ammonia present, so that the sulphur and ammonia cycles are linked. In simple models this link is not explicitly included, but in comprehensive models it should be explicitly modelled.

4.6 EXAMPLES OF SIMPLE AND COMPREHENSIVE MODELS

In the following sections a more detailed account will be given of the atmospheric transport processes influencing acid deposition. Reference will be made to two programmes of model development. One typifies the simple practical approach, which works closely with empirical data, and the other illustrates the comprehensive approach, which works from first principles. In explaining the basis of the models the main factors influencing acid deposition will be described in more detail.

4.6.1 The EMEP model

The EMEP model is the basic tool by which the long-range transport of air pollutants in Europe has been evaluated within the EMEP programme. EMEP stands for the 'Co-operative Programme for Monitoring and Evaluation of the Long-Range Transmission of Air Pollutants in Europe'. The programme, which started in 1977, is the natural focus for European collaboration on acid rain and involves all countries with concerns over the issue. It is now organised under the Convention on Long-Range Transboundary Air Pollution, drawn up in 1979 under the auspices of the United Nations Economic Commission for Europe. Its objective was to provide governments with information on the deposition and concentration of air pollutants and on the quantity of material crossing national boundaries, so that strategies for emission reduction could be developed as necessary. EMEP is essentially concerned with atmospheric processes, such as how much of the emission from one region is deposited in another. Consideration of whether such deposition is potentially damaging, or not, is left to other research programmes.

Information on current air pollution levels and wet deposition over Europe is tackled by EMEP using a largely observational programme. Source–receptor relationships are investigated using modelling.[8,9] Both aspects have developed over the years as the range of pollutants has widened from the earlier interest in sulphur species. Not all aspects can be tackled given existing resources and the intention has been to make practical use of available meteorological data and measurements, which could be performed on a routine basis. The modelling of first, sulphur species, and more recently, the nitrogen cycle, has followed this aim.[10]

4.6.2 The Regional Acid Deposition Model (RADM)

The National Acid Precipitation Assessment Program (NAPAP) was set up by the US Congress in 1980 to assess acidic deposition causes, effects and control strategies. In September 1990, NAPAP published its final assessment which consisted of two parts. The first consists of a series of reports documenting the present state of science. The second is an integrated assessment based on a number of key questions. Apart from questions concerning the relationship between acidic deposition/air pollutant concentrations and effects, which are not the concern of atmospheric transport studies, these questions are centred on the relationship between emissions and acidic deposition/air pollutant concentrations and the sensitivity of the relationship to change.

The atmospheric model used to undertake this part of the analysis is called the Regional Acid Deposition Model (RADM). It incoporates the most important contributing processes and has been under development since 1983.

A somewhat similar model—the Acid Deposition and Oxidant Model (ADOM)—has been developed in parallel with RADM. Comments regarding RADM apply in the main to ADOM as well, so no distinction is made in the following discussion.[5] In order that the treatment of each of its processes is at least at the level of the present state of science, few concessions to simplicity have been made and the model is consequently complex, consisting of a set of modules, each corresponding to a different process. Before RADM can be used for assessment purposes an evaluation of its suitability is necessary, the key element of which is a comparison of its predictions against a detailed field study consisting preferably of both ground-level and airborne measurements. Such an evaluation programme is currently under way.

By comparing and contrasting the development of the EMEP and RADM models, the key elements of the long-range transport of atmospheric acidity will become more apparent. This is the intention of the following sections, which are divided according to each of the main processes contributing to an integrated model (Fig. 4.1).

4.7 ATMOSPHERIC TRANSPORT

4.7.1 Lagrangian and Eulerian framework

The basis of modelling the long-range transport of air pollutants is to track the movement of air parcels over long distances. The EMEP model uses a Lagrangian (moving) co-ordinate system in which properties of the atmosphere are assigned to individual air parcels that move through space in time. The RADM model uses an Eulerian (fixed) co-ordinate system in which properties of the atmosphere are assigned to points in space at each given time, without an attempt to identify individual air parcels from one time to another. Conceptually, it is easier to think about source–receptor relationships in a Lagrangian framework, but it is simpler to consider chemical reactions in an Eulerian framework.

In neither case is it possible to track precisely the movement of air parcels over 'long distances' (~ 1000 km).[11] The general movement can be followed, but to obtain good agreement between calculated and observed daily average concentrations at a monitoring site, which in the case of wet deposition should be related to the duration, intensity and past history of the precipitation, requires quite accurate estimates of air-parcel trajectories.

4.7.2 Statistical models

When calculating annual average concentrations such precision is not required and this is the reason that so-called 'statistical models' have been a useful interpretative tool since the early 1970s. In statistical models, transport is simulated via one or more long-term average wind roses, which represent the speed and directional frequencies of winds within sectors at a single level in the atmosphere. Wet deposition in statistical models is typically dependent on the average duration and frequency of wet and dry periods rather than on actual occurrences and amounts.

4.7.3 Meteorological data

Neither RADM nor the EMEP model is a statistical model. In the EMEP model the movement of air parcels is followed using the wind field on a chosen meteorological surface in the atmosphere. In the latest version of the model the wind is taken from a numerical weather prediction model, which combines, in an optimal way, information from measurements for a certain point in time with information on the meteorological situation over the previous few days, taking into account the physical laws governing atmospheric dynamics. The representative wind is taken to be at a level of about 700 m above the surface. Other meteorological properties, such as cloud cover and temperature, are part of the numerical weather prediction model and are used in the model of nitrogen transport, since they can strongly influence chemical reactions.

The EMEP model is run using a grid of square cells in the horizontal, of sides 150 km. It is centred on north-western Europe and would contain most of the nitrogen and sulphur source areas. The model makes use of the precipitation averaged over each grid element every 6 h, where the precipitation over land is taken directly from observation, while that over the sea is taken from the numerical weather prediction model. All information from the EMEP model is provided as averages over 150×150 km grid squares in order to limit the amount of computation. The resolution is adequate for the purpose for which the model was designed, i.e. long-range transport over distances of 1000 or 2000 km.

The meteorological information required to run RADM is obtained from a meso-scale meteorological model,[12] which provides three-dimensional fields of wind, pressure, temperature, humidity, cloud and surface precipitation rate that are used by the other modules of RADM. The meso-scale model predictions are based on solving the primitive equations of atmospheric flow and precipitation formation.

The predictions are updated every 12 h with actual meteorological data to ensure that the predictions are constrained never to deviate far from observations, but provide, in a meteorological sense, the best interpolation between observations. The meteorological model provides data at 15 levels between the surface and a level 15 km above the surface. The current version of the model is applied with six vertical layers, whose approximate heights are 0–150 m, 150–550 m, 550–1300 m, 1·3–3·8 km, 3·8–7·8 km and 7·8–15 km. The lowest three layers are typically within the daytime mixing layer of the atmosphere which contains the highest pollution concentrations, and in which rapid vertical mixing takes place. The remaining layers are in the free troposphere.

Thus, the RADM considers transport throughout the depth of the troposphere, whereas earlier models only follow air parcels in the mixing layer where the highest concentrations occur. The EMEP model allows for exchange of material between a well-mixed boundary layer at the ground and the free troposphere, but does not explicitly treat transport in the free troposphere.

It needs to be stated that atmospheric transport models are subject to development and change. For example, the number of vertical layers and their positioning are flexible parameters in RADM, as is the choice of horizontal grid. Presently, runs are performed using a domain of 35×38 grid cells where each cell has a dimension of 80×80 km. The domain covers north-eastern North America and just like the EMEP model is nested within a wider meteorological domain. Computational storage requirements limit the choice of grid size and layers, but are not the only consideration.

For assessment purposes RADM is required to provide estimates of long-term average (annual or seasonal) depositions or concentrations. RADM is an episodic model requiring a detailed set-up of meteorological inputs for 3–5 day episodes. Resource constraints mean that it is not possible at present to directly simulate a long-term (multi-season or multi-year) period. In contrast, by limiting its field of application the developers of the EMEP model have been able to calculate concentration and deposition patterns over Europe for the period 1979–1988 and will continue to extend their period of assessment as monitoring continues.

4.7.4 Horizontal resolution in models

To limit the amount of computation, all information in both models is provided as averages over grid squares; thus the models do not provide information on a finer scale than about 100 km in the horizontal and a more detailed model would be required to study variations on depositions over distances significantly less than 100 km (Fig. 4.2).

As parcels of air, which can be thought of as having horizontal dimensions of order 100×100 km and vertical dimensions of order 1 km, are moved by the wind in Lagrangian models, they pick up additional pollutants as they cross source areas. In Eulerian models the action of the wind leads to exchange of pollution between grid cells, while the sources emit

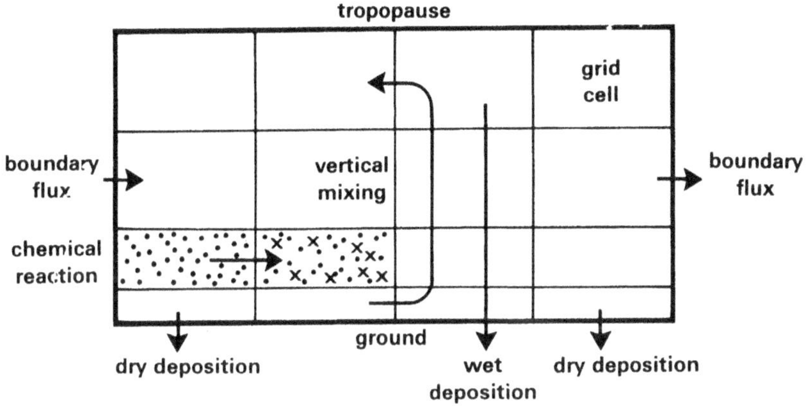

Fig. 4.2. Schematic diagram of the numerical treatment of processes in comprehensive models.

pollution into the grid cells near the surface in which the sources are located. A major, but well-recognised, problem with Eulerian models is that the movement of air between neighbouring grid cells cannot be described accurately. Any approximate treatment of the movement of air between fixed cells used to obtain numerical solutions of the basic equations involves some 'numerical diffusion', or artificial spreading. Generally, a numerical method is chosen which uses relatively little computer time, considering its accuracy, and does not generate negative concentrations. This difficulty with Eulerian models is offset by the ease with which changes in concentrations caused by chemical reactions and the input of fresh emissions can be treated within a fixed grid.

4.8 ATMOSPHERIC REMOVAL MECHANISMS

As air parcels are moved around by meteorological processes, pollutants emitted directly into the atmosphere are subject to loss processes. For the sulphur cycle, SO_2 is the primary emission and is removed by dry deposition at the surface, chemical transformation to sulphate and wet removal in precipitation. A small fraction of the sulphate emission is emitted directly as sulphate, but most of the airborne sulphate is generated by gradual conversion of SO_2 and is depleted by wet and dry loss mechanisms.

4.8.1 Dry deposition

Dry deposition is an important removal mechanism because it operates throughout the travel path of a given emission, while it is in the lowest layers of the atmosphere, or mixing layer, in which pollution is generally confined for most of its lifetime in the atmosphere.

Since on average there is an interval of a few days between periods of rain, much of the travel occurs in dry conditions. There may be cloud present, particularly low cloud or fog in the mixing layer. This can affect the chemistry of SO_x and NO_x. There has recently been much interest in the mechanism of cloud deposition, the deposition of cloud droplets containing sulphur or nitrogen compounds onto hillsides.[13–15]

Both the EMEP and RADM models recognise that the deposition velocity, the parameter describing the rate of removal, is a spatially and temporally varying quantity and approximations to its value over a grid element are used to take into account the main factors which influence its variation. It is inversely proportional to the sum of three characteristic resistances, representing

(1) the resistance to atmospheric transport to the surface by turbulent mixing;
(2) the resistance to transport across layers adjacent to surface elements, such as leaves, by diffusive processes; and
(3) the resistance at the surface itself which depends on the state of the surface.

The deposition velocity can usually be thought of having an upper bound of magnitude 0·01 or 0·02 m s^{-1}, which applies only when the atmospheric resistance is important. Surrogate data which can be easily specified, such as season, latitude or land-type classification, or time of day, solar isolation or surface wetness, averaged over grid cell elements (80×80 km or 150×150 km according to the model) can be adjusted in each grid cell as the numerical computation proceeds. These surrogate data allow broad categories of deposition velocity behaviour to be defined. As it is a single-layer model, the EMEP model introduces an effective deposition velocity characteristic of a height 50 km above the surface and dependent on atmospheric stability to take acount of the atmospheric resistance over this depth. In principle the multi-layer structure of RADM should automatically take into account the atmospheric resistance to surface removal and is only limited by practical considerations regarding the vertical resolution of the model. The main problem is the lack of observational data on deposition velocity[16] for all chemical species of interest. Measurement of deposition velocity is a difficult, non-routine procedure and variability in results cannot necessarily be explained by a broad categorisation of differences in conditions under which the measurements have been made.

4.8.2 Loss of material through top of mixing layer

In the EMEP model the height of the mixed layer at 12.00 GMT every day is estimated from an analysis of radiosonde data. Any one air parcel starts at 12.00 GMT with a height equal to that derived from observations. Subsequently, the upper boundary of the vertically well-mixed air parcel carrying the pollution follows the vertical motion of the air derived from the underlying numerical prediction model. After 24 h the mass of the air parcel is redefined following the next analysis of mixing layer depth. If the height of the air parcel has grown higher than the analysed mixing layer depth, that part sticking into the free atmosphere is assumed lost, together with the pollution it carried. If the opposite is true the height of the air parcel is redefined assuming that the pollution has been diluted with background tropospheric air. This procedure either defines a loss of material to the free troposphere which is then lost from the system or the addition of tropospheric background pollution, the origin of which is not directly attributed to any of the sources treated in the model (though it may have originally come from them, having been transported to the free troposphere through an earlier exchange cycle).

The meso-scale meteorological model underlying the RADM generates vertical wind motions of reasonable magnitude, which are used directly in the RADM to transport material throughout the troposphere. Such vertical winds could not be derived from observations. In addition, vertical exchange by diffusion takes place using a formulation, which gives maximum mixing within the mixing layer, i.e. near the surface, and decreases above the mixing layer. This treatment falls naturally within an Eulerian grid cell model and is the Eulerian equivalent of a well-mixed vertical layer in the Lagrangian treatment. The six layers used in the current version of RADM are too coarse to accurately take into account the depth of the boundary layer. Instead, a representative eddy diffusivity dependent on boundary layer height is used to separate the mixing layer, where more rapid vertical mixing occurs, from the free troposphere where less rapid vertical mixing occurs. Radiosonde measurements in the north-eastern US are taken near sunrise and sunset and therefore do not provide adequate information about actual maximum daytime mixing depths.

4.8.3 Background sulphur and nitrogen sources

It is clear in the EMEP model that transport of material from the troposphere into the mixing layer provides one source of background pollution in sulphur or nitrogen budgets. Transport of material across lateral boundaries of the model domain provides another source. This material has originated outside the area of study. It may be material which has originated within the domain of the model and has been carried along a trajectory, which has crossed the edge of the domain and later returned.

Since the domain of the model has been chosen to include all major combustion sources in north-western Europe and to have as few as practicable sources in grid cells near the boundaries of the domain, one does not expect these background contributions to be a large fraction of the sulphur and nitrogen directly attributed to sources within the study region. In the EMEP model parcels of air are followed for 96 h along trajectories to specified receptor points. One expects only a very small fraction of sulphur or nitrogen to remain airborne for more than 96 h, but any that does is another potential source of background material.

Analysis from earlier versions of the EMEP model[8] have shown that it was not possible to entirely neglect background contributions. It has not been possible from analysis of observations to assess the background contribution in detail, according, for example, to the direction or latitude from which an air mass originates. Instead, constant fixed values of the concentrations of SO_2 and sulphate are assumed to exist in air fed into the model calculations. In addition, precipitation, which scavenges much of the troposphere as well as the mixing layer near the surface, is assumed to contain a background sulphate concentration which is added to the calculated levels of sulphate in precipitation.

In the RADM the problem of specifying tropospheric background concentrations does not arise. Instead, the model requires concentrations of trace gases at its boundaries, in principle at heights throughout the troposphere. These are not available, so that values typical of clean air conditions are chosen from available measurements. Aircraft measurements have been made in order to extend knowledge of the typical concentrations of the many chemical species calculated in the model at different heights in the atmosphere. Data from these flights can be used to establish seasonal trends in critical chemical variables to guide in the setting of chemical concentrations at RADM boundaries. The choice of lateral boundaries in the RADM, like those in the EMEP model, has been made so as to place them as far as is practicable from the major sulphur and nitrogen emission regions of the north-eastern US and Canada, so that the level of acidic deposition will be largely determined by physical and chemical processes which occur within the model domain.

The other problem faced by the developers of RADM is that as one is only able to run the model for limited periods, an artificial way is used to generate initial concentrations of the trace gases at the start of a RADM simulation. Careful choice of episodes ensures that the uncertainty in knowledge of initial conditions is minimised by ensuring that the sulphur and nitrogen contributing to, say, a notable wet deposition event are emitted mainly within the period of study and not before.

4.8.4 Sector analysis

The significance of background concentrations of sulphate in precipitation in Europe is seen at monitoring sites in extremely rural areas near the western edge of the continent, such as on the West Coast of the British Isles and Norway. Useful information as to the origin of material collected at these monitoring sites can be obtained using a technique known as 'sector analysis'. In this technique observations are grouped into sectors according to the general direction of the airflow as determined from calculations of air-parcel trajectories (which may be derived from the EMEP modelling activities). The dependence of observed concentration or deposition on transport direction should reflect the position of major source areas. Trajectories of air parcels representing a column of polluted air are tracked back for 96 h from a monitoring station. If the majority of the trajectory path lies within a certain sector the measured data for the day concerned are allocated to that sector. If this is not the case, i.e. trajectories are somewhat curved, the data are grouped in a ninth sector, which therefore includes all material of 'undecided' origin. The purpose of the procedure is to make a fairly strict allocation to ensure that one is fairly certain of the possible source areas, within about 1500 km of the monitoring site, which may be contributing to the measured concentration.

If there are a number of source areas at various distances within a sector, this analysis is not very revealing. However, if there are no combustion sources in a certain sector this indicates something about the background concentration of sulphate (or nitrate or ammonium) in precipitation, which arises either from natural emissions or from very-long-range transport (greater than 96 h travel time). The sulphate concentration in precipitation is always greater than zero even in unpolluted sectors. For example, Birkenes[17] is a long running site in southern Norway where a significant fraction of the wet deposition is associated with the apparently clean north-western sector and Atlantic trajectories. At

Eskdalemuir in southern Scotland the sector with the greatest fraction of wet deposition is associated with westerly winds made up from a large number of heavy rain events containing rather lower concentrations of sulphate. At Valentia in south-western Ireland, there is significant wet deposition associated with all the sectors covering Atlantic trajectories. In all three cases trajectories associated with sectors containing major source areas of Central Europe and the British Isles make contributions and have particularly high average sulphate concentration in precipitation, but they do not dominate the annual wet deposition of sulphate.

The sector analysis supports one of the major conclusions of the EMEP sulphur and nitrogen models that sulphur and nitrogen of indeterminate origin, arising from the way background levels are specified, make major contributions in western edges of the continent particularly in high rainfall areas. In particular, 50% of the sulphur budget of Norway is predicted to arise from sulphur of indeterminate origin. If important, background sulphur would contribute towards the fluxes across the boundaries of the grid domain required as data to run the RADM. So far there has been insufficient evaluation of RADM results to establish if it could be important. Since RADM is essentially a model which describes episodes rather than one which directly produces sulphur budgets it may be difficult to resolve these questions. Considerable information on the flux of material carried at different levels in westerly winds out across the Atlantic from north-eastern North American sources has been obtained from the Western Atlantic Ocean Experiment (WATOX).[18]

4.8.5 Natural emissions

The quantification and attribution to sources of this background sulphur therefore remains one of the major unresolved problems in our understanding of the sulphur cycle over Europe. One explanation could be the widespread release of natural sulphur compounds, which, though of low emission rate per unit area, could amount to substantial emissions on a global scale. Recently, most interest has focused on dimethyl sulphide (DMS).[19] It is produced primarily in seawater as a consequence of enzyme reactions involving phytoplankton and zooplankton, particularly during phytoplankton 'bloom' periods. In the atmosphere oxidation of DMS occurs by various mechanisms and it has been estimated that emissions from the North Sea could contribute as much as 30% of the atmospheric sulphur acidity over southern Scandinavia during the late spring and early summer phytoplankton blooms.[20] The unpredictability of the time and location of such blooms makes confirmation of the model calculations difficult. Such natural emissions have not therefore been included in the EMEP model. If they do exist they would be represented, albeit approximately, by the empirical background concentrations used.

The RADM presently includes only man-made sulphur emissions which can be justified in terms of the dominance of industrial sources, since the model's use is mainly directed towards simulating pollution episodes caused by sources within the model domain. Work on natural emissions is directed elsewhere towards hydrocarbon emissions from vegetation and NO emissions from soil.

4.8.6 Very-long-range transport

The other main potential source of background sulphur (or nitrogen) apart from natural emissions is material that has travelled very long distances in the atmosphere, such as from north-eastern North America to Europe.[21] Since wet and dry removal processes control the degree of large-scale transport by removing contaminants from the atmosphere and preventing further transport, very long-range transport is likely to be most effective for contaminants carried above the mixing layer (to avoid dry deposition) in precipitation-free regions.

Some of the processes which can lift material from the mixing layer into the free troposphere have already been discussed. Dynamic processes connected with the formation of precipitation such as convective activity or systematic vertical motion will promote this transfer, so part of the material entering a rain system may pass through it into the free troposphere, while another part is deposited within precipitation. Only more elaborate multi-layer regional transport models, able to resolve meteorological features, such as extra-tropical cyclones over global scales are suitable to address these problems.[22] The RADM has been used to look at the way precipitation systems promote mixing throughout the troposphere, but has not been run in a form which allows transport over periods of many days and distances of 5000 km or more to be investigated. In the EMEP programme an alternative transport model from the standard EMEP model has been developed.[23] Although major incursions of pollutants from Europe and North America occur over the Arctic, the contribution of North American sources to European deposition has not so far been demonstrated to be of major significance.[24,25]

4.9 CHEMICAL REACTIONS

4.9.1 Gas-phase oxidation

Chemical reactions and transformations of SO_x and NO_x play a vital role and greatly complicate any simple description of the removal of sulphur and nitrogen by wet deposition. In simple terms SO_2 is oxidised to sulphate in non-precipitating conditions. The sulphate usually forms part of a particle or droplet. The gaseous chemical processes thought to be largely responsible in daylight hours for this oxidation are photochemical reactions, promoted by high levels of solar radiation. Sunlight drives the chemistry of the atmosphere breaking key molecules into fragments which then react with other species. The key reactive fragment formed is the hydroxyl radical. This radical initiates the gaseous oxidation of SO_2. It is produced by the photochemical dissociation of reactive compounds and free radical reactions in the atmosphere. Since the only role SO_x in these reactions is to recycle radicals, the gas-phase oxidation process is almost linear and more SO_2 leads to proportionally more sulphate production.

In the EMEP model gas-phase oxidation is described by a linear transformation rate which is assumed to vary seasonally reflecting higher solar radiation levels in summer and lower ones in winter. Given that the oxidation rate is not the major rate for removing SO_2 from the atmosphere this is quite acceptable. In the RADM[26] a chemical reaction scheme is used which simulates the behaviour of reactive compounds under sunny conditions and therefore includes the behaviour of NO_x, volatile organic compounds and the stable and reactive products, intermediate species and radicals into which they are broken down when subject to solar radiation. The model cannot include every chemical species or reaction which may exist in the atmosphere, which would require thousands of chemical species and reactions. However, the gas-phase module has been designed to reproduce the major features of gas-phase chemistry when tested against smog chamber and more extensive, explicit mechanisms. It uses over 20 representative organic species to simulate approximately and efficiently the full chemical system. The reason for including detailed gas-phase chemistry in the RADM model is not to attempt especially accurate calculations of airborne sulphate formation. Stable oxidising species, such as ozone and hydrogen peroxide, are important products from the cycle of photochemical reactions. They play a major, if not dominant, role in the oxidation of SO_2 in cloud droplets. The detailed gas-phase chemistry is therefore primarily required to provide estimates of the production of these oxidants in clear air.[27] The absorption of these oxidants in cloud droplets may well strongly influence the overall rate of removal of SO_2 in rain systems. The detailed treatment of gas-phase chemistry also automatically takes account of the rather complicated system of NO_x species and reactions.

The EMEP nitrogen model is an attempt to couple the basic EMEP trajectory model with a nitrogen scheme which preserves the main features of the NO_x chemistry. Even so it contains six NO_x species—NO, NO_2, HNO_3 (nitric acid gas), PAN, particulate nitrate (excluding ammonium nitrate), and ammonium nitrate (NH_4NO_3). The oxidising species involved in these reactions, such as OH, CH_3COO_2 and ozone are prescribed functions of latitude, time of year and time of day taken from a tropospheric chemistry model.[28] Any attempt to develop a scheme for calculating these oxidants as the model calculations proceeded would effectively lead to the development of a complex gas-phase module along the lines introduced into the RADM. Complex gas-phase schemes have been introduced into the EMEP model for particular applications.[29]

4.9.2 Oxidation in clouds

Clouds may be present in the mixing layer without precipitation occurring. Such cloud will influence the solar radiation levels in the lowest layer of the atmosphere and therefore the rate at which photochemical reactions take place. The influence of cloud cover on the nitrogen reactions treated in the EMEP nitrogen transport model is included in a simple way. In the RADM a more complete radiation scheme is applied. The treatment shows that reaction rates can be enhanced relative to clear sky conditions.

The other aspect of clouds is that the SO_2 dissolved in cloud droplets may be oxidised to sulphate and assessment of this process is the main difficulty faced by atmospheric transport models. The first problem is describing the clouds in physical (meteorological) terms because of the vast differences in the scale of natural phenomena and scales that are used in the models. Cloud droplets are typically of 10 μm radius, while individual convective clouds (10 km) are smaller than the grid size of the models. Clouds do not represent a fixed reservoir of water in which reactions proceed to a state of chemical and physical equilibrium. Even in low-level stratus cloud droplets are formed dynamically when a moist air parcel rises sufficiently for a droplet to form and may then later evaporate in a descending air parcel. Continual exchange of air and pollutants by mixing takes place within individual clouds. This has the effect of introducing fresh material into the cloud from outside.

For example, hydrogen peroxide (H_2O_2) is thought to be a very effective oxidising species of SO_2 in water droplets. The amount of oxidation of SO_2 caused by H_2O_2 in a cloud may depend upon the amount of H_2O_2 entrained rather than on the rate of reaction of SO_2, as the SO_2 concentration in polluted air is generally in excess of the H_2O_2 concentration. H_2O_2 is generated in photochemical reactions in the gas phase, so that its formation is favoured in daytime fair-weather anticyclonic conditions, rather than in night-time cloudy disturbed cyclonic weather. As a reactive gas, H_2O_2 is thought to be readily deposited on surfaces at the ground. With limited measurements of H_2O_2 concentration, the comprehensive Eulerian model is one way of assessing the degree of coupling between its photochemical cycle and its removal in cloud systems.[26]

Experimental studies of cloud processes are hard to perform. Investigations at Great Dun Fell in the UK of the impingement of artificial SO_2 releases onto a ridge immersed in cloud have enabled controlled field studies to be undertaken. It has been demonstrated that H_2O_2 is rapidly destroyed in cloud with entrainment of outside air being the major replenishment mechanism rather than in-situ production. On occasions the increase in oxidised SO_2 was more than that which could be produced by the available H_2O_2. The extra amount could be accounted for by reaction with ozone.

Ozone is a ubiquitous trace gas in the atmosphere which occurs at concentrations greater than most polluted SO_2 concentrations, subject to regional dispersion, apart from levels in concentrated SO_2 plumes. However, it is much less soluble in water than H_2O_2. Moreover, the overall rate of oxidation of SO_2 in the presence of ozone and water droplets decreases as the acidity of the droplets increases, as inevitably happens as the oxidation process produces sulphuric acid in the droplets. Ammonia will dissolve readily in cloud droplets to form ammonium in solution which reduces the acidity of droplets. The presence of ammonia is likely to promote oxidation of SO_2 by ozone.

SO_2 oxidised to sulphate may be removed from the atmosphere when the droplet is captured and washed out by other droplets during the formation of rain. This is discussed more fully in the next section. More probably, since most cloud is present in the absence of rain, the droplet will re-evaporate and the oxidised SO_2 will remain airborne as a small sub-micrometre size particle containing sulphate. Such particles are deposited more slowly than the equivalent SO_2 and are more likely to form the nucleus for the formation of further cloud droplets and eventually after a number of cycles be removed in the rain.

The role of clouds in the nitrogen cycle has not been regarded as of equal importance as in-cloud sulphur reactions, because it has been generally assumed that gas-phase oxidation of NO_2 under various conditions, is the main oxidation route. Wet deposition of nitrate largely occurs through the subsequent removal of oxidised nitrogen species by precipitation. In the EMEP model clouds are not considered and the average, seasonally dependent oxidation rate of SO_2 is assumed to contain some contribution from in-cloud oxidation.

With a horizontal grid cell size of 80×80 km and three layers typically in the atmospheric boundary layer it would be difficult to specify accurately the amount, or even the fraction, of cloudiness in the lower atmosphere. Ideally, the RADM requires a scheme which would enable the extent of fair-weather cloud to be quantified in terms of readily available meteorological observations, which could be easily implemented. For initial model simulations cloudiness has been related to relative humidity. It would be very difficult to test this part of the model against observations. Fortunately, oxidation in fair-weather cloud may not be a strong influence on the concentrations of sulphate and nitrate in precipitation. Chemical processes in cloud during the formation of precipitation are vital to a proper understanding of the composition of rain.

4.10 WET DEPOSITION

In simple models and in the EMEP model, the removal of sulphur species (and of nitrogen species that are scavenged by rain) is dealt with in a straightforward manner. It is assumed that when significant rain is present the air parcels are depleted of airborne material, the rate of depletion being the same throughout the depth of the polluted layer. In the EMEP model, and some others, the rate of depletion is directly proportional to the (average) rate of precipitation. Material removed from the atmosphere is assumed to have been transferred to the precipitation. If precipitation falls on a receptor site, the wet deposited material is added to that deposited during earlier precipitation events to produce seasonal or annual totals of wet deposition. The procedure is complicated in the case of EMEP by the requirement to attribute the sulphur or nitrogen wet deposition according to their sources in 28 European countries or to background input, as appropriate.

The simple approach ensures a mass balance between material carried along in the air and material transferred to rain. If the removal rate is chosen appropriately, agreement with the observed average deposition can be obtained. For SO_2 the rate of removal is considerably higher than it would be if it were caused solely by washing out SO_2 dissolved in cloud or rain droplets. It appears very likely that the cloud oxidation processes connected with the photochemical-generated oxidants discussed earlier, play an important role in rain systems.

To consider these, the main physical and dynamical processes associated with rain formation should be described. These are very complex to formulate mathematically because of the complicated dynamics and the interaction between gaseous and liquid (and ice) phases of water. Water droplets grow (or evaporate) at various rates in different parts of the system. The major mechanism for growth to a size sufficient to fall as raindrops is by the interception of small droplets (accretion). No attempt to describe these processes could be made in even the most comprehensive Eulerian model.

However, certain key elements are allowed for in a multi-layer tropospheric model.[30,31] Precipitation events are associated with the ascent of water vapour out of the atmospheric boundary layer and the ascending air parcels carry with them material released in the boundary layer. As a mechanism for redistributing, mixing and diluting pollutants, rain systems can play an important role. During the ascent of moist air, cloud droplets grow by condensation and chemical reactions can take place within the droplets. The time available for the reactions to take place depends on the vertical velocities carrying moist air upwards.

Calculations which couple cloud dynamics with (simple) microphysics of cloud formation together with chemical reactions in the cloud have been made[32,33] and have endeavoured to include two-dimensional flow effects and the main chemical reactions. These methods ensure that a mass balance of water and chemical species is maintained along realistic airflow, cloudflow and rainflow lines in the precipitation system. The cloud physics and chemistry are closely linked. The rate of condensation of water vapour onto cloud droplets is related to updraught velocities which determine the time available for chemical reactions to take place. It is not clear that the simplified treatment of wet deposition currently used in comprehensive models takes all these interactions into account.

However, these approaches are not directly transferable to the RADM, which works with grid size dimensions too coarse to stimulate either frontal or convective rain. The approach taken in RADM is to use concentrations from RADM as the initial concentration to feed into a scavenging module. This module then allows concentrations to change along the vertical direction. The detailed treatment depends on the type of cloud, which may cover a variable fraction of each grid cell. Air is exchanged between layers in a way consistent with broad-scale meteorological features, such as precipitation rate and vertical profiles of temperature and moisture. The chemical changes produced by aqueous chemical reactions in the module are then fed back into the basic RADM in terms of grid averaged concentrations.[34]

Though an improvement on single-layer models, the way that dissolved gases and captured particles are removed from the atmosphere by precipitation in RADM cannot be considered to be a simulation of the process, even if it is interpreted as a kind of grid average. Detailed comparison of the predictions of concentrations in precipitation with observations is necessary to evaluate the procedure, and although this is under way at the present time it has not been completed.

4.11 CONCLUSIONS

The purpose of improving atmospheric transport models is to be able to perform more accurate assessments of the contribution of sources to the concentration or deposition at receptor sites, and to be able to extrapolate with confidence from existing conditions, on which the performance of the model can be tested, to other situations in which realistic changes in the present-day emissions of primary pollutants are considered. With this goal in mind, one can review the preceding sections and identify the main processes that should be included in comprehensive models if one goes beyond what can be achieved with simple models. Reservations about the use of simple models are implicit in the assumptions used to derive them and the simplicity of the models allows these to be clearly seen. With more complex models it is not so easy to see what assumptions are being made, so that they need to be comprehensible, as well as comprehensive.

Although comprehensive models go far beyond what has been attempted with simple models they are still limited in the following ways.

(1) To undertake calculations large datasets describing meteorological conditions, rates of emissions of numerous chemical species and information concerning geophysical characteristics of the region of study must be specified. As well as being demanding of resources, those data inevitably contain errors which will propagate through each module of the model, so that the predictions of the model will

contain some inaccuracies. There is no proven method ensuring that errors are contained. Only by evalution of the predictions against the observations can systematic biases be avoided.

(2) Comprehensive models effectively consist of a set of 20 or so coupled differential equations from which average concentrations in grid cells of 80×80 km by several hundred metres are calculated. Computing considerations limit the number and size of grid cells and the duration for which the model can operate. Even with perfectly accurate datasets the computing restraints, in one way or another, introduce errors arising from the imprecise specification of the following processes:

(i) the artificial conditions at the start of a calculation;
(ii) boundary conditions at the edges of the region studied;
(iii) fine-scale variations in dry deposition within grid cells, plus cloud deposition, and fair-weather cloudiness within boundary layers;
(iv) the coupled physical and chemical processes involved in wet deposition; and
(v) numerical errors arising in the computation.

In comprehensive models it is not possible to describe accurately plumes of width much less than 80 km, and this introduces errors in the chemical processes. There is a possibility that errors would multiply and lead to unrealistic predictions. The early tests of the comprehensive models indicate that this does not happen and suggest that these errors are bounded. One must therefore be optimistic about the potential of the comprehensive model, as an assessment tool for investigating responses of the deposition and concentrations to changes in emission. The reservations must be that the models have not demonstrated high accuracy when calculating concentration in precipitation and cannot be expected to. Furthermore, there are practical constraints on running the models, which even now limit the usefulness of the model. Attempts to overcome the five process errors mentioned will only exacerbate this problem. Computationally efficient methods would be required.

It may therefore be argued that the two kinds of models reviewed in this chapter represent alternative compromises in reaching effective assessment methods beyond that possible by simple models. One is with us now and has been used extensively. The comprehensive model shows potential, but may be limited in its practical usefulness by its heavy demand on data and computing resources. However, the comprehensive model,

RADM, has been applied in the US National Acid Precipitation Assessment Program, the results of which were published in September 1990 and it will be used in further regulatory studies. In summary, it has been used to draw the following general conclusions (Mahoney, J. R., 1990, pers. comm). 'Changes in emissions are expected to cause slightly less than proportional changes in average deposition levels. Regional differences in emission changes are important. In general, the influence of emission controls decrease as distance increases away from the control region.' Such conclusions are a useful confirmation of on the conclusions that have been drawn from simpler modelling techniques.

REFERENCES

1. Fisher, B. E. A., A review of the processes and models of long-range transport of air pollutants. *Atmos. Environ.*, **17** (1983) 1865–80.
2. Clark, T. L., Voldner, E. C., Dennis, R. L., Seilkop, S. K., Alvo, M. & Olson, M. P., The evaluation of long-term sulphur deposition models. *Atmos. Environ.*, **23** (1989) 2267–88.
3. Klug, W. & Lüpkes, C., *A Comparison between Long-Term Interregional Air Pollution Models*. Institut für Meteorologie, Technische Hochschule, Darmstadt, Germany, 1985.
4. Chang, J. S., Brost, R. A., Isaksen, I. S. A., Madronich, S., Middleton, P., Stockwell, W. R. & Walcek, C. J., A three-dimensional Eulerian acid deposition model: physical concepts and formulation. *J. Geophys. Res.*, **92** (1987) 14,681–700.
5. Venkatram, A., Karamchandani, P. K. & Misra, P. K., Testing a comprehensive acid deposition model. *Atmos. Environ.*, **22** (1988) 737–47.
6. Buijsman, E., Maas, J. M. & Asman, W. A. H., Anthropogenic NH_3 emissions in Europe. *Atmos. Environ.*, **21** (1987) 1009–22.
7. Asman, W. A. H. & Janssen, A., A long-range transport model for ammonia and ammonium for Europe. *Atmos. Environ.*, **21** (1987) 2099.
8. Eliassen, A., The OECD study of long-range transport of air pollutants: long range transport modelling. *Atmos. Environ.*, **12** (1978) 479–87.
9. Eliassen, A., Aspects of Lagrangian modelling. In *Air Pollution Modelling and its Application III,* ed. C. De Wispeldere. Plenum, New York, USA, 1984, pp. 3–20.
10. Iversen, T., Saltbones, J., Sandnes, H., Eliassen, A. & Hov, O., *Airborne Transbandary Transport of Sulphur and Nitrogen over Europe: Model Descriptions and Calculations*, Norwegian Meteorological Institute Report EMEP MSC-W 2/89. DNMI, Blindern, Norway, 1989.
11. Sykes, R. I. & Hatton, L., Computation of horizontal trajectories based on the surface geostrophic wind. *Atmos. Environ.*, **10** (1976) 925–34.
12. Anthes, R. A. & Warner, T. T., Development of hydrodynamic models suitable for air pollution and other mesometeorological studies. *Monthly Weather Rev.*, **106** (1978) 1045–78.

13. Dollard, G. J. & Unsworth, M. H., Field measurements of turbulent fluxes of wind-driven fog drops to a grass surface. *Atmos. Environ.*, **17** (1983) 775–80.
14. Dollard, G. J., Unsworth, M. H. & Marve, M. J., Pollutant transfer in upland regions by occult precipitation. *Nature,* **302** (1983) 241–3.
15. Hill, T. A. & Choularton T. W., A model of sulphate production in a cap cloud and subsequent turbulent deposition onto the hill surface. *Atmos. Environ.,* **20** (1986) 1753–71.
16. Fowler, D., Transfer to terrestrial surfaces. *Phil. Trans. R. Soc.* (Lond)., **B305** (1984) 281–97.
17. Schaug, J., Hanssen, J. E., Nordop, K., Ottar, B. & Pacyna, J. M., *Summary Report from the Chemical Co-ordinating Centre for the Third Phase of EMEP*, Norwegian Institute for Air Research Report EMEP-CCC 3/87. DNMI, Blindern, Norway, 1987.
18. Gallaway, J. N. & Whelpdale, D. M., WATOX-86 Overview and Western North Atlantic Ocean S and N atmospheric budget. *Global Biogeochem. Cycles,* **1** (1987) 261–81.
19. Turner, S. M., Malin, G. & Liss, P. S., Dimethyl sulphide and dimethyl sulphonisopropionate in European coastal and shelfwaters. In *Biogenic Sulfur in the Environment,* American Chemical Society Symposium Series, 393, ed. E. S. Saltzman & W. J. Cooper. ACS, Washington, USA, 1989, 183–200.
20. Fletcher, I. S., North Sea dimethyl sulphide emissions as a source of background sulphate over Scandinavia: a model. In *Biogenic Sulfur in the Environment*, ACS Symposium Series 393, ed. E. S. Saltzman & W. J. Cooper. ACS, 1989, pp.489–50.
21. Levy, H., The long-range transport of sulphur and nitrogen compounds. In *The Long-Range Transport of Natural and Contaminant Substances*, ed. A. H. Knap. Kluwer, Dordrecht, The Netherlands, 1989, pp. 231–57.
22. Levy, H. & Moxim, W. J., Examining the global impact of local/regional air pollution: the role of global chemical transport models. In *Air Polution Modelling and Its Application VII*, ed. H. van Dop. Plenum, New York, USA, 1989, pp. 139–55.
23. Iversen, T., Tarrason, L. & Eliassen, A., *Transatlantic Sulphur Transport: Preliminary Estimates from a Three-Dimensional Eulerian Model*, Norwegian Meteorological Institute Report EMEP/MSC-W 1/89. DNMI, Blindern, Norway, 1989.
24. Whelpdale, D. M., Eliassen, A., Galloway, J. N., Dovland, H. & Miller, J. M., The transatlantic transport of sulphur. *Tellus,* **40B** (1988) 1–15.
25. Fisher, B. E. A., Transfer of sulphur between continents via the mid-latitude free troposphere. In *Air Pollution Modelling and its Application VII*, ed. H. van Dop. Plenum, New York, USA, 1989, pp. 195–204.
26. Chang, J.S., The Regional Acid Deposition Model and Engineering Model. National Acid Precipitation Assessment Program Report SOS/T4, US Government Printing Office, Washington, 1990.
27. Karamchandani, P. K., Kuntasal, G. & Venkatram, A., Results from a comprehensive acid deposition model: predictions of regional ozone concentration. In *Air Pollution Modelling and its Application VII*, ed. H van Dop. Plenum, New York, USA, 1986, pp. 321–30.

28. Isaksen, I. S. A. & Hov, O., Calculations of trends in the tropospheric concentration of O_3, OH, CO, CH_4 and NO_x. *Tellus,* **39B** (1987) 271–85.
29. Eliassen, A., Hov, O., Isaksen, I., Saltbones, J. & Stordal, F., A Lagrangian long-range transport model with atmospheric boundary layer chemistry. *J. Appl. Meteorol.,* **21** (1982) 1645–61.
30. Walcek, C. J. & Taylor, G. R., A theoretical method for computing vertical distributions of acidity and sulphate production within cumulus clouds. *J. Atmos. Sci.,* **43** (1986) 339–55.
31. Walcek, J. & Berkowitz, C., Modelling of wet scavenging and subgrid-scale vertical transport by clouds in a tropospheric chemical model. In *Air Pollution Modelling and its Application VII*, ed. H. van Dop. Plenum, New York, USA, 1988, pp. 267–75.
32. Fisher, B. E. A., The transport and removal of sulphur dioxide in a rain system. *Atmos. Environ.,* **16** (1982) 775–83.
33. Hales, J. M., A generalised multi dimensional model for precipitation scavenging and atmospheric chemistry. *Atmos. Environ.,* **23** (1989) 2017–32.
34. Middleton, P., Chang, J. S., Corral, J. C. del, Geiss, H. & Rosinski, J. M., Comparison of RADM and OSCAR precipitation chemistry data. *Atmos. Environ.,* **22** (1988) 1195–1208.
35. Dennis, R. L., Evaluation of regional acidic deposition models. In *Acid Deposition: State of Science and Technology*, ed. P. M. Irving. National Acid Precipitation Program, Government Printing Office, Washington, 1991, pp. 57–63.

Chapter 5

Deposition of Atmospheric Acidity

BRUCE B. HICKS
Atmospheric Turbulence and Diffusion Division, NOAA/ARL, PO Box 2456, Oak Ridge, TN 37831, USA

5.1 INTRODUCTION

5.1.1 The forms of acid deposition

A 'wet deposition collection bucket' is an easily understood experimental device that is designed to collect precipitation in a way that does not jeopardize the chemical integrity of the collected sample. Moreover, samples of precipitation can be collected with assurance that the presence of the collection vessel is not, in itself, interfering with the deposition that it is intended to measure. This is not the case for dry deposition, a much more gradual and largely diffusive flux of pollutants from the air, which occurs all the time. This difference has great repercussions that limit the understanding of acid deposition in general. In practice, the knowledge of wet deposition is presently being refined, but an appreciation of the complexities of dry deposition is only just starting. To a considerable extent, it is the measurement difficulty that is limiting investigations of dry deposition.

The distinction between wet and dry deposition is somewhat pragmatic. The most common form of wet deposition sampler is arranged to protect the collection vessel from dry deposition when it is not raining. The use of a 'dry bucket' that is open only when it is not precipitating is an obvious extension of this overall approach. Hence, a convenient separation exists between wet deposition (that which is collected by a wet-only bucket) and dry deposition. As the science has developed, the definition of dry deposition has been refined. Dry deposition from the atmosphere is more than just the collection in a dry bucket: as currently defined, dry deposition is the gravitational settling of large particles not associated with falling precipitation, and the turbulent exchange of small particles and trace gases to the surface.

The factors that govern wet deposition are the occurrence of precipitation and the presence of pollution in air, primarily at cloud altitude. For dry deposition, the critical factors are the concentrations in air very near the surface and the ability of the surface to 'capture' pollutants that come in contact with it. Concentrations in air are a common factor; no form of deposition can occur unless the pollutants are available in the air. The focus of much recent research has been on the development of methods for quantifying deposition when air concentrations are known. The natural variability of the processes that cause both wet and dry deposition is such that there can be no single 'constant of proportionality' that relates either wet or dry deposition rates to air concentrations, yet such concepts pervade much of the literature (and indeed gain substantial credibility if used to address long-term averages). In the case of wet deposition, the link with air concentrations is through a scavenging ratio (defined as the ratio of concentrations in precipitation to concentrations of the same substance in air). In the case of dry deposition, the nature of the surface has no strong influence on the rate of deposition of large particles; the flux is determined by air concentrations and the fall velocity—a known function of particle size. For trace gases, the uptake rate may depend on such factors as the wetness of the surface, and the kind of plants that grow there. Although the concept of gravitational settling is not appropriate for gas exchange, a 'deposition velocity' has relevance. Thus, just as scavenging ratios can be used to estimate wet deposition from air-quality data, so dry deposition rates can be estimated if an appropriate deposition velocity (V_d) is known. Then, the flux delivered to the surface by dry deposition can be estimated as $F_d = V_d \times [C]$, where $[C]$ is the concentration in air to which the vegetation is exposed. V_d depends on a wide range of environmental variables, as well as on the characteristics of the pollutant being deposited. V_d may vary from zero to many cm s^{-1}; this will be discussed later.

For high-elevation locations, a third category of deposition (sometimes called 'occult') must be considered; this corresponds to the interception and impaction of cloud and fog droplets.

Table 5.1 summarizes the various chemical species that are of importance in the context of acidic deposition (of both gases and particles).

Table 5.1
The distribution of chemical species of relevance to acid deposition, among wet and dry deposition[a]

	Wet	*Trace gases (dry)*	*Fine particles (dry)*	*Coarse particles (dry)*
Sulfur species	SO_3^{2-} SO_4^{2-}	SO_2	Most SO_4^{2-}	Some SO_4^{2-}
Nitrogen species	NO_3^- NH_4^+	HNO_3 NH_3 NO_x	Most NO_3^- Most NH_4^+	Some NO_3^- Some NH_4^+
Oxidants		O_3 H_2O_2		
Carbon species	Organic acids	Volatile organic compounds	Organic acids	Graphitic materials
Trace metals	Some		Some	Dominant
Crustal materials	Some		Some	Dominant
Oceanic materials	NaCl Some SO_4^{2-}		Some NaCl	Most NaCl

[a]The dry component is partitioned into trace gas, fine particulate, and coarse particulate fractions.

5.1.2 Deposition in remote areas

The common belief that the atmosphere is made up of nitrogen (N_2) and oxygen (O_2), with trace quantities of other gases is rather misleading. In fact, water vapor is present in large quantities in the lower atmosphere (often accounting for more than 2% by volume). After water vapor, the most prevalent gaseous component of air is carbon dioxide (CO_2), present in slowly but steadily increasing quantities, soon to reach about 0·04%.

Even in remote areas, rainfall contains trace quantities of chemicals scavenged from the atmosphere. CO_2 is dissolved, for example, until the weak solution of carbonic acid (H_2CO_3) that results comes to a pH of about 5·6 (after which further dissolving of CO_2 is impaired). The value of pH of 5·6 is sometimes taken to be representative of precipitation acidity in remote areas; however, other chemicals exist that will complicate matters considerably. The dominant chemicals are often organic, leading to concentrations of formic (HCOOH) and acetic acid (CH_3COOH) in precipitation. Nitrogen oxides (NO_x) are generated by biological activity at the surface and by lightning, as well as by combustion processes that occur naturally. NO_x species oxidize in air to form nitrates, which then acidify precipitation. Sulfur dioxide (SO_2) is exuded from surface fissures,

Table 5.2
Examples of differences in the composition of precipitation at different locations—a remote island (Amsterdam Island, Indian Ocean), a remote inland area (Katherine, Northern Territory, Australia), an eastern inland site in North America (Charlottesville, Virginia), and central Sweden[a]

	Amsterdam Island (Indian Ocean)[b]		*Katherine (Australia)*[b]	*Virginia (USA)*[c]	*Central Sweden*[d]
	Total	*Excluding sea salt*			
Ca^{2+}	8·6	—	1·5	3·5	2·5
Mg^{2+}	45·9	0	1·0	2·0	1·0
Na^+	206·5	4·8	3·3	6·0	3·0
K^+	4·4	0	0·8	0·9	0·5
NH_4^+	1·8	1·8	2·8	10·6	10·0
H^+	8·8	8·8	15·5	50·4	26·5
SO_4^{2-}	29·2	4·9	3·0	47·5	30·5
NO_3^-	1·3	1·3	3·6	21·8	13·5
Cl^-	237·7	2·5	6·1	4·2	4·5
$RCOO^-$	2·2	2·2	10·5	10·9	
pH	5·06	5·01		4·30	4·55

[a]Concentrations are volume-weighted mean values, expressed in μeq liter^{-1}. Note that none of the values listed is more than an approximation to precipitation quality. These data refer to different periods, with different sampling protocols, and hence close comparison is not warranted.
[b]Data from Refs 2 and 3.
[c]Data for Charlottesville, Virginia, are as listed in tabulations of Ref. 4; RCOOH data are from Ref. 5.
[d]Data for central Sweden (Bredkalen) from Ref. 1.

and especially from volcanoes; this leads to the presence of sulfates in precipitation. In addition, other sulfur species (hydrogen sulfide (H_2S) and dimethyl sulfide (DMS), primarily) are emitted by biological processes, particularly from wetlands and the ocean. The products of atmospheric reactions involving all of these substances are potentially acidifying, and add to the natural background acidity of precipitation in remote areas that is associated with CO_2 and the incorporation of organic acids. If no neutralizing agents are scavenged along with the species discussed so far, then the pH of precipitation in remote areas would be expected to be less than 5·6.

In oceanic regimes, sea salt dominates the chemistry of precipitation; it acts as a neutralizing agent (the pH of seawater is about 8·1). Results obtained in a coastal transact of wet deposition in Sweden[1] verify the expectation that the wet deposition of sea salt is greatest at sea and diminishes rapidly with inland distance. For other trace chemicals, the concentrations found in rainfall collected at offshore locations were much the same as would be expected by linear extrapolation of concentrations measured on land; however, because the amount of precipitation is much less at sea (by about 30%), the wet deposition of trace chemicals is less.

Table 5.2 lists the average composition of precipitation collected on a remote island (Amsterdam Island) in the southern Indian Ocean. After removal of the contribution of sea salt, the remaining chemistry is dominated by sulfuric (H_2SO_4), nitric (HNO_3), organic acids (HCOOH and CH_3COOH), and H_2CO_3, in that order. The volume-weighted mean pH was 5·06, considerably less than the value associated with absorption of CO_2 alone. Thus, acid rain is not a symptom of industrialization. Inspection of the remote site data of Table 5.2 reveals several important characteristics.

(1) The chemistry of precipitation at sea is dominated by sea salt, but there is little net effect on the acidity.
(2) The pH at remote sites can be substantially less than the value of 5·6 that is anticipated by consideration of dissolved CO_2 alone.
(3) The contribution to the total hydrogen ion (H^+) concentration by organic acids is largest for the inland remote location, where these acids could account for 60–70% of the H^+ concentration.

5.1.3 Deposition in polluted areas

Table 5.2 includes details of the chemical composition of precipitation, derived from tabulations of the US National Atmospheric Deposition Program for Charlottesville, Virginia, and for three locations at a similar distance from the coast, in central Sweden. The data show the anticipated influence of industrial sources.

(1) The inland areas that are not far removed from sources display sulfate (SO_4^{2-}), nitrate (NO_3^-), and ammonium ion (NH_4^+) concentrations that are an order of magnitude greater than observed at the remote sites.
(2) Base cation concentrations are also elevated. The effect on total acidity is the consequence of interactions involving all of these ionic species, and hence is not easy to generalize.

High-quality network data on wet acid deposition are now becoming available. Early information (i.e. before the 1970s) was of questionable quality, since bulk deposition collectors were then favored; these permit contamination of the sample by dry depositing materials that are of local origin such as twigs, insects, and soil particles. In addition, early measurements of pH are often questioned; pH is difficult to measure because of the low ionic strengths of precipitation samples.

To minimize the effects of contamination, it is now accepted that precipitation samples should be collected using wet-only collectors that remain closed until precipitation is detected, coupled with quick analysis. Research networks, which require the highest quality data, therefore normally report 'event' data, collected as soon as possible after the end of each period of uninterrupted precipitation. For practical reasons, most 'event' networks have adopted a daily sampling protocol. On the other hand, routine monitoring studies have mostly adopted a longer sampling period, often 1 week.

Figure 5.1 presents two alternative depictions of the pattern of H^+ deposition for the USA, derived by combining the daily observations reported by a number of research networks,[6] and from the weekly data of the national routine monitoring programs of the US and Canada.[7] The diagrams both refer to the same year: 1983. The expected similarities are evident, but it is also clear that considerable differences occur (especially in the more remote areas where strong acids are likely to be less important). However, both analyses show that the area most affected by wet acid deposition in North America is in the Ohio/Pennsylvania/New York region.

Two important points should be made immediately. First, different sampling protocols lead to different interpretations of the pattern of acid deposition. The causes of such differences have been examined at considerable length (e.g. Ref. 8) and will not be discussed further, here. It is

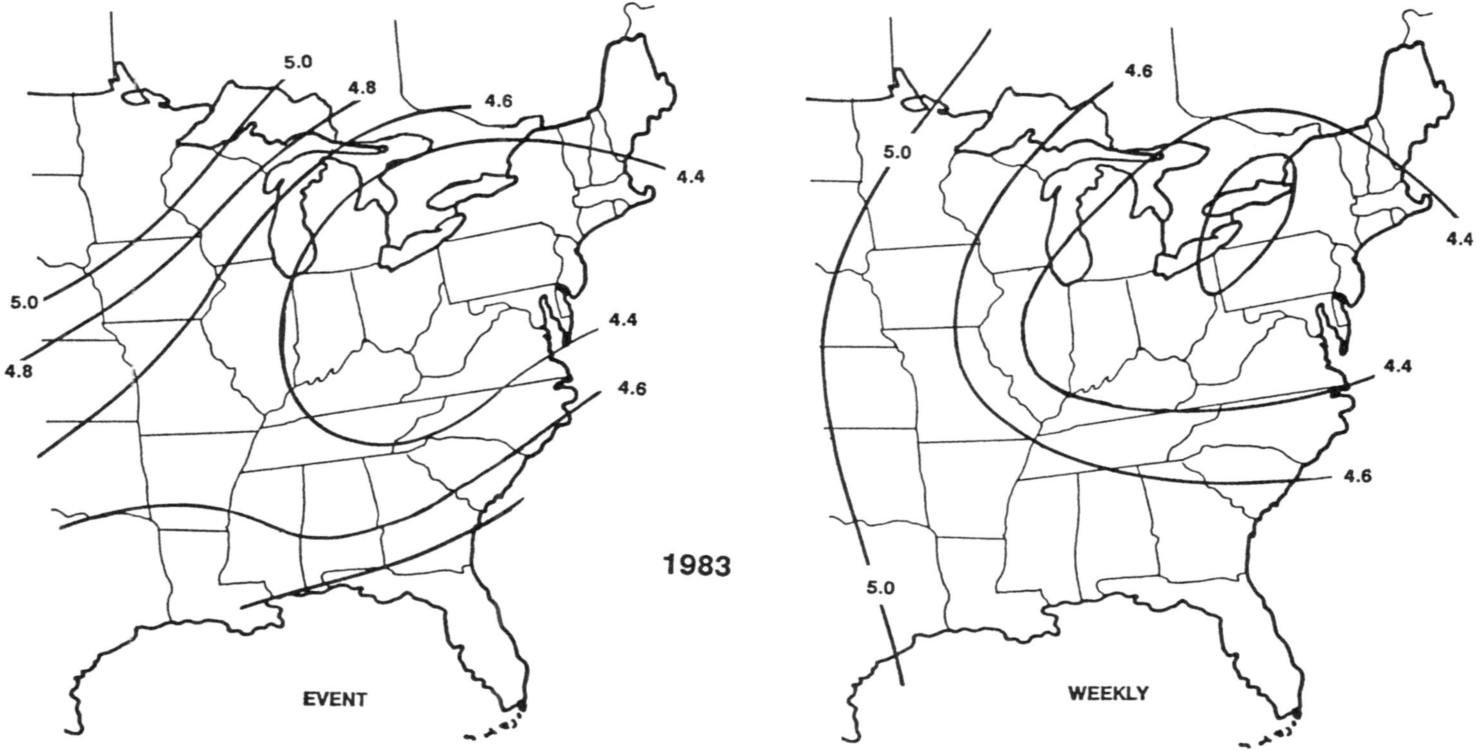

Fig. 5.1. The pattern of acid rain (expressed in pH units) across North America in 1983, as derived from event data (left) and from weekly averages (right).

sufficient to note that to derive high-quality wet deposition data it is necessary to eliminate dry deposition from the sample and to conduct the chemical analysis as soon as possible after sample collection. Secondly, proper depiction of the phenomenon requires that databases cross national boundaries, and therefore that there be close co-operation (and intercomparison) among all sampling networks. In North America, the sampling protocols of the US and Canada differ, and hence it has not been easy to construct a clear picture of the large-scale spatial characteristics of acid deposition.

A major goal of related research has been to demonstrate the cause-and-effect link with emissions of SO_2 from upwind industrial areas. Such links are obviously complicated by meteorological and air chemical considerations, and also by the tacit assumption that acid deposition and SO_2 emissions are intimately linked; Tables 5.1 and 5.2 show that much of the perceived problem is not due to sulfur compounds alone, but also to nitrogen species and sometimes to organic acids. Moreover, representations of deposition such as are shown in Fig. 5.1 represent wet deposition only; a similar plot for dry deposition is not yet possible.

5.2 PROCESSES CONTROLLING DEPOSITION

It is useful to consider a hypothetical situation of a landscape exposed to a mix of pollutants, both gaseous and particulate, some acidic and some not. If precipitation occurs, it will scavenge pollutants from the air through which it falls at rates that depend on the chemical and physical nature of the pollutants, as well as on storm characteristics. In many cases, the precipitation system will draw air from the more polluted mixed layer of the lower atmosphere, and expose this air to relatively efficient in-cloud scavenging mechanisms. Thus, wet deposition depends on both the amount of airborne material available to be scavenged and the efficiency of the precipitation scavenging process. If precipitation does not occur, then particles carried by the air will still slowly deposit as a result of gravity, and trace gases will be taken up by reactions that occur if they come in contact with the surface. The relative importance of these various deposition mechanisms depends on the relative efficiency of the chemical, physical, and biological processes that combine to cause wet and dry deposition.

5.2.1 Dry deposition of gases and diffusing particles

Budget studies have demonstrated that dry deposition of SO_2 delivers about as much sulfur to the surface as does wet deposition of sulfate,[9,10] but it is also known that in some locations and for some other chemical species dry deposition can greatly exceed wet, and vice versa. The uncertainty regarding dry deposition is therefore viewed as a major problem confronting all attempts to quantify acid deposition. Dry deposition rates are intimately linked to air concentrations. A central problem that is the cause for much of the uncertainty about dry deposition is the relationship between air concentrations (C) and the flux (F_d) from the air to the surface over which the air passes. As mentioned above, the relationship between C and F_d defines the deposition velocity, $V_d = F/C_a$, a highly variable quantity that is the subject of most contemporary research on dry deposition.

Among the mechanisms contributing to dry deposition are gravitational settling, turbulent exchange, Brownian and molecular diffusion, interception and impaction, and biological mechanisms that influence stomatal aperture. It is frequently not possible to specify, in advance or with confidence, which of the many contributing processes will be the most important cause of dry deposition from the atmosphere in a given set of circumstances. However, two extreme cases can be described with confidence.

(1) Large particles deposit under the influence of gravity. These large particles have sufficient inertia so that they do not respond to the rapid changes in air velocity associated with atmospheric turbulence. They fall at a terminal speed at which the acceleration force (gravity) is counterbalanced by frictional retardation (drag). This terminal speed is thus a strong function of the size and mass of the particle, and also depends on the viscosity of the air (and therefore on altitude) and on the shape of the particle; spherical and streamlined particles such as raindrops fall more quickly than particles with complicated shapes.

(2) At the opposite extreme, gas molecules and very small particles are not significantly affected by gravitational settling. Moreover, these trace gases and minute particles have sufficiently low mass (and therefore low inertia) that they follow air motions faithfully, and are therefore transported towards the surface mainly by turbulent diffusion. After delivery to the vicinity of receptor surfaces, they can then be diffused to the surface and captured.

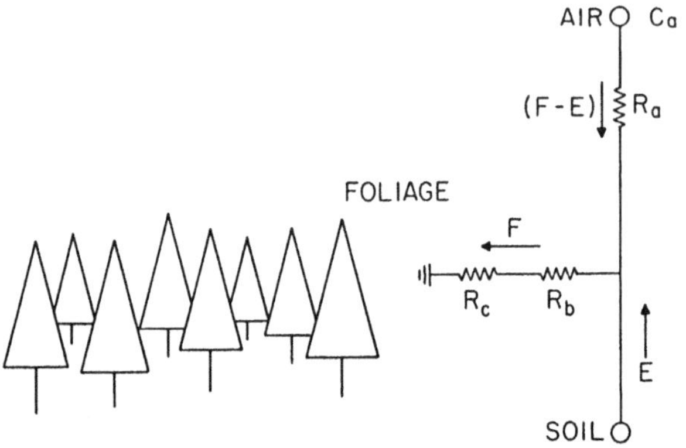

Fig. 5.2. The resistance analogy to dry deposition, showing the roles of the aerodynamic resistance (R_a), the canopy resistance (R_c), and the intermediate boundary layer resistance (R_b). Note that a flux from the surface (E) is shown to emphasize a limitation of the standard resistance model: it works best when the flux is always downwards.

The intermediate case of small particles that are sufficiently large to display significant inertia effects while being too small to be dominated by gravity presents an especially intriguing scientific problem.

Discussion of how the many potential important processes affect dry deposition is simplified by considering a conceptual model of the contributing mechanisms. Figure 5.2 illustrates the most common model; the diagram closely resembles an electrical circuit, in which the flux of material from the atmosphere is likened to an electrical current, and a series of resistances represent the roles of turbulent aerodynamic processes (the resistance R_a), surface effects within the canopy (R_c), and intermediate viscous boundary layer factors (R_b). The downwards flux of some specific chemical species from the atmosphere (F) is the dry deposition of interest here, and in general the possibility of emission (E) of this same species will need to be considered.

Table 5.1 lists the trace gases that are of greatest concern. Some of these are highly reactive and soluble species that are removed efficiently upon contact with any surface, such as HNO_3; others are less reactive and poorly soluble species such as nitrogen dioxide (NO_2). The intent of much current research is to identify the processes that control the exchange from the atmosphere to the surface of chemicals such as these, and to formulate the processes that are most important. For trace gases, surface uptake is frequently controlled by the ability of the surface to absorb the specific chemical species. For HNO_3, this is not a major problem; this particular trace gas is so reactive that molecules are collected immediately upon contact with all vegetative (and soil) surfaces that have been measured so far. Less reactive but soluble trace gases tend to be transferred to vegetation through stomata, or through the leaf cuticle when stomata are closed. For snow and ice, the chemical composition of the surface has been shown to be important, and the presence of surface films or breaking waves is known to modify deposition to liquid water surfaces. Thus, the surface deposition of trace gases tends to be controlled by the available concentration and by the ability of the surface to absorb (or capture) the material in question.

For submicrometer particles, the list of potentially contributing factors is much longer than for either trace gases or large particles. Consequently, the greatest uncertainties about dry deposition concern the case of aerosols. There is a paucity of field data, so laboratory studies form the foundation for many of the theoretical models that have been developed. As mentioned above, recent studies have confirmed that ammonium nitrate (NH_4NO_3) is a relatively rare species, which coexists with NH_3 and HNO_3 (both gases) when concentrations of each are adequately high.[11]

Dry deposition of gases and small particles is a daytime phenomenon, because at night the air is sufficiently stratified that turbulent exchange is quite inefficient. Hence, even if a surface will readily take up a pollutant that comes into contact with it, the dry deposition rate at night is minimal because the air cannot deliver pollutants to the surface. This is the reason that it is fundamentally incorrect to extend early point-value estimates of deposition velocity, made during the daytime, as if they applied all the time. A clear example of the error is the frequent misuse of the value of $1\ cm\ s^{-1}$ as a deposition velocity for SO_2, whereas in fact this is only typical of values for a small portion of the day. Figure 5.3 gives an example of the time sequence of dry deposition (of ozone, in this example) obtained in recent studies of eddy fluxes. A strong diurnal cycle is obvious. NO_2 fluxes are expected to behave similarly, since the controlling processes are the same as for ozone (except for the complexity of the chemistry for NO_2).

Air–surface transfer of gaseous, liquid, and solid atmospheric pollutants, in general, involves all of physics, chemistry, and biology.[13,14] Figure 5.4 illustrates the variety of mechanisms contributing to dry deposition. The diagram is not exhaustive, such mechanisms as electrostatic attraction ('electrophoresis', in the jargon of the trade—literally an electrical forcing mechanism) and radiative pressure ('photophoresis') are omitted. These forces are generally small, and their effects are limited to particles. Stefan flow is worthy of passing attention, since its role extends to trace gases. Stefan flow is the net movement of air away from an evaporating

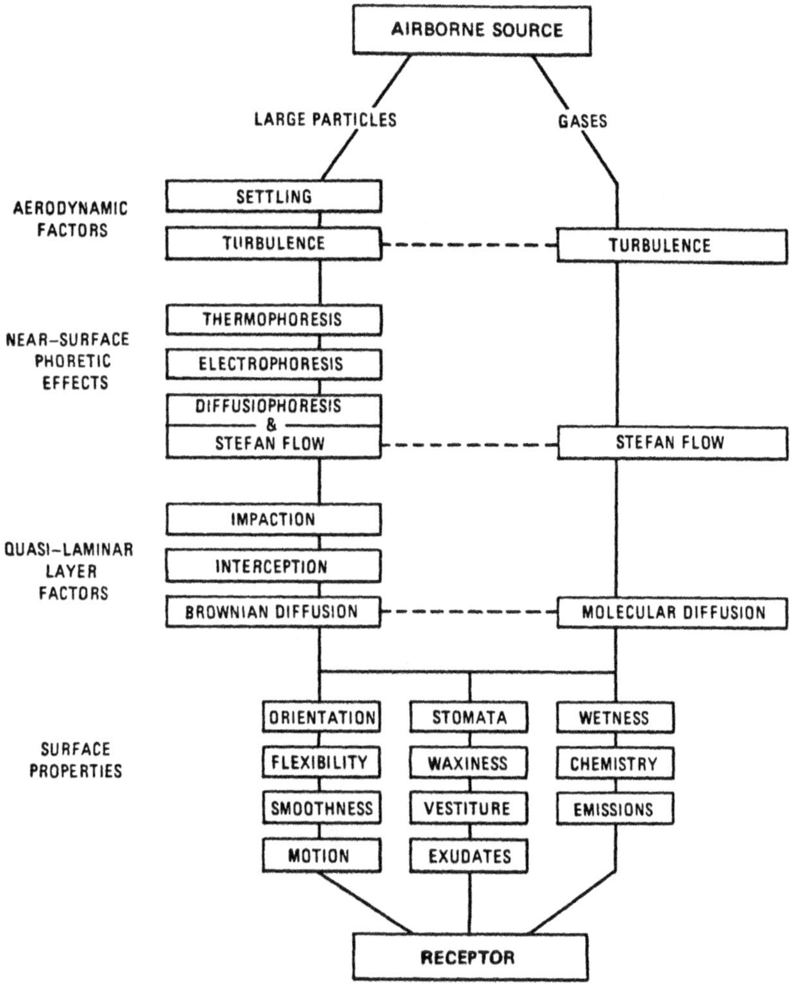

Fig. 5.3. A representation of the sequence of mechanisms influencing the deposition from the atmosphere of trace gases (the right-hand side) and diffusing particles (the left-hand side). Not all processes are identified; in fact, this representation is a simplification. (Adapted from Ref. 12).

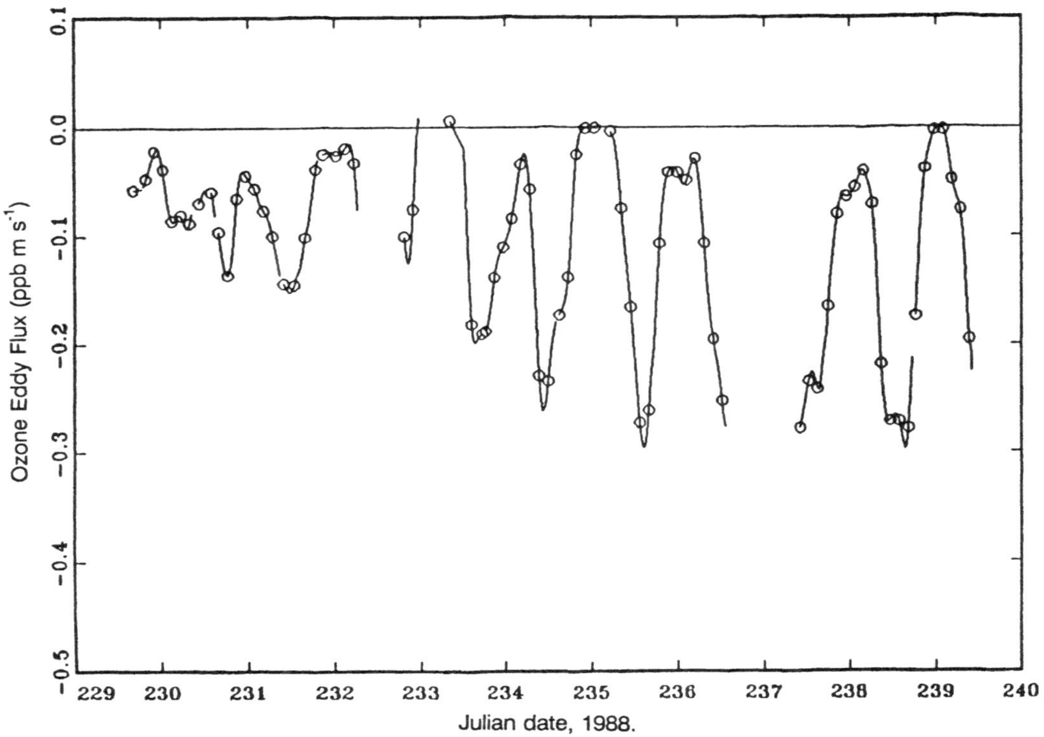

Fig. 5.4. A time sequence of ozone fluxes, measured by eddy correlation during a recent field study conducted in central Pennsylvania. Fluxes towards the surface (deposition) are plotted as negatives, following micrometeorological convention. Ozone is selected for illustration here because its deposition appears to be controlled by similar processes as for NO_2, which is much harder to measure. (Reproduced with permission from R.T. McMilleno.)

surface, due to a displacement of the air by the evaporating molecules. Its consequences are also usually small. Until recently, the poor quality of experimental measurements has not warranted detailed consideration of Stefan flow, but new technology provides measurement precision such that there is now a fairly common need to take Stefan flow into account. Recent work has underlined the importance of several additional complicating factors. In particular, surface fluxes of some gases can be bidirectional. Some plants can be sinks for some trace gases and sources of others, and the roles may be different for other kinds of vegetation growing in different situations.

The chemical affinity of different surfaces for different atmospheric trace species is an underlying complexity. Material surfaces (stone, brick, metals, etc.) react with different trace gases in ways that are strong functions of the chemical and physical nature of their surface, and especially of the existence (or otherwise) of water microlayers resulting from condensation. Trace gases that are soluble in water appear to be efficiently exchanged with plant tissue inside stomata, but other trace gases can be readily absorbed by the cuticle. Highly reactive trace gases, such as HNO_3 may deposit efficiently to whatever natural surface is available.[15,16] A common factor in many such considerations is the presence of water in some form. It is now clear that the availability of water plays a strong part in the dry deposition process. Obviously, the deposition to open expanses of water presents a special case of considerable scientific interest (see Ref. 17). A water body residing in a landscape will generally not receive as much dry deposition per unit areas as its surroundings; the water is smoother and often cooler than the air in daytime. Hence, flow across it will tend to be stratified during that time of the day (midday) when dry deposition would otherwise be greatest. Hence, dry deposition directly to the surfaces of inland lakes does not appear to be a critically important issue.

5.2.2 Dry deposition of gravitationally settling particles

As mentioned earlier, gases such as HNO_3 will deposit to the surfaces of particles in the air, and these particles will then deposit and carry nitrates with them. The relative importance of this mechanism is still difficult to judge. It seems likely, however, that the large-particle component to the total nitrogen deposition will be greatest in areas with the greatest concentrations of HNO_3 and the most large particles. As yet, there is little information on the rate of deposition of nitrates by large particles. It is known, however, that the deposition velocity for gaseous nitrate is more than an order of magnitude greater than that for particles, especially for the accumulation size range particles that offer the greatest surface area in the particle size distribution. Hence, in order for the large-particle contribution to nitrate deposition to equal that of trace gas deposition, it is necessary for concentrations of nitrate associated with particles to exceed HNO_3 concentrations (in units of mass of NO_3 per unit mass of air) by at least an order of magnitude.

5.2.3 Wet deposition

If clean air (i.e. containing no particles) cools, then its relative humidity (RH) will increase until the air becomes saturated (RH = 100%), and soon afterwards droplets will start to form. If there are particles in the air, then these will provide sites on which the water will condense. The particles are 'nuclei' for the droplets that form. Some particles are hygroscopic: they form droplets when the humidity reaches some critical level below 100% RH. A major consideration concerning the impact of deposition of cloud and fog droplets (see below) is that it is the sulfate and nitrate aerosol particles that tend to nucleate first. After these most-available nuclei become droplets, the air continues to cool, water continues to be condensed onto the droplets that are now growing in size, and the concentration of pollutants in these droplets will then tend to be decreased by dilution.

It is useful to distinguish between two extremes of precipitation scavenging and wet deposition processes. On the one hand, large-scale synoptic systems generate precipitation in stable layers aloft, and deposit rain over wide areas as a consequence. The rain droplets are generated in relatively clean air, although perhaps contaminated by pollution from long-range transport. As these droplets fall, they scavenge trace gases and particles from the polluted layers of air below the clouds. As rain proceeds, the mixed layer is gradually cleansed. As it is cleansed, the concentrations of material in precipitation collected at the surface gradually decrease. Thus, the signature of such stable precipitation systems is a gradually decreasing concentration of pollutant in the precipitation collected at the surface, throughout the precipitation period. In this instance, the precipitation normally extends for many hours, and sometimes for days. Such systems are characteristic of regions at high latitudes, where strong convective activity is rare.

At the other extreme, strong convection causes air to rise from the surface, to penetrate through the top of the mixed layer, and to generate clouds from which rain can fall. These convective cells are highly active processors of air from the mixed layer. They inhale air from the mixed layer, scavenge pollutants from it, transfer these pollutants to hydrometeors, and then deposit them at a high rate and with great efficiency. As a rule of thumb, an average convective cell processes approximately six times its own volume of mixed layer air. Concentrations of pollutant in precipitation collected through the lifetime of a typical convective cell are likely not to have the same sort of signature as in the case of stratiform systems. As a simple extreme case, a convective cell might scavenge material at a constant rate through its lifetime, such that the concentration in precipitation collected at the surface might remain constant.

Whereas stratiform systems are primarily high latitude and cold weather phenomena, convective cells occur mainly at low latitudes and in warm weather. Depending upon the geographical location, and the synoptic regime, one can usually judge as to which kind of system dominates. Meteorological records classify conditions accordingly, using an index of convective instability. Consideration of wet deposition therefore links directly with standard meteorological observing routines.

Stratiform systems cleanse material from the mixed layer; the net result is that extended rainfall will tend to dilute material scavenged from the mixed layer, so that an inverse relationship between the concentration in precipitation and the amount of precipitation would be expected. On the other hand, a convective cell could have a constant amount of pollution removed per unit precipitation through the lifetime of an event. Hence, there should be little relationship between concentration in rain and the amount of rainfall for this particular circumstance.

Two different methods have become popular for relating the deposition of some chemical in rain to the concentration of the same chemical in air. In the first, precipitation is supposed to scavenge material from the air through which it falls, at a rate that is proportional to the amount in the air at that time. The process that is imagined is like that of radioactive decay, and a similar form of exponential relationship results, defining a *scavenging rate*. In the second, the average concentration in precipitation (C_r) is assumed to be related to the average concentration in air $[C]$, and a *scavenging ratio* (S_r) is then defined as the constant of proportionality. The flux of material delivered to the surface via wet deposition is then computed as $F_w = S_r \times [C]$. In practice, both the scavenging rate and ratio depend on many different environmental variables, as well as on the chemical and physical form of the pollutant being scavenged.

It is clear that high-altitude clouds have access to pollutants from more distant sources than do convective clouds that ingest air from the mixed layer. It follows that the neutralizing capacities associated with wind-blown surface dust are less available to clouds at high altitude. It is equally clear that the highest concentrations of pollutants in liquid water will exist in clouds (and perhaps in fogs), and that the greatest acidities should be expected in the case in which a mountain-top is exposed to the passage of clouds that are above the layer where neutralizing dust is readily available. The deposition of fog and cloud droplets to mountain-top environments will be discussed later.

In practice, precipitation at any location is rarely either always convective or always stratiform. Instead, the situation is expected to be somewhere in between. Work conducted using radioactivity has provided clear evidence that the relationship between the concentration and precipitation and the amount of precipitation varies widely over these two extremes, with some measurements supporting the theory that convectively dominated situations will show little dependence of concentration in rain on the amount of precipitation, whereas at high latitudes where stratiform systems dominate a strong dependence is found.

Table 5.3 summarizes results obtained in the trial operation of a research network intended primarily to provide data on dry deposition and on the various methods by which it can be computed. Values are presented of representative scavenging ratios for a range of relevant chemical species, and of the slope of the best-fitting relationship of the general kind $S_r = a \times h^b$ (where h is precipitation amount and a and b are fitted by linear regression). The data show that the value of b is typically about 0·4, for the three stations that are considered. These three stations are all located in the inland eastern portion of the US. A broader survey of the dependence of C_r on h was reported recently;[18] the value of the slope b changes as expected, from values indicative of dominance by dilution in areas where stratiform rain is dominant, to values in agreement with those listed in Table 5.3 for areas where strongly convective precipitation is frequent. A general conclusion is inescapable—that great care should be exercised when estimating wet deposition from air concentration data, since the form of the relationship depends on the climatological regime.

5.2.4 Throughfall

It is sometimes important to distingush between deposition from the atmosphere and deposition to the ground. The difference arises, in particular, in the case of a forest. Two different but equally important questions can then be asked concerning (i) the deposition incident to the canopy, and (ii) the deposition incident to the ground surface beneath the

Table 5.3
Sample results relating to scavenging ratios and their dependence on precipitation quantity, derived from operations of the National Oceanic and Atmospheric Administration CORE/satellite network (see Fig. 5.8)[a]

Site	*Species*	*b (One week)*	*1 week* S_r *(for h = 20 mm)*	*Dry/total (annual)*
Oak Ridge, TN				
	Total S	0·37±0·15	310±19%	0·46±0·04
	Total N	0.46±0·11	1470±14%	0·40±0·04
	$[NH_4^+]$	0·00±0·18	140±24%	0·17±0·04
	$[Cl^-]$	0·45±0·13	60300±17%	—
	$[Na^+]$	0·56±0·15	6170±19%	0·04±0·01
	$[Ca^{2+}]$	0·71±0·25	6350±34%	0·09±0·01
	$[Mg^{2+}]$	0·64±0·17	6950±22%	0·06±0·01
	$[K^+]$	0·33±0·14	490±19%	0·16±0·02
Argonne, IL				
	Total S	0·22±0·06	230±8%	0·40±0·03
	Total N	0·25±0·06	845±8%	0·45±0·03
	$[NH_4^+]$	0·28±0·11	500±16%	0·18±0·03
	$[Cl^-]$	0·11±0·09	7820±13%	0·02±0·01
	$[Na^+]$	0·34±0·09	1900±13%	0·07±0·01
	$[Ca^{2+}]$	0·33±0·11	2250±16%	0·07±0·01
	$[Mg^{2+}]$	0·27±0·10	1900±14%	0·07±0·01
	$[K^+]$	0·38±0·11	660±16%	0·17±0·02
State College, PA				
	Total S	−0·11±0·22	230±9%	0·38±0·03
	Total N	0·24±0·12	1340±8%	0·43±0·01
	$[NH_4^+]$	0·09±0·19	290±12%	0·29±0·03
	$[Cl^-]$	0·56±0·20	650±14%	—
	$[Na^+]$	0·49±0·24	605±12%	0·21±0·03
	$[Ca^{2+}]$	0·39±0·14	515±20%	0·29±0·05
	$[Mg^{2+}]$	0·04±0·12	340±18%	0·33±0·04
	$[K^+]$	0·09±0·16	350±12%	0·26±0·04

[a]Precipitation data are derived from operations of the National Atmospheric Deposition Program (NADP) at the same (or very near) locations.

canopy. If the focus is on the first of these two questions, then an appropriate experimental response for the wet deposition component would be to use precipitation collectors erected in clearings or above the canopy. For dry deposition, above-canopy micrometeorological methods could be used, or an assessment made on the basis of air concentration data and an assumption concerning an appropriate deposition velocity. For the case of sulfur and nitrogen species, the dry deposition is then largely due to uptake by foliage of trace gases (primarily SO_2, NO_2 and HNO_3). On the other hand, an experimental answer to the second question might be better obtained using precipitation collectors mounted beneath the canopy. This 'throughfall' is then influenced both by the concentrations of pollutants in the incident precipitation and by the washing of pollutants previously dry deposited to the vegetation. The two deposition quantifications are clearly different; the amount delivered by throughfall to soil beneath a vegetation canopy is not directly and immediately tied to the sum of wet deposition and dry deposition incident to the canopy itself. The major difference is the amount of pollutant that is intercepted and stored in the vegetation, which might be delivered to the ground only when leaves fall or trees die.

Reference 19 reports on a study of the removal of previously dry-deposited sulfur from foliage, by throughfall. Isotopically labeled sulfur was used as a tracer; the results indicate that about half of the SO_2 deposited to foliage is available for removal by throughfall. The remainder is delivered to the surface over a far longer time-scale. The amount of information that is presently available on this subject is very limited, and care should be taken in extrapolating this particular result to circumstances other than that studied by Lindberg and Garten.[19]

Wet deposition is notoriously irregular, both in space and in time. The effects of passage through a canopy make throughfall even more irregular. Reference 20 reports that many more samplers are required to obtain a representative sample of throughfall chemistry for some chemical species (e.g. potassium) than for others (e.g. nitrate).

5.2.5 Cloud and fog droplet deposition

In situations when atmospheric mixing is minimal, radiative cooling can be an effective mechanism by which airborne particles become associated with small droplets and eventually deposit to the surface. Radiative fogs are caused at night, near the surface where nocturnal cooling is the greatest. Radiative heat losses also cause clouds to form high in the atmosphere, but at heights such that precipitation from them is not of concern in the present context. Air that is caught in a strong updraft will cool rapidly as it expands with increasing altitude. This is the convective cloud mechanism discussed above. Air that is forced to rise as one air mass comes into contact with another at a frontal zone will also tend towards becoming saturated, and clouds may then form. Air flowing from a warm region into a cooler environment will cool, and if cooled sufficiently condensation will occur on whatever particles were carried with the inflowing air mass and on the particles that were in the air into which it mixed. The list of mechanisms could continue, but it is adequate here to point out that fogs tend to occur in humid environments, on cool nights, and in low-lying areas. In contrast, clouds generated by more dynamic mechanisms can be intercepted by mountains. Furthermore, the presence of a mountain causes changes in air flow, which can cause orographic clouds to form and hover around the peak.

Large particles in the atmosphere fall according to the law of gravity, with a velocity limited by viscosity. This limiting velocity is that determined by the Stokes–Cunningham relationship. Raindrops and falling hydrometeors of all kinds obey the same laws. The deposition of fog and cloud droplets is therefore of special interest, since these droplets are sufficiently small that their sedimentation rate is slow, whereas the impaction and interception by foliage and other obstacles in their path can be large. At this time, there are limited capabilities to estimate the deposition rates of pollutants and cloud and fog droplets. Models of the processes involved have been developed,[21] but techniques to test the predictions of these models are as yet in their infancy. Fog and cloud droplet deposition occurs primarily at high altitudes, where cloud droplets can be highly enriched in pollutants, and where cloud base is frequently penetrated by mountains. It has been estimated that the deposition of cloud droplets can be a dominant source of base cations and of sulfur and nitrogen compounds at mountain-tops.

The mechanisms of cloud droplet deposition are precisely the interception, impaction, and gravitational settling processes already considered in the context of dry deposition. As in the case of the dry deposition of submicrometer aerosol particles, the lack of a device for routinely measuring the flux from the atmosphere imposes the need to use indirect methods. In most studies conducted so far, the deposition rates have been computed from measurements of the concentration of some chemical species (sulfate, for example) in cloud liquid water, from which the flux to the surface is computed using a deposition velocity model. In this special case, multiple-process models are again appropriate.[21,22] The cloud droplet total deposition velocities predicted by the Lovett model[21] are about 7% of the wind speed. The total deposition velocity is made up of contributions by both impaction (and interception) and sedimentation. At low wind speeds sedimentation is comparable to impaction. The model predicts that impaction is the dominant mechanism throughout the depth of the canopy but that sedimentation becomes increasingly important towards the floor of the forest. In very light winds sedimentation will become more important.

Field verification of such a model is presently lacking. At this time, throughfall methods appear especially promising, and form the foundation for a comprehensive program currently investigating exposure conditions of mountain forests.[23] A summary of the related contributions to total deposition will be found in Ref. 24.

5.3 TIME- AND SPACE-SCALES OF DRY AND WET DEPOSITION

5.3.1 Dry deposition

5.3.1.1 Time variability of dry deposition

Dry deposition is characterized by a strong diurnal cycle, regardless of whether the controlling processes are biological or meteorological. In practice, the diurnal cycle is such that very little deposition occurs at night, no matter what the condition of the surface. This is because the atmosphere becomes stably stratified at night, and transfer across the lowest layers of atmosphere in contact with the surface becomes diminishingly small. A direct repercussion of this is that extension of daytime measurements as if they were representative of long-term averages is most certainly incorrect. Thus, findings of 1·0 cm s^{-1} for the deposition velocity of SO_2 in daytime experiments cannot be extended to long-term averages without taking the diurnal cycle into account. Field experiments conducted so far indicate that a 1·0 cm s^{-1} deposition velocity for SO_2 translates into approximately 0·4 cm s^{-1}, applied to long-term average concentration data.

Overriding the diurnal dominance of the frequency domain of dry deposition is the episodic nature of air pollution outbreaks. Dry deposition rates are intimately linked to air concentrations, which are periodic and couple to synoptic conditions. An air pollution episode is typically several days in duration, and combines with the diurnal variation to yield a net result as illustrated in Fig. 5.3.

5.3.1.2 Space variability of dry deposition

The processes that cause dry deposition are strongly influenced by surface conditions and by meteorological mixing. The resistance to transfer across the air surface interface is largely controlled by biological factors, that are specific to the biological species with which the surface is covered. A change from an area with one biological species to another will cause a change of deposition. Thus, land-use categories have become favored tools for classifying dry deposition aerial variability.

Terrain complexity has an additional role. It is not yet possible to compute the turbulent exchange across the lowest layers of the atmosphere in circumstances in which the transfer mechanisms are dominated by local topographic factors. The presence of a mountain, for example, will modify the large-scale stable stratification that would otherwise limit deposition from occurring at night. Downwind of a mountain, it is quite likely that areas will exist where deposition will continue around the clock, relatively undiminished by the effects of stratification. At this time, such discussion is largely conjectural; capabilities to investigate the matter using experimental techniques have yet to be developed. As an interim measure, it has been proposed that dry deposition rates should be bounded in such circumstances, between limits that correspond to the standard stratified condition on the one hand and to the assumption of rapid transfer through the atmosphere associated with terrain induced turbulence, on the other hand.[25]

Dry deposition is strongly affected by variability in the kind of surface. Edge effects are likely to be important whenever the surface is not spatially homogeneous. The edges of forests, for example, are known to receive much more dry deposition (per unit horizontal area) than uniform areas planted with the same trees. The magnitude of the edge effect can be great. Throughfall studies (e.g. Ref. 26; spruce) have demonstrated that dry deposition rates are greatest to trees at the edges of stands, and decrease (approximately exponentially) to values representative of horizontally uniform stands over a distance of about several tree heights downwind of the edge. The plots presented in Ref. 26 indicate that the edge 'amplification factor' can be as much as a factor of 10, but varies with the chemical species. Comparison with results presented elsewhere leads to the conclusion that the edge amplification factors depend on the architecture of the forest. For example, Ref. 27 reports edge deposition 'amplifications' that are in the range 2–5. It is clear that such results permit a first-order attempt to quantify the consequences of vegetation complexity on dry deposition from the atmosphere. The critical factor appears to be the patchiness of the surface, expressed in terms of the horizontal size of individual stands relative to their height above the surroundings, with an obvious additional role of the transparency of the canopy to the wind.

5.3.2 Wet deposition

5.3.2.1 Time variability of wet deposition

The fundamental difference between stratiform precipitation patterns and convective precipitation has already been emphasized. For a stratiform cloud, a slow exponential decay of concentrations as rain proceeds is expected, in close analogy to behavior underlying radioactive decay. This sub-event variability is observed in some circumstances, but is often complicated by factors that lie outside the normal set of considerations that can be addressed objectively. Figure 5.5 is an example showing a time sequence of concentrations of a few selected ions in precipitation collected during a major field study conducted in central Ohio in 1981 (the OSCAR experiment).[28,29] The variability through the passage of a single precipitating cell is fairly well simulated using a complicated precipitation chemistry model (identified as RSM in the diagram); however, it is evident that the model overestimates the flux of H^+ by a considerable amount. This is likely a consequence of the omission by the model of effects related to base cations, of local origin and therefore not easily considered in a regional numerical model.

Figure 5.5 shows data from a precipitating convective cell; its complexity deserves some attention. It is informative to consider the situation in which a circular precipitation cell traverses across a single observing station. The time sequence observed at this single measurement location corresponds to a cross-section of precipitation from the cell as it passes overhead. In this instance, the record of sub-event precipitation chemistry may well appear as a single-peaked curve, if the station only observes the edge of the storm as it passes overhead. However, if the convective cell passes directly over the observing site, then the record may well have two peaks. If the sampling interval is not short enough, then the appearance may well be one of a maximum concentration at the leading edge, followed by a complicated behavior through the remainder of the storm. The use of a single station to derive information concerning the nature of the cloud is therefore rather tenuously based, since the characteristic features observed by any single station are determined by how a cloud passes across that point.

5.3.2.2 Space variability of wet deposition

The problem of how to interpret the signature of a single event using observations of a single station has caused many researchers to employ networks of sampling stations, with fine time resolution so as to look at the

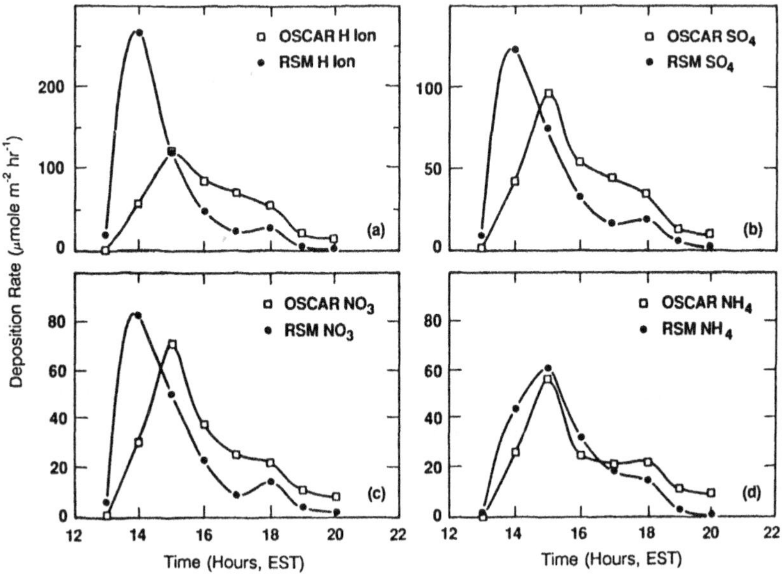

Fig. 5.5. Time cross-sections of precipitation chemistry of a single rain cell, showing a comparison with predictions of the RSM model. The data were obtained in the 1981 OSCAR field experiment (see Refs 28 and 29.)

spatial variability as well. Figure 5.6 is an example of a convective case, also drawn from the OSCAR experiment.[28,29] The deposition footprints left by a single precipitation cell are different for different chemical species. This sub-event variability is such that explanation of its details is beyond contemporary capabilities.

5.3.2.3 Variability of precipitation

The previous sections have focused on the variability of wet deposition of chemical species. In fact, precipitation itself is among the most variable of all natural processes. At this time, it is not possible to predict the amounts of rain that will fall, where it will fall, or when, with confidence. The density of stations used to monitor precipitation for hydrological purposes is a clear indication of a variability in space that must be confronted. In practice, measurement of the amount of water falling over a landscape is exceedingly difficult, cannot yet be predicted, and can be derived only from dense networks of precipitation gages.

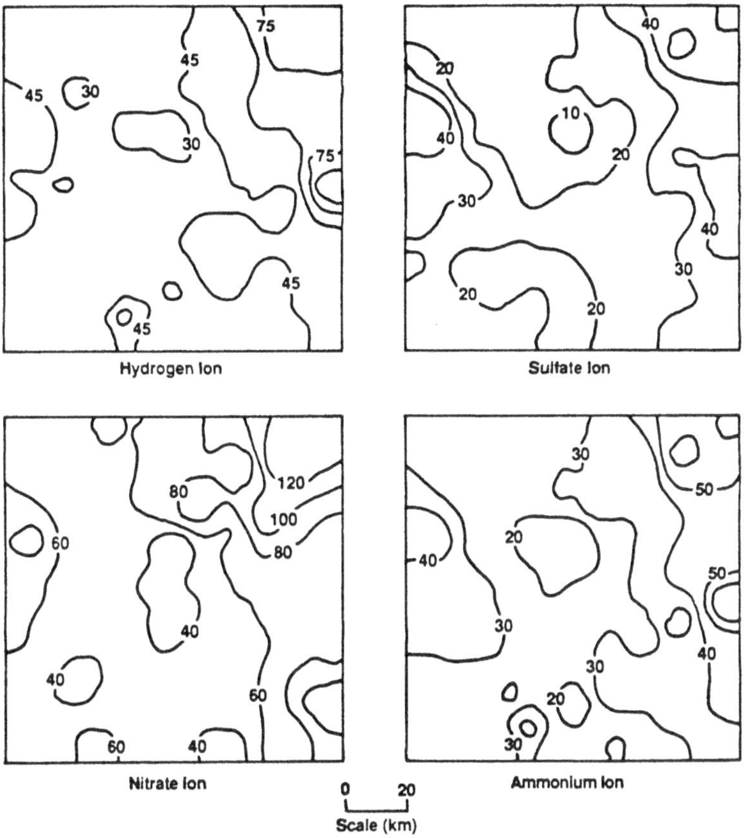

Fig. 5.6. Wet deposition 'footprints' left by the rain cell of Fig. 5.5. Deposition isopleths are in units of μmol m^{-2}.

5.4 TECHNIQUES FOR MEASURING DEPOSITION

5.4.1 Dry deposition

Dry deposition is notoriously difficult to measure directly. A recent workshop addressed the matter of how to measure dry deposition, both for intensive research purposes and for routine monitoring.[30] Table 5.4 summarizes the merits of different classes of measurement methods. For any particular situation there is no 'best' method for all pollutants, nor is there a 'best' method for all situations for any specific pollutant. Instead,

Table 5.4
A summary of the applicability of different methods for quantifying dry deposition[a]

	Surface sampling techniques	*In-air methods*
Applicability	Mainly for particles	Mainly for trace gases
Most direct methods	Snowpack accumulation Surface washing (including throughfall/stemflow)	Eddy correlation
Computational methods	Calibrated watersheds	Gradients, Variances; Similarity Methods
Indirect methods	Surrogate surfaces	Inferential concentration monitoring

Adapted from Ref. 30.
[a]Note that in-air methods are capable of addressing the deposition of small particles, although with considerable uncertainty. Likewise, ongoing research programs are exploring the use of some surface sampling methods for trace gas deposition.

as yet there is a limited range of measurement techniques from which it is only sometimes possible to select an optimum approach.

5.4.1.1 Buckets and surrogate surfaces

Early work using so-called dry buckets has been largely discredited for the chemical species of contemporary interest; the collection efficiency of a bucket or any surrogate collector is substantially different from that of a natural landscape, and hence extending from bucket observations to landscapes presents significant difficulties. Furthermore, analysis of dry bucket data collected by the US National Atmospheric Deposition Program has indicated that every site is affected by biological matter and local dust that are not components of the deposition that it is desired to measure.[12]

Figure 5.7 shows the distribution of contamination indices derived from analysis of NADP data. In this plot, a rating of 8 indicates that a site was affected all the time, by soil particles for the upper part of the figure, and by guano for the lower portion.

5.4.1.2 Surface accumulation methods—throughfall

Other techniques (particularly throughfall methods) are especially suited to the provision of particle deposition data, for which micrometeorologi-

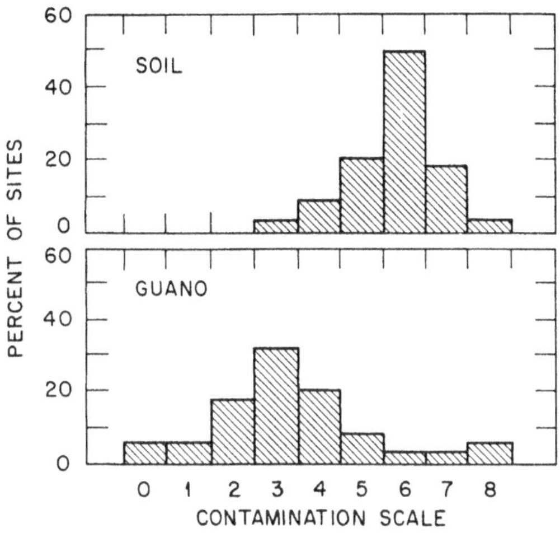

Fig. 5.7. Results of an objective analysis of the contamination of dry buckets in the US NADP network by local soil particles and by bird droppings. An index of 8 indicates that the site in question was always affected.

cal techniques are less suited. At the National Acid Precipitation Assessment Program (NAPAP) workshop on dry deposition, wash-off methods (especially throughfall) were judged to be the best developed of the techniques which measure the accumulation of pollutants at the surface. Other techniques require some special surface for their application, e.g. snowpack accumulation[31] and foliar extraction methods.[32] Techniques based on the use of a calibrated watershed[33,34] were viewed as promising, but appear limited in the ways in which they can be applied. Some new surrogate surfaces also appear promising; they may provide a standardized method for measuring large-particle deposition. Table 5.5 presents a summary of technique characteristics. All of the methods work best for particles, and provide averages over time-scales longer than for the micrometeorological methods that will be discussed later (see Table 5.6).

Special considerations arise when the need to assess the deposition to the ground under a vegetation canopy, such as at the floor of a forest. The amount and chemical composition of precipitation reaching the ground is different from what is measured above the canopy or in a clearing in the forest. The combination of precipitation that falls directly through the canopy without being intercepted and that which is intercepted but drips off the canopy is called throughfall; that which runs down the canopy and

Table 5.5
Characteristics of surface-accumulation methods for measuring dry deposition

Snowpack accumulation	Limited to winter Deposited materials migrate rapidly when snow starts to melt Time resolution typically days or weeks
Surface washing (structures as well as foliage)	Highly variable from surface to surface, and hence many samplings are required Time resolution: days (for artificial washing) to seasons (for some throughfall protocols) Space resolution determined by the size of the sampling array
Calibrated watersheds	Effects of soils often hard to assess Utility depends on the precipitation pattern Time resolution depends on the kind of soil, usually months to years Space resolution determined by the size of the watershed
Surrogate surfaces	Cannot be easily related to natural surfaces, except for coarse particles Time and space resolution depend on the sampler and the sampling protocol

flows down tree trunks is called stemflow. In fully leafed forests, stemflow typically amounts to 10–20% of the throughfall.

The amount of water reaching the soil (i.e. throughfall plus stemflow) may be greatly different from the amount of incident precipitation. Depending on the density of the canopy and on the intensity and duration of the precipitation event, a portion of the precipitation (and sometimes all of it) is intercepted by the canopy. Of this intercepted precipitation, some evaporates during the precipitation event, some remains on the canopy elements (leaves, needles, twigs, branches, etc.), some drips off the elements and falls to the ground, and some runs down the branches, stems, and trunks, eventually also reaching the soil. The chemical composition of water reaching the soil may also be substantially different from the incident precipitation. During the time the precipitation is in contact with canopy elements, an exchange of materials can occur with the biological material of the canopy and with chemicals that were previously deposited to the canopy. There may be some neutralization of the acidic deposition, addition of pollutants that were previously dry deposited to the canopy and washed off by the precipitation, retention of nitrogen compounds in the canopy, and leaching of some chemicals from plant tissue (especially potassium). Intercepted water that remains on the canopy eventually evaporates leaving a residue of non-volatile materials on the canopy. The residual material can then influence stemflow and throughfall in subsequent precipitation events.[35]

For some purposes, it is necessary to quantify the average deposition of specific chemical species to the floor of a forest over short time intervals, typically by rainfall event. In other applications, the purpose is to derive dry deposition information from long-term throughfall measurements. Depending on the purpose, substantially different measurement protocols may be required. Both kinds of study require measurement of incident precipitation, throughfall, and stemflow. If the goal is to determine the input of specific chemical species to the floor of a forest on an event basis, then the spatial irregularity of throughfall becomes a major problem. In the past, workers have used large numbers of simple collectors, or sometimes troughs, to impose the necessary spatial averaging. The number of collectors required is obviously a function of the architecture of the canopy. At this time, the number of collectors to be used can best be determined by statistical analysis of trial data.

On the other hand, if the intent is to derive information on the dry deposition of material to the canopy between precipitation events, then it is clear that a large number of events must be considered. There is then a reduced need for a large number of sub-canopy collectors, and an increased emphasis on long-term continuity of the experimental procedures and observance of standardized measurement procedures. Statistical methods can be used to separate the dry deposition and canopy exchange contributions to throughfall chemistry.[35]

Hence, the specific needs for throughfall data determine the way in which the data must be obtained. In all cases, determining the exchange of materials between the canopy and intercepted water is difficult. In particular, available data[35] indicate that ion mobility within plant tissue governs the rate of canopy exchange for different ions: SO_4^{2-}, K^+, Ca^{2+}, NO_3^-.

5.4.1.3 In-air methods—micrometeorology

Micrometeorological field investigations are intended to identify the processes controlling deposition to natural surfaces, and to provide parameterizations of the relevant processes. Such studies have the unique benefit of

Table 5.6
Characteristics of in-air methods for measuring dry deposition

Eddy correlation	Requires fast-response wind and concentration sensors, typically with time constant less than 1 s Requires intensive data reduction and analysis Present chemicals: SO_2, O_3, NO_2, NO, CO_2, submicrometer particles, SO_4^{2-} Time resolution: 15–60 min Space resolution: depends on the height of measurement; typically 1 km for towers, 10–100 km for aircraft
Gradients	Chemical concentration sensors must resolve differences smaller than 1% Interpretation requires accurate knowledge of atmospheric turbulent diffusion, often not available Present chemicals: SO_2, O_3, CO_2, HNO_3, submicrometer particulate species Time and space resolution: as for eddy correlation
Similarity methods	When based on concentration variance, the sign of the flux is lost When based on gradients, the sensors must resolve concentration differences smaller than 1% Time and space resolution: as for eddy correlation
Inferential methods (including concentration monitoring)	Require the availability of schemes to derive accurate deposition velocities from site data Time resolution: either much less than or much greater than one day Space resolution: typically 1 km

examining the deposition phenomenon in an integrated manner; however, the number of chemical species that can be considered is limited. Table 5.6 summarizes the features of different in-air methods for deriving dry deposition data.

5.4.1.4 Concentration monitoring and inferential approaches

Inferential and concentration monitoring measurements have a special attraction, because they require only that air concentrations be measured and that sufficient supporting meteorological and surface data be obtained to permit an estimate to be made of the relevant deposition velocity. In the concentration monitoring approach, the emphasis of field programs is on measuring air concentrations well, from which assumptions are made about the relevant value of V_d in order to yield an estimate of the deposition rate to the surface. The inferential approach is designed to reduce the uncertainty concerning the estimation of V_d—a sub-network of intensive research stations measures V_d using whatever experimental method is appropriate, and the routine concentration monitoring network is expanded to measure those atmospheric and surface properties that are known to control V_d. These supplementary data are then used to extrapolate values of V_d from locations where intensive measurements are made to locations where simpler observations only are available. The methods involved in applying the inferential method are thoroughly discussed elswhere.[36] Figure 5.8 shows the array of stations deployed acrosss North America to test the nested network philosophy of the inferential approach to dry deposition measurement.

5.4.2 Wet deposition

5.4.2.1 Direct measurement

Abilities to measure wet deposition to a landscape are sometimes perceived to be well developed, but reality may often be different. The discussion above focuses on the difficulties in evaluating dry deposition to a selected area; in the case of wet deposition, the spatial variability of the precipitation process itself imposes difficulties that are difficult to overcome. In general, the capabilities of existing methods are not clear. Since wet deposition varies greatly in both space and time, interpreting data obtained at a limited number of sampling locations is always difficult. It is obvious that a single precipitation-sampling station provides no information on the representativeness of its own data, and it is clear that only two or three sampling points will generate only marginal statistics. In ecosystem-input studies, it seems necessary to consider carefully the limitations imposed by sampling wet deposition at only a few locations. The additional complications associated with throughfall serve to emphasize the need for caution.

5.4.2.2 Coupling precipitation chemistry and quantity data

Measurements made using bulk deposition collection vessels have not been discussed here, but clearly such vessels will be susceptible to precisely the same difficulties that exist with the use of collection devices for measuring small particles and trace gas deposition. Furthermore, bulk devices are normally open vessels, in which precipitation samples are not protected from chemical reactions with airborne trace constituents after collection. Bulk devices should not be used if it is possible to evaluate dry

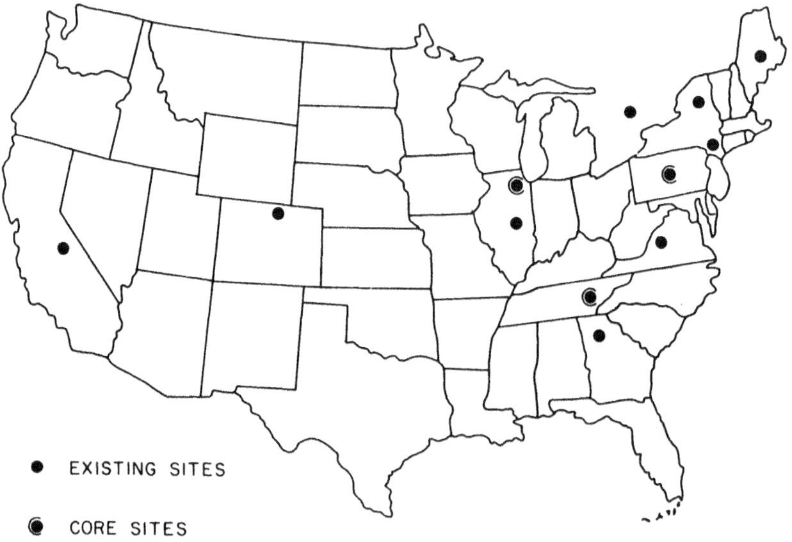

Fig. 5.8. The nested network of sites being operated as part of the NAPAP dry deposition research program. CORE sites are emphasized; these are where intensive measurement techniques are employed to provide reference data for calibrating simpler methods employed elsewhere.

and wet deposition independently, using techniques carefully tailored to satisfy the different requirements of each separate measurement.

5.4.3 A nested network approach

The variability of the precipitation process imposes the conclusion that computation of deposition of trace chemicals to a landscape should not be based solely upon those few precipitation chemistry measurements that are available, but should also take into account the broader base of precipitation quantity data derived from meteorological networks. A nested network operation appears to be optimal. In concept, every station could be reporting quantities related to wet deposition and dry deposition. In practice, the costs involved are usually prohibitive. Consequently, a subset of special stations for more intensive measurement should be selected (also at random) in each quadrant. The procedure that will be discussed is one in which a small sub-network of locations is especially instrumented to conduct intensive studies, supported by a wider network of locations at which simpler measurements are made. The overall philosophies of this nested network operation have been presented elsewhere.[38]

If air concentration data cannot be obtained, then some other method must be employed to come up with the dry deposition data. The use of ratios of dry to wet deposition has already been discussed; it was shown that the variability involved in the dry/wet deposition ratio is such that short-term application is not warranted.[36] However, when constructed over seasonal or annual periods, the dry/wet ratio appears to have far more meaning. In practice, computation of the dry/total ratio appears more attractive, since this quantity is necessarily constrained to lie between zero and unity. Estimates derived in the trial operation of a dry deposition measurement network have been combined with the corresponding wet deposition data to produce the average wet/total ratios that are listed in Table 5.3.

Inspection of the dry/total ratios listed in Table 5.3 reveals that for sulfur and nitrate deposition (expressed as a mass of total sulfur and mass of total nitrogen deposited), the dry/total ratio is in the range 0·4–0·5. This is not the case for the base cations, Ca^{2+}, Mg^{2+}, K^+ and Na^+. It is clear from inspection of the table that the particles behave completely differently from the sulfur and nitrogen species dominated by trace gases. This is not unexpected, since radioactive fallout studies conducted during the 1950s and 1960s indicate approximately a 10:1 ratio of wet to dry deposition, a result which was supported by budget studies as well as by detail to deposition studies to specific surfaces. Thus, the dominance of wet deposition for particle transfer that is expected on the basis of radioactive fallout studies is supported by the field data obtained in the trial network operation that forms the basis for much of the discussion here. However, approximately equal quantities of sulfur and nitrogen are deposited by wet and dry deposition.

Thus, substantially different techniques need to be employed in order to compute or estimate dry deposition components using a dry/wet ratio applied to wet deposition measurements. As a first order of approximation, the wet deposition estimates of nitrogen and sulfur compounds can be doubled in order to derive an estimate of the total deposition. On the other hand, the wet deposition estimates need only be increased by approximately 10% in order to derive total deposition estimates for the trace metals (base cations) described earlier.

5.5 CONCLUSIONS AND RECOMMENDATIONS

To estimate total deposition to a landscape, it is necessary to take dry and wet components into account. Collection of bulk samples should be avoided, because of changes that occur after collection of precipitation and because of the inability to relate the dry component measured by a bulk collector to the dry deposition to the surrounding area. Estimates of dry and wet deposition need to be based upon similar data sources, and need to have similar levels of confidence associated with them. For example, it seems scarcely worthwhile to improve the wet deposition estimate of NO_3^- deposition if the dry deposition estimate for NO_x is crude.

Difficulties arise with both the wet and dry component. For the wet component, the problem is the spatial variability of the precipitation process. For the dry deposition component the problem is the strong diurnal cycle and the dependence upon the vegetation species and its water status. Techniques are presently available to reduce the effects of each of these complexities, but complete confidence in these methods has yet to evolve.

A precipitation network is a key ingredient. Without this component, it is necessary to assume that precipitation measured at a single point is representative of the area. This is known to be an invalid assumption. A better assumption, based upon statistical analysis conducted so far, is to employ a network of simple precipitation gages in order to extend a limited database of precipitation chemistry observations. In this way, the effects of the variance of the precipitation amount can be reduced substantially.

If the precipitation chemistry data are the only chemical data available, then estimates of dry deposition can be based upon them, using a dry/total (or dry/wet) ratio which is computed on the basis of seasonal data. This approach does not yield statistically meaningful information on a short-term basis. In order to derive shorter term information, air concentration data are necessary. If such air concentration information is available, then dry deposition rates can be computed from it provided appropriate measurements are made of meteorological and surface properties that would lead to an evaluation of the deposition velocity for each chemical species concerned.

A minimum set of observations at a target area would involve, therefore, a precipitation network, a sub-network where precipitation chemistry data are obtained, and air-quality data obtained at this sub-network of intensive sites. In addition, measurements should be made of those properties that control the deposition velocity. Using such a database, independent estimates of wet and dry deposition can be obtained, from which total deposition can be computed by simple addition.

Some problems that can arise in field operations concern firstly the lack of a precipitation network, and secondly the application of incorrect assumptions concerning the dry to wet ratio.

(1) If it is necessary to assume that a single measurement of precipitation is representative of an area, then the error resulting is likely to be substantial, and to dominate any estimate of the total deposition. This source of error is simple to eliminate, is inexpensive to address, and is inexcusable to overlook.

(2) It is well known that the deposition rates of sulfur and nitrogen compounds are about the same for wet and dry processes. However, this approximate equality extends only to areas where there is sufficient precipitation, and where the vegetation is such that dry deposition can occur without severe effects of drought or water stress. It is yet unknown how the deposition ratio might differ in a desert environment, in periods of drought, or when conditions are dominated by torrential rains. Moreover, it is inappropriate to extend this equality to other chemical species, especially those whose dry deposition is not dominated by gaseous components. For base cations in particular, wet deposition must be expected to dominate dry.

All of the above depends upon the measurement of some index of local chemistry. This index of local chemistry is either precipitation chemistry, air chemistry, or both. If local chemical data are lacking, then model outputs can perhaps be used. However, the use of such a model to predict deposition to a landscape requires that a complicated model be used to account for local as well as distant sources. In this regard, it would not be adequate to use a long-range transport model alone, nor to rely entirely upon a model that handles local sources. Instead, sources at all ranges must be considered. In particular, nitrogen more than sulfur compounds are substantially influenced by local sources as well as distant. The situation for base cations appears even more locally influenced.

ACKNOWLEDGEMENT

This report is based on work carried out as a component of Task Group

II (Atmospheric Chemistry) of the US National Acid Precipitation Assessment Program.

REFERENCES

1. Granat, L., *A Comparative Study of Precipitation Chemistry at Inland, Coastal and Island Sites in the Bothnian Bay Area*, Report CM-73. Department of Meteorology, University of Stockholm, Sweden, 1988.
2. Galloway, J. N. & Gaudry, A., The composition of precipitation on Amsterdam Island, Indian Ocean. *Atmos. Environ.*, **18** (1984) 2649–56.
3. Likens, G. E., Keene, W. C., Miller, J. M. & Galloway, J. N., Chemistry of precipitation from a remote, terrestrial site in Australia. *J. Geophys. Res.*, **92** (1987) 13299–314.
4. National Atmospheric Deposition Program, *NADP/NTN Annual Data Summary, Precipitation Chemistry in the United States, 1985*. Colorado State University, CO, USA.
5. Keene, W. C. & Galloway, J. N., Organic acidity in precipitation of North America. *Atmos. Environ.*, **18** (1984) 2491–7.
6. Eynon, B. P., Statistical analysis of precipitation chemistry measurements over the eastern United States. Part II: Kriging analysis of regional patterns and trends. *J. Appl. Meteorol.*, **27** (1988) 1334–43.
7. NAPAP, *Annual Report, 1984, of the US National Acid Precipitation Assessment Program*. Washington, DC, USA, 1984.
8. Sisterson, D. L., Wurfel, B. A. & Lesht, B. M., Chemical differences between event and weekly precipitation samples in Northeastern Illinois. *Atmos. Environ.*, **19** (1985) 1453–69.
9. Shannon, J. D., A model of regional long-term average sulfur atmospheric pollution, surface removal, and net horizontal flux. *Atmos. Environ.*, **15** (1981) 689–701.
10. Galloway, J. N., Church, T. M. & Wolff, G. T., The flux of S and N eastward from North America . *Atmos. Environ.*, **18** (1985) 2595–607.
11. Brost, R. A., Delany, A. C. & Huebert, B. J., Numerical modeling of concentrations and fluxes of HNO_3, NH_3, and NH_4NO_3 near the surface. *J. Geophys. Res.*, **93** (1988) 7137–52.
12. Hicks, B. B., The acid deposition phenomenon and its effects, A-7, Dry deposition processes, Critical Assessment Review Papers, Volume 1, Atmospheric Sciences, EPA-600/8-83-016 AFG, 7-1–7–70, USEPA, NC, USA.
13. Sheih, C.-M., Wesely, M. L. & Hicks, B. B., Estimated dry deposition velocities for sulfur over the eastern United States and surrounding regions. *Atmos. Environ.*, **13** (1979) 1361–8.
14. Hicks, B. B. & Matt, D. R., Combining biology, chemistry, and meteorology in modeling and measuring dry deposition. *J. Atmos. Chem.*, **6** (1988) 117–31.
15. Huebert, B. J. & Robert, C. H., The dry deposition of nitric acid to grass. *J. Geophys. Res.*, **D1, 90** (1985) 2085–90.

16. Duyzer, J. H., Bouman, A. M. M., Aalst, R. M. van & Dierderen, H. S. M. A., *Assessment of Dry Deposition Fluxes of NH_3 and NH_4^+ over Natural Terrain,* Report Number P 87/034. Netherlands Division of Technology for Society, TNO, Delft, The Netherlands, 1987.
17. Slinn, W. G. N., Hasse, L., Hicks, B. B., Hogan, A. W., Lal, D., Liss, P. S., Munnich, K. O., Sehmel, G. A. & Vittori, O., Some aspects of the transfer of atmospheric trace constituents past the air–sea interface. *Atmos. Environ.,* **12** (1978) 2055–87.
18. United States Geological Survey, *Distribution and Variability of Precipitation Chemistry in the Conterminous United States, January through December, 1983*, USGS Open File Report 87-558, US Geological Survey, VA, USA.
19. Lindberg, S. E. & Carter, C. T. Jr., Sources of sulphur in forest canopy throughfall. *Nature,* **386** (1989) 148–51.
20. Kostelnik, K. M., Lynch, J. A. & Corbett, E. S., *Assessing Spatial Variability of Throughfall Chemistry Beneath a Mixed Hardwood Forest Canopy*, Report ER 8717. Environmental Resources Research Institute, The Pennsylvania State University, University Park, PA, USA, 1987.
21. Lovett, G. M., Rates and mechanisms of cloud water deposition to a subalpine balsam fir forest. *Atmos. Environ.,* **18** (1984) 361–71.
22. Slinn, W. G. N., Predictions for particle deposition to vegetative canopies. *Atmos. Environ.,* **16** (1982) 1785–94.
23. Mohnen, V. A., *Exposure of Forests to Gaseous Air Pollutants in Clouds*, MCCP Management Center Report 87/1. W. R. Fleming & Associates, Albany, NY, USA, 1987.
24. Lovett, G. M., Reiners, W. A. & Olson, R. K., Cloud droplet deposition in subalpine balsam fir forests: Hydrological and chemical inputs. *Science,* **218** (1982) 1303–4.
25. Hicks, B. B. & Meyers, T. P., Measuring and modeling dry deposition in mountainous areas. In *Acid Deposition at High Elevation Sites*, ed. M. H. Unsworth & D. Fowler. Kluwer, Dordrecht, The Netherlands, pp. 541–52.
26. Beier, C., Atmospheric deposition of N and S compounds in Denmark. PhD thesis, Technical University of Denmark, Denmark, 1988.
27. Hasselrot, B. & Grennfelt, P., Deposition of air pollutants in a wind-exposed forest edge. *Water Air Soil Pollut.,* **34** (1987) 135–43.
28. Easter, R. C., The OSCAR experiment. In *Deposition Both Wet and Dry*, ed. B. B. Hicks. Butterworth, Boston, MA, USA, pp. 97–120.
29. Albritton, D., Fehsenfeld, F., Hicks, B., Miller, J., Liu, S., Hales, J., Shannon, J., Durham, J. & Patrinos, A., Atmospheric processes. In *NAPAP Interim Assessment, Volume III: Atmospheric Processes and Deposition*. US National Acid Precipitation Assessment Program, Washington, DC, USA, 1987, Ch. 4.
30. Hicks, B. B., Wesely, M. L., Lindberg, S. E. & Bromberg, S. M., eds, *Proceedings of the NAPAP Workshop on Dry Deposition*. Harpers Ferry, WV.
31. Dovland, H. & Eliassen, A., Dry deposition on a snow surface. *Atmos. Environ.,* **10** (1978) 783–5.
32. Lindberg, S. E. & Lovett, G. M., Field measurement of particle dry deposition rates to foliage and inert surfaces in a forest. *Environ. Sci. Technol.,* **19,** 238–44.

33. Eaton, J. S., Likens, G. E. & Bormann, F. H., The input of gaseous and particulate sulfur to a forest ecosystem. *Tellus,* **30** (1978) 546–51.
34. Dillon, P. J., Jeffries, D. S. & Scheider, W. A., The use of calibrated lakes and watersheds for estimating atmospheric deposition near a large point source. *Water Air Soil Pollut.,* **18** (1982) 241–58.
35. Lovett, G. M. & Lindberg, S. E., Dry deposition and canopy exchange in a mixed oak forest as determined by analysis of throughfall. *J. Appl. Ecol.,* **21** (1984) 1013–28.
36. Hicks, B. B., Baldocchi, D. D., Meyers, T. P., Matt, D. R. & Hosker Jr, R. P., A preliminary multiple resistance routine for deriving dry deposition velocities from measured quantities. *Water Air Soil Pollut.,* **36** (1987) 311–30.
37. Hicks, B. B., Hosker, R. P. & Womack, J. D., Comparisons of wet and dry deposition derived from the first year of trial dry deposition monitoring. In *The Chemistry of Acid Rain, Sources and Atmospheric Processes* ed. R. W. Johnson *et al.* American Chemical Society, Washington, DC, USA, 1987, pp. 196–203.
38. Hales, J. M., Hicks, B. B. & Miller, J. M., The role of research measurement networks as contributors to Federal assessments of acid deposition. *Bull. Amer. Meteorol. Soc.,* **68** (1987) 216–25.

Chapter 6

Measurement Methods for Atmospheric Acidity and Acid Deposition

JED M. WALDMAN
Department of Environmental and Community Medicine, UMDNJ–Robert Wood Johnson Medical School, 675 Hoes Lane, Piscataway, NJ 08854, USA

&

J. WILLIAM MUNGER and DANIEL J. JACOB
Department of Earth and Planetary Sciences, Division of Applied Science, Harvard University, 29 Oxford Street, Cambridge, MA 02138, USA

6.1 INTRODUCTION

Production of atmospheric acidity from fossil-fuel combustion and industrial activities constitutes a priority environmental problem. The deleterious effects of acid deposition on aquatic biota, vegetation, and materials are well established. Health effects from inhalation of acidic aerosol have been suspected for some time and are currently being investigated. Sulfuric acid (H_2SO_4) and nitric acid (HNO_3) are the main acids responsible for the environmental effects of atmospheric acidity. Organic acids (e.g. formic (HCOOH) and acetic acids (CH_3COOH)) can be major contributors to the total burden of atmospheric acidity, but they are less important in terms of environmental effects because they are rapidly consumed by bacteria. Other acids (e.g. hydrochloric (HCl), nitrous (HNO_2), sulfonic acids (—SO_2OH)) are usually of minor importance, although they can make significant contributions to atmospheric acidity in particular regions of the globe. Assessment of acid deposition requires accurate measurements of the concentrations of acids in the atmosphere, of the concentrations of acid precursors (sulphur dioxide (SO_2) and nitrogen oxides (NO_x)), and of the acid fluxes to the earth's surface. The object of this chapter is to summarize the current state of knowledge regarding these measurement techniques.

H_2SO_4 in the atmosphere is almost exclusively present as an aerosol because of its extremely low vapor pressure. In contrast, most other acids (HNO_3, HCl, organic acids) are mainly present in the gas phase, except under cloudy conditions. Acids in the atmosphere can be neutralized by various bases, principally ammonia (NH_3) and carbonate soil dust (e.g. calcium carbonate ($CaCO_3$)). NH_3 in its pure form is present in the gas phase but, as it neutralizes acidity, becomes incorporated into the aerosol phase. In particular, it condenses on H_2SO_4 aerosol and combines with gaseous HNO_3 to produce an ammonium sulfate ($(NH_4)_xH_{2y-x}(SO_4)_y$) or ammonium nitrate (NH_4NO_3) aerosol, respectively.

Acidic and basic compounds in the atmosphere are generally water soluble and, therefore, are readily dissolved into cloudwater, fogwater, and rainwater. Precipitation is a major mechanism for removal of acids from the atmosphere, but deposition of gases and aerosols in the absence of rainfall (dry deposition) is also important. A thorough understanding of atmospheric acidity thus requires ambient sampling of the gas, aerosol, and droplet phases, as well as measurements of wet and dry deposition fluxes.

Most of the historical understanding of atmospheric acidity has been gained from rainwater composition measurements, principally because the collection and analysis of rainwater samples is relatively straightforward. However, the past few decades have seen the development of several new methods for collection and analysis of gas-phase, aerosol, and cloudwater samples. New approaches have also been designed for evaluating dry deposition fluxes. Some of the measurement approaches have become universally accepted, while others are still in a developmental stage.

The focus in this chapter is to review the general considerations for collection, storage, and analysis of samples. Methods for sampling precipitation, cloudwater, particles, gases and dry deposition are presented first. General aspects of sample storage and analytical techniques are discussed next. These include analyses of collected samples as well as continuous (i.e. real-time) determinations of acid components. Commercial monitors for measuring concentrations of acid precursors are also briefly discussed; these techniques are not presented in detail because this information is readily available in the instrument manuals or in technical notes.

6.2 SAMPLING METHODS

In this section, the sample collection techniques used for components of atmospheric acidity are presented. Because these components occur in different phases (gaseous, aerosol, droplet and precipitation), a variety of methods are necessary. Some methods allow real-time determination of species. Others require the collection and preservation of samples for later analysis. Although differing in their collection schemes, many of these latter methods yield similar sample aliquots for analysis. For example, aqueous aliquots are obtained not only from precipitation and cloud droplet collection, but also from aerosol filter extraction, gas impinger solution, and gas adsorbent elution. Thereafter, chemical analyses of these aliquots are often similar for components of interest, such as acidity, ionic species, etc.

6.2.1 Precipitation

Precipitation samples (rainwater and snow) are routinely collected for a number of purposes at sites throughout the world. The concern herein is the determination of chemical composition of rain and snow and its contribution to acidic deposition. Nearly all precipitation collectors use some sort of open vessel, such as a funnel or bucket. Nonetheless, there is a wide variety in the design of the collectors and the materials used to meet the needs of individual studies. In this section, some of the considerations that are important in choosing a collector and devising a sampling protocol are summarized. Details on collector construction are readily available in the literature (*cf.* Ref. 1).

The collector surface needs to be chemically inert for the sample integrity to be maintained. Plastic surfaces are generally recommended for samples used to determine the major inorganic compounds.[1] Prior washing with a laboratory soap and copious distilled water is a necessary standard practice; collection of sample 'blank' is recommended after the final rinse. For determination of trace-metal concentrations, more thorough washing procedures include a strong acid rinse.[2] Plastics may leach metals, so collectors made of glass are sometimes used in studies of metal deposition when very low levels are expected. Plastic surfaces are generally avoided in investigations of organic consituents in precipitation; instead, glass or metal collection materials are used.

Two basic types of samplers have been historically used: bulk samplers and wet-only samplers. Bulk precipitation samplers are continually open and collect both rainwater (and snow) and dry deposited material (particles and absorbed gases). Wet-only precipitation samplers are designed to open only during precipitation events. Bulk samplers are generally no longer used, since the amount of dryfall included in a bulk sample is not well known, making it difficult to interpret results of this method.[3]

Wet-only precipitation samples may include the precipitation from several individual storms over fixed intervals, single storms, or sub-fractions of a storm. These are referred to as composite, event, and serial samples, respectively. Composite samples are adequate for determination of the total wet deposition of conservative species at a site. The National Atmospheric Deposition Assessment Program (NADAP) has utilized a weekly composite schedule. However, to relate changes in composition to environmental factors such as storm track, precipitation intensity, or wind direction, event samples are required. Moreover, concentrations of constituents change over the course of individual storms. The first few millimeters of rainfall usually contain the highest species concentrations, which can only be measured with a serial sampling. A further problem in composite sampling is that the concentrations of non-conservative species such as acidity and alkalinity may not equal the volume-weighted average values taken from the individual events.

In order to determine the wet deposition flux or 'chemical loading' at a site, it is important to accurately measure the volume of precipitation as well as its composition. Spatial variability in both precipitation amount and composition must be considered if the results are to be extrapolated to a larger area, such as a watershed or geographical region. In cases where loading is to be calculated, the amount of precipitation collected in the sampler should be compared to a standard meteorological gage. This serves as a reference for comparison to sampling networks which use different collectors. The potential for even standard gages to underestimate precipitation amount under certain conditions has been pointed out.[4] The magnitude of the error increases with higher wind speed and with smaller drop sizes.

At sites where precipitation is accompanied by strong winds, it may be difficult to accurately determine the amount of rainfall (or snow) by standard measurements. Discrepancies in collector and gage amount may indicate differences in efficiency related to the aerodynamics of the collector or a bias against low precipitation intensities. For example, collectors that rely on some form of moisture sensor to open their cover may not open reliably at low precipitation intensities. Furthermore, spatial variability in precipitation is evident over a range of scales. Variability, expressed as the ratio of the standard deviation to the mean for several samplers, is typically at least 10% within a 5 km radius; it can be more than 30% for a 50 km radius.[3]

Siting requirements for the NADAP network of precipitation sampling sites are outlined in Ref. 5. Collectors are to be located in an area that typifies the region and minimizes local sources. They should be 10 km away (or 20 km downwind) from nearby industrial facilities or population centers. Roadways must not be within 100 m; pastures must be not closer than 20 m; and parking lots should be at least 700 m away from the collector. Data have been compared for two pairs of nearby precipitation stations.[6] In each pair, one violated the NADAP siting requirements in a variety of ways, primarily in proximity to roadways and urban areas. Weekly variations in concentrations were considerable—often as high as 50%. However, no significant differences were found in the weighted quarterly and yearly precipitation deposition between pairs.

6.2.2 Cloudwater and fogwater

6.2.2.1 Collector design criteria

Cloud droplets range in size from 1 to 100 μm, with mass median diameters 10–30 μm; most of the cloudwater is present in the size range 5–50 μm.[7] Particles of such size are difficult to collect efficiently because their inertia may prevent them from following the air streamlines converging towards the inlet of a sampler. Furthermore, water droplets in the atmosphere are in a fragile thermodynamic balance with the ambient humidity, which is very sensitive to perturbations by a sampling device. Conventional methods for collecting aerosol samples may, therefore, lead to biases caused by anisokinetic sampling or changes in temperature or pressure. Cloudwater sampling is further complicated by the relatively large sample volumes needed for routine wet chemical analysis. For a typical cloud liquid water content of $0{\cdot}1$ g m^{-3}, collection of 10 ml of cloudwater in 1 h would require a sampling rate of $1{\cdot}7$ m^3 min^{-1}.

In addition to cloud droplets, clouds also contain some interstitial aerosol, consisting of particles which did not grow upon cloud formation (being either too small or too hydrophobic). These particles account for only a minute fraction of the total particle mass in clouds (droplet plus aerosol), but they can contain, in some cases, a substantial fraction of the total ionic loading. The fraction is found to be largest in fogs, where supersaturations are typically lower, and aerosol concentrations are higher, than in clouds.[8,9] Because interstitial aerosol is microphysically and chemically different from the cloud droplets, it is generally considered desirable to exclude it from the cloudwater sample.

Most sampler designs are aimed at reliable size-selective collection in clouds—retaining only the liquid water droplets while excluding interstitial particulate material. This is generally done by impaction on a sampling surface, which relies upon the increasing inertia of particles with their size. The collection efficiency of droplets is determined by (i) the geometry of the impaction surface, (ii) the collector Stokes number (St), and (iii) the Reynolds number (Re). The dimensionless numbers are defined below:

$$\mathrm{St} = \frac{D^2 \varepsilon_\mathrm{p} u}{9 \mu W} \tag{6.1}$$

$$\mathrm{Re} = \frac{D u \varepsilon_\mathrm{a}}{\mu} \tag{6.2}$$

where D is droplet diameter, ε_p and ε_a are the droplet and air densities, respectively, u is the velocity of air flowing past the impaction surface, μ is the viscosity of air, and W is the characteristic width of the impaction surface. Collection efficiencies are a function of St and Re,[10] and they have been measured for a variety of impaction surface geometries.[11] For most surfaces (cylinders, spheres, plates), droplets with $\mathrm{St} > 1$ are efficiently collected over the range of typical Reynolds numbers. Higher St values are required for impaction on slotted rods.[12]

Aspiration of air at the inlet of a sampling device perturbs the flow streamlines if the sampling velocity, u_s, differs from the local wind velocity, u_a, i.e. anisokinetic sampling.[11] Large cloud droplets may not be sampled efficiently if their inertia prevents them from following the flow streamlines converging towards the inlet (or diverging from the inlet). The inlet sampling efficiency is a function of (i) the inlet St (computed from eqn (6.1) where W is instead the width of the inlet), (ii) the ratio $u_\mathrm{s}/u_\mathrm{a}$, (iii) the orientation of the inlet into the wind, and (iv) the inlet geometry. Extensive experimental data for inlet sampling efficiencies are tabulated in Ref 13. For a sampler oriented into the wind, inlet losses are maximum for sampling velocities two to five times the ambient wind velocity; under those conditions, a sampler oriented into the wind can lose over 30% of particles with an inlet St greater than 0·5. In a crosswind, losses can be considerably larger.

Changes in temperature or pressure within or near the sampling device must be sufficiently small to avoid condensation or evaporation of the droplets during their approach to the impaction surface. This effect can be computed by solving the appropriate equations for mass transfer and droplet trajectory.[12] Thermodynamic perturbations of the sample may also occur after impaction due to temporal variations in the humidity of the air flowing by the impaction surface. To avoid such perturbations the droplets should be sheltered away from the impaction surface immediately following collection.

6.2.2.2 Current collector designs

A number of ground-based collectors have been developed over the past few decades for sampling fogs or clouds intercepting mountain slopes (Table 6.1). Reference 12 gives a review of historical devices, including the passive collector designs. Using six recently developed instruments, a performance intercomparison was conducted in 1984 at a mountain site above Los Angeles.[14] No significant differences in chemical composition were found between samples collected with the rotating arm collector,[12] the linear jet impactor,[15] and the screen mesh collector[16] (see Fig. 6.1).

More recently, the Cloud Water Project active cloudwater collector was developed to allow remote (i.e. battery) operation.[17] It employs a bank of Teflon strings through which air is drawn at 7·6 m s^{-1} with a 12 V DC fan. The Cloud Water Project collector has been used widely, allowing identical protocols at sites throughout North America.[18] The collectors described above share the problem that they require an on-site operator. Subsequent efforts have improved further on the Cloud Water Project design, providing higher collection efficiencies, increased sample size, the ability to collect rime ice, and automated start-up and sample changer.[19,20] The Caltech active strand collector, shown in Fig. 6.2, can be considered state-of-the-art.[21] In this design, air is drawn across a bank of inclined Teflon strings (0·5 mm diameter) at a velocity of 9 m s^{-1}. Cloud droplets impact on the strings and flow by aerodynamic drag and gravity to a Teflon trough underneath. With a square inlet (400 cm^2) oriented into the wind, efficient collection is achieved for droplet diameters 4–100 μm. A downward-facing inlet is added in precipitating clouds to avoid sampling of raindrops.

Table 6.1
Current designs for active, ground-based cloudwater collectors[a]

Design	*Impaction surface*	*Collection rate* ($m^3\ min^{-1}$)	*Lower size cut* (μm)	*Alias*[b]	*Reference*
Rotating arm	Slotted rods Teflon-coated	5	10–20	RAC	12
Linear jet	Roller plates Teflon-coated	1·2	5[c]	DRI	15
Fan-drawn	Mesh in pipe Polypropylene	1·5	5	GGC	16
Fan-drawn	Screen mesh Polypropylene	1·5	?[d]	ASRC	136
Fan-drawn	String Teflon	21	5·2[c]	CWP	17
Fan-drawn	String Teflon	21	3·5[c]	CASC	19,20

[a]Refer to Ref. 12 for review of historical devices and passive collector designs.
[b]RAC—Caltech rotating arm collector; DRI—Desert Research Institute; GGC—Global Geochemistry Corp.; ASRC—Atmospheric Science Research Center; CWP—Cloud Water Project; CASC—Caltech active strand collector.
[c]Most designs will also collect drizzle or rain during periods of precipitation; DRI, CWP and CASC use inlet shields to prevent collection of precipitating drops.
[d]Not reported or cannot calculate from theory.

Under conditions where no electrical power is available, cloudwater samples have been collected with passive ground-based devices (e.g. Refs 17 and 22). Virtually all use some kind of plastic strings or screen mounted vertically in the wind. Droplets impact on the strings as they are advected past, and flow by gravity to a collection reservoir at the bottom. A major problem with these devices is that collection rates depend on wind speed; in calm or low wind, no sample will be collected. In addition, collection efficiencies vary greatly with the ambient wind velocity. Depending on the diameter of the strings, high winds may be required for collection of representative samples. With 0·5 mm diameter strings, wind speeds over 5 m s^{-1} are required to collect efficiently droplets larger than 5 μm.[19,20]

Two airborne devices have been used extensively to sample clouds from aircraft: a jet impactor[23] and the Mohnen slotted-rod impactor.[24] The latter instrument is by far the more popular. It consists of a vertical assembly of cylindrical rods (0·95 cm diameter), with slots in the front

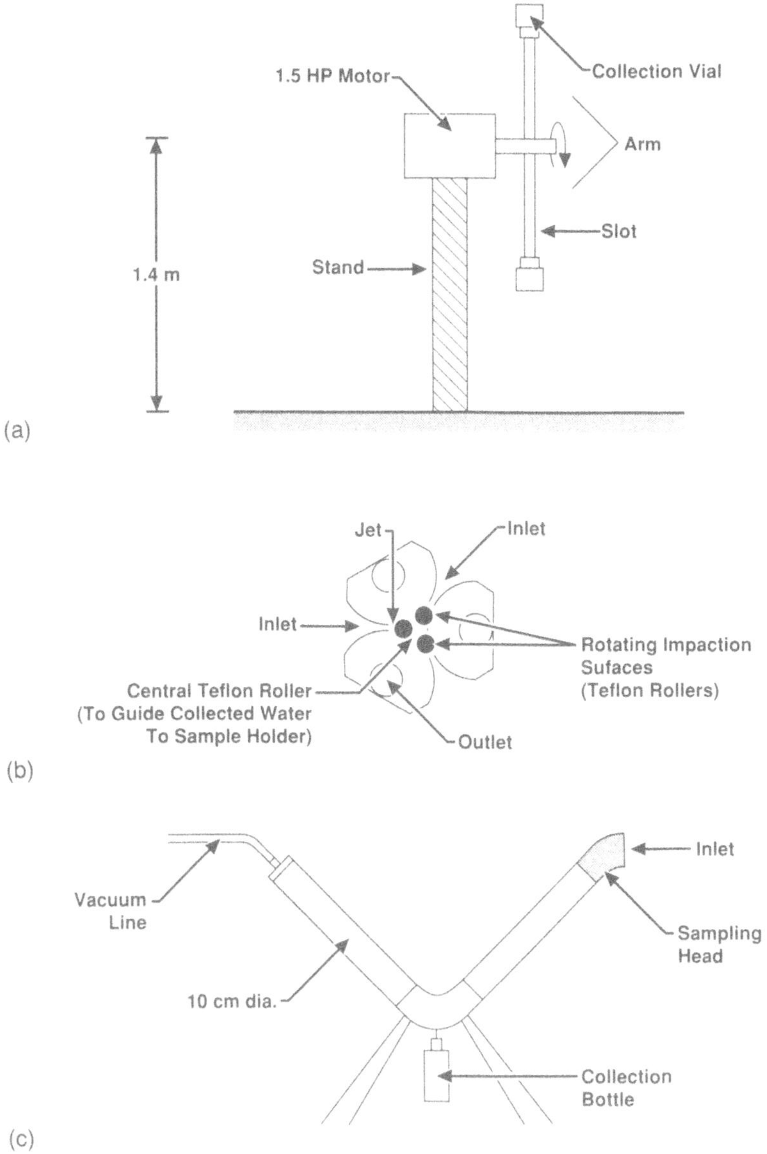

Fig. 6.1. Fogwater collector designs. (a) Rotating arm (Caltech RAC); (b) linear jet (DRI); (c) screen mesh (GGC). See Table 6.1 for further descriptions and figure citations.

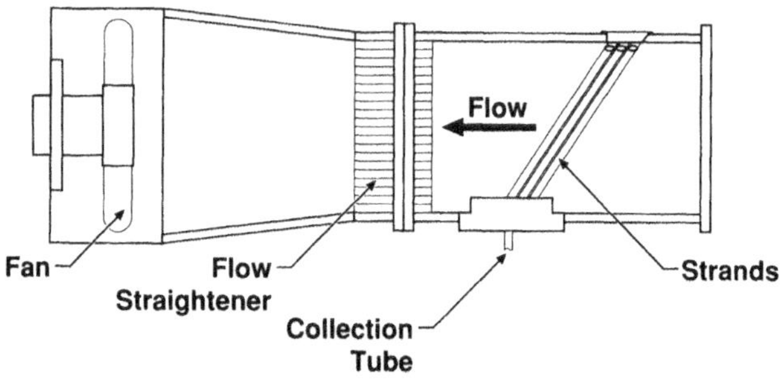

Fig. 6.2. Diagram of the Caltech active strand collector. (Reproduced with permission from Ref. 19.)

0·19 cm wide and 0·64 cm deep. The assembly extends out of the fuselage; cloud droplets deviate from the flow streamlines around the rods, enter the slots where they impact the surface, and flow down to collection vials below the assembly. Theoretically, the Mohnen slotted-rod device should collect efficiently droplets larger than 3 μm diameter.[24] However, laboratory and field tests indicate that the collection efficiency is actually less than 50% over the range of cloud droplet sizes, and varies with droplet size in an unclear manner.[25] Factors affecting the collection efficiency include the recirculation of air within the slots, details in the machining of the rods, and location of the fuselage. It has been concluded that cloudwater samples collected with the Mohnen slotted-rod device may not be representative of the ambient cloudwater.[26] Further work is needed to design and evaluate reliable airborne collectors for cloudwater sampling.

6.2.3 Aerosols

There are a few techniques allowing real-time determination of aerosol species (see Section 6.3.5). However, most methods for chemical analysis of aerosol originate with collection of the particulate material onto a substrate, frequently a filter. These may incorporate some fractionation or pre-separation according to size. Some methods use impingers or impaction plates rather than filters. The most critical issues with respect to aerosol sampling are the inlet, the collection medium, and sampling artifacts.

6.2.3.1 Inlets

High-volume aerosol samplers (known as 'Hi-vols') are used to monitor compliance with the National Ambient Air Quality Standard for Particulate Matter.[27] The initial standard was set on total suspended particulate material (TSP), and the peaked roof inlet was used. The federal standard was amended to regulate particulate material < 10 μm aerodynamic diameter or PM-10,[28] and size-selective inlets were developed for the Hi-vol.[29,30] In either configuration, Hi-vols draw 1·4 m^3 min^{-1} (40 cfm) of air onto a 20×25 cm (8×10 in) filter; enough mass can be collected in 24 h to allow net determination on a laboratory balance (±1 mg).

Most acid-containing aerosols are generally found in the fine particle regime, frequently less than 1 μm diameter.[31,32] These small aerosols are less prone to anisokinetic sampling biases than coarse particles or cloud droplets. Size selectivity for acid aerosol samplers is nevertheless important because the coarser fraction tends to be enriched with an alkaline component (e.g. carbonate). Artifact neutralization may thus occur on the filter substrate or when the filter is analyzed. Jet impactors, elutriators, and cyclones can all be used effectively as size-selective inlets to remove coarse particles from the sample stream.

Although Hi-vol data are frequently reported to identify regional patterns for sulfate, nitrate and ammonium (notwithstanding the biases given above), samples for direct analyses of aerosol acidity should not be taken with the TSP or PM-10 sampler. More suitable are samplers which segregate the fine mode. For example, the dichotomous sampler gives simultaneous samples of the two size classes: fine—$d_p \leq 2{\cdot}5$ μm, and coarse—$2{\cdot}5 \leq d_p \leq 10$ μm. The sampler uses a low flow rate (16·7 liters min^{-1}) and a virtual impaction separation scheme to collect the two fractions (Fig. 6.3).

6.2.3.2 Filter substrate

A number of filter substrates are commonly used for particulate sampling (Table 6.2). A portion of the filter can be extracted for major ion determinations, or collected material can be analyzed directly on the filter. Glass fiber (GF) is highly alkaline and absorbs SO_2 and other acid gases.[34,35] Absorption of SO_2 can produce significant artifact sulfate.[36] Nitrate artifacts can also be a problem with GF filters.[37,38] Quartz and Teflon substrates have been shown to be relatively free of these sulfate and nitrate artifacts.

Measurements of aerosol acidity have been made using cellulose acetate,[39] quartz,[40] and Teflon-membrane filters.[41] In the cases of quartz and cellulose, the filters require an elaborate acid wash to eliminate alkalinity.[42] In recent years, Teflon (polytetrafluoroethane (PTFE)) membrane

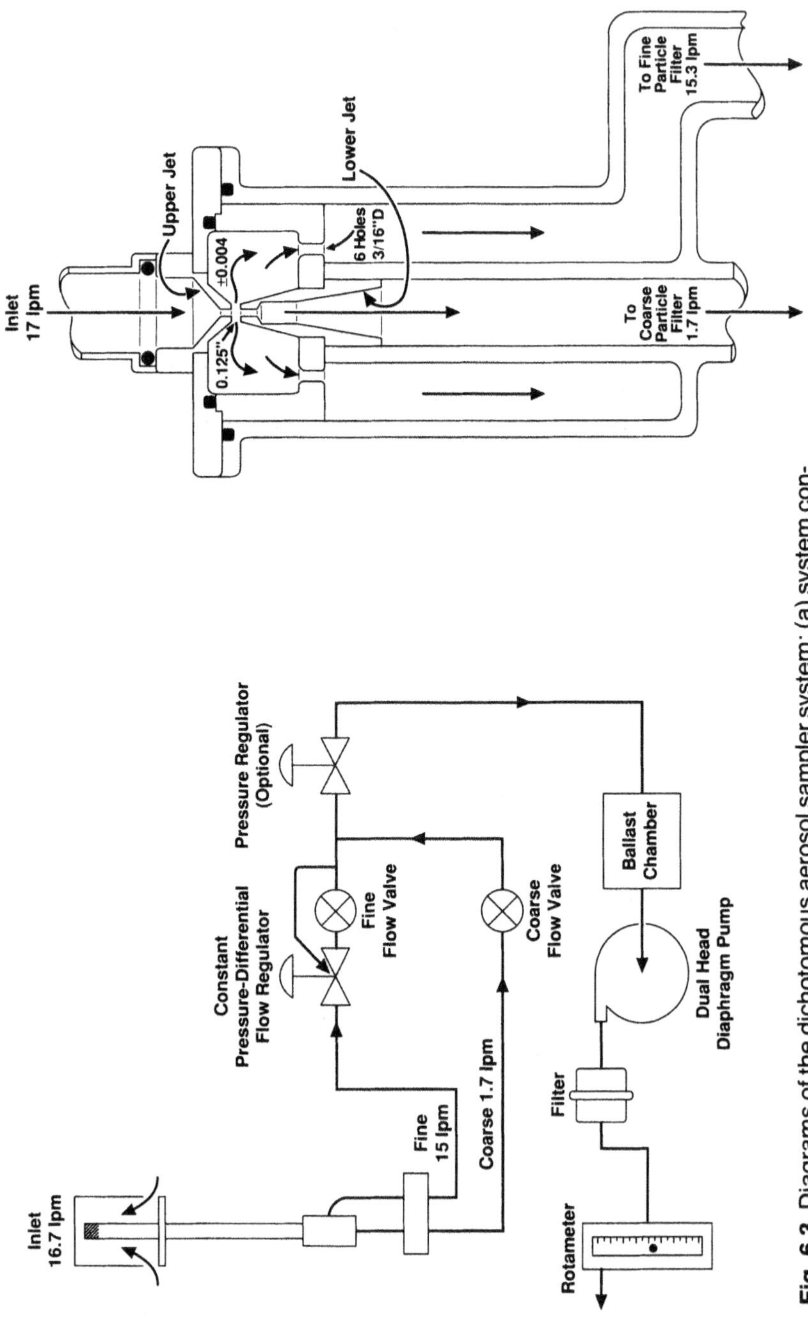

Fig. 6.3. Diagrams of the dichotomous aerosol sampler system: (a) system configuration; (b) virtual impactor. (Reproduced with permission from Ref. 33.)

Table 6.2
Properties of filters used for particulate samples[a]

Filter	*Composition*	*Density (mg cm^{-2})*	*Surface reactivity*	*Efficiency*[b] *(%)*
Teflon	Polytetrafluoro-ethane membrane ($(CF_2)_n$) (PTFE)	0·5	Neutral	99
Whatman 41	Cellulose $(C_6H_{10}O_5)_n$	8·7	Neutral[c,d]	58
Whatman GF/C	Glass fiber	5·2	Basic (pH 9)[e]	99
Gelman Quartz	Quartz fiber	6·5	pH 7[c]	98
Nucleopore	Polycarbonate $(C_{15}H_{14}-CO_3)$	0·8	Neutral	93
Millipore	Cellulose acetate/nitrate	5·0	Neutral[c,d]	99

[a]Adapted from Ref. 33.
[b]For $d_p > 0{\cdot}03$ μm at $V_{face} = 10$ cm s^{-1}.
[c]Requires acid wash to remove alkaline sites.
[d]Reacts with HNO_3(g).
[e]Reacts with SO_2(g).

has become the substrate of choice, having the advantages of being chemically inert and providing very low blank levels for the species of interest (sulfate, nitrate, ammonium, and acidity). However, besides the greater expense compared to most other filters, the chief drawbacks to PTFE are (i) the relatively large pressure drop across the filter, which limits the flow rate for sample collection, and (ii) a membrane (vis à vis fibrous) substrate more easily overloads, which constrains the sample duration and may enhance particle-to-particle reactions on the filter surface.

6.2.3.3 Sample artifacts

Neutralization by ammonia of particulate acidity on the filter can be an important problem if the atmosphere switches from an acidic to an alkaline regime over the course of collection. Since low ammonia concentrations are *sine qua non* whenever ambient aerosol contains free acidity, this concern is less important for short sampling durations during which atmospheric conditions remain somewhat stable. However, particular attention to this problem must be paid when using longer sampling durations, e.g. 12 h or more. An acid-coated diffusion denuder can be used to scavenge $NH_3(g)$ in the air stream and protect the filters from neutralization (see the next section).

6.2.4 Gases

Whenever possible, gases are analyzed directly by continuous or semi-continuous analytical instruments. Direct read-out analyzers are commercially available for SO_2 and NO_x, the most important acid precursors. These instruments are included in most air-quality surveillance networks, such as required by federal and state clean air laws. The principles of these gas analyzers are discussed in Section 6.3.4, while the general aspects of sample collection are given herein.

At monitoring stations, sampling inlets need to be located above, and away from, nearby obstructions. For both gases and particulates, the location of an inlet should allow sampling of a continuously renewing air parcel. Zones of stagnant air or building exhaust are avoided, for example. The deposition of reactive gases can lead to localized concentration gradients in the surface layer above the ground surface or near buildings and foliage. Hence, care should be taken to locate the inlet above this surface layer. Site selection criteria for the air pollutant monitoring set the acceptable range of 3–5 m above ground level for the height of sampler inlets.[43]

Chemically inert inlets and sampling lines are essential. A glass inlet with a manifold and bypass pump are frequently used to minimize the transit time from inlet to detectors. From the manifold, PTFE or stainless-steel tubing is most often used for sampling inorganic gases. Materials such as rubber or nylon are avoided because they react with the sample or are degraded by it. Biases for a number of inlet designs and materials with respect to HNO_3, the most absorbent of acid gases have been compared.[44] Uncoated stainless-steel surfaces caused substantial losses. Glass led to some loss, while PTFE coating was demonstrated to allow unbiased collection.

When concentrations of the gas species of interest are below instrument detection limits or a continuous analyzer is not available, collection in a suitable substrate allows gaseous components to be concentrated over a period of minutes, hours or even days. Sensitivity is enhanced but at the expense of time resolution in measurements. A number of approaches are utilized to make time-integrated samples. These entail adsorption or absorption into (i) a solution, (ii) a chemically treated filter or, (iii) a porous solid or chemically treated surface. These substrates are generally base-treated to collect the acidic gases; treating with an acid gives good retention of NH_3. Methods for collection with bags or other containers are not generally suitable for acidic gases as they tend to react with surfaces.

Liquid sorbents can be used in a bubbler or spray chamber. A particularly efficient chamber for the collection of HNO_3 and organic acids, which are highly soluble species and can be collected in distilled water, has been developed.[45] Collection of SO_2 can also be achieved by including hydrogen peroxide (H_2O_2) or formaldehyde (HCHO) in the scrubbing solution, which oxidizes or stabilizes dissolved SO_2, respectively.

SO_2 can also be quantitatively collected using filters impregnated with carbonate.[46] The same method has been used to collect organic acid vapors, such as CH_3COOH and HCOOH.[47] Nylon, sodium carbonate (Na_2CO_3) and sodium-chloride-impregnated filters can also be used to collect HNO_3.[48,49] Filters impregnated with citric or oxalic acid can be used for NH_3 collection.[50,51,52] In each case, an inert filter (often Teflon) is used to remove particulate material prior to the absorbent filter.

However, there are some interferences with these methods. While the gases mentioned will not adsorb to the Teflon membrane, some of the gas, especially HNO_3, can be depleted before passing through the prefilter by condensing on any material previously collected.[37] This can be minimized by using short sampling intervals. Under atmospheric conditions rich in NH_4NO_3, a positive interference can occur by dissociation into HNO_3(g) and NH_3(g). Carbonate-treated substrates may convert some portion of NO_2 and peroxyacetyl nitrate (PAN) to HNO_3. A positive artifact has also been reported in the use of base-impregnated filters for carboxylic acid sampling.[47] A fraction of SO_2 may be collected on nylon filters, which is a problem if it is to be collected on a downstream sampler.

Several approaches have been used for concurrently sampling gaseous along with aerosol components of atmospheric acidity. This is a difficult problem because components can shift between phases and because the action of sampling may alter the thermodynamic equilibria, as described above. Three of the current methods are the denuder difference method, the annular denuder system, and the transition flow reactor system.

6.2.4.1 Denuder difference method (DDM)

The DDM uses the collection of two parallel samples to calculate HNO_3 and aerosol NO_3^-.[53] An open nylon filter collects both the gas and particulate nitrate species. A second nylon filter is placed behind a diffusion denuder which removes only HNO_3. The concentration of gaseous HNO_3 is calculated by difference.[54]

6.2.4.2 Annular denuder system (ADS)

The ADS is also based on selectively removing chemically reactive gases to the walls of a coated tube, as in the DDM. Both methods take advantage of the higher diffusivities of gases compared to aerosols under laminar flow conditions. In the ADS approach, the concentrations of gases are determined by extracting the denuder and measuring the adsorbed quantity directly.

Gas removal efficiencies of cylindrical diffusion denuders have been computed.[55] The sampling rate must be kept low to maintain laminar flow ($Re < 2000$). The configuration of an annular tube can be optimized to provide great increases in the wall surface area, in the flow rate for the same Re, and in the efficiency for the same dimensions.[56,57] With prior removal of gases, aerosols can be collected downstream.

Configured with a pre-separator (to remove coarse particles), one or more denuder tubes, and a filter pack, applications of the ADS have become widespread in recent years. Using various chemical coating solutions on the walls of a diffusion denuder tube, collection of a wide range of gaseous compounds can be performed. Ambient measurements have been made for HNO_2, HNO_3, SO_2, NH_3, HCHO, CH_3COOH, HCOOH, nicotine and other vapors. The collection efficiencies of ADS samplers have been tested in a number of studies.[58,59,60] A frequently used configuration includes a jet impactor, three denuder tubes, and a two-stage filter holder (Fig. 6.4(a)). The tubes are coated with Na_2CO_3, Na_2CO_3, and citric acid, respectively. Aerosol is collected on a Teflon filter.

One problem of the ADS is that the removal of acidic and basic gases in the denuder can alter the thermodynamic equilbrium of the $H_2SO_4:HNO_3:NH_3:H_2O$ mixture in the air sample downstream. Under certain conditions, this might cause volatilization of NH_4NO_3 aerosol from the filter.[62] Frequently, a nylon or carbonate-coated filter is added behind the Teflon filter to capture volatilized nitrate. The possibility that some of the nitrate lost is actually HNO_3 has been examined by using a second (acid-coated) back-up filter to capture NH_3 and thus verify the stochiometry of the volatilized aerosol. Its magnitude is not found to be important except under wintertime conditions.[60]

6.2.4.3 Transition flow reaction system (TFRS)

The TFRS consists of a cyclone (removing coarse particles), two reactive liners which collect constant fractions of HNO_3 and NH_3, followed by three filters (see Fig. 6.4(b)).[51] As with the ADS and most simple filter-pack approaches, a Teflon filter is used first in-line to collect particulate

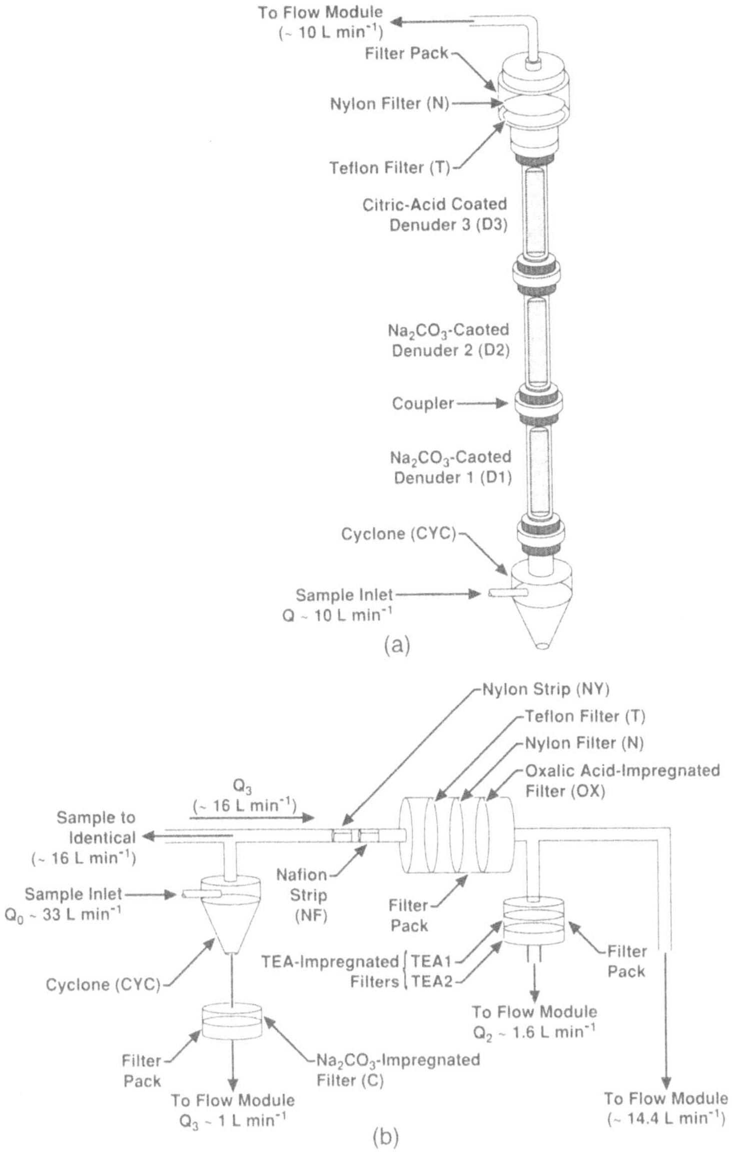

Fig. 6.4. Diagrams of sampling systems for gaseous and aerosol components of atmospheric acidity: (a) thc annular denuder system (ADS); (b) transition flow reactor (TFRS) system. (Reproduced with permission from Ref. 61.)

material. It is followed downstream by a nylon filter then an oxalic-acid-impregnated filter which, in this case, capture HNO_3 and NH_3 passing the two reactive liners. The system also samples SO_2 and NO_2 using triethanolamine-impregnated (TEA) filters. The gaseous concentrations are calculated using an algorithm based on the collection efficiencies of the reactive liners and the chemical speciation on the individual filters.[51]

The ADS and TFRS were evaluated together for measurements of selected gases and aerosols.[61] Both samplers were shown to provide good sensitivity and precision (below 1 $\mu g\ m^{-3}$) on daily (22 h) samples for most of the measured components. However, the results suggested some bias in HNO_3, fine aerosol NO_3^-, H^+ and SO_2 measurements by the TFRS. These appear due to the differences in fundamental design of the TFRS and ADS samplers; the ADS approach (upstream removal of HNO_3 and NH_3) appears to eliminate the uncertainty in the TFRS measurement caused by artifact HNO_3 volatility and acid neutralization of the filter pack. Lower SO_2 measurements by the TFRS were the result of only 80–90% recovery from TEA-impregnated filters. Finally, 7-day samples showed good agreement with composite daily values, indicating that the long-term concentrations could accurately be made by ADS samplers.

6.2.5 Dry deposition

Dry deposition is the removal of airborne material to surfaces near the ground. Unlike wet precipitation, in which relatively large droplets or flakes fall under the predominant influence of gravity, dry deposition involves turbulent transport of aerosol and gases to the surface layer. This is followed by some mechanism of uptake at the surface, either impaction or diffusion (aerosol) or adsorption/absorption (gases). These complex interactions of aerodynamic and surface characteristics make it inherently more difficult to quantify these pathways. There are three conceptual approaches to dry deposition measurement: (i) direct collection on surrogate or natural surfaces, (ii) flux measurements by eddy correlation or profile techniques, and (iii) indirect estimation using atmospheric concentration monitoring and estimated deposition velocities. Examples of the three different approaches are given in Table 6.3. References 63 and 64 provide useful reviews on the status of dry deposition measurement methods for gases and particles reported in the current literature (see also Chapter 5).

Table 6.3
Dry deposition measurement techniques

Direct (surrogate surface)	*Flux (eddy correlation)*	*Concentration (gradient)*
Buckets	Flame photometric detector	Filter pack
Teflon	Optical particle counter	Denuder difference
Petri dishes	Laser particle spectrometer	Annular denuder method
Filter media		Transition flow reactor
Leaf washes		
Synthetic grass		

6.2.5.1 *Direct collection*

Surrogate surfaces, such as open buckets, flat plates, Petri dishes, synthetic foliar surfaces have all been used to measure dry deposition. Collection on surrogate surfaces is strongly dependent on the surface used and its position.[65] Because the surrogate surfaces are neither aerodynamically nor chemically similar to many natural surfaces, they will not experience the same deposition. As an alternative, deposition to vegetation can be measured by washing the accumulated deposition off leaves.[66] Clearly, this is only valid for chemical species that are neither leached from nor absorbed by the leaves. Unfortunately, there is not a direct relationship between deposition to surrogate surfaces and deposition to natural surfaces as determined by foliar washing.[67] Surrogate surface methods can provide estimates of seasonal and regional variation in deposition, but should be used with caution to estimate overall fluxes to a site.[68]

It is possible to directly measure dry deposition to snow by periodical sampling of snow layers or by comparing the content of a snowpack just before it begins to melt to the sum of individual storms.[69,70] Dry deposition to dew has also been directly measured by collecting the accumulated dew on a Teflon sheet.[71]

6.2.5.2 *Flux measurement*

The flux of material across a plane above the surface may be determined by micrometeorological methods. The eddy correlation method uses the correlation between rapid fluctuations in vertical wind speed and species concentrations to calculate the vertical flux of material. Sensors that can detect variations at 1 Hz or better are necessary. Measurements must be made at a distance above the canopy, such as on a tower projecting above the canopy or from an airplane.

Fluxes can also be derived from measured concentration gradients by making the approximation that the flux of a species is proportional to its vertical concentration gradient. The proportionality constant is known as the eddy diffusivity or turbulent transfer coefficient. This approximation is justified in the presence of small eddies, but frequently fails when the eddies are large, as within a canopy or high above it.[72] A similarity approximation can be used to estimate fluxes from the ratio of gradients for two species when the flux of one of them, such as water vapor, has been measured by some other means.[73]

Both eddy correlation and the flux–gradient relationship are limited to sites with sufficiently simple terrain and surface homogeneity to allow well-developed profiles and uniform eddies to develop. Eddy correlation and flux–gradient relationships have been used extensively to determine fluxes of gaseous species (e.g. Refs 74 and 75) and fine particles (e.g. Refs 76, 77, 78 and 79). HNO_3 and particulate nitrate gradients have been measured by Huebert and Robert;[25] the investigators point out the complicating effect of gas-to-particle interconversion in the surface layer.

6.2.5.3 Estimation with deposition velocities and concentration monitoring

One of the important applications for the gas/aerosol systems (ADS and TFR) described above is direct monitoring of acid deposition components. Dry deposition can be indirectly estimated from measured concentrations and assumed deposition velocities. Deposition velocity is defined by the following empirical relationship:

$$J = -V_d \times C \tag{6.3}$$

where J is the deposition flux of a species, V_d is its deposition velocity over the surface of interest, and C is its concentration at a particular height. The convention is upwards to be the positive direction (hence downward or depositional flux is negative). Since the deposition is gradient-driven near the surface, the values of V_d are valid only with respect to fixed height. Deposition velocity is a complex function of atmospheric stability, wind speed, surface characteristics and the chemical reactivity of the species.

Considerable research effort has been made to determine deposition velocities for the species involved in acid deposition, as a function of meteorological variables and surface types (cf. Ref. 80). Reference 79 presents preliminary results from a dry deposition network using measured aerosol and gas-phase concentrations and assumed V_d values. Like the direct flux measurements, the indirect deposition measurements require a reasonably simple terrain and homogeneous surface. Care must be taken that the sites where concentration measurements are made are representative of the surrounding area, and that appropriate values for V_d are used. Dry deposition measurement methods are most reliable for smooth terrain such as agricultural fields or unbroken forest, with minimal topography. As yet there is no reliable method to directly determine dry deposition to patchy forests or ecosystems in mountainous terrain. Unfortunately, these conditions are typical of the regions generally considered to be most sensitive to acid deposition.

6.2.6 'Occult' precipitation

Several pathways for pollutant removal fall outside the nominal categories of 'wet' and 'dry' deposition. These include dew, fog, and intercepted cloud. Deposition accompanying these meteorological conditions is sometimes called 'occult' precipitation, since it is not recorded by standard precipitation collectors. Early research[81,82,83] and recent work[71,84,85,86] demonstrated the importance of these pathways under certain conditions.

The collection methods for 'occult' precipitation are frequently based on the premise of a surrogate surface collector. Therein, the design issue is to accurately mimic natural deposition. Dew collectors are designed to duplicate the radiant cooling of ground surfaces.[71] Fog deposition to open terrain may be sampled using flat collectors that capture sedimenting droplets.[6,86] In some cases, the volume of liquid is often so small, even on large plates, that a quantitative extraction of the material deposited on the surface is required. Fluxes of fog droplets to open pastures have also been estimated using the gradient method.[87,88]

Wind-driven fog and intercepted clouds can provide substantial inputs of moisture and chemical deposition in mountainous regions. The major process governing deposition is generally impaction of droplets on plant surfaces, such as pine needles, leaves and branches, so it is important to correctly mimic the natural geometry, if surrogate surfaces are used.[84] Droplet size, wind speed, and relative humidity governed the deposition to a given forest canopy.[84]

6.2.7 Remote sensing

At present, it is not possible to measure most acid species in the atmosphere by remote sensing, i.e. determinations at a distance, *without* bringing samples into an instrument. Light Direction and Ranging (LIDAR) techniques have been used extensively to measure the vertical distributions of total aerosol from aircraft.[89] In addition, both HNO_3 and HCl absorb infra-red (IR) radiation and could, in principle, be detected by spectroscopy. Remote sensing of the troposphere is complicated by interferences from aerosols, clouds, and water vapor. In the case of HNO_3 and HCl, an additional complicating factor is the presence of high mixing ratios in the stratosphere above.[90] However, these difficulties are not insurmountable; Ref 91 proposes that Fourier transform IR (FTIR) measurements of tropospheric HNO_3 and HCl be conducted from the Earth Observing Satellite (EOS) to be launched in the late 1990s.[91] Detection limits and the extent of vertical resolution have not been established yet, but concentrations as low as 0·1 ppb are likely to be measurable.

6.3 ANALYTICAL METHODS

As noted in Section 3.2, most techniques for monitoring components of acid deposition generate, or ultimately produce, water solution or aqueous aliquots. Procedures for the storage and analysis of such aliquots are discussed in this section. Table 6.4 gives a summary list of the various wet chemical techniques standardly used, along with their detection limits. In the case of precipitation samples collected near population centers, limits of detection for major species are rarely an issue. Likewise, cloudwater samples tend to be several times more concentrated than rainwater. However, for remote precipitation samples, sensitivity of analyses is important because component concentrations are generally low. Furthermore, sensitivity can limit aerosol determination by requiring longer periods for filter collection.

6.3.1 Sample storage

Precipitation or cloudwater samples are inherently unstable. A variety of processes can occur following their collection, which will lead to changes in the chemical composition of the sample. These include:

- evaporation of water,
- degassing of volatile constituents,

Table 6.4
Analytical methods for major components of acid deposition

Solution	
Acidity	
Hydrogen	pH measurement
Strong/total acidity	Base-addition titration
Anions	
Strong acid anions (SO_4^{2-}, SO_3^{2-}, NO_3^-, NO_2^-, Cl^-, Br^-, F^-)	Ion chromatography
Weak acid anions ($HCOO^-$, CH_3COO^-, etc.)	Ion-exclusion chromatography
Cations	
NH_4^+, Na^+, K^+	Ion chromatography
NH_4^+	Colorimetry of fluorescence
Ca^{2+}, Mg^{2+}, Al^{3+}	Atomic absorption spectrocopy
Gaseous	
Sulfur dioxide	West–Gaeke reaction Fluorescence analysis Flame photometric detection (total S)
NO_x and NO_y	Chemiluminescent analysis FTIR analysis Tunable diode laser spectroscopy
Acidic gases	
HCl, SO_2, HNO_3, HNO_2	Alkaline substrate and extraction
Basic gases	
NH_3	Acidic substrate and extraction
Aerosol and particulates	
Soluble material	Filter capture and extraction FTIR analysis
Metals	X-ray fluorescence analysis

- absorption of soluble gases,
- adsorption on container walls,
- leaching of material from the container walls,
- precipitation or dissolution of particles,
- microbial alteration of the sample, and
- chemical reaction (oxidation/reduction).

Ideally, samples would be analyzed immediately after collection to ensure chemical integrity. However, this is often impractical, so some form of storage is required. Selection of an inert sample container minimizes the problems with adsorption and leaching. Polyethylene has been found to be suitable for most of the major inorganic ions in precipitation. Tightly sealing containers with a minimum headspace are necessary to avoid evaporation and exchange of gases. Samples being stored for trace-metal analysis are usually acidified to prevent metals from adsorbing to the walls. Filtration of precipitation and cloudwater samples using inert membrane (e.g. 0·22 μm pore size) is recommended to remove insoluble components, such as dust or fly ash. These can interfere with chemical analysis and may also slowly interact with the water solution and lead to spurious biases if not removed shortly after collection.

Compounds that are easily oxidized (e.g. S(IV)) or reduced (e.g. H_2O_2) must be stabilized immediately upon collection. Microorganisms are found in rainwater, and, if given an opportunity they will grow in the sample. Species that are nutrients for their growth (e.g. organic and nitrogenous compounds) will be most affected.[92] Over short time periods, refrigeration is adequate to inhibit microorganisms in the sample. For long-term storage or shipment from distant sampling stations, freezing or some form of biocide is necessary. A variety of preservatives have been compared and it was found that chloroform or toluene had the least adverse affect on the sample, although all the preservatives altered the sample composition to some extent.[3] Chloroform has come to be the preservative of choice for samples to be shipped from remote sites.[93] Any preservative used must be of high quality to avoid contamination, and its blank value should be determined.

Aerosol filter samples are likewise unstable. Acid and bases on filters may react with one another or with the filter media (see Section 6.2.3). Volatile species, such as NH_4NO_3 and ammonium chloride (NH_4Cl), may evaporate, while microbial activity can consume organic acids. If rapid extraction and analysis is not possible, refrigeration or freezing is used to minimize losses. Neutralization of aerosol acidity on filters is a serious problem because NH_3 vapor is a ubiquitous contaminant in the field and laboratory, and the reaction is very rapid.[94] Collected filters should be handled and stored exclusively in NH_3-free hoods and containers.

6.3.2 Acidity

Acids and bases, in the Brönsted definition, are compounds which can donate and accept protons, respectively.[95] Strong acids are those with equilibria which strongly favor loss of one or more protons, e.g. $HA \rightleftarrows H^+ + A^-$; weak acids are less prone to dissociate, unless a strong base is added. A measure of acid strength is 'K_a' or acid dissociation constant; for convenience, the term $pK_a (= -\log_{10}(K_a))$ is frequently used. Strong acids are generally those with $pK_a < 3$; pK_a values for HNO_3 and H_2SO_4 are below zero, i.e. $K_a > 1$.

As highlighted earlier, the concept that acidity exists in various phases is important to the 'acid deposition' question. Deposition from each phase contributes to the total flux, and components can transfer among phases. Hence, acidity is some measure of the amount of proton which can be donated or potentially donated in any phase. However, the convention of a pH scale is only meaningful for solutions, and generally this means an aqueous medium. 'Acidity' in the gaseous or aerosol phase represents a sum of individual compounds which are measured in their respective phases. The net acidity, however, is measured in solution, following elution or extraction of the sample.

Measurements of sample acidity are performed with a pH electrode and pH meter in a liquid aliquot. While this seems a relatively straightforward measurement, there are a number of ways in which it can be made in error or biased. As with any analytical method, accuracy is largely dependent upon the standards used. Standard buffers are commercially available and should be appropriately certified. Buffer contamination, instrument miscalibration, electrode damage or other operational problems are not uncommon. Stirring of dilute rainwater samples while taking a pH reading is found to cause inaccuracies from the streaming potential; this can be overcome by addition of an electrolyte such as potassium chloride (KCl). Samples should be adjusted for constant ionic strength, since a slight difference in measured pH value will result for the dilute, low ionic strength sample compared to the standard buffer range (e.g. 0·01 M). Also, because pH is a logarithmic scale, biases will increase with lower pH. For example, −0·02 unit bias will give +5 μmol litre^{-1} at pH 4·0 and +50 μM at pH 3·0.

Acidity in precipitation samples is frequently naturally buffered, and dissolved carbon dioxide (CO_2) is a prominent weak acid which can bias pH measurements. This is particularly a problem in samples containing net alkalinity; pH > 6·0 indicates that substantial carbonate is present in a system open to atmospheric CO_2 (340 ppm). To remove dissolved CO_2(g), sample aliquots are purged with nitrogen (N_2), helium (He) or another inert gas.

Gran's graphical method has been widely applied to determine strong and weak acidities using base-addition or coulometric titrations; details of the algorithm can be found elsewhere.[95,96] If weak acidity is a small fraction of the total sample acidity, it can be difficult to measure it by titration. The lower pH values of many clouds and fogs makes titration for weak acidity imprecise. Instead, determinations of individual weak acid compounds are necessary.

6.3.2.1 *Aerosol acidity*

No reference method exists for determination of aerosol acidity. Flame photometric detectors (FPD) have been used to determine H_2SO_4 and ammonium sulfates (see following section), although they do not allow a measure of the particulate acidity. For filter samples, H_2SO_4 can be specifically extracted with benzaldehyde and analyzed,[97] although there are interferences from sulfites and bisulfate salts.[98]

Determination of the aerosol acidity can be performed by filter extraction and titration.[99,100] As for precipitation samples, titration with strong base according to Gran[101] allows the strong and weak acidity to be measured.[96] Dissolution of filter samples in 0·05–0·10 m mol litre^{-1} mineral acid solution improves the precision of the determination.[102] This acid-addition method causes removal of dissolved carbonate and suppresses the contribution from other weak acids to the measured acidity. With the addition of the strong acid 'blank', the pH measurement alone is found to provide accurate results for calculating aerosol strong acidity.[60]

In the handling of filters, post-sampling neutralization can be a serious problem. NH_3 is a common contaminant in the field and laboratory. In fact, levels in exhaled breath are usually many times ambient concentrations.[103] As a precaution, many investigators isolate filters during handling in a glove box or hood which has an NH_3 scrubber (e.g. Ref. 60). Another important problem is that some acidic aerosol components can volatilize from the filter surface. Early measurements of organic acids appear to have been biased because the aerosol fraction was not adequately stabilized.[47]

6.3.3 Ionic species

Ion chromatography (IC) has been perhaps the greatest single asset to the analysis of acid precipitation. All of the major ions present in precipitation can be determined at low concentrations by this method.[104,105] Both suppressed and unsuppressed column techniques are routinely used for precipitation analysis of anions. Atomic absorption spectrometry is used for metallic cations,[106] although technological improvements in cation analysis are making IC determinations of major cation species (Na^+, K^+, NH_4^+, Ca^{2+} and Mg^{2+}) more practical and economical, as well.[107] For NH_3, the indophenol colorimetric method,[108] a fluorescence method[109] or the ion-specific (NH_4^+) electrode are methods used in lieu of IC.

Concentrations of S(IV) (i.e. dissolved SO_2 and its adducts in solution) are not routinely determined in precipitation samples, although studies of SO_2 oxidation rates in the field often analyze for it. Without special care, oxidation of S(IV) to sulfate during sample storage will rapidly occur. Addition of HCHO is found to be effective at retarding S(IV) oxidation in samples.[110] The subsequent reaction with pararosaniline to produce a colored product is sensitive to S(IV) concentrations on the order of 1 μ mol litre^{-1}. Even greater sensitivity was achieved with a fluorometric flow-injection method.[111] Naturally occurring hydroxymethanesulfonate (HMSA), the sulfonic acid anion formed from reaction of S(IV) with HCHO can be analyzed directly by ion-pairing chromatography.[112]

Satisfactory calculation of an ion balance in a sample is a necessary, although not sufficient, condition of complete and unbiased analysis. The ion balance is usually given as the ratio of total cation equivalents to total anion equivalents per liter, or vice versa (equivalents = moles×ion charge). Determination errors in major species or the omission of a major component will lead to a distorted ion balance. However, errors in minor species, evaporation of water, chemical reactions (oxidation/reduction), and many of the concerns for sample preservation listed above *cannot* be precluded by a favorable ion balance.

6.3.4 Gas species

SO_2, and NO_x are most frequently measured with continuous analyzers. The US EPA 'Reference Method' for SO_2 is the West–Gaeke bubbler sampler, which stabilizes SO_2 with pararosaniline.[113] The bubbler technique gives a time-integrated value. However, SO_2 fluorescence analyzers are commercially made and are a US EPA-approved 'Equivalent Method'. SO_2 molecules are excited by a zinc or cadmium lamp, causing fluorescent emissions (220–400 nm), which are measured. The sensitivity offers a limit of detection of ~ 5 ppb.[114]

The continuous FPD offers slightly greater sensitivity (~ 2 ppb) and can be used to measure SO_2 and other sulfur-containing gases and particulates.[115] Sulfur compounds are detected in a H_2-rich flame by chemiluminescence emission of excited S_2 molecules at 394 nm. An inert filter is used to remove sulfur-containing particulate compounds. Water vapor is a negative interference, increasing the FPD background current.

Determinations of SO_2 are possible if the other sulfur-containing gases are under 5%. To separate SO_2 and the other sulfur-containing gases at ppb levels, a gas chromatography column is used.[116] In either case, detector response is strongly affected by gas flows of the H_2 and sample air. In aircraft, monitoring requires modified mass flow controller to ensure accuracy.[117]

The chemiluminescent gas-phase reaction between NO and ozone (O_3) is the basis of the EPA Reference Method for NO_2:[113]

$$O_3 + NO \rightarrow NO_2^* + O_2 \tag{6.4}$$

$$NO_2^* \rightarrow NO_2 + h\nu \tag{6.5}$$

where light is emitted between 590 and 3000 nm. Analyzers use a catalytic oven (e.g. molybdenum) to convert NO_2 to NO. Dual-channel or single-channel (with alternating sampling) instruments are used to determine NO only and NO_x ($NO+NO_2$) such that NO_2 is found by difference. These configurations are used in routine monitoring stations, and the level of detectability is generally in the 1–5 ppb range.

Reference 118 describes a modified chemiluminescent instrument that is sensitive to NO at sub-ppb levels.[118] Furthermore, its rapid response allows it to be used for eddy correlation measurements of NO fluxes. A photolysis cell utilizing a high intensity xenon lamp converts a *known* fraction of NO_2 to NO, allowing determination of both NO and NO_2.

Other odd nitrogen compounds can be quantitatively converted to NO over a heated gold catalyst in the presence of H_2 or CO, to give total NO_y (= $NO_x+HNO_2+HNO_3+PAN+RONO_2$).[119] In principle, HNO_3 could be measured with a NO_x analyzer using the same approach as NO_2 (e.g. molybdenum converter) where one channel contains a nylon or akaline filter to remove HNO_3, and the determination is made by difference.[114] However, this approach has not been adequately tested to ensure that only HNO_3 is removed and that it is not lost in the inlet to the analyzer.

FTIR has made IR spectrometry useful in atmospheric studies. It has greater speed and sensitivity than conventional dispersive IR spectroscopy and eliminates some of the interferences caused by ambient levels of water vapor and CO_2; these are conquered using simultaneous wavelength sampling, a long pathlength ($\geq$ 1 km), and computer data acquisition/processing.[120] Measurements of gaseous concentrations in the 1–10 ppb range are possible for HNO_3 and PAN, as well as for NH_3 and organic species such as HCHO.[121]

The tunable diode laser spectrometer (TDLS) is also based on IR spectrometry. While FTIR employs a continuous wavelength source, TDLS uses a laser light source with smaller wavelength range (~ 300 cm^{-1}) and a very narrow line width. Atmospheric studies have monitored a wide variety of species: SO_2, O_3, PAN, N_2O, NH_3, CH_4, HCl, CO and H_2O_2.[122]

Acidic gases (SO_2, HNO_3 and HCl) can also be collected on an alkaline medium and analyzed (for SO_4^{2-}, NO_3^- and Cl^-) as previously described. Filters, diffusion (e.g. annular) denuder tubes, and impingers can provide very sensitive time-integrated ambient measurements: < 1 ppb for 12 h or more.[32] Interferences in SO_2 methods are generally minor. Hydrolysis of NO_2 and PAN in alkaline media including a nylon filter can bias HNO_3 determinations. In the annular denuder method, NaCl-coated tubes have been shown to give unbiased collection of HNO_3.[123] A semi-continuous analyzer for HNO_3 and NH_4NO_3 aerosol in air uses magnesium-sulphate-coated denuders in series (two at ambient, one at 150°C) to collect the vapor and aerosol constituents, respectively.[124] After 30 min of sampling, the denuders are heated to 700°C, volatilizing NO_x which is then monitored with a NO_x analyzer. An analogous thermodenuder approach allows highly sensitive semi-continuous determinations of NH_3.[125]

Organic acids cannot be directly determined in the gas phase; instead, they are collected and concentrated for analysis. Several common methods for the collection and analysis of formate and acetate, the most common organic acids in the atmosphere, have been compared.[47] Artifact formation was a serious problem for methods based on collection on an alkaline substrate. The method apparently most free from artifacts was a mist chamber in which the acids are collected by absorption in a spray of distilled water.[45] Ion-exclusion chromatography and normal-ion chromatography are suitable for the analysis (cf. Ref. 107).

6.3.5 Aerosol samples

Continuous analyzers for aerosol components have not reached the state of sensitivity, selectivity, and economy achieved for many gaseous species. Because of its ease, economy and versatility, collection of samples on filters is used far more frequently. The use of different extraction solutions is indicated for different ions and sampling media. Aqueous extraction is generally adequate for major ions.[126] Wetting of the substrate is not a problem for fibrous filters; however, concern has been raised that water alone may be insufficient to fully wet hydrophobic PTFE membranes frequently used.[127] Often, a small amount of a wetting solution (e.g. ethanol) is applied directly to the Teflon surface prior to filter extraction.[128] With surface wetting and ultrasonication, complete dissolution and recovery of filter-borne solutes can be achieved.[50]

A measurement method for aerosol sulfates has been developed using semi-continuous FPDs.[129,130,131] The instrument is configured to analyze sulfur, and SO_2 is first removed with a diffusion denuder. Since H_2SO_4 is volatile at temperatures below other sulfate aerosols (120°C), a programmed temperature step allows H_2SO_4 to be quantified by difference. Unfortunately, it is not possible with this method to differentiate other acidic sulfate aerosols from non-acidic sulfate, i.e. NH_4HSO_4 versus $(NH_4)_2SO_4$.

X-ray fluorescence (XRF) allows multi-element analysis of particulate material directly on filters *without* extraction.[132] Samples are irradiated by a narrow X-ray column. Sample fluorescence is detected with either a wavelength or an energy dispersive spectrometer. Because the response is reduced by particle penetration into collection media, membrane-only filters are recommended.[133] Analogous material thin-section standards are required to quantitatively interpret XRF responses. Total particulate sulfur can be measured with a minimum detection limit of $\sim$ 20 ng cm^{-2}; this converts to $< 0{\cdot}1$ μg m^{-3} for most aerosol samples. A recent review of XRF techniques and detection limits is given in Ref 115.

FTIR spectroscopy has been applied to semi-continuous determinations of all major aerosol components in air.[134,135] Particulate material is collected on a zinc selenium surface for several minutes and then analyzed. Peaks can be resolved to differentiate H_2SO_4, NH_4HSO_4 and $(NH_4)_2SO_4$, as well as NH_4NO_3. This may become the state-of-the-art for acid aerosol analyzers, but in the meanwhile its expenses and humidity-related interferences hold it back from wider field application.

6.4 SUMMARY

It is important to recognize similarities and difference among techniques for measurements of atmospheric acidity. Many of the same chemical components exist in the various phases (e.g. NO_3^- in droplets, aerosol and gas phases). Transfer of species between phases occurs naturally, but it can also be caused by alterations in thermodynamic equilibria caused by sampling techniques. Collection of rainwater, snow, cloudwater, dew. etc. yields aqueous aliquots which require similar concern regarding contamination, preservation and analysis. Techniques for time-integrated samples of gaseous or aerosol components generally produce samples suitable for similar analytical determinations. The analyses of individual analytes carry further constraints on the preservation and storage of samples. Accuracy of analyses must be checked with traceable standards and intercomparison with independent techniques. The correct ionic balance of samples is a necessary, but not sufficient conditions of complete and error-free results.

Certain materials are unsuitable as collection media or for construction of sampling apparatus. For example, SO_2 and HNO_3 will react with alkaline substrates, leading to biases in the gas (or the aerosol product) determinations. Weak acids are also prone to artifact formation on certain collection media. Hence, care in selecting materials is essential.

Size selective inlets are essential for acidic aerosol collection. While acids are found almost exclusively in the fine aerosol mode, coarse particulate material is frequently alkaline. This will neutralize the samples' acidity if the fractions are collected together. Gaseous NH_3 also has the capability to cause negative artifacts in acidic aerosol samples, and diffusion denuders are generally used to protect collected material during sampling.

REFERENCES

1. Galloway, J. N. & Likens, G. E., Calibration of collection procedures for the determination of precipitation chemistry. *Water, Air and Soil Pollution,* **6** (1976) 241–58.
2. Ross, H. B., The importance of reducing sample contamination in routine monitoring of trace metals in atmospheric precipitation. *Atmos. Environ.,* **20** (1986) 401–5.
3. Galloway, J. N. & Likens, G. N., The collection of precipitation for chemical analysis. *Tellus,* **30** (1978) 71.
4. Rodda, J. C. & Smith, S. W., The significance of the systematic error in rainfall measurement for assessing wet deposition. *Atmos. Environ.,* **20** (1986) 1059–64.
5. Bigelow, D. S., *NADP/NTN Site Section and Installation*, Resource Ecology Laboratory Report. Colorado State University, Fort Collins, CO, USA, 1984.
6. Artz, R. S. & Rolph, G. D., Evaluation of precipitation chemistry siting criteria using paired stations from northern Maine and southeastern Texas. *Atmos. Environ.,* **23** (1989) 1033–50.
7. Pruppacher, H. R. & Klett, J. D., *Microphysics of Clouds and Precipitation*. Reidel, Dordrecht, The Netherlands, 1978.
8. Waldman, J. M. & Hoffmann, M. R., Pollutant behavior in fog. In *Sources and Fates of Aquatic Pollutants,* Advances in Chemistry Series 216, ed. S. J. Eisenreich. Ann Arbor Science, Ann Arbor, MI, 1987, pp. 79–129.
9. Brink, H. M. ten, Schwartz, S. E. & Daum, P. H., Efficient scavenging of aerosol sulfate in liquid-water clouds. *Atmos. Environ.,* **21** (1987) 2035–52.

10. Israel, R. & Rosner, D. E., Use of a generalized Stokes number to determine the aerodynamic capture efficiency of non-Stokesian particles from a compressible gas flow. *Aerosol Sci. Technol.,* **2** (1983) 45–51.
11. Friedlander, S. K., *Smoke, Dust, and Haze.* J. Wiley, New York, USA, 1977.
12. Jacob, D. J., Wang, R. F. & Flagan, R. C., Fogwater collector design and characterization. *Environ. Sci. Technol.,* **18** (1984) 827–33.
13. Davies, C. N. & Subari, M., Aspiration above wind velocity of aerosol with thin-walled nozzles facing and right angles to the wind direction. *J. Aerosol Sci.,* **13** (1982) 59–71.
14. Hering, S. V., Blumenthal, D. L., Brewer, R. L., Gertler, A., Hoffman, M. R., Kadlecek, J. A. & Pettus, K., Field intercomparison of five types of fogwater collectors, *Environ. Sci. Technol.,* **21** (1987) 654–63.
15. Katz, U. &. Miller, D. F., *Development and Evaluation of Cloud Water Collectors,* Final Report. Desert Research Institute, Reno, NV, USA, 1984.
16. Brewer, R. L., Gordon, R. J., Shepard, L. S. & Ellis, E. C., Chemistry of mist and fog from the Los Angeles urban area. *Atmos. Environ.,* **17** (1983) 2267–70.
17. Daube Jr, B. C., Kimball, K. D., Lamar, P. A. & Weathers, K. C., Two new ground-level cloud water sampler designs which reduce rain contamination. *Atmos. Environ.,* **21** (1987) 893–900.
18. Weathers, K. C., Likens, G. E., Bormann, F. H., Bicknell, S. H., Bormann, B. T., Daube Jr, B. C., Eaton, J. S., Galloway, J. N., Keene, W. C., Kimball, K. D., McDowell, W. H., Siccama, T. G., Smiley, D. & Tarrant R. A., Cloudwater chemistry from ten sites in North America. *Environ. Sci. Technol.,* **22** (1988) 1018–26.
19. Collett Jr, J. F., Daube, B. C., Munger, J. W. & Hoffmann, M. R., Cloudwater chemistry in Sequoia National Park. *Atmos. Environ.,* **23** (1989) 999–1007.
20. Collett Jr, J. F., Characterization of cloudwater and precipitation chemistry and deposition at elevated sites in Central and Southern California. PhD thesis, California Institute of Technology, Pasadena, CA, USA, 1989.
21. Jacob, D. J., Waldman, J. M., Haghi, M., Hoffmann, M. R. & Flagan, R. C., Instrument to collect fogwater for chemical anlaysis. *Rev. Sci. Instrum.,* **56** (1985) 1291–3.
22. Falconer, R. E., & Falconer, P. D., Determination of cloud water acidity at a mountain observatory in the Adirondack Mountains of New York State. *J. Geophys. Res.,* **85** (1980) 7465–70.
23. Hegg, D. A. & Hobbs, P. V., Cloud water chemistry and the production of sulfates in clouds. *Atmos. Environ.,* **15** (1981) 1597–1604.
24. Winters, W., Hogan, A., Mohnen, V. & Barnard, S., *ASRC Airborne Cloud Water Collection System.* Publication No. 728. ASRC-SUNYA, Albany NY, USA, 1979.
25. Huebert, B. J. & Robert, C. H., The dry deposition of nitric acid to grass. *J. Geophys. Res.,* **90** (1985) 2085–90.
26. Huebert, B. J. & Baumgardner, D., A preliminary evaluation of the Mohnen slotted-rod cloudwater collector. *Atmos. Environ.,* **19** (1985) 843–6.

27. US Environmental Protection Agency, National primary and secondary ambient air standards. Appendix B: Reference method for the determination of suspended particulates in the atmosphere. *Federal Register,* **36** (84) (1971) Part II.
28. US Environmental Protection Agency, Revisions to the National Ambient Air Quality Standards for Particulate Matter. Appendix J: Reference method for the determination of particulate matter as PM_{10} in the Atmosphere (40 CFR Part 50). *Federal Register,* **52** (1987) FR 24664.
29. McFarland, A. R., Ortiz, C. A. & Bertch Jr, R. W., A 10-micron cutpoint size selective inlet for hivol samplers. *J. Air Pollut. Control Assoc.,* **34** (1984) 544.
30. Wedding, J. B. & Weingand, M. A., The Wedding ambient aerosol sampling inlet (D_{50} = 10 μm) for the high volume sampler. *Atmos. Environ.,* **19** (1985) 535.
31. Tanner, R. L., Garber, R., Marlow, W., Leaderer, B. P. & Leyko, M. A., Chemical composition of sulfate as a function of particle size in New York summer aerosol. *Ann. NY Acad. Sci.,* **322** (1979) 99–114.
32. Koutrakis, P. Wolfson, J. M., Spengler, J. D., Stern, B. S. & Franklin, C. A., Equilibrium size of atmospheric sulfates as a function of the relative humidity. *J. Geophys. Res.,* **94** (1989) 6442–8.
33. Stevens, R. K., Modern methods to measure air pollutants. In *Aerosols: Research, Risk Assessment and Control Strategies*, ed. S. D. Lee, T. Schneider, L. D. Grant & P. J. Verkerk. Lewis, Chelsea, MI, USA, 1986, pp. 69–96.
34. Pierson, W. R., Hammerle, R. H. & Brachaczek, W. W., Sulfate formed by interaction of sulfur dioxide with filters and aerosol deposits. *Anal. Chem.,* **48** (1976) 1808–11.
35. Appel, B. R., Wall, S. M., Haik, M., Kothny, E. L. & Tokiva, Y., Evaluation of techniques for sulfuric acid and particulate strong acidity measurements in ambient air. *Atmos. Environ.,* **14** (1980) 559–63.
36. Coutant, R. W., Effect of environmental variables on collection of atmospheric sulfate. *Environ. Sci. Technol.,* **11** (1977) 873–8.
37. Spicer. C. W. & Schumacher, P. M., Particulate nitrate: laboratory and field studies of major sampling interferences. *Atmos. Environ.,* **13** (1979) 543–52.
38. Appel, B. R., Wall, S. M., Tokiwa, Y. & Haik, M., Interference effects in sampling particulate nitrate in ambient air. *Atmos. Environ.,* **13** (1979) 319–25.
39. Commins, B. T. & Waller, R. E., Observations from a ten-year study of pollution at site in the city of London. *Atmos. Environ.,* **1** (1967) 49–68.
40. Morandi, M. T., Kneip, T. J., Cobourn, W. G., Husar, R. B. & Lioy, P. J., The measurements of H_2SO_4 and other sulfate species at Tuxedo, NY with a thermal analysis flame photometric detector and the analysis of simultaneously collected quartz filter samples. *Atmos. Environ.,* **17** (1983) 843–8.
41. Tanner, R. L., Cederwall, R., Garber, R., Leahy, D., Marlow, W., Meyers, R., Phillips, M. & Newman, L., Separation and analysis of aerosol sulfate species at ambient concentrations. *Atmos. Environ.,* **11** (1977) 955–67.

42. Holland, J. C., Harrell, R. M., Appel, B. R. & Kothny, E. L., Measurement of the alkalinity of filter media. *J. Air Pollut. Control Assoc.*, **39** (1989) 59–62.
43. US Environmental Protection Agency, *Site Selection for the Monitoring of Photochemical Air Pollutants*, Report EPA 450/3-78-013. Office of Air Quality Planning and Standards, Research Triangle Park, NC, USA, 1978.
44. Appel, B. R., Povard, V. & Kothny, E. L., Loss of nitric acid within inlet devices for atmospheric sampling. In *Proceedings of the 1987 EPA/APCA Symposium of Measurement of Toxic and Related Air Pollutants*. APCA, Pittsburgh, PA, USA, 1987, pp. 158–67.
45. Talbot, R. W., Stein, K. M., Harriss, R. C. & Cofer, W. R., Atmospheric geochemistry of formic and acetic acids at a mid-latitude temporate site. *J. Geophys. Res.*, **93** (1988) 1638–52.
46. Kelly, T. J., Tanner, R. L., Newman, L., Galvin, P. J., & Kadlecek, J. A., Trace gas and aerosol measurements at a remote site in the northeast US. *Atmos. Environ.*, **18** (1984) 2565–76.
47. Keene, W. C., Talbot, R. W., Andrae, M. O., Beecher, K., Berresheim, H., Castro, M., Farmer, J. C., Galloway, J. N., Hoffmann, M. R., Li, S. M., Maben, J. R., Munger, J. W., Norton, R. B., Pszenny, A. A. P., Puxbaum, H., Westberg, H. & Winiwarter. W., An intercomparison of measurement systems for vapor and particulate phase concentrations of formic and acetic acids. *J. Geophys. Res.*, **94** (1989) 6457–71.
48. Forrest, J. D., Tanner, R. L., Spandau, D., D'Ottavio, T. & Newman, L., Determination of total inorganic nitrate utilizing collection of nitric acid on NaCl-impregnated filters. *Atmos. Environ.*, **14** (1980) 137–44.
49. Anlauf, K. G., Fellin, P., Wiebe, H. A., Schiff, H. I., Mackay, G. I., Braman, R. S. & Gilbert, R., A comparison of three methods for measurement of atmospheric nitric acid and aerosol nitrate and ammonium. *Atmos. Environ.*, **19** (1985) 325–33.
50. Russell, A. G. & Cass, G. R., Acquisition of regional air quality model validation data for nitrate, sulfate, ammonium ion and their precursors. *Atmos. Environ.*, **18** (1984) 1815–27.
51. Knapp, K. T., Durham, J. L. & Ellestad, T. G., Pollutant sampler for measurements of atmospheric acidic dry deposition. *Environ. Sci. Technol.*, **20** (1986) 633–7.
52. Quinn, P. K. & Bates T. S., Collection efficiencies of a tandem sampling system for atmospheric aerosol particles and gaseous ammonia and sulfur dioxide. *Environ. Sci. Technol.*, **23** (1989) 736–9.
53. Forrest, J. D., Spendau, J., Tanner. R. L. & Newman, L., Determination of atmospheric nitrate and nitric acid employing a diffusion denuder with a filter pack. *Atmos. Environ.*, **16** (1982) 1473–85.
54. Spicer, C. W., Howes, J. E., Bishop, T. A., Arnold, L. H. & Stevens, R. K., Nitric acid measurement methods: an intercomparison. *Atmos. Environ.*, **16** (1982) 1487–1500.
55. Gormley, P. & Kennedy, M., Diffusion from a stream flowing through a cylindrical tube. *Proc. Royal Irish Acad.*, **52A** (1949) 133–69.
56. Ferm, M., Method for the determination of atmospheric ammonia. *Atmos. Environ.*, **13** (1979) 1385–93.

57. Possanzini, M., Febo, A. & Liberti, A., New design of a high-performance denuder for the sampling of atmospheric pollutants. *Atmos. Environ.*, **17** (1983) 2605–10.
58. Vossler, T. L., Stevans R. K. & Baumgardner, R. E., A study of the performance of annular denuders and preseparators . In *Proceedings of the 1987 EPA/APCA Symposium of Measurement of Toxic and Related Air Pollutants*. APCA, Pittsburgh, PA, USA, 1988, pp. 168–73.
59. Waldman, J. M., Lioy, P. J., Meulen, A. van der, Reijnders, H. & Hayes, C., Operation of annular denuder systems for atmospheric acidity measurements in Bilthoven, The Netherlands. In *Proceedings of the 1987 EPA/APCA Symposium of Measurement of Toxic and Related Air Pollutants*. APCA, Pittsburgh, PA, USA, 1988, pp. 691–8.
60. Koutrakis, P., Wolfson, J. M., Slater J.L., Brauer, M., Spengler, J. D., Stevens, R. K. & Stone, C. L., Evaluation of an annular denuder/filter pack system to collect acidic aerosols and gases. *Environ. Sci. Technol.*, **22** (1988) 1463–8.
61. Sickles, J. E., Hodson, L. L., Rickman, E. E., Saeger, M. L., Hardison, D. L., Turner, A. R., Sokol, C. K., Estes, E. D. & Paur, R. J., Comparison of the annular denuder system and the transition flow reactor for measurements of selected dry deposition species. *J. Air Pollut. Control Assoc.*, **39** (1989) 1218–24.
62. Slater, J. L., Koutrakis, P., Keeler, G. J. & Wolfson, J. M., Determination of aerosol strong acidity losses due to interactions of collected particles: results from laboratory and field studies. In *Proceedings of the 1988 EPA/APCA Symposium on Measurement of Toxic and Related Air Pollutants*. APCA, Pittsburgh, PA, USA, 1988, 176–81.
63. Hicks, B. B., Wesley, M. L., Lindberg, S. E. & Bromberg, S. M., *Proceedings of the NAPAP Workshop on Dry Deposition*. NOAA/ATDD, Oak Ridge, TN, USA, 1986.
64. Nicholson, K. W., The dry deposition of small particles: a review of experimental measurements. *Atmos. Environ.*, **22** (1988) 2653–66.
65. Davidson, C. I., Miller, J. M. & Pleskow, M. A., The influence of surface structure on predicted particle dry deposition to natural grass canopies. *Water Air Soil Pollut.*, **18** (1982) 25–43.
66. Lindberg, S. E., Harriss, R. C. & Turner, R. R., Atmospheric deposition of metals to forest vegetation. *Science*, **215** (1982) 1609–11.
67. Bytnerowicz, A., Miller, P. R. & Olszyk, D. M., Dry deposition of nitrate, ammonium and sulfate to a *Ceanothus Crassifolius* canopy and surrogate surfaces. *Atmos. Environ.*, **21** (1987) 1749–57.
68. Hosker, R. P. & Lindberg, S. E., Review: atmospheric deposition and plant assimilation of gases and particles. *Atmos. Environ.*, **16** (1982) 889–910.
69. Cadle, S. H., Dasch, J. M. & Kopple, R. V., Wintertime wet and dry deposition in northern Michigan. *Atmos. Environ.*, **20** (1986) 1171–8.
70. Dovland, H. & Eliassen, A., Dry deposition on a snow surface. *Atmos. Environ.*, **10** (1976) 783–5.
71. Pierson, W. R., Brachaczek, W. W., Gorse, R. W., Japar, S. M. & Norbeck, J. N., On the acidity of dew. *J. Geophys. Res.*, **91** (1986) 4083–96.

72. Denmead, O. T. & Bradley, E. F., Flux-gradient relationships in a forest canopy. In *The Forest-Atmosphere Interaction*, ed. B. A. Hutchinson & B. B. Hicks. Reidel, Dordrecht, The Netherlands, 1985, pp. 421–42.
73. Stull, R. B., *An Introduction to Boundary Layer Meteorology*. Kluwer Academic, Dordrecht, The Netherlands, 1988.
74. Garland, J. A., The dry deposition of sulphur dioxide to land and water surfaces. *Proc. Royal Soc.* (Lond), **A354** (1977) 245.
75. Sheperd, J. G., Measurements of the direct deposition of sulfur dioxide onto grass and water by the profile method. *Atoms. Environ.,* **8** (1974) 69–74.
76. Garland, J. A. & Cox, L. C., Deposition of small particles to grass. *Atmos. Environ.,* **16** (1982) 2699–702.
77. Nicholson, K. W. & Davies, T. D., Field measurements of the dry deposition of particulate sulfate. *Atmos. Environ.,* **21** (1987) 1561–71.
78. Sievering, H., Small-particle dry deposition under high wind speed conditions: eddy flux measurements at the Boulder Atmospheric Observatory. *Atmos. Environ.,* **21** (1987) 2179–85.
79. Hicks, B. B., Matt, D. R., McMillen, R. T., Womack, J. D., Wesley, M. L., Hart, R. L., Cook, D. R., Lindberg, S. E., Pena, R. G. de & Thonson, D. W., A field investigation of sulfate fluxes to a deciduous forest. *J. Geophys. Res.,* **94** (1989) 13003–11.
80. Walcek, C. J., Brost, R. A., Chang, J. S. & Wesley, M. L., SO_2, sulfate and NHO_3 deposition velocities computed using regional land use and meteorological data. *Atmos. Environ.,* **20** (1986) 9949–64.
81. Oberlander, G. T., Summer fog precipitation on the San Francisco peninsula. *Ecology,* **37** (1956) 851–2.
82. Yoshida, A. & Kuroiwa, D., Wind force on a conifer tree and the quantity of fog water thereby captured. In *Studies on Fog in Relation to Fog-Preventing Forest*, ed. T. Hori. Tanne Trading Co., Sapporo, Japan, 1953, pp. 261–78.
83. Kerfoot, O., Mist precipitation on vegetation. *Forest. Abst.,* **29** (1968) 8–20.
84. Lovett, G. M., Rates and mechanisms of cloud water deposition to a subalpine balsam fir forest. *Atmos. Environ.,* **18** (1984) 361–71.
85. Goodman, J., The collection of fog drip. *Water Resources Res.,* **21** (1985) 392–4.
86. Waldman, J. M., Depositional aspects of pollutant behavior in fog. PhD thesis, California Institute of Technology, Pasadena, CA, USA, 1986.
87. Dollard, G. J. & Unsworth, M. H., Field measurements of turbulent fluxes of wind-driven fog drops to a grass surface. *Atmos. Environ.,* **17** (1983) 775–80.
88. Harvey, M. J. & McArthur, A. J., Pollution transfer to moor by occult deposition. *Atmos. Environ.,* **23** (1989) 1073–82.
89. Browell, E. V., Remote sensing of tropospheric gases and aerosols with an airborne DIAL system. In *Optical Laser Remote Sensing*, ed. D. K. Killinger & A. Mooradian. Springer, New York, 1983, pp. 138–47.
90. Farmer, C. B., Raper, O. F. & Norton, R. H., Spectroscopic detection and vertical distribution of HCl in the troposphere and stratosphere. *Geophys. Res. Lett.,* **17** (1976) 869–72.

91. Beer, R. & Glavich, T. A., Remote sensing of the troposphere by infrared emission spectroscopy. *J. SPIE,* **1129** (1989) 42–51.
92. Herlihy, L. J., Galloway, J. N. & Mills, A. L., Bacterial utilization of formic and acetic acid in rainwater. *Atmos. Environ.,* **21** (1987) 2397–402.
93. Keene, W. C., Galloway, J. N. & Holdren, Jr, J. N., Measurement of weak organic acidity in precipitation from remote areas of the world. *J. Geophys. Res.,* **88** (1983) 5122–30.
94. McMurry, P. H., Takano, H. & Anderson, G. R., Study of the ammonia (gas)–sulphuric acid (aerosol) reaction rate. *Environ. Sci. Technol.,* **17** (1983) 347–52.
95. Stumm, W. & Morgan, J. J., *Aquatic Chemistry*, 2nd edn. John Wiley, New York, USA, 1981.
96. Brosset, C., A method of measuring airborne acidity: its application for the determining of acid content on long-distance transported particles and in drainage water from spruces. *Water Air Soil Pollut.,* **6** (1976) 259–75.
97. Leahy, D., Siegel, R., Klotz, P. & Newman, L., The separation and characterization of sulfate aerosol. *Atmos. Environ.,* **9** (1975) 219–29.
98. Eatough, D. L., Isatt, S., Ryder, I. & Hansen, L. D., Use of benzaldehyde as a selective solvent for sulfuric acid: interferences by sulfate and sulfite salts. *Environ. Sci. Technol.,* **12** (1978) 1276–9.
99. Junge, C. & Scheich, G., Determination of the acid content of aerosol particles. *Atmos. Environ.,* **5** (1971) 165–75.
100. Brosset, C. & Ferm, M., Man-made airborne acidity and its determination. *Atmos. Environ.,* **12** (1978) 909–16.
101. Gran, G., Determination of the equivalence point in potentiometric titrations, Part II. *The Analyst,* **77** (1952) 661–71.
102. Stevens, R. K., Dzubay, T. G., Russwurm, G. & Rickel, D., Sampling and analysis of atmospheric sulfates and related species. *Atmos. Environ.,* **12** (1976) 55–68.
103. Larson, T. V., Covert, D. S., Frank, R. & Carlson, R. J., Ammonia in the human airways: Neutralization of inspired acid sulfate aerosols. *Science,* **197** (1977) 161–3.
104. Mulik, J. D. & Sawicki, E., eds, *Ion Chromatographic Analysis of Environmental Pollutants*. Ann Arbor Science, Ann Arbor, MI, USA, 1979.
105. Weiss, J., *Handbook of Ion Chromatography*. Dionex Corp, Sunnyvale, CA, USA, 1986.
106. Cresser, S. M. & Ebdon, L. D., Atomic spectrometry update—environmental analysis. *J. Anal. Atom. Spectrom.,* **1** (1986) 1R–7R.
107. Tarter, J. G., ed., *Ion Chromatography*. Marcel Dekker, New York, USA, 1987.
108. Harwood, J. E. & Huyser, D. J., Some aspects of the phenol–hypochlorite reaction as applied to ammonia analysis. *Water Res.,* **4** (1970) 501–15.
109. Rapsomanikis, S., Wake, M., Kitto, N. & Harrison, R. M., Analysis of atmospheric ammonia and particulate ammonium by a sensitive fluorescence method. *Environ. Sci. Technol.,* **22** (1988) 948–52.
110. Dasgupta, P. K., DeCesare, K. & Ullrey, J. C., Determination of atmospheric sulfur dioxide without tetrachloromercurate(II) and the mechanism of the Schiff reaction. *Anal. Chem.,* **52** (1980) 1912–22.

111. Dasgupta, P. K. & Yang, H. C., Trace determination of aqueous sulfite, sulfide and methanethiolate by fluorimetric flow injection analysis. *Anal. Chem.*, **58** (1986) 2839–44.
112. Munger, J. W., Tiller, C. T. & Hoffmann, M. R., Determination of hydroxymethanesulfonate in fog water. *Science*, **231** (1986) 247–9.
113. US Environmental Protection Agency. *Air Quality Criteria for Particulate Matter and Sulfur Oxides*. Report EPA-600/8-82-029, Environmental Criteria and Assessment Office, Research Triangle Park, NC, USA, 1982.
114. Finlayson-Pitts, B. J. & Pitts Jr, J. N., *Atmospheric Chemistry: Fundamentals and Experimental Techniques*. John Wiley, New York, USA, 1986.
115. Lodge, J. P., ed., *Methods of Air Sampling and Analysis*, 3rd edn. Lewis, Chelsea, MI, USA, 1989.
116. Stevens, R. K., Mulik, J. D., O'Keeffe, A. E. & Krost, K. J., Gas chromatography of reactive sulfur in gases in air at the part per billion level. *Anal. Chem.*, **43** (1971) 827–31.
117. Garber, R. W., Daum, P. H., Doering, R. F., D'Ottavio, T. & Tanner, R. L., Determination of ambient aerosol and gaseous sulfur using a continuous FPD. III. Design and characterization of a monitor for airborne application. *Atmos. Environ.*, **17** (1983) 1381–5.
118. Ridley, B. A. & Howlett, L. C., An instrument for nitric oxide measurements in the stratosphere. *Rev. Sci. Instrum.*, **45** (1974) 742–46.
119. Fahey, D. W., Eubank, C. S., Hubler, G. & Fehsenfeld, F. C., Evaluation of a catalytic reduction technique for the measurement of total reactive odd-nitrogen NO_y in the atmosphere. *J. Atmos. Chem.*, **3** (1986) 435–68.
120. Griffiths, P. R., Recent commercial instrumental developments in Fourier Transform Infrared Spectrometry. *Adv. Infrared Raman Spectrosc.*, **10** (1983) 277.
121. Tuazon, E. C., Winer, A. M. & Pitts Jr, J. N., Trace pollutant concentrations in a multiday smog episode in the California south coast air basin by long path length Fourier Transform Infrared Spectroscopy. *Environ. Sci. Technol.*, **15** (1981) 1232–7.
122. Cassidy, D. T. & Reid, J., High-sensitivity detection of trace gases using sweep integration and tunable diode lasers. *Appl. Opt.*, **21** (1982) 2527–30.
123. Febo, A., DeSantis, F. & Perrino, C., Measurement of atmospheric nitrous and nitric acids by means of annular denuders. In *Physico-Chemical Behavior of Atmospheric Pollutants*, Air Pollution Research Report 2. Reidel, Dordrecht, The Netherlands, 1986, pp. 121–5.
124. Klockow, E., Niessner, R., Malejczyk, M., Kiendl, H., Berg, B. Vom, Keuken, M. P., Wayers-Ypelaan, A. & Slanina, J., Determination of nitric acid and ammonium nitrate by means of a computer-controlled thermodenuder system. *Atmos. Environ.*, **23** (1989) 1131–8.
125. Keuken, M. P., Slanina, J. & Wayers-IJpelaan, A., Determination of ammonia, nitric acid and ammonium nitrate aerosols by means of automated thermodenuder systems. Paper presented at the 1988 Annual Meeting of the American Association for Aerosol Research, Chapel Hill, NC, USA, 1988.
126. Ferek, R. J., Lazrus, A. L., Haagenson, L. & Winchester, J. W., Strong and weak acidity of aerosols collected over the northeastern United States. *Environ. Sci. Technol.*, **17** (1983) 315–24.

127. Derrick, M. & Moyers, J., Precise and sensitive water soluble ion extraction method for aerosol samples collected on polytetrafluoroethylene filters. *Analytical Lett.*, **14** (1981) 1637–52.
128. Wolfson, J. M., Determination of microgram quantities of inorganic sulfate in atmospheric particulates. *J. Air Pollut. Control Assoc.*, **30** (1980) 688–90.
129. Huntzicker, J. J., Hoffman, R. S. & Ling, G. S., Continuous measurement and speciation of sulfur-containing aerosols by flame photometry. *Atmos. Environ.*, **12** (1980) 83–8.
130. Cobourn, W. G., Husar, R. B. & Husar, J. D., Continuous in situ monitoring of ambient particulate sulfur using flame photometry and thermal analysis. *Atmos. Environ.*, **12** (1978) 89–98.
131. Tanner, R. L., D'Ottavio, T., Garber, R. & Newman, L., Determination of ambient aerosol sulfur using a continuous flame photometric detection system. I. Sampling system for aerosol sulfate and sulfuric acid. *Atmos. Environ.*, **14** (1980) 121–7.
132. Dzubay, T. G., ed., *X-ray Fluorescence Analysis of Environmental Samples*. Ann Arbor Science, Ann Arbor, MI, USA, 1977.
133. Davis, D. W., Reynolds, R. L., Tsou, G. C. & Zafonte, L., Filter attenuation corrections for the X-ray fluorescence analysis of atmospheric aerosol. *Anal. Chem.*, **49** (1977) 1990–3.
134. Cunningham, P. T. & Johnson, S. A., Spectroscopic observation of acid sulfate in atmospheric particulate samples. *Science*, **191** (1976) 77–9.
135. Johnson, S. A., Groczyk, D. G., Kumar, R. & Cunningham, P. T., *Analytical Techniques for Ambient Sulfate Aerosols*. Argonne National Laboratory Report ANL-81-12, Argonne, IL, June 1981.
136. Fuzzi, S., Castillo, R. A., Jiusto, J. E. & Lala, G. G., Chemical composition of radiation fogwater microphysics. *J. Geophys. Res.*, **89** (1984) 7159–64.

Chapter 7

Field Studies of Cloud Chemical Processes Leading to Acid Deposition

T. W. CHOULARTON
Department of Pure and Applied Physics, UMIST, PO Box 88, Manchester, M60 1QD, UK

7.1 INTRODUCTION

Despite occupying only about 15% of the volume of lower trophosphere, clouds play a very important role in the chemistry of the atmosphere. The clouds play the major role in the upward transport of trace gases emitted from the earth's surface. They readily entrain material from the free troposphere and so bring these gases into intimate contact with this rich source of oxidants. Clouds act as reactors for chemical species in the atmosphere, particularly for soluble species. Precipitating clouds deposit soluble gases and particulate matter back to the surface. They reflect photochemically active ultraviolet (UV) radiation from the sun.

This chapter concentrates on the contribution of field studies to the understanding of cloud chemical processes and the very important wet deposition processes. These will include measurements made on the ground, either in hill cap clouds or radiation fogs or by aircraft in cloud systems. In addition, modelling work will be discussed as appropriate.

A comprehensive review of the work undertaken will not be attempted rather a highly selective path will be followed reflecting the author's view of the most important steps forward. Particular emphasis will be given to the role of clouds in the aqueous-phase oxidation of sulphur dioxide (SO_2) to sulphuric acid (H_2SO_4) the major constitutent of acid rain.[1] The role of clouds in the chemistry of oxides of nitrogen (NO_x) will also be discussed. The discussion of wet deposition processes will focus on the role of hill cap clouds in providing a major increase in wet deposition in regions of complex topography. Clouds also play an important role in the control of the important tropospheric oxidants, ozone (O_3) and hydrogen peroxide (H_2O_2). After a short review of the relevant chemistry being investigated follows a discussion of the distribution of the main oxidants in the atmosphere and the possible role of clouds in influencing the oxidising properties of the atmosphere.

7.2 THE FORMATION OF OXIDES OF SULPHUR AND NITROGEN IN CLOUD

Observations indicate that the deposition of H_2SO_4 and nitric acid (HNO_3) are not always highly correlated with the production of their gaseous precursors.[2] For example, the wet deposition of sulphate in Western Europe between 1970 and 1980 did not reflect the known increase in sulphur emissions.[3] This implies that factors such as the availability of oxidants can control the rate of strong acid production. Hence, it is very important that these processes are quantified in the real atmosphere.

7.2.1 Sulphur chemistry

The principal oxidants for dissolved SO_2 S (IV) in cloud are H_2O_2, O_3 and oxygen catalysed by Fe (III) and Mn (II). Of these the reaction with O_3 leads to the highest conversion rate for a cloudwater pH above about 5·5. The rate is, however, inversely proportional to the square of the hydrogen ion (H^+) concentration and so decreases rapidly at low pH.[4] The rate of oxidation by H_2O_2 increases linearly with increasing H^+ concentration,[5] but the corresponding decrease in SO_2 solubility with increasing acidity leads to an overall conversion which remains roughly constant with pH.

Over the normally observed pH range for cloudwater (3–6) it is therefore expected that the oxidation of S(IV) in aqueous solution will be dominated by the O_3 reaction above pH 5·5 and by the acid-catalysed H_2O_2 reaction at lower pH, provided this oxidant is available in sufficient quantities. The reaction with oxygen is considered to be of little importance below pH 7, although the metal-catalysed reactions may be important above pH 5 at night-time[1] if the catalyst is available in sufficient concentration.

In addition, it has been suggested that the hydroxyl (OH) and hydroperoxyl (HO_2) free radicals, scavenged by droplets from the gas phase, might also be important oxidants in midsummer, provided that their accommodation coefficients for collection by water droplets are larger than 0·0001.[6]

7.2.2 Nitrogen chemistry

The relative importance of nitrogen compounds as contributions to regional-scale pollution is increasing due to a reduction in sulphur emissions and an increase in the emissions of NO_x. The atmospheric chemistry of nitrogen species is, however, much less well understood than sulphur chemistry.

One of the most important processes in the atmospheric chemistry of nitrogen dioxide (NO_2) is the conversion of NO_2 to nitric acid (HNO_3). The deposition velocity of NO_2 is small, hence the atmospheric lifetime of NO_x is strongly influenced by the rate of oxidation to HNO_3, which is rapidly removed either from its gaseous form by dry deposition, or by wet deposition after entering aqueous solution.

The production of HNO_3 during daylight hours is mainly controlled by the gas-phase reaction of the OH radical with NO_2. This mechanism is well understood. However, further possible reactions of NO_2 involving multiphase processes are very poorly quantified. The production of HNO_3 in cloud droplets is a possible pathway and this reaction may proceed through the gas-phase production of NO_3 and N_2O_5 followed by the capture of these molecules in cloud droplets. This latter process is well known but there is a lack of field evidence of its effectiveness. It should be more effective at night because of the removal of nitrogen trioxide (NO_3) and N_2O_5 by photolysis during the day. Evidence for this may be obtained by studying the rate of production of NO_3^- in cloudwater and the NO_y/NO_x ratios in the gas phase during the change from day to night or visa versa.

It is also been suggested that nitrous acid (HNO_2) is formed by heterogeneous processes involving NO_2. This can be detected by cloudwater measurements of NO_2^- which itself may be a source of NO_3^- by oxidation in droplets.

7.3 THE DISTRIBUTION OF OXIDANTS IN THE ATMOSPHERE

Gas phase measurements of H_2O_2, SO_2 and O_3 in the lower troposphere over the eastern US in October and November 1984 have been reported.[7] It was found that in general gas phase H_2O_2 concentrations in air near the surface were much less than in air aloft in the free troposphere. Maximum H_2O_2 concentrations were observed just above cloud tops. A latitudinal gradient also existed both near the surface and aloft with higher concentrations in the south. The H_2O_2 concentrations varied from 0 to 4·1 ppbv (1 ppbv = 1 part in 10^9 by volume). O_3 concentrations showed no variation with altitude and were typically 35 ± 10 ppbv. The O_3 concentration is usually in excess of the SO_2 concentration in the atmosphere and the SO_2–O_3 reaction is not usually oxidant limited. SO_2 concentrations were generally highest near the surface. Similar results have been obtained by the UMIST aircraft flying in the vicinity of Great Dun Fell (Cumbria, UK) during April. These measurements were made concurrently with the ground-based field experiments described below.[8]

H_2O_2 is a photo-oxidant which is produced from HO_2 radicals which in turn can derive from OH reactions during the photo-oxidation of hydrocarbons. HO_2 is also produced through formaldehyde (HCHO) photolysis. The most likely explanation for the enhanced H_2O_2 concentrations near cloud top is the enhanced photochemistry produced by scattered radiation either just above or within the upper portions of the cloud. The details of this are not clear. Field evidence for the in-cloud production of H_2O_2 has been presented.[9] The aqueous H_2O_2 measurements in cloudwater at Whiteface Mountain[10] and in stratus clouds[11] exhibit a seasonal trend with maximum concentrations in the summer months, consistent with its photochemical origin.

In an important modelling paper, it has been shown that clouds also tend to reduce the concentrations of O_3, OH, nitric oxide (NO) and formaldehyde (HCHO),[12] further emphasising the need for a proper understanding of cloud chemical processes and their proper simulation in global-scale tropospheric chemistry models.

7.4 AIRBORNE MEASUREMENTS OF CLOUD CHEMICAL PROCESSES

Most cloud chemistry occurs in clouds forming in the free atmosphere and not in contact with the ground. Clearly it is most important to understand the physical and chemical properties of these clouds. These have to be studied by aircraft, unfortunately it has proved extremely difficult to obtain spatially and temporally consistent measurements of the chemical dynamical and microphysical properties of the clouds. This has led to great difficulties in the quantitative interpretation of the data. Nevertheless, these studies have at least led to an identification of the most important processes controlling the chemistry of clouds which can be addressed through ground-based field projects (described below) or through laboratory and modelling studies.

A series of airborne experiments designed to investigate the cloud physics and chemistry of tropospheric clouds has been performed by the Cloud and Aerosol Research Group at the University of Washington, WA, USA. Their early studies focused on orographic wave clouds. These form when air is forced to rise over mountains. As the air is forced to rise it cools adiabatically and if the air stream becomes supersaturated a cloud will form, the droplets forming on small hygroscopic particles in the atmosphere called cloud condensation nuclei (CCN). These particles consist predominantly of sulphates, chlorides and nitrates of both natural and anthropogenic origin. These wave clouds have the advantage that an air stream flows through the cloud and in principle it is possible to make measurements in the same air-mass before entering and after leaving the cloud. The cloud is essentially stationary but individual droplets have a relatively short lifetime as they grow on the leading edge of the cloud, travel through the cloud and evaporate at the trailing edge.

These clouds have been used for an investigation of the production of sulphate in cloud.[13,14] The technique adopted was to expose filters from an aircraft to the air entering and leaving the mountain wave cloud. The masses of sulphate were determined by later analysis of the filters by ion chromatography. The sulphate production rate in the cloud was then determined from the increase in the sulphate mass on the downwind side of the cloud and an estimate of the transit time of the air in the cloud. The derived sulphate production rates were analysed as a function of other measured variables such as cloud liquid water content, cloudwater pH, SO_2 and O_3 concentrations and temperature. Major uncertainties exist in the data principally arising from difficulties in determining the trajectory of the droplets (and hence the transit time through the clouds). Difficulties also arise in determining whether the filter measurements were always representative of the air entering and leaving the cloud.

A total of 28 case studies were performed and of these 16 showed significant sulphate production. The mean value of the sulphate production in the wave clouds was $0{\cdot}9\pm0{\cdot}4\ \mu g\ m^{-3}$. This implies a very high sulphate production rate as the transit time through the clouds was only a few minutes. No firm evidence is available to establish which oxidation mechanism is responsible. The production rate is broadly consistent with laboratory studies[5] of oxidation by H_2O_2 if the ambient concentration was around 0·3 ppbv (a reasonable value). If this was the dominant oxidant the sulphate production rate would be expected to decline raipidly with time as the H_2O_2 was used up.

Measurements of the scavenging and production of sulphate and nitrate in cumulus, stratocumulus and stratus clouds have been reported.[15] The procedure adopted was to obtain measurements of the size distributions of cloud droplets and the interstitial aerosol. Measurements were also made of the cloud liquid water content, temperature, vertical air velocity and the chemical composition of the cloud droplets just above the cloud base. Measurements were made of the size distribution of aerosol and CCN, vertical air velocity, temperature and the chemical composition of the aerosol and atmospheric trace gases just below the cloud base. A scavenging coefficient for sulphate (F_s) was defined as the ratio of the sulphate concentration in the cloudwater to the sulphate concentration in the aerosol below the cloud base.

The values of F_s obtained in this way varied from 0·1 to 11. The values greater than unity were attributed to the production of sulphate by the aqueous-phase oxidation of SO_2. From the data it was deduced that the proportion of aerosol sulphate incorporated into the cloudwater (nucleation scavenged) as the droplets formed on the aerosol particles at cloud base was $0{\cdot}7\pm0{\cdot}2$. The sulphate production was estimated to be $1{\cdot}0\pm0{\cdot}3\ \mu g\ m^{-3}$, very similar to the value obtained for wave clouds in the earlier work.[13,14]

The results obtained for nitrate in this study were particularly interesting. They measured a total scavenging coefficient for nitrate (F_N) defined as the nitrate concentration in cloudwater divided by the nitrate (particulate and gas) concentration measured just below the cloud base. The values of F_N obtained varied from $0{\cdot}4\pm0{\cdot}3$ to 38 ± 25. The values greater than unity imply an aqueous phase production of nitrate. They attributed this to the formation of soluble higher oxides of nitrogen such as N_2O_5 from the insoluble NO_2 reacting with O_3 in the interstitial air, the soluble compound then being absorbed into the cloud droplets.

Recently, Hegg *et al.* have turned their attention to the much more complex dynamical and microphysical conditions which occur in frontal rainbands. A comparison has been made between field data obtained by aircraft and radar from a cold frontal rainband studied during the Genesis of Atlantic Lows Experiment (GALE)[17] and a rainband chemistry model described in Ref. 16. Both physical and very comprehensive chemical measurements were available from the aircraft.

The model employed was a two-dimensional kinematic model with bulk-parameterised cloud microphysics. The model included representations of cloudwater, cloud ice, snow, graupel and rain and a physical parameterisation of the processes by which water is transferred between these categories. The chemical species represented in the model were H_2O_2, sulphate, HNO_3, O_3, ammonium (NH_4^+) and peroxyacetyl nitrate (PAN). Hence the aqueous-phase oxidation of SO_2 to sulphate by O_3 and H_2O_2 was included. No production of nitrate was therefore included in the model and no chemical transformations were allowed to take place in the ice phase.

It was found that the model results were in broad agreement with the observations. The sulphate deposition was underestimated by a factor of two by the model due either to too low a nucleation scavenging rate or an underestimate of the below-cloud scavenging of sulphate. The nitrate aloft was also underestimated due to the lack of any nitrate production in the model. The model predicted and observed precipitation rates agreed well and due to substantial below-cloud evaporation the precipitation efficiency was only 8%. The model indicated that sulphate produced by oxidation made a larger contribution to wet deposited sulphate than nucleation scavenging and that H_2O_2 and O_3 made roughly equal contributions to the oxidation. The scavenging efficiencies of all the chemical species examined were low (2·5–5%). These low values were attributed to the hydrometeors growing predominantly by the deposition of water vapour onto ice crystals and to the low value of the precipitation efficiency in the rainband.

The results described above are indicative of the likely role of H_2O_2 as an oxidant in natural clouds as suggested by laboratory experiments. Further, very strong evidence was presented in Ref. 18 which found a very strong anti-correlation between concentrations of SO_2 and aqueous-phase H_2O_2 in stratocumulus clouds. This strongly suggests that a fast reaction is occurring between these species.

7.5 FIELD STUDIES OF THE CHEMISTRY OF RADIATION FOGS

Radiation fogs provide a rather different chemical environment to free tropospheric clouds and hill cap clouds. This is because they form in very stagnant air at the surface in which anthropogenic pollutants build up to very high levels. The stagnant conditions also favour prolonged droplet lifetimes. Consequently, the dominant chemical processes may be rather different.

For a detailed description of the microphysics of radiation fogs the reader is referred to Ref. 19 for a field study and to Ref. 20 for a model. Radiation fog occurs when the sky is free from cloud and the air very still. The ground surface cools to space by long-wave emission through the atmospheric window. The air immediately above the surface in turn cools by radiating to the surface outside the window. As cooling occurs the relative humidity increases and hygroscopic aerosol particles grow. These continue to cool by radiative cooling to the ground and to space and cloud droplets form as the particles are activated. The shallow fog then progressively shields the ground as it deepens and the liquid water content increases and the main radiative heat sink is transferred to the fog top. This destabilises the layer below thc fog top and turbulence develops. This enhances the flux of fog droplets to the surface and also entrains air into the fog from above its top. The main radiative heat sink is the droplets themselves near the fog top. The droplets are, therefore, colder than the surrounding air which may remain slightly undersaturated as the droplets grow. It is in this framework that the chemical processes take place.

The most comprehensive field studies of the chemistry of radiation fog are those conducted in the Po Valley (e.g. Ref. 21). Detailed measurements included the aerosol and droplet size distributions, the concentrations of trace substances in the aerosol and liquid phase, gas phase concentrations and meteorological conditions near the surface. A number of fogs between 1985 and 1987 were studied. These detailed measurements showed that most of the sulphate and nitrate in the fogwater came from nucleation scavenging. Evidence was found for the aqueous phase oxidation of S(IV) to sulphate. An excess of nitrate was also found which was attributed to the scavenging of gas phase HNO_3. The loss of material to the ground by fogwater deposition (a very important pathway for delivering phytotoxic pollutants to vegetation in high concentrations) and the entrainment of material from above the fog top played a central role in determining the evolution of the concentrations of the chemical species in the fogwater. They concluded that in order to interpret the data quantitatively, particularly in establishing the importance of various chemical conversion processes vertically resolved measurements within and above the fog layer are crucial.

Recently, under the auspices of the Ground-Based Cloud Experiment (part of EUROTRAC), still more comprehensive measurements have been made in the Po Valley including profile information through the fog from a 50 m tower. These data are still being interpreted and the results are awaited with interest.

Probably one of the most important next steps in this study would be to combine the observation programme with a detailed model of the chemistry and microphysics of radiation fog. An example of a chemical scheme which could be developed and applied to a detailed radiation fog model is that of Ref. 22. A chemical model of a night-time urban fog was developed using input chemistry relevant to southern California. It was suggested that the extreme acidity of the fog droplets could be attributed to the condensation and growth of droplets on acidic condensation nuclei or the oxidation of dissolved S (IV). In these urban conditions the oxidation of S (IV) by oxygen catalysed by Fe (III) and Mn (II) was important as well as H_2O_2 and O_3, the two oxidants found to be most important in free tropospheric clouds and hill cap clouds. It was also found that the formation of the methanesulphonate ion (HMSA) was significant. Nitrate formation was mainly due to the scavenging of gas phase HNO_3.

The very high concentrations of damaging pollutants in urban fogs, together with the rather different chemistry to other clouds (due to their highly polluted environment) make such fogs an important topic for future field studies and modelling work.

7.6 FIELD EXPERIMENTS ON ATMOSPHERIC CHEMISTRY ON GREAT DUN FELL

In this section an account is given of a series of field experiments conducted over a five year period at the UMIST field station on Great Dun Fell in Cumbria. The main objectives of these experiments have been to investigate the pathways by which SO_2 is oxidised to sulphate in cloudwater. This is the main pathway for the production of acid rain. In parallel with the experiments on the chemical processes experiments have been conducted to investigate the deposition of sulphate and other dissolved species by rainfall and by the direct deposition of cloudwater to vegetation when the wind-driven cloud is in direct contact with the surface.

The series of experiments have been among the largest field experiments ever conducted in the area of cloud chemistry. They have involved collaboration between the scientists from the Departments of Pure and Applied Physics and Chemistry at UMIST, the School of Environmental Sciences at the University of East Anglia, the Environmental Sciences Division of AEA Technology Harwell, the Institute of Terrestrial Ecology and the Environment and Technology Division of National Power.

7.6.1 The geography of the Great Dun Fell area

Great Dun Fell is situated on the north-west to south-east oriented ridge of the Pennine chain in Cumbria, in the north of England. The summit of Great Dun Fell rises to 847 m above sea level. To the south-west is the Eden Valley, typically 200 m above sea level in this area. To the north-east is Teesdale with an extensive area of moorland typically 600 m above sea level. The summit of Great Dun Fell is enveloped by cloud for an average of part of 250 days per year. This together with the ridge structure of the topography makes the site ideal for cloud chemistry studies. The prevailing airflow from the south-west frequently blows from Eden Valley over the ridge forming a cap cloud on the ridge summit.

7.6.2 Cloud physics research at Great Dun Fell

The atmospheric physics group at UMIST has been performing field studies on Great Dun Fell since 1975. The early work focused mainly on the microphysics of the hill cap cloud and the airflow over the site. These early studies focused mainly on the effects of the mixing of dry air, from above the top of the cap cloud into the cloud, on the evolution of the cloud droplet spectrum. This is a process which can have a major effect on the efficiency of rain production in clouds as well as affecting the cloud liquid water content and droplet size distribution as a function of height above the cloud base. This work has been published in a series of papers (eg. Refs 23 and 24).

A model of the airflow at the site was published as Ref. 25 and received experimental verification in a series of field experiments reported in Ref. 26. This work showed that the airflow over the site did indeed approximate very well to flow over a two-dimensional ridge. In addition it was demonstrated that the airflow frequently remained attached to the surface throughout as the wind flowed up the hill over the summit and down into Teesdale. This ability to use the site as essentially a flow-through reactor using equipment on the surface at a number of sites on the hillside and in the valley has been particularly important in the development of the site for cloud chemistry investigations.

7.6.3 The cloud chemistry experiments on Great Dun Fell

In order to investigate the aqueous phase oxidation of SO_2 to sulphate measurements have been made at three sites. The first site (site 1) was located upwind of the hill in the Eden Valley. This is a house, owned by UMIST near the village of Dufton, called Wharleycroft. This site was used to measure the atmospheric trace gas concentrations in the airstream entering the cloud along with the chemical composition and size distribution of the atmospheric aerosol on which the cloud droplets would form.

The second major site (site 2) was located on the south-west face of Great Dun Fell approximately 650 m above sea level. This site was based on a mobile laboratory and was equipped to measure the chemistry, liquid water content and droplet size distribution of the hill cap cloud just above the cloud base. These measurements were repeated at the summit (site 3). At this site UMIST has a permanent laboratory, rented from the Civil Aviation Authority who have a major radar installation on the summit of Great Dun Fell. Limited measurements were also made at a number of other sites on the slopes of Great Dun Fell during some of the campaigns.

7.6.4 Details of the measurements

Figure 7.1 shows the distribution of equipment for the major campaigns on Great Dun Fell. In addition to the measurements shown in this Figure, during campaigns involving UMIST and National Power a series of experiments was performed in which an artificial release of SO_2 was made from a 20 m tower at an upstream site. Sulphur hexafluoride (SF_6) was also released to act as a tracer gas.

During the later campaigns a Cessna aircraft owned by UMIST was flown in the area and made vertical profile measurements of H_2O_2 and O_3.

To summarise is a typical experiment as shown in Fig. 7.1. At the upstream site 1, measurements were made of gas phase H_2O_2, O_3 and SO_2. Samples of atmospheric aerosol were collected for later analysis for the major anions and cations. At site 2 on the south-west slope of Great Dun Fell, measurements were made of O_3 in the gas phase, H_2O_2 in the aqueous phase (as it is a very soluble gas). Samples of cloudwater were collected for later analysis for the major dissolved anions and cations. Mea-

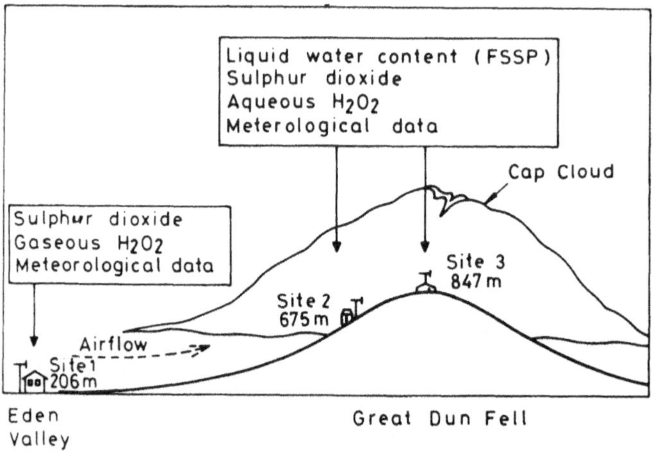

Figure 7.1. The distribution of equipment during an experiment to investigate the oxidation of S(IV) at Great Dun Fell.

surements were also made of the cloud droplet size distribution and liquid water content. These measurements were all repeated at the summit site 3 with the addition of measurements of gas phase SO_2. In addition, standard meteorological parameters of wind speed, wind direction, temperature and humidity were made at all three sites. Additional cloudwater samples were collected at a number of other sites on the slope of the hill.

The details of the instruments used to make the measurements will not be reproduced here. The reader is referred to papers describing the work for these details (readily available examples include Refs 8 and 27–29).

7.6.5 Results from the field studies

A substantial number of case studies have been performed over the years, some involving an artificial release of SO_2, others relying on background concentrations. The studies have also been performed at different times of the year. Consequently, different individual case studies have more clearly highlighted different aspects of the results. In the following sections are presented examples of results obtained from individual case studies to draw out the main findings from the experiments.

7.6.5.1 The reaction of SO_2 with H_2O_2 in aqueous solution

To begin with, the results of a SO_2 release experiment performed during October 1986 are considered.[30] A strong anti-correlation was found between the aqueous phase H_2O_2 and gas phase SO_2 concentration measured at the summit site, strongly suggesting that the SO_2 is removing the H_2O_2 by reaction. The concentrations of H_2O_2 were, however, always small and even if all the H_2O_2 present was used in oxidising dissolved SO_2 the extra sulphate produced would be insignificant. The pH of the cloudwater was around 4·0 during this case study, too small for any significant contribution of the sulphate from the oxidation of dissolved SO_2 by O_3 in the time available as the air flowed up the hill (a few minutes). Consistent with this no sulphate production was detected in the cloudwater. These results are typical of experiments conducted during the autumn and winter months.

Experiments conducted during the spring will now be considered. Figure 7.2 shows data gathered on the 20 April 1988. The figure illustrates many of the features common to many of the field studies conducted at Great Dun Fell. This graph shows aqueous phase concentrations of H_2O_2 from sites 2 and 3 which have been converted to gas phase equivalents using Henry's law. The gas phase concentration of H_2O_2 measured at site 1 is shown. In addition, the trace of SO_2 concentration measured at the

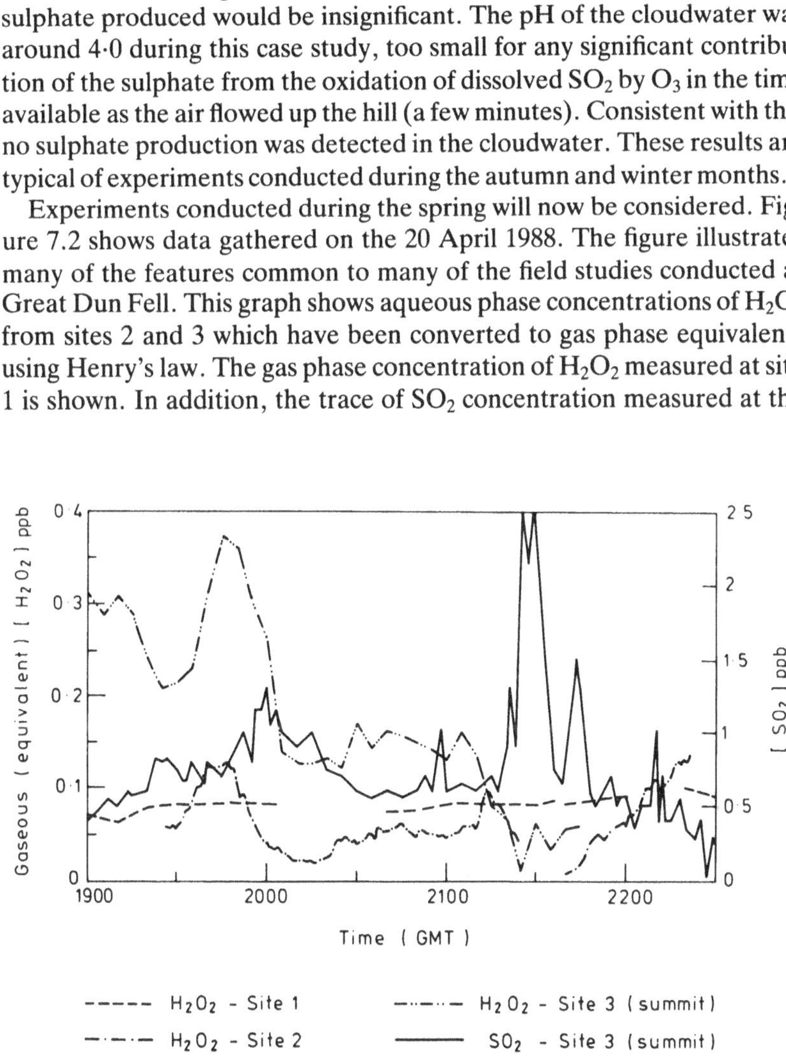

Figure. 7.2. Time histories of gas phase equivalent H_2O_2 measured at sites 1, 2 and 3 on 20 April 1988. The time history of SO_2 is also shown at the summit site 3. (Reproduced with permission from Ref. 8.)

summit site 3 is plotted. No deliberate release of SO_2 was performed during this experiment.

Once again it can be seen that a strong covariance exists between the SO_2 concentration and the H_2O_2 at the in-cloud sites 2 and 3. The input H_2O_2 concentration measured at site 1 was constant through the experiment. So very strong evidence exists for a reaction between the SO_2 and H_2O_2 in solution. It is also immediately apparent that the concentrations of H_2O_2 are very much higher than during the experiments conducted in the autumn, quite high enough for a significant amount of sulphate production. It is generally found that H_2O_2 concentrations are a factor of 10–100 times higher in the spring than in the autumn months.

It can also be seen that in conditions where the SO_2 concentration was low the concentrations from sites 1 (the valley) and 2 (the hillside) were very similar. The concentration at site 3 (the hill summit) was, however, much higher than the other two sites. This was attributed to the effects of the entrainment of H_2O_2-rich air from above the top of the cap cloud measured by the aircraft. In the hill cap cloud the effects of entrainment were generally largest close to the hill top. This is the region where downward entrainment of air from the free troposphere into the cloud is a maximum due to Kelvin–Helmholtz instability induced in the stable layer just above the cloud top as the airstream was forced to rise over the range of hills.

Figure 7.2 shows that at both sites 2 and 3 detectable concentrations of H_2O_2 remained in the cloudwater despite the presence of SO_2. If the transit time of air in clouds between the sites is known then the rate of removal of H_2O_2 may be compared with the rates expected from the laboratory studies described above. These comparisons have been performed for several case studies presented in several papers,[8,27,28] using the hill cap cloud chemistry model[31] for comparison. Considerable scatter in the results was obtained but a consistent finding was that the rate of depletion of H_2O_2 was much faster than predicted using recent laboratory results. The experiments in which artificial releases of SO_2 were performed generally had much higher concentrations of SO_2 which tended to totally deplete the H_2O_2 concentration. Hence, these were not able to confirm the very fast reaction rates detected in the other experiments.

7.6.5.2 The quantification of sulphate production in the cloud

Sulphate production, attributable to the oxidation of SO_2 in aqueous solution, was detected in most of the experiments conducted during the spring months. The most detailed comparison between the field observations and the prediction of the hill cap cloud chemistry model was reported in

Ref. 29, which describes an SO_2 release experiment performed on 30 April 1986. The plume entered a deep cap cloud which on this occasion was not significantly affected by entrainment. During the period when the SO_2 plume reached the hill summit the sulphate concentration in the cloudwater roughly doubled and the H_2O_2 concentration was reduced to below the detection limit.

It was apparent that the H_2O_2 depletion could only account for about 40% of the sulphate produced at the hill summit. Comparison with the model predictions showed that oxidation by O_3 at the rate expected from laboratory measurements[4] could account for the remainder of the sulphate produced.

The effectiveness of O_3 as an oxidant in the Great Dun Fell cloud during the spring months and not during the autumn is due to the higher cloudwater pH in the spring. This is probably due to higher ammonia (NH_3) concentrations, emitted from the hillside due to biological activity, during the spring.

7.7 WET DEPOSITION FROM HILL CAP CLOUDS

A large section of this chapter has been devoted to the use of a hill cap cloud as a captive cloud system for studying chemical and physical processes in the field. These clouds do, however, act as very important pathways for wet deposition in regions with even quite low hills. These two processes are the seeder–feeder effect and the direct turbulent deposition of cloudwater.

7.7.1 The seeder–feeder process

The cap clouds which frequently envelop low hills usually contain large concentrations of the major ions: sulphate, nitrate, NH_4^+, H^+, chloride (Cl^-) and sodium (Na^+). These either originate from nucleation scavenging or from chemical transformations within the droplets as discussed above. It has been observed at Great Dun Fell that the concentrations of these ions in cloudwater at the hill summit on average exceed those in rain by factors of between 2 and 3.[32]

This low-level cloud is readily scavenged by raindrops falling from higher cloud layers increasing the rainfall on and near the hill summit by the seeder–feeder effect.[33] The observed high concentration of the ions in the cap cloudwater lead to an increase in their concentration in rainfall with altitude and an increase in wet deposition. At Great Dun Fell the deposition of these ions was observed to increase between the valley floor and hill summit by a factor of four. This was associated with a factor of two increase in rainfall. It should be stressed that this effect only occurs in regions of relatively low hills (500–1000 m) typical of northwestern Europe. Measurements at high elevations in the Alps of France, in Austria and in western North America show generally smaller concentrations of the major ions in rainfall at high elevations (> 300m).[34,35]

Models of this process have been developed.[36] Both the modelling and field studies have been extended to larger scale regions of complex topography,[37] which show that the wet deposition in large scale regions of complex topography is very much larger than had previously been thought. These results have been applied to the major regions of elevated topography in the UK.[38] The results suggest that wet deposition is considerably higher in areas such as the Lake District, Welsh mountains and Scottish Highlands than had previously been considered.[39] Table 7.1 summarises the results. The Review Group on Acid Rain[40] has now published a new report with much modified wet deposition maps of the UK to take account of these effects.

7.7.2 Turbulent deposition of cloudwater

The direct turbulent deposition of cloud water from wind-driven cloud in contact with the ground close to hill tops can be an important pathway for atmospheric pollutants to be deposited to vegetation. This deposition process is commonly referred to as 'occult' deposition.[41,42] As discussed above it has been shown that at a hill-top site, the concentrations of major ions such as sulphate are typically two or three times larger in cloudwater than in rainwater measured at the same site.[32]

Recently detailed measurements have been made of the deposition of cloudwater to vegetation. Measurements were first made using a micrometeorological technique.[43,44] Gradient measurements of cloud liquid water content were made to determine liquid water fluxes to a moorland site. More detailed measurements of the cloud droplet deposition velocity as a function of droplet size, over moorland on Great Dun Fell, have been made.[45] These measurements were made by the gradient technique using Knollenberg forward-scattering spectrometer probes to measure the number, concentration and size distribution of cloud droplets at two levels above the ground. Similar measurements have been made over a forest canopy on Dunslair heights in southern Scotland, using the eddy correlation technique.[46] These measurements have shown that the cloud droplet deposition velocity is a strong function of droplet size, but that the

Table 7.1 The peak annual deposition rates for four ion species in mountainous areas during the period 1986–1988

	Snowdonia	*Lake District*	*York Moors*	*West Highlands*	*Cairngorms*
Grid co-ordinates	270 350	330 510	470 490	230 710	330 770
Precipitation (mm)	2 900	2 940	900	3 220	1 380
Acid deposition (mg m^{-2})	93	132	82	129	69
Sulphate deposition (g S m^{-2})	3	3·6	1·5	2·8	1·1
Nitrate deposition (g N m^{-2})	1·15	1·43	0·72	1·15	0·58
Ammonium deposition (g N m^{-2})	1·56	2·62	0·85	1·14	0·44
Percentage increase in deposition	73	76	41	73	50

deposition velocity for liquid water in a wind-driven cloud is close to that for momentum and considerably greater than the deposition velocity of submicrometer aerosol.

This can, therefore, be an important pathway for deposition close to hill tops, particularly when they are forested. The forest increases the surface roughness, the intensity of the turbulence and hence the rate of deposition to the canopy. Estimates have been made of the importance of this process in wet deposition to elevated regions in the UK.[47] It is suggested that the wet deposition of major ions is increased by about 30% in the major high-level regions of the UK (e.g. the Lake District, Welsh mountains, etc.).

7.8 DISCUSSION

In this chapter a brief description has been presented of recent advances in the understanding of cloud chemical processes relevant to acid deposition through field work. The most successful approaches have involved a combination of field investigation with assistance from detailed numerical models to aid the interpretation, the results in turn leading to better physical models. Substantial progress has been made in understanding the deposition processes both in simple cloud systems such as hill cap clouds and in the complex dynamical and microphysical situations associated with frontal systems. The main points to be brought out from the work described may be summarised as follows.

(1) Particulate sulphate, nitrate and chloride is incorporated into cloud droplets at an efficiency that often approaches 100% as the droplets form. This can lead to a very high efficiency of wet deposition if precipitation growth is dominated by coalescence and accretion. This efficiency is much reduced, however, if the growth of precipitation is dominated by vapour diffusion growth of ice crystals. This last process tends to make frontal systems rather inefficient at wet-depositing material incorporated into their cloud systems.

(2) In most cloud systems the dominant oxidants for S(IV) are H_2O_2 and O_3. The exception is probably radiation fog forming in highly polluted urban areas where oxidation by oxygen catalysed by Fe(III) and Mn(II) can be important.

Oxidation by H_2O_2 is very fast and usually limited by the availability of H_2O_2. Aircraft measurements have shown that H_2O_2 is often present in abundance in the free troposphere. Hence, entrainment of this air into cloud systems can be an important source of extra oxidant. The entrainment rate may frequently be the limiting step in determining the rate of sulphate production.

Oxidation by O_3 is strongly pH dependent and is slow at low pH. Although this oxidant is abundant throughout the atmosphere it is only likely to be of major importance in sulphate production when NH_3 gas or some other neutralising agent are present in sufficient concentrations.

(3) Through the seeder–feeder effect and direct turbulent deposition of cloud droplets hill cap clouds make an important contribution to wet deposition in complex topography. The importance of these processes has only recently been recognised.

7.9 FUTURE WORK

In order to understand the complex physical and chemical systems in and around clouds closely co-ordinated field, laboratory and modelling studies are required. The use of hill cap clouds as natural laboratories needs to be combined with more difficult aircraft-based studies of frontal and convective precipitation systems. Particular problems which remain outstanding at the time of writing are as follows.

(1) The pathways leading to the formation of nitrate in cloudwater through gas and possibly aqueous phase chemical processes. This leads to important questions surrounding the nitrogen cycle and the regulation of the oxidant concentrations in the atmosphere.
(2) The ice phase plays the major role in the generation of precipitation in most storm systems. To date, ice has been assumed to play a passive role in cloud chemistry. This needs investigation.
(3) Clouds play a central role in determining the radiative balance of the globe. Small changes in cloud structure can result in substantial changes in surface temperature.[47] Chemical processes in and around clouds generate the number and change the activity of CCN. These in turn are important in determining the droplet size and distribution and hence the radiative properties of the clouds. The role of clouds in determining the CCN spectrum is not well understood.

REFERENCES

1. Calvert, J.G., Lazrus, A., Kok, G.L., Heikes, B.G., Walega, J.G., Lind, J. & Cantrell, C.A., Chemical mechanisms of acid generation in the atmosphere. *Nature,* **317** (1985) 27–35.
2. Lee, I.Y. & Shannon, J.D., Indications of non-linearities in processes of wet deposition. *Atmos. Environ.,* **19** (1985) 143–9.
3. Granat, L., Sulphate in precipitation as observed by the European atmosphere chemistry network. *Atmos. Environ.,* **12** (1978) 413–24.
4. Maahs, H.G., Kinetics and mechanism of the oxidation of S(IV) by ozone in aqueous solution with particular reference to SO_2 conversion in nonurban tropospheric clouds. *J. Geophys. Res.,* **88** (1983) 10721–32.
5. Penkett, S.A., Jones, B.M.R., Brice, K.A. & Eggleton, A.E.J., The importance of atmospheric ozone and hydrogen peroxide in oxidising sulphur dioxide in cloud and rainwater. *Atmos. Environ.,* **13** (1979) 123–37.
6. Chameides, W.L. & Davis, D.D., The free radical chemistry of cloud droplets and its impact upon the composition of rain. *J. Geophys. Res.,* **87** (1982) 4863–77.
7. Heikes, B.J., Kok, G.L. & Lazrus, A.L., H_2O_2, O_3 and SO_3 measurements in the troposphere over the Eastern United States during the fall. *J. Geophys. Res.,* **92** (1987) 915–31.
8. Gallagher, M.W., Downer, R.M., Choularton, T.W., Gay, M.J., Stromberg, I.M., Mill, C.S., Radojević, M., Tyler, B.J., Bandy, B.J., Penkett, S.A., Davies, T.J., Dollard, G.J. & Jones, B.M.R., Case studies of the oxidation of sulphur dioxide in a hill cap cloud using ground based and aircraft based measurements. *J. Geophys. Res.,* **95** (1990) 18517–37.
9. Heikes, B.G., Lazrus, A.L., Kok, G.L., Kunen, S.M., Gandrud, B.W., Gitlin, S.N. & Sperry, P.D., Evidence for aqueous phase hydrogen peroxide synthesis in the troposphere. *J. Geophys. Res.,* **87** (1982) 3045–51.
10. Lazrus, A.L., Gandrud, B.W., Gitlin, S.N., Heikes, B.W., Kok, G.L., Lind, J.A., Shetter, R.E., Walega, J.G. & McLaren, S.E., Hydrogen peroxide at an Adirondack mountain site. *J. Geophys. Res.,* **97** (1987) 915–56.
11. Kelly, T.J., Daum, P.H. & Schwartz, S.E., Measurements of peroxide in cloudwater and rain. *Geophys. Res.,* **90** (1985) 7861–71.
12. Lelieveld, J. & Crutzen, P.J., Influences of cloud photochemicals on tropospheric ozone. *Nature,* **343** (1990) 227–33.
13. Hegg, D.A. & Hobbs, P.V., Cloudwater chemistry and the production of sulphate in clouds. *Atmos. Environ.,* **15** (1981) 1592–604.
14. Hegg, D.A. & Hobbs, P.V., Measurements of sulphate production in natural clouds. *Atmos. Environ.,* **16** (1982) 2663–8.
15. Hegg, D.A., Hobbs, P.V. & Radke, L.F., Measurements of scavenging of sulphate and nitrate in clouds. *Atmos. Environ.,* **18** (1984) 1939–46.
16. Rutledge, S.A., Hegg, D.A. & Hobbs, P.V., A numerical model for sulphur and nitrogen scavenging in narrow cold frontal rainbands. 1. Model description and discussion of microphysical fields. *J. Geophys. Res.,* **91** (1986) 14385–402.

17. Hegg, D.A., Rutledge, S.A., Hobbs, P.V., Barth, M.C. & Hertzman, O., The chemistry of a mesoscale rainband. *Quart. J. Royal Meteorol. Soc.*, **115** (1989) 867–86.
18. Daum, P.H., Schwartz, S.E. & Newman, L., Studies of the gas and aqueous phase. In *Composition of Stratiform Clouds. Precipitation Scavenging, Dry Deposition and Resuspension*, Vol. 1. Elsevier, New York, USA, 1983, pp. 31–52.
19. Roach, W.T., Brown, R., Caughey, S.J., Garland, J. & Readings, C.J., The physics of radiation fog. 1. A field study. *Quart. J. Royal Meteorol. Soc.*, **102** (1976) 313–33.
20. Brown, R., A numerical study of radiation fog with an explicit formulation of the microphysics. *Quart. J. Royal Meteorol. Soc.*, **106** (1980) 781–802.
21. Fuzzi, S., Nordini, G. & Facchini, M.C., Heterogeneous processes in the Po Valley radiation fog. *J. Geophys. Res.*, **93** (1988) 11141–51.
22. Jacob, D. & Hoffmann, M.R., A dynamic model for the production of H^+, NO_3^-, and SO_4^{2-} in urban fog. *J. Geophys. Res.*, **88** (1983) 6611–22.
23. Baker, M.B., Blyth, A.M., Carruthers, D.J., Caughey, S.J., Choularton, T.W., Conway, B.J., Fullarton, F., Gay, M.J., Latham, J., Mill, C.S.M., Smith, M.H. & Stromberg, I.M., Field studies of the effect of entrainment upon the structure of clouds at Great Dun Fell. *Quart. J. Royal Meterol. Soc.*, **108** (1982) 899–916.
24. Choularton, T.W., Consterdine, I.E., Gardiner, B.A., Gay, M.J., Hill, M.K., Latham, J. & Stromberg, I.M., Field studies of the optical and microphysical characteristics of clouds enveloping Great Dun Fell. *Quart. J. Royal Meteorol. Soc.*, **112** (1986) 131–48.
25. Carruthers, D.J. & Choularton, T.W., Airflow over hills of moderate slope. *Quart. J. Royal Meteorol. Soc.*, **108** (1982) 603–24.
26. Gallagher, M.W., Choularton, T.W. & Hill, M.K., Some observations of airflow over a large hill of moderate slope. *Boundary Layer Meteorol.*, **42** (1987) 229–50.
27. Chandler, A.S., Choularton, T.W., Dollard, G.J., Gay, M.J., Hill, T.A., Jones, A., Jones, B.M.R., Morse, A.P., Penkett, S.A. & Tyler, B.J., A field study of the cloud chemistry and cloud microphysics at Great Dun Fell. *Atmos. Envion.*, **22** (1988) 683–94.
28. Chandler, A.S., Choularton, T.W., Dollard, G.J., Gay, M.J., Gallagher, M.W., Hill, T.A., Jones, B.M.R., Penkett, S.A., Tyler, B.J. & Bandy, B., A field study of the oxidation of SO_2 in a cap cloud at Great Dun Fell. *Quart. J. Royal Meterol. Soc.*, **115** (1989) 397–420.
29. Clark, P.A., Gervat, G.P., Hill, T.A., Marsh, A.R.W., Chandler, A.S., Choularton, T.W. & Gay, M.J., A field study of the oxidation of SO_2 in cloud. *J. Geophys. Res.*, **95** (1990) 13985–95.
30. Gervat, G.P., Clark, P.A., Marsh, A.R.W., Choularton, T.W. & Gay, M.J., Controlled chemical kinetic experiments in cloud: a review of the CERL/UMIST Great Dun Fell project. In *Acid Deposition at High Elevation Sites*, ed. M.H. Unsworth & D. Fowler, Kluwer Academic Press, Dordrecht, 1988, pp. 283–98.

31. Hill, T.A., Choularton, T.W. & Penkett, S.A., A model of sulphate production in a cap cloud and subsequent turbulent deposition onto the hill surface. *Atmos. Environ.*, **20** (1986) 1763–71.
32. Fowler, D., Cape, J.N., Leith, I.D., Choularton, T.W., Gay, M.J. & Jones, A., The influence of altitude on rainfall composition at Great Dun Fell. *Atmos. Environ.*, **22** (1988) 1355–62.
33. Bergeron, T., On the low-level redistribution of atmospheric water caused by orography. In *Proceedings of the International Conference on Cloud Physics,* 1965, pp. 96–100.
34. Duncan, C.L., Welch, E.B. & Ausserer, W., The composition of precipitation at Snoqualmire Pass and Stevens Pass in the central cascades of Washington. *Water Air Soil Pollut.*, **30** (1986) 217–29.
35. Puxbaum, H., Vitovec, W. & Kovar, A., Chemical composition of wet deposition in the Eastern Alpine region. In *Acid Deposition Processes at High Elevation Sites,* ed. M.H. Unsworth & D. Fowler, Kluwer Academic Press, Dordrecht, 1988, pp. 419–30.
36. Hill, T.A., Jones, A. & Choularton, T.W., Modelling sulphate deposition onto hills by washout and turbulence. *Quart. J. Royal Meteorol. Soc.*, **113** (1987).
37. Dore, A.J., Orographic Enhancement of Rainfall and Snowfall, PhD thesis, University of Manchester, Manchester, UK, 1990.
38. Dore, A.J., Choularton, T.W., Fowler, D. & Stortom-West, R., Field measurements of wet deposition in an extended region of complex topography. *Quart. J. Royal Meteorol. Soc.*, **116** (1990) 1193–212.
39. Barret, C.F., Atkins, D.H., Cape, J.N., Fowler, D., Irwin, J.E., Kallend, J.E., Martin, A., Pitman, M.J., Scriven, R.A. & Tuck, F., *Acid Deposition in the UK.* Warren Spring Laboratory, Stevenage, UK, 1983.
40. Review Group on Acid Rain, *Acid Deposition in the UK,* 2nd Report. Warren Spring Laboratory, Stevenage, UK, 1990.
41. Nagel, J.F., Fog precipitation on Table mountain. *Quart. J. Royal Meteorol. Soc.*, **82** (1958) 452–60.
42. Kerfoot, O., Mist precipitation on vegetation. *Forest. Abstr.*, **29,** (1968) 8–20.
43. Dollard, G.J. & Unsworth, M.H., Field measurements of turbulent fluxes of wind driven fogs to a grass surface. *Atmos. Environ.*, **17** (1983) 775–80.
44. Dollard, G.J., Unsworth, M.H. & Harvey, M.J., Pollutant transfer in upland regions by occult precipitation. *Nature,* **302** (1983) 241–3.
45. Gallagher, M.W., Choularton, T.W., Morse, A.P. & Fowler, D., Measurements of cloud droplet deposition at a hill site. *Quart. J. Royal Meteorol. Soc.*, **114** (1988) 1291–303.
46. Beswick, K.M., Hargreaves, K., Gallagher, M.W., Choularton, T.W. & Fowler, D., Size resolved measurements of cloud droplet deposition velocity to a forest canopy using the eddy correlation technique. *Quart. J. Royal Meteorol. Soc.*, **117** (1991) 623–45.
47. Slingo, A., Sensitivity of the Earth's radiation budget to changes in low clouds. *Nature,* **343** (1990) 49–51.

Chapter 8

History of Atmospheric Acidity

PETER BRIMBLECOMBE
School of Environmental Sciences, University of East Anglia, Norwich, NR4 7TJ, UK

8.1 INTRODUCTION

Acid rain undoubtedly became a key environmental issue in the 1980s. It created as much public and media interest as other major environmental issues such as nuclear power and lead in petrol. Its rapid rise into public view gave an impression that it was a problem that had arisen suddenly, fully formed, within a decade. Yet the issue had already achieved considerable international prominence at the UN Conference on the Human Environment in Stockholm in 1972,[1] and indeed its history extends back far longer.

Much about our society and the way it perceives environmental issues can be learned by a study of the history of atmospheric acidity, but there are also important scientific reasons for thinking in historical terms. In particular it is useful to know about the pre-industrial base-line composition of the atmospheres, the long-term changes in atmospheric acidity and the development of scientific ideas surrounding rainfall chemistry.

The importance of scientific and environmental histories is receiving wider note, and there are already a number of papers on the history of acid rain (e.g. Refs 2 and 3). However, there is already a little evidence of the mythologisation that is apparent in other aspects of air pollution history,[4] i.e. an Anglocentric scientific world has attributed the discovery of 'acid rain' to Angus Smith in the 1870s,[5] despite the fact that Ducros described 'pluie acide' in 1845.[6]

There are a number of problems in writing a history of acid rain and a few deserve a mention here.

(1) In some circles acid rain has become a synonym for air pollution. It is preferable to regard acid deposition as the consequence of the dispersion of air pollutants over large areas. However, inclusion of the history of building damage in this chapter is not in keeping with this view, because in the past building damage arose much as a result of localised urban air pollution.
(2) As the focus of this book is on atmospheric acidity, less attention is paid than might be wished to issues such as ozone, or volatile organic compounds, although they are linked with the production of acids.
(3) It is difficult to decide where history ends, but by and large the discussion of the most recent decades has been kept disproportionately brief (despite previous criticism of this approach, e.g. Ref. 7), because they are proper topics for other chapters in this book.

8.2 EARLY OBSERVATIONS

8.2.1 Rainfall composition

There are many early observations of rainfall composition. Ehrenberg's[8] researches suggest that observations of blood rain (rain containing reddish dust; for modern accounts, see Refs 9 and 10) were made more than 3000 years ago. The use of such observations in prognostication has meant that they have been well recorded throughout history, even during the Dark Ages.[11] Classical writers also commented on other aspects of rainfall composition. In Pliny's *Natural History XXXI.29,* agricultural damage caused by unusually saline rains in the Bosphorus is reported. Similar observations can be found from northern Europe more recently, i.e. the English scientist William Derham noticed that in the aftermath of the great storm of 1703 the grass had become so salty that the sheep refused to eat it.[12]

There have also been many debates over which types of water are best for health or agriculture. Vetruvius VIII.2–4 argued that rain was very healthy, although Pliny's *Natural History XXXI.21* points out that rain is full of dust. He was supported in this by the wondrous medieval polymath Hildegard of Bingen in *Causae et Curae.*[13] Sir Kenelm Digby thought that rain was responsible for delivering the essential nitre to plants and regarded equinoxial rains as having rather special properties.[14] The discovery of animals in rain (even fish and frogs, e.g. Ref. 15) led to a belief that life might originate in clouds (e.g. Douglas, 16th century; Ref. 16), although sceptics such as William of Conches suggested strong winds could carry creatures aloft.[13]

8.2.2 Long-range transport of pollutants

In areas remote from sources of pollution, the pollutant concentrations are typically low. This has meant that the initial effects of acid rain may be imperceptible and escape observation. Despite this there are some remarkable early observations of long-range transport of air pollutants. In the late 16th century, there were complaints that smoke from extensive agricultural burning in England damaged the vine crops in France.[17] In the late 18th century, polluted rain fell in Colonial Detroit.[18] In England, country folk said they could smell the smoke from London almost 100 miles distant.[19] In the north of England the terms, 'moorgrime' and 'moorgroime' arose in the dialects of Yorkshire and Lancashire. These referred to black sooty accumulations found in the wool of sheep grazing on upland pastures. Fleeces are somewhat greasy and act as good collectors of droplets from fog and low cloud, so it is possible that the shepherds were observing the effects of pollution transported from industries many tens of kilometers distant.[20]

An early description of long-range transport relevant to the acid rain issue comes from the Norwegian playwright Ibsen's play *Brand*, written in 1865:

> Worse times, worse sins,
>   through the night of future flashes,
> Britain's suffocating coal dust is
>   slowly descending over the countryside,
> Soiling all that is green,
>   strangling all that strives to grow,
> Creeping low and mixed with poison,
>   stealing sun and light from the green valley,
> Pelting down as a rain of ashes
>   over the long sinced doomed village.

Ibsen (e.g. in *Enemy of the People*) was interested in environmental matters and such literary observations were paralleled by notes in scientific literature. In the 19th century, Brøgger in Norway described dirty snows;[21] in Britain, Frankland's studies of the quality of English rainfall led him to believe that nowhere was it free from soot.[22] Angus Smith, so often (and wrongly) credited with the discovery of acid rain, became interested in air composition as Britain's first Alkali Inspector. Reference 23, like many, notes that Smith's work[24] was overlooked by subsequent investigators. It would be wrong to overemphasise the obscurity of Smith's work, which was known to the major bibliographical compilers (e.g. Refs 25 and 26). Indeed, there was no difficulty in discovering Smith's book when preparing a paper on historical changes in atmospheric trace components.[27]

Some particularly well-documented examples of black rainfall are those observed on the Scottish coast some 30 km north of Aberdeen in the years 1862 and 1863. The Reverend James Rust, Minister of Slains wrote a book on the remarkable black rain that fell during some convective storms.[28] Rain fell from an 'immense threatening cloud' that appeared 'as black as a plume from a steamer'. The falls bespattered clothes that were out to dry. Water that had been set aside for shaving the following morning curdled the lather from the soap into a mottled black mass. Sufficient meteorological information is available from this period to construct weather maps and determine the trajectories for air on previous days. The back-trajectory for the event of 14 January 1862 (Fig. 8.1) indicates that the source of the air from which the black rain fell could have been industrial Yorkshire or the Midlands. The air trajectories for other black rain incidents recorded by Rust also indicate that the air probably passed over the industrial areas to the south.[29] Thus, under certain atmospheric conditions, long-range transport of significant quantities of pollutants may have been possible a century ago despite the relatively low fuel consumption.

8.2.3 Damage to materials and ecosystems

Damage to materials, especially buildings, has been well known since the earliest times. Roman authors only comment on the blackening of buildings, but enhanced rates of weathering of stone and corrosion of iron by air pollutants receive comment by the 17th century[17,24] and architects such as Wren and Hawksmoor were aware of the problems of air pollution at this time.[30] There are a number of studies of long-term damage (i.e. many

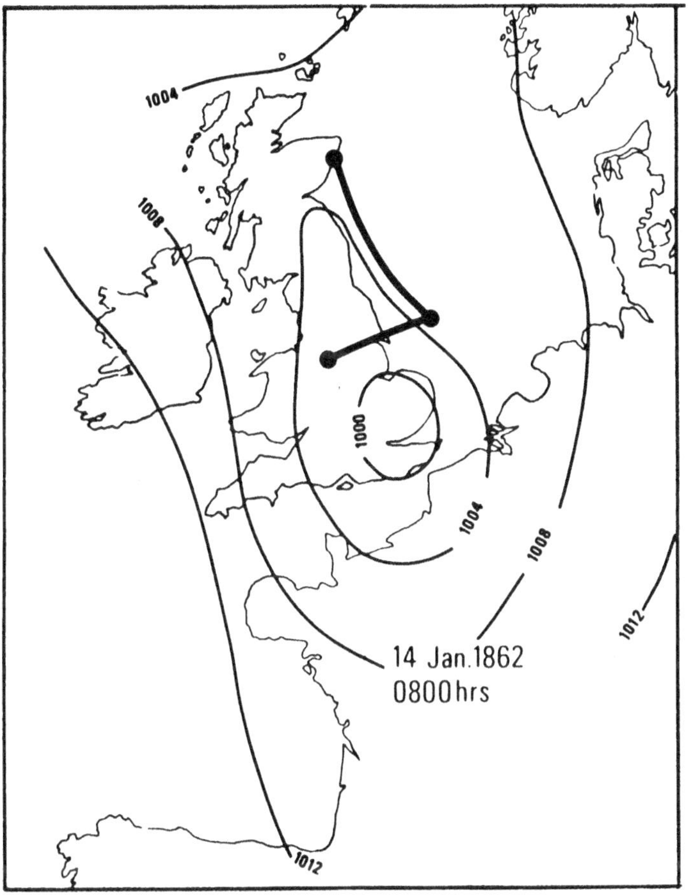

Figure 8.1. Back-trajectory for the fall of black rain in Scotland 14 January 1862.

hundreds of years) to individual structures: Trajan's column,[31] York Minster[32] and Westminster Abbey.[30] Studies of the history of large ecclesiastical buildings in England make it clear that contemporary observers attributed damage to very localised pollution sources; often adjoining buildings. Furthermore, there is much less evidence of damage to buildings in non-urban settings, so the term 'acid rain' needs to be appropriately qualified when talking about damage to buildings in the past.

However, long-term records of damage to buildings at remote sites are not necessarily easy to come by or interpret.[32]

In near-urban localities there are early descriptions of damage to plants from air pollutants. An example of this is found in the records of the English town of Beverley in the 15th century where fruit trees were damaged by the emissions of kilns.[33] By the 17th century there were numerous comments about the damage in London.[17,34] As the sensitivity of plants towards the air pollutants varies among species, it is possible to estimate the concentration of pollutants at a site by examining the changes in plant diversity. Lichen are very sensitive to air pollutants and records of the distribution of lichen in Epping Forest begin some two centuries back. These were used to estimate the changing sulphur dioxide (SO_2) concentrations there over two centuries.[35]

It is often rather difficult to find early descriptions of detrimental effects of acid rain on ecosystems in remote regions. However, it is clear from reports that some lakes and tarns in southern Norway began to lose their fish populations by 1890 due to acidification;[36] however, in Britain much of the blame for declining fish stock is placed on the discharge of effluents directly into the rivers. In the 1920s there were measurements of surface water pH in Norway[37] and its relationship to trout production,[38,39] and there were even some experiments in adding lime.[40] Interestingly, there is an awareness of the fact that observations of acid rain damage were first made by fishermen in more popular accounts of the history of acid rain. However, they refer to large changes occurring in the late 1960s.[41]

8.3 DEVELOPMENT OF RESEARCH ON PRECIPITATION COMPOSITION

8.3.1 Early analyses

The earliest chemical analysis of rainfall seems to have been made in Copenhagen by Olavus Borrichius in 1674 and repeated by Hjarne at the beginning of the 18th century.[42] Marggraf detected calcium chloride ($CaCl_2$) and nitrate in rain and snow in 1749–1750,[43] followed by Bergman.[44] Hermbstaedt detected organic matter in rain,[45] and both Mulder and Zimmermann found chlorine and several other substances in rain.[46,47]

The analytical determination of chloride in rain, as distinct from the descriptions in classical writings, was a very significant development. The atomic theorist John Dalton showed chloride concentrations in Manchester rainfall to be higher when the wind blew more directly from the sea.[48] The incorporation of sea-spray in the rainfall collected at coastal sites was thought to be a problem in determining rainfall amount, but a British Rainfall Association study in the 1860s showed even at coastal stations, sea-spray made only a small contribution to the total rainfall catch.[49]

Ducros, a pharmicist in Nimes, studied acid rain in 1842 and concluded that it contained nitric acid (HNO_3), although he regarded the source as a natural one.[6]

The Rivers Pollution Commission also had some interest in rainfall composition for a brief period in the 1860s. They were concerned about rainfall caught for drinking purposes. The analyses justified their concern, as even at sites over 25 miles from any large town, a surprising degree of contamination was to be found.[22]

8.3.2 Agricultural research

The early studies mentioned above were sporadic and it was agricultural scientists who made the first systematic long-term observations of rainfall composition. Their earliest measurements were concerned with the chlorine content in rainfall, probably because it was relatively easy to determine. The first analysis of a full year's rainfall in the British Isles was for chloride (Cl^-), and it was made at a near coastal location at Penicuick in Scotland in 1843.[50] Unfortunately, the analysis is not reliable. Some theorised that plants such as the marigold, that had a marine ancestry, ought to benefit from the addition of salt to the soils in which they grew. This idea was soon discarded and it became evident as time went on that rain-borne sea salt had little effect on most crops except right at the coastal margin and there the effect was probably harmful.[51]

In the 17th century, Sir Kenelme Digby and John Evelyn thought that nitrogenous compounds in rainfall contributed to plant growth,[52] and by the mid-19th century much active research centred around this issue. Justus von Liebig thought that rainfall made a significant contribution to the soil nitrogen and claimed that some 27 kg were deposited on every hectare in a year, sufficient, it would have seemed, for many crops.[53] Work in England undertaken by the young agricultural chemist J. T. Way suggested that the quantities of nitrogen brought down in the rain were too small to supply the needs of most agricultural crops.[24] The momentum of this early work was such that rainfall continued to be analysed by many agricultural stations well into the present century.

At Rothamsted in England a large rain gauge of a thousandth of an acre area was used for many years and the station continues to monitor rainfall composition. There are early analyses for Cl^-, ammonia (NH_3) and NO^-_3 in rainfall. The sulphate analyses are more limited although the few that are available are interesting to compare with the modern determinations. Between 1881 and 1887 some 7·8 kg of sulphur were deposited on each hectare at Rothamsted. In the measurements made between 1955 and 1966 this had increased to 12·2 kg ha^{-1} a^{-1}, no doubt due to the increased amounts of sulphur present in the air over the British Isles. Unfortunately, it is not easy to be certain about the reliability of the early sulphur analyses.

For the cases of the Cl^-, NH_3 and NO^-_3 one is much more fortunate, because there are monthly analyses in addition to annual totals. In Fig. 8.2(a) there is a nearly perfect overlap between the modern figures for Cl^- deposit and those obtained from the early data. It is not simply that the amount of Cl^- deposited has remained constant, but the seasonal distribution has not varied either. The lack of change is likely at a site like Rothamsted where the main source of Cl^- in rainfall should be from sea salt driven far inland. Although high-Cl^- coals could add to the Cl^- concentration in rain at some sites, even today the anthropogenic contribution is much smaller than this natural source of Cl^-. Particularly evident in the seasonal distribution is the higher deposit of Cl^- in the winter months which comes from storms.

If the seasonal distribution of NO^-_3 deposit is examined (see Fig. 8.2(c)), a difference between the modern and old values is evident. Not only is more deposited each year in modern times, but the distribution has changed. Last century there was little seasonal variation, but nowadays there is a peak in NO^-_3 deposit in spring. There has been a large increase in the emission of nitrogen oxides (NO_x) over the last century. The measured concentrations of NH_3 in rainfall ought to have increased also,[54] but there is no evidence of this in the observations (Fig. 8.2(b)). It may well be that the NH_3 was lost from, or transformed within, the bulk samples. However, NO^-_3 and NH_3 observations taken together suggest, at least, that nitrogen deposition increased over the period.

One should not think that these were the results of a solitary research programme. Agricultural stations all over Europe made measurements (e.g. Ref. 55), and similar, but more widely scattered measurements come from North America (e.g. Refs 56–60). The results add to a picture of rising amounts of nitrate deposit since the earliest measurements late last century. The early precipitation data was well summarised by Eriksson.[25,61]

Relatively little work was undertaken by non-agricultural scientists, so

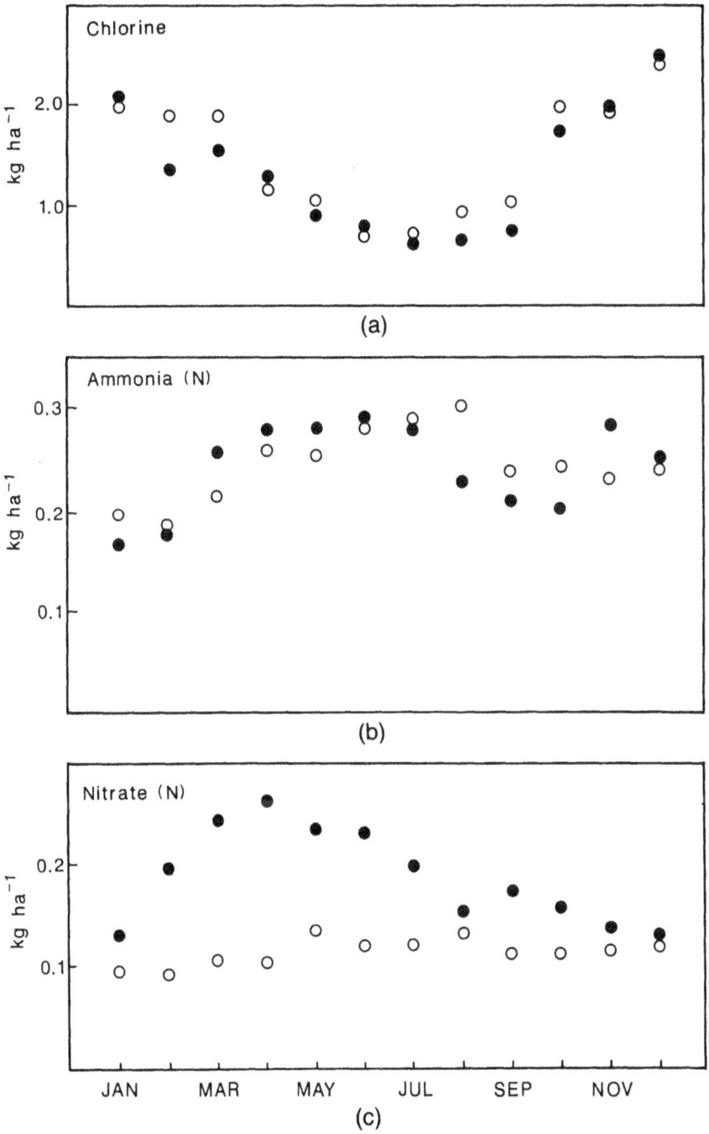

Figure 8.2. Seasonal patterns of wet deposition of anions in rainfall at Rothamsted. (○) 1877–1915. (●) 1955–1966. (Reproduced with permission from Ref. 27.)

there is no particular interest in pollution at remote sites. The few who were particularity concerned with air pollution focused most of their attention in towns (e.g. Refs 62–64). There is evidence of concern about dispersion from the chemical industry. A particularly early example of this would be the chlor-alkali industry, where the large quant-ities of hydrogen chloride (HCl) caused great damage to vegetation over large areas and resulted in the passage of the Alkali Act 1863 in the UK.

Smelters also caused much damage to vegetation through the release of smoke and SO_2. There are historical accounts of these in North America[65,66] and Germany.[67] It was noted that smelter smoke is a neglected area of acid rain history.[66] However, because the effects of smelter smoke (or even the very acidic emissions from the alkali industry) are most noticeable locally, they have often been treated as part of the history of general air pollution. The difficult distinction between the history of acid rain and that of air pollution will undoubtedly remain to vex us.

8.3.3 Origins of modern studies

As has been seen most early work on precipitation analysis arose from an interest in the nutrient budgets of soils, apart from a few air pollution studies and hydrological analyses. This work came to an end early this century and analyses of precipitation were less common until a network of stations was set up in Scandinavia (broadening later to much of Europe) in the years following World War II. The network of stations ultimately became known as the European Air Chemistry Network (EACN). In some cases the measurements recommenced at the very sites of much earlier measurements. At Rothamsted, for instance, it is possible to build up a set of rainfall analyses that run in a broken series back to the middle of last century. As was seen in Fig. 8.2 it is possible to assemble useful datasets that show the variation in rainfall composition over long periods. A number of subsequent European networks developed after the EACN and these—Organisation for Economic Co-operation and Development (OECD) (1973–1974); Cooperative Programme for Monitoring and Evaluation of Long Range Transmission of Air Pollutants for Europe (EMEP) (1978–1982); World Meteorological Organisation (WMO) (1979——)—are described briefly in Ref. 71.

At the same time was there as a reawakening of interest in the contribution that rain made to agricultural nutrients, Gorham began his studies of precipitation input to lakes.[67,68] He found that the acids introduced into precipitation by industrial processes caused a progressive loss of alkalinity in surface waters and a rise in the free acidity of soils. In addition to this work, C. E. Junge, the atmospheric chemist, made a series of studies of the composition of precipitation in North America,[69] the most significant measurements made on that continent since the agricultural measurements much earlier in the century.

Concern about the 'acid rain' problem in its modern sense was initiated by Oden, who argued that the long-term increases in the deposit of acidity would have profound ecological consequences. His claims provoked widespread public and political interest, which culminated with the initiation of many major studies of the problem.[70] It was argued that acid rain changed the pH of surface waters, led to a decline in fish stocks, leached toxic minerals from soils, decreased forest growth and accelerated damage to materials. Norway, for instance, showed its concern in 1972 by establishing the Effect of Acid Precipitation on Forests and Fish (SNSF) Project to examine the effects of acid precipitation on forests and fish. The Organization for Economic Co-operation and Development initiated a two year study in 1973 that firmly established the notion of long-distance transport of air pollutants.[71] A 'Convention on Long-Range Transboundary Air Pollution' was signed in 1979.

In North America there was also considerable interest in the work being done in Scandinavia. The experimental catchment at Hubbard Brook was well suited to deposition studies and provided an excellent base for work. However, large national programmes took some time to develop in the US. Perhaps the best known of these was the National Acid Deposition Program started in 1976 to co-ordinate long-term monitoring of precipitation. In the same year the MAP3S program was initiated by the US Department of Energy. It was concerned with understanding the transport, transformation, and fate of pollutants released from energy-related activities.

Needless to say even the scientific research programmes have not been able to avoid the political background to acid rain. At conferences there were always stories about difficulties in publishing data, over-cautious refereeing and the problems inherent in volatile media interest. The source of scientists, funding became almost as important as their results. An example of sensitivity towards funding sources was typified by the Surface Water Acidification Programme funded by the Central Electricity Generating Board and National Coal Board of the UK. While some discovered funds were difficult to apply for, others found it morally impossible to accept those offered.[72,73]

8.4 TRENDS IN PRECIPITATION ACIDITY

8.4.1 Evolution of the atmosphere

At contemporary carbon dioxide (CO_2) partial pressures (350 ppm) the pH value of water in equilibrium with the CO_2 would be about 5·6. In past geological time such as on the Archaean earth where CO_2 pressures may have been as high as 1 atm (101 kPa), then carbonic acid would have represented a greater acidifying influence (e.g. Ref. 74).

Further interest in acid rain in the geological past has been stimulated by research into the possibility of meteorite impacts at the Cretaceous–Tertiary boundary. The heating of the earth's atmosphere due to the shock of large meteorites (bolides) would lead to the production of enormous quantities of NO_x and the subsequent formation of HNO_3. Some evidence for increased weathering following the impact of the putative Cretaceous–Tertiary bolide(s) is suggested by high ^{87}Sr:^{86}Sr ratios in seawater at that time.[75]

8.4.2 Pre-industrial atmospheric acidity

A fairly continuous history of the input of materials into the precipitation that has fallen in remote areas can be obtained from ice cores.[77] Snow laid down in polar regions stores information about atmospheric composition in annual layers. Drilling programmes have obtained samples from deep within ice caps of Greenland and Antarctica. A good marker of potentially acidifying pollution in these remote regions is the presence of non-sea-salt or excess sulphate. The non-sea-salt sulphate arises from the oxidation of reduced sulphur compounds: these would include SO_2 from human activities and volcanoes, but some would come from the oxidations of organo-sulphur compounds such as dimethyl sulphide (DMS) produced from phytoplanktonic activity in the oceans. Further acidification can occur through the production of methane sulphonic acid (MSA) another product of DMS oxidation. MSA is regarded as a indicator of biologically derived acidity. It has been shown that during the last ice age ($>$ 15 000 years ago) MSA concentrations were two to five times higher than today.[78] Some may be due to snow accumulation decreases, but it is probably an indication of enhanced transport and biological activity.

Figure 8.3 shows the concentration of non-sea-salt sulphate in ice layers of various ages in both the Antarctic and Greenland. In the Antarctic no clear trends in the excess sulphate can be seen, but there are peaks in the record that are the result of large volcanic eruptions. Gradual increases in atmospheric sulphate that have resulted from human activity are not apparent in the snows of the Antarctic, but Greenland which is closer to the heavily industrialised areas of North America, Europe and

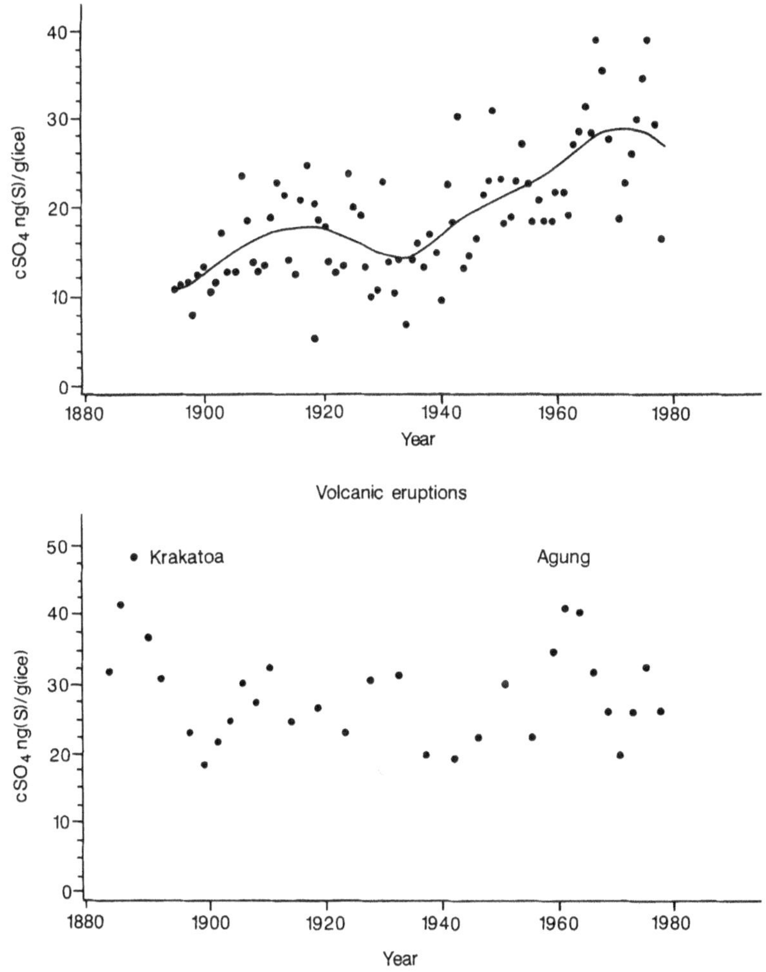

Figure 8.3. (a) Annual sulphate concentration in shallow core from Dye 3, South Greenland fitted with a spline curve. (From Ref. 70, reproduced by permission of author.) (b) Annual sulphate concentration at Dome C, central east Antarc-tica. (Reproduced by permission from Ref. 79.)

the Far East shows clear trends. The sulphate concentrations in the ice from the Dye 3 drilling site seem to have increased nearly three-fold over the last century. Studies of the nitrate concentrations of Greenland ice show an increase in this anion also. Nitrate in snow is expected as the result of the transport and oxidation of NO_x from combustion processes into remote regions.

Volcanic acids are also found in ice cores where there is a particularly good record of major eruptions such as that of Agung (1963), Santa Maria (1902), Krakatoa (1883) and Tambora (1815). The depositional record gives an indication of the vast quantities of acid released by eruptions. In the case of the Tambora eruption some 150 Tg S were produced.

Other indications of the importance of naturally derived acids have come from recent studies (see Ref. 80) in the Northern Territory of tropical Australia. These have indicated rainfall events with pH values as low as 3·5. The rainwater shows only traces of nitrate and sulphate attributable to anthropogenic sources and most of the acidity arises from organic acids probably of natural origin.

8.4.3 History of emissions and acidity

The changes in sulphur emission for Europe since 1840 are shown in Fig. 8.4. The small pie diagrams illustrate the distribution of the emissions within the countries of Europe. In the early part of last century it is clear that Britain made a larger contribution to the anthropogenic sulphur budget than continental Europe. Gradually, other countries increased their dependence on fossil fuels, so not only did the total European emissions increase, but there was a shift in the distribution. The established industrial giants, (including Britain and Germany) retained their dubious lead in the emission of sulphur until recently. The Eastern Bloc countries (i.e. growth of the former USSR emissions given, for instance, in Ref. 83), often with high-sulphur coals, have now become much more dominant as emitters. Emissions have declined in many Western European nations, either through economic recession and fuel change (e.g. Britain) or more significantly by active control policies (e.g. the former FRG, although smaller countries also show impressive and hard-won decreases).

The transport and deposition of sulphur emissions can now be determined by computer modelling. These take as input the strength of the sulphur sources in various countries and account for the dispersion and subsequent removal. These can be compared with numerous measurements of the amount of sulphur deposited in the various parts of Europe

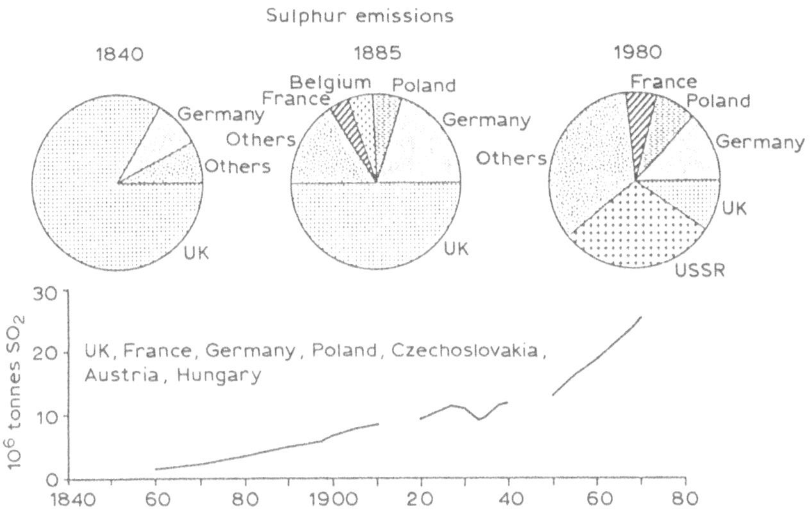

Figure 8.4. Annual sulphate emissions from Europe since 1840 with pie diagrams to illustrate relative proportions from various countries. (Adapted from Refs 81 and 82.)

to check the performance of models. It is possible to estimate the amount of pollutant that is transported into the various countries of Europe. Many countries face the problem that a large fraction of their sulphur budget arises from foreign sources, so difficult to curb within the framework of international law.

Sulphur emissions in the past can be estimated from fuel statistics and a knowledge of its composition, enabling models to extend our knowledge of sulphur deposition in the past or estimate what changes future differences in emission patterns will bring about. Figure 8.5 shows the estimated deposition patterns for sulphur in the year 1840.[82,84] As might be expected the distribution centres about the British Isles (responsible for half the European sulphur emission at this time). This should be compared with that of 1970, which illustrates the much changed distribution pattern. In addition to the change in distribution, the contours on the map indicate the vastly increased magnitude of deposition.

These modelling exercises based on emission data from the 1840s suggest that relatively little sulphur pollution would have reached Scandinavia from other European sources.[85] Yet it is known from observations made in Scandinavia at the time that pollution episodes were

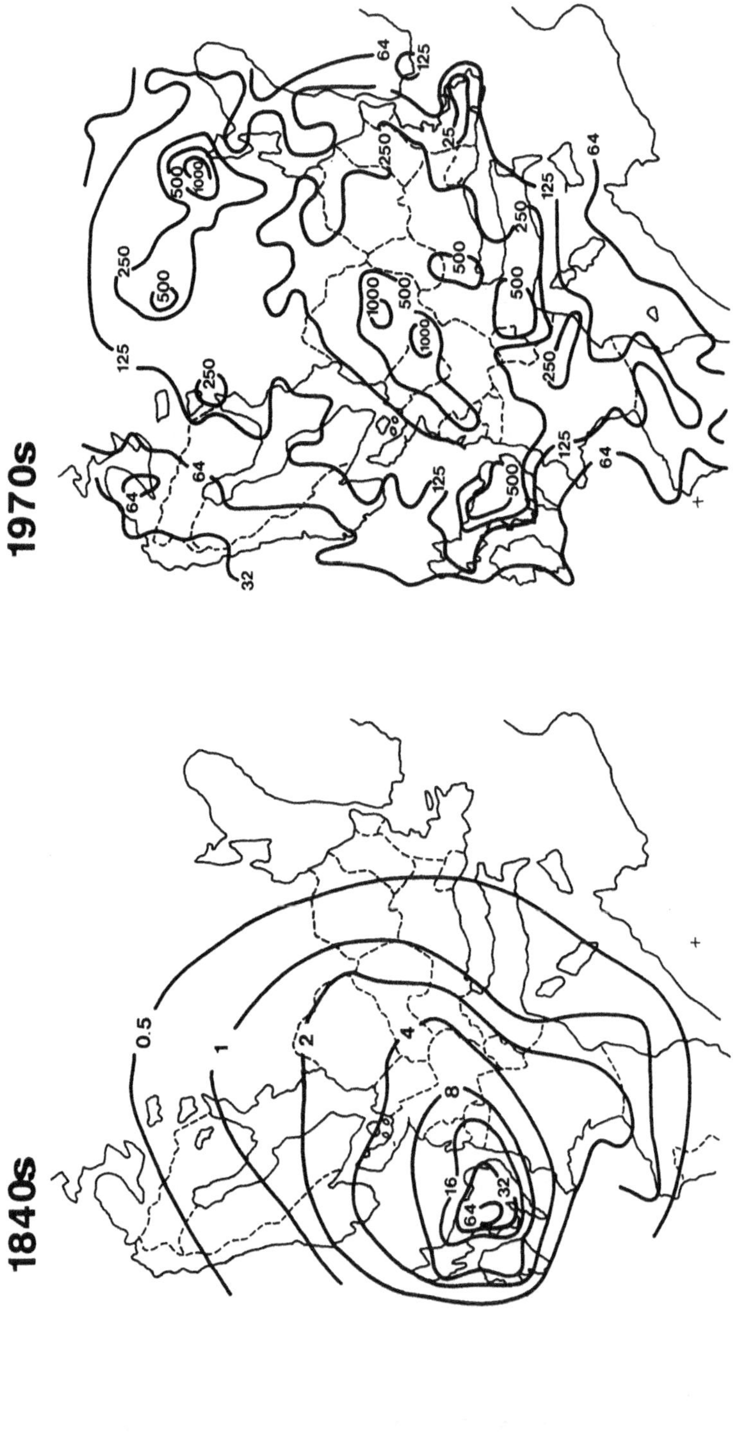

Figure 8.5. Modelled deposition of sulphur (mg S m^{-2} $month^{-1}$) over Europe for the 1840s and 1970s. (Reproduced with permission from Ref. 84.)

occasionally observed. It is important to remember that deposition determined in such modelling exercises is an average. Unusual meteorological conditions may give rise to a polluted event that has a much higher than average pollutant load. It is these extreme conditions that may have been noticed in the past. They have significance today too, because isolated but very acidic falls of rain may have profound environmental consequences.

The historical emissions of sulphur within the US have also been determined and show a distinct bias, with higher quantities emitted along the north-eastern border with Canada at the turn of the century. As in the European setting the emissions become more distributed with time, but the east still predominates (Fig. 8.6).

The acidity of rainfall has usually been attributed to sulphur emissions and the subsequent formation of sulphuric acid (H_2SO_4), but other acids, especially HNO_3, need to be considered. Figure 8.6 shows the estimated emission of NO_x from North American sources over the last century. HNO_3 appears to make an increasingly important contribution to the acidity of rainfall in both Europe and North America (e.g. Refs 87 and 88).

Rather more surprising is the acidification caused by the alkaline gas NH_3. The effect of NH_3 on increasing the solubility of SO_2 and hence its oxidation to H_2SO_4 has long been known, but more recently the nitrification of NH_3 has been recognised as a source of acidification (e.g. Ref. 89). There is particular concern in the Netherlands where the NH_3 emission density currently perturbs the nitrogen cycle. A study has been undertaken of the historical concentration and deposition of NH_3 for the period 1870–1980,[54] and the modelling suggests substantial increases in the deposition of NH_3. These might be as high as four-fold in some areas. Historical rainfall analyses at Rothamsted and Hungary do not show the predicted increase.[90,55] The trends may well be masked by loss of NH_3 during bulk sampling by earlier workers.

There seems to have been a general increase in the ozone concentration of the European troposphere over the last century. Measurements made at Montsouris, France 1876–1905 and Arkona, Germany 1965–1984 suggest that the average ozone concentration changed (see Fig. 8.7) from 11 ppb in the earlier measurements to about 21 ppb in the later series.[91] In addition to being phytotoxic, ozone becomes involved in the oxidation of sulphur and NO_x so is an important consideration in the formation of acid rain.

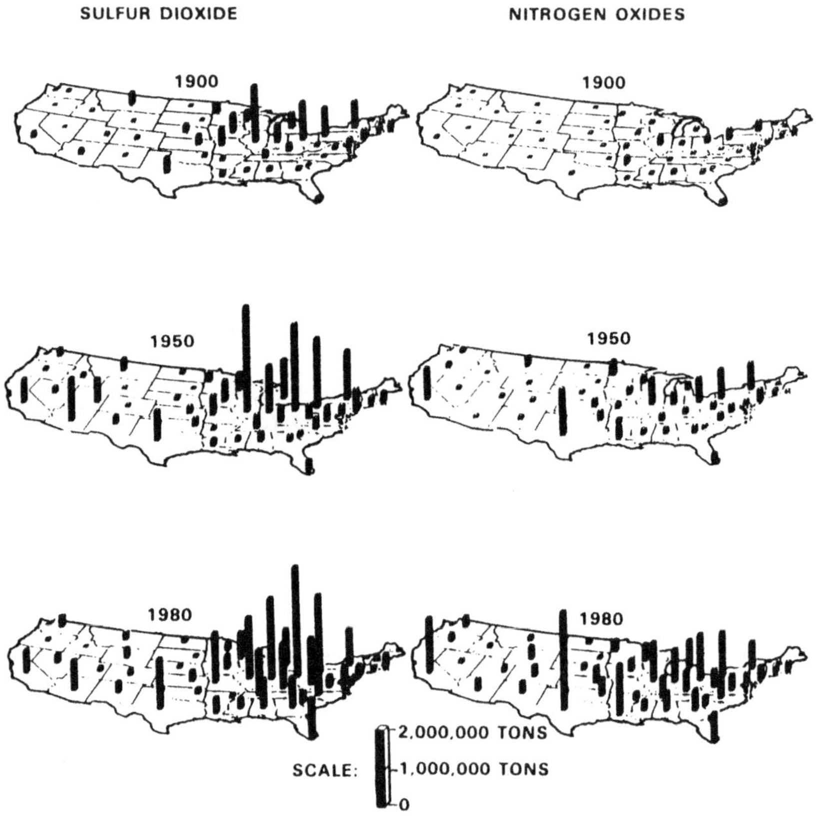

Figure 8.6. Sulphur and nitrogen oxide emissions for 1900, 1950, 1980. (Reproduced with permission from Ref. 86.)

8.4.4 Analysis of the precipitation record

The development of precipitation monitoring networks from the late 1950s onwards has led to the accumulation of an increasingly large dataset. The dataset, though large, has not always been easy to interpret. Quite early on questions were raised about the reliability of networks on the basis of analytical quality.[92] As time has gone on such questions have not necessarily become easier to resolve. Confounding factors such as change in climate and network density and distribution contribute to the difficulties.[93]

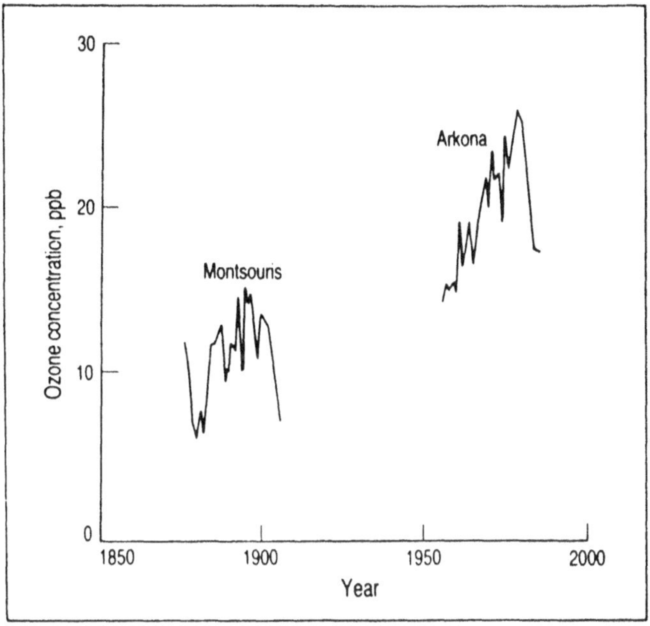

Figure 8.7. Near-surface ozone concentrations measured at Montsouris France 1876–1905 and Akrona, German 1956–1984. (Reproduced with permission from Ref. 91.)

It has always been particularly important to determine the long-term trend in precipitation chemistry, yet it is that task that has often proved particularly difficult. For example, an evaluation was undertaken of the EACN network for the period 1955–1982 and the shortcomings of the dataset were overcome to show that there had been important increases in sulphate concentrations in Scandinavia from the late 1950s to the late 1960s, but some declines in the 1970s were suggested.[94] The long-term record at Hubbard Brook in the US has shown a decline for many species including sulphate over the period 1963–1982.[95] However, declining trends at individual stations could conceal the broadening distribution of stations recording acidified precipitation.

8.4.5 Biological indicators

Biological indicators of pollutants have been treated in a report by Burton.[96] Perhaps the best-known use of biological indicators in determining past acidification has been the use of diatom assemblages in lake sediments. The European and North American work has been summarised. In the UK changes in the diatom assemblage are evident from the mid-19th century, these changes being found a little later in Norway and much later in Sweden and Finland. Acidification in German lakes may have taken place over the last 30–40 years. Small lakes in the Austrian Alps showed no trends to lower pH values in recent times. When all the evidence is placed together it strongly supports the belief that acid deposition is responsible for the changes in European surface waters. In North America, although the picture was rather more complicated, lakes found to have the earliest and largest pH decreases are in the areas of highest acid loading. It has been concluded that acidification is the most likely cause for the decline of fisheries in the eastern US,[98] and the decline in fish survival in south-central Pennsylvania streams has been traced.[99]

Dendrochronology has been used to examine the historic changes in the health of trees. At some sites in Germany the tree-ring analysis shows that decrease in tree vitality began in the 1950s.[100,101] North American trends have been summarised.[102] There has been widespread red spruce decline in the north-eastern US since 1986, but it should be remembered that there were red spruce declines in the 1870s and 1880s when air pollution concentrations were much lower.[103]

Although it is not necessarily easy to establish long-term changes in soil pH because of changing methodology, land use, etc. there is good evidence for the declining pH of soils for this century. A set of soils in south-western Sweden that had first been examined in 1927 have been resampled, and a decrease of 0·3–0·9 pH units in spruce and hardwood was found. A large number of Austrian soils where the first pH measurments were made in 1958–1970 have been examined.[105] The decline in pH was attributed to deposition on the biomass. The observed changes in pH and exchangeable cations is reviewed elsewhere.[106]

8.4.6 Evidence of a reversal

Various changes under way in Europe (economic recessions, increases in efficiency, fuel changes and legislative controls) have caused a recent decline in the overall emission of sulphur.[167,108] Attempts have been made to establish trends using the three networks EACN, OECD and EMEP.[71] It was found that SO_2 and SO_4^{2-} concentrations did not vary in parallel with sulphur emission data across Europe. This was attributed to changes in emission height and contamination of the samples.

However, other work has been able to show declining amounts of a number of rainfall contaminants. Most noticeable have been the declines in sulphate observed in the northern British Isles,[109] the Netherlands[110] and Sweden.[111] Decreases in sulphate and nitrate were found,[110] but changes in H^+ were more difficult to distinguish statistically. Downward trends in sulphur in precipitation from the Atlantic and Norwegian Sea

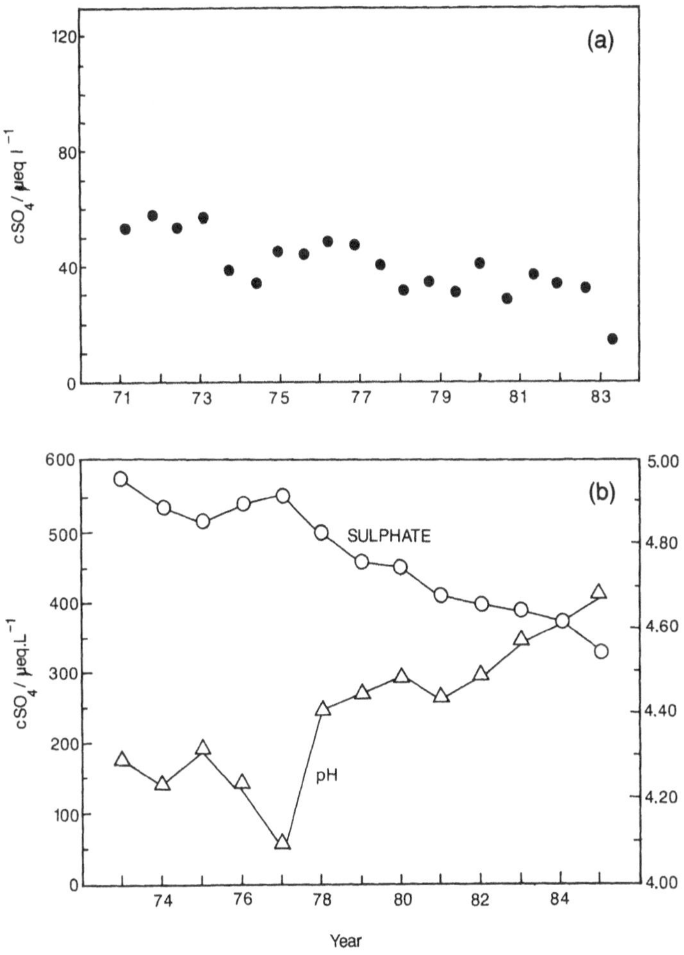

Figure 8.8. (a) Average volume-weighted sulphate ion concentration in precipitation at Hubbard Brook. (Reproduced with permission from Ref. 114.) (b) Average sulphate ion concentration and pH in Clearwater Lake, Canada. (Reproduced with permission from Ref. 115.)

sectors have been noted, but not decreases in the UK sector despite lower sulphur emissions.[112] A suggested explanation involved increased oxidants. In Sweden, decreases in sulphate have been found,[87,111] but nitrate appears to be relatively constant.

In North America there has been some reduction in emissions and, downward trends were found for the SO_2 release in 32 American states,[113] the effect of recent decreases in the emission of sulphur in the north-eastern and mid-western US on the precipitation quality at Hubbard Brook have been examined (see Fig. 8.8).[114] Although there is a decrease in both SO_4^{2-} and H^+ over the period 1963–1983 the correlation is not particularly good. However, one needs to consider the increases in contribution from HNO_3 that were particularly significant up to 1975. Similar results were found for central Ontario where decreases in sulphur emissions have led to a gradual decrease in deposited sulphur.[115] Changes in nitrate deposition are more difficult to discern, although it has been argued that in Britain the decreases in the wet deposition of acids and non-marine sulphate have occurred while nitrate deposition has increased.[116]

Wet deposition of acids and heavy metals in sites in Germany has tended to decrease over the period in the polluted regions, but no striking trend for the acid concentration has been found in recent years.[117]

In 1987, 305 of the lakes in Norway that were surveyed for water chemistry had previously been sampled in the 1974–1975 period. The pH values appeared to have changed little over the period. The 1986 sulphate concentrations were significantly lower than those of the previous decade especially in areas with the highest deposition. Nitrate concentrations were found to be increasing in the lake waters of southernmost Norway.[118] On a more localised scale, the declining sulphur emissions from the smelters of the Sudbury Basin have led to a rapid decrease in the concentrations of sulphate and a rise in pH in acidic lakes.[119] With declining emission in some areas there are indications of improvement, with the dominance of acidiphilic members of the rotifer community declining.[120]

8.4.7 The future

While there are indications that in some parts of the developed world decreases in acid rain are in evidence, projections of the future sulphur loadings in the absence of reduction in Europe are disturbing.[84] These expected decreases may not be paralleled in the developing world where energy use is likely to increase dramatically (hence the Scientific Committee on Problems of the Environment (SCOPE) review). The rapidity of increases in emissions of NO_x and sulphur oxides (SO_x) are the greatest in the tropics,[121] where industrialisation is not necessarily coupled with efficient pollution control (see, for instance, Ref. 122). Acid rain may be increasing in India,[123,124] and in China the concentrations of sulphates are found to be extremely high. Interestingly enough, the rainfall is not quite as acidic as might be expected because of the high concentrations of neutralising calcium present in the air.[125]

8.5 URBAN CONTEXT

As explained in the introduction, some of the problems attributed to acid rain are rather localised. However, as many urban air-pollution problems are treated within the context of acid rain some discussion is necessary here.

Air pollution has been a feature of urban life for a very long time. Many classical writers dealt with the subject and there were laws concerning it in Mediterranean countries 2000 years ago.[126,127] The early use of coal within the British Isles led to an awareness of the problems of air pollution in the Middle Ages.[127] Air and rain in cities began to be analysed last century and the studies of Smith are most famous because of his early detection of acid rain.[5] Less well known are the more numerous analyses of Russell in London, who, despite discovering an acidic sample in his test run, failed to find acid rainwater during the systematic programme using methyl orange or litmus as indicators.[62] This may be rather similar to the current situation in China,[125] where alkalis appear to raise the pH of rainfall samples with a large amount of pollutant sulphate.

8.5.1 Trends in air pollution

It has been concluded that by and large air quality has become worse (in terms of smoke and haze) except in large US cities where there have been some improvements since 1970.[128] Reference 128 concludes that visibility data if carefully used can be a surrogate for air pollution measurements. However, many argue that the air of cities in the developed world has improved lately, in terms of the primary pollutants measured.[129]

Past air pollution in cities has largely been the result of combustion processes. The increasing utilisation of fossil fuels that dominated modern urban history was particularly rapid in the London of the 19th century. This increasing energy consumption has generally led to an increased production of air pollutants. Many other European cities last century saw rapid industrialisation and parallel increases in air pollution. There are few measurements of air pollution old enough to show that increase, but in the records of sulphate deposit in London there are some hints of a peak at the turn of the century.[130] London fog frequency may well be an indicator of air pollution and this seemed to reach a peak in London about the same time.[131]

However, despite increases in the production of major air pollutants a number of important changes have tended to lower the concentration of air pollutants in cities that achieved an early industrial development. These would include the following (in an approximately chronological order).

(1) *Dilution effects:* This century the dispersion of sources of air pollutants has often lowered the pollutant source strength. Thus, while cities have produced more pollutants as they evolved, they have also expanded in size which has had a diluting effect on pollutants. This expansion can be particularly rapid as an efficient urban transport system develops.
(2) *Electrification:* The consumption of energy as electricity within a city can grow without the city increasing the production of pollutants. The pollutants from electricity generation may be produced beyond the boundaries of the city. This might be regarded as another mechanism of dispersion of pollution sources.
(3) *Fuel changes:* In some cities there has been a transition from fuels such as coal, which produce large amounts of soot and SO_2 to oil or gas. Such changes are not necessarily part of control policies and in the 20th century transport-based consumption of petroleum fuels has probably been most important in altering urban air quality.
(4) *Control methods:* There has been an evolution in the law and technology of air pollution control.

The earliest of these effects (i.e. dilution) has been at work for the longest and is in evidence in cities with a long history of air pollution problems. The concentrations of SO_2 and soot are frequently found to be decreasing.[129] By contrast legislative effects can often only be seen in the last decade or so and here lead may be the best example. There are notable exceptions to such trends, particularly in cities of the developing world which use increasing amounts of fuel and in some cases high-sulphur coal.

While classical air pollutants are often on the wane, others may be increasing. Particularly notable has been concern over petroleum-generated hydrocarbons, ozone and the NO_x. These compounds have come to play a bigger role in the chemistry of the urban atmosphere and may be seen to reflect a shift from stationary to mobile sources of air pollution. Los Angeles is the archetype of a city affected by a photochemical smog. The investigations of Chandler's detective Philip Marlowe took place in this atmosphere every bit as charactersitic of a city as the London fogs penetrated by Sherlock Holmes. The fact that the air pollution in Los Angeles was termed smog (SMoke+fOG) is an indicator of the confusion that accompanied this newer form of pollution. The Los Angeles smog was first observed in the 1940s and despite some attempts to ameliorate the situation by shutting down large industrial plants the actions were based on inadequate understanding.[132,133] It was not until the work of Haagen-Schmidt that the smog was seen as being something quite different to the classical smoke and SO_2 situation.[134] Photochemical smogs are now recognised in many cities, even those in northern Europe where it was thought that lack of sunlight would result in inactive photochemistry.

8.5.2 Damage to buildings

Damage to buildings from air pollutants, apart from a mere besooting by smoke, does not really become apparent prior to the extensive use of fossil fuel in cities. However, the adoption of coal in London led to frequent complaints about the corroding quality of the coal smoke. The aggressive nature of coal smoke was attributed to the sharpness of its atoms.[135] There was much damage to St Paul's Cathedral and Westminster Abbey in London of the 1600s. Keepe wrote of Westminster Abbey that 'you rather behold the skeleton of a church . . . exposed, as also to the continual smokes of the sea-coal which are of a corroding and fretting quality, which have added more furrows to her declining years.[136]

It was not only large public buildings that suffered; soiling by coal smoke caused economic loss to ordinary householders. In the London of the 18th century the rate of darkening to the paintwork of some houses was so rapid that repainting needed to be frequent,[137] and in other cities dyes faded in the polluted air.[138]

Despite improvements in the overall quantities of pollutants in much urban air there is a general feeling that these improvements in air quality have not been matched by an equal improvement rate of degradation of our cultural heritage. Possible explanations are as follows.[139]

(1) The total pollutant load in the urban atmosphere may well have decreased, but some components are probably still increasing, e.g. ozone and NO_x, or catalysts such as iron, manganese and vanadium

which could promote the sulphation of building stones. The impact of the recent changes in urban composition on the urban fabric is not well understood and the role of oxidised species such as nitrate is not clear (e.g. Ref. 140).

(2) The corrosion rates may actually be decreasing, but the time resolution or accuracy of the measurements is low.
(3) Building materials possess some kind of 'memory' and the damage appearing now represents pollutant deposition that occurred in the past.[140]
(4) It is a psychological or sociological effect, e.g. the cumulative damage rather than the rate of damage is perceived.

8.5.3 Effects on health

Concern about the links between air pollution and health have been expressed since classical times. The earliest attempt at a statistical correlation was undertaken by John Graunt in 1662,[141] and there is some evidence of increased mortality during foggy periods in 17th century London.[24] To what extent these effects can be linked to acids is not certain even today. However, the influence of H_2SO_4 was strongly suspected in the classical air pollution events at Meuse, France (1930), Donora, USA (1948), and in London (1952). An association of the health effects with acid aerosols seems reasonable because of the strong synergism that exists between SO_2 and smoke, i.e. SO_2 appears to have a greater effect on health in the presence of smoke.

The large numbers of excess deaths that resulted from major pollution episodes of the first half of the 20th century (most particularly that in London in 1952) had a profound influence on the evolution of modern air pollution control legislation. The UK Clean Air Act of 1956 was a far-reaching outcome of the 1952 London smog. However, despite its undoubted importance it must not be forgotten that its attention was directed towards smoke only and not SO_2 or acidic substances. Other improvements to air quality as a result of the passage of this act were incidental.

It is only in recent years that attention has begun to shift away from the simple gaseous pollutants to a clear interest in the role of acid particles (see Chapter 9). However, the question of the relevance of making acid particles a criteria pollutant is still being argued,[142] although it will no doubt gain momentum.

8.6 HISTORY OF CONTROL

The direct attacks on the pollution problem have usually been technological (see Chapter 14).

8.6.1 Fuel type

One of the simplest air pollution controls, at least at the outset, is to change to a less-polluting fuel. This was a solution pressed upon the limeburners in 13th century London who were using coal. They were required to return to the traditional brushwood fuel. Later, when coal had been widely adopted as a fuel in London, the less smoky anthracite might be adopted as a fuel in neighbourhoods where sensitivity to air pollution was high. This practice appears to have been common in English cities last century.[24]

The sulphur emissions can be lowered by using fuels of a lower sulphur content. In particular low-sulphur coals are increasingly desirable, but they are not always geographically distributed in a way that makes their use economic. There is a broadening base of legislation to specify the allowable sulphur content of fuels. Transitions to gas which is very low in sulphur content have often been seen as explanations for improving urban air quality.

Fuel changes can also benefit nitrogen oxide emissions. Although NO_x is formed from nitrogen in the air via the Zeldovich cycle, the nitrogen present in fuel can increase the production of nitric oxide (NO).

8.6.2 Chimney heights

Tall chimneys capable of increasing the dispersion of smoke have a long history. As early as the 14th century there are indications that legislators required chimneys to be high enough to disperse the emissions properly.[24] In the 19th century the problem became more serious as there was no strict relationship between the emissions of a particular factory and the chimney height requirement.[143] At the end of the century in areas such as the Ruhr where the problem of smelter smoke was severe, giant chimneys became the preferred solution.[67] Smelters in North America were not long in following this approach.

The 'tall stack' policy of Britain's Central Electricity Generating Board has been a much criticised use of the dispersive method of controlling air pollution. Historically it arose at a time when there was a great deal of pressure to reduce air pollution in cities. Power generation shifted to bigger stations far from urban centres. The great dispersive power of such chimneys was hardly welcomed by the countries downwind.

8.6.3 Preprocessing the fuel

Attempts to render coal less offensive began in Elizabethan England. John Thornburgh the Dean of York undertook experiments to desulphurise coal, but no details of this method survive.[24] In Stuart England experiments to char coal were by no means infrequent. These activities were not so much to reduce air pollution as to broaden the use of coal which had not proved a successful fuel in many industries. However John Evelyn, the diarist and author of the air pollution abatement tract *Fumifugium*,[17] was quick to realise that the charring might be an approach to reducing smoke.

The practice of washing coal to remove dirt became important at the beginning of the century.[144] However, it is not particularly effective as a method of reducing sulphur. The sulphur in coal is largely present either as pyrite or organosulphides. Pyritic sulphur can be removed by physical processes such as gravity separation based on the high density of pyrite. Other processes such as bacterial or atmospheric oxidation have not made a great impact.

Gasification of coal was first applied on an industrial scale in 1926 (the Winkler generator); however, the methods, while yielding a sulphur-free fuel,[145] are very wasteful on coal (30–40% loss).

8.6.4 Controlling the combustion process

The development of the steam engine led to an early interest in lowering the production of smoke. Smoke, it was realised, represented wasted fuel and also coated the boilers with an insulating layer.[146] Careful control of inflowing air was deemed the best control and a regular spate of books (e.g. Refs 147 and 148), patents and often ineffectual devices or 'smoke consumers' increased to a virtual deluge by the end of the century. It is possible to reduce sulphur emissions by adding lime to the furnace, but such processes have yet proved rather inefficient. A somewhat more successful use of controlled combustion has been in the area of developing furnaces that produce rather low amounts of NO_x.

8.6.5 Flue gas purification

One of the currently favoured approaches to the acid rain problem is removing the pollutants from the flue gas. Grit arrestors and precipitators have been common for some time.[149] Finer particles can be removed by electrostatic precipitators which have a long history.[150] The principle was demonstrated in the laboratory setting in 1824,[150] with the first successful commercial precipitator dating from 1907.

The principle of flue gas desulphurisation has been under investigation for more than a century.[151] The construction of large power stations in cities of the UK and the attendant public outcry during the 1920s prompted the construction of scrubbers at Battersea,[152] Swansea and Fulham. Catalytic oxidation of SO_2 on the surface of platinum, copper, iron and vanadium was of importance to the H_2SO_4 industry. Dry absorption of SO_2 by injecting lime directly into the bed of the furnace was developed by Donath in 1915,[145] and although implemented on a large scale in the 1960s was not very successful. Wet limestone techniques received little attention in the first half of the century.

The success of flue gas desulphurisation (FGD) in the past is not entirely clear. It has been noted that the scrubber at Battersea removed 80–90% of the SO_2 from the flue gases for most of its working life.[153] However, the loss of plume buoyancy and reduction in dispersion led to complaints locally.[154] However, despite local complaints, there can be no doubt that such systems lower the total emission of SO_2. Modern FGD systems have made great advances over the early attempts and form the basis of much current control.

8.6.6 Politics and legislation

It may be that the earliest laws about the trans-national movements of air pollutants were made at the beginning of the 17th century,[17] but it is just as likely that this was merely a piece of wishful thinking by the proto-environmentalist John Evelyn. In the present century, the most familiar case is the investigation from 1929 to 1941 of damage caused by the Trail smelters in Canada to crops in the US.[155] There were, no doubt, a wealth of other incidents along the borders of European countries. Some US interstate pollution cases are summarised in Ref. 159. Another complex interstate situation pertains in the Baltic States of the former USSR and the Baltic Sea, which also suffer from acid deposition.[156]

The Trail Smelter Arbitration has had great influence on international law. The wording of this decision was reflected in that of the Declaration of the UN Conference on the Human Environment in Stockholm in 1972.[1] The OECD became an important forum for discussing international pollution problems in the early 1970s. In 1979 the UN Economic Commission for Europe convened a meeting that adopted the convention and resolution on Long-Range Trans-boundary Air Pollution (LRTBAP) which came into force in 1983.[1] The convention commits the signatories (which include Eastern European countries) to the principles of reducing air pollution, but it is rather lacking in clear objectives and timetables for reduction. In 1983 a number of countries committed themselves to a protocol to the LRTBAP convention, requiring a 30% reduction in sulphur emissions. Some countries (including the UK) have still not formally agreed to the protocol, while others have aimed at greater reductions. The policies of nations are naturally sensitive to the international politics and further reductions in emissions seem inevitable.

The importance of societal features (inefficient automobile engines, cheap fuel, expanding consumerism, modern technology and the conglomeration of manufacturing activity into trans-national corporations) in shifting air pollution control from national to an international frame of reference has been reviewed:[157] Thus caution must be used when reducing trans-boundary acidification problems to the 'tall stack' policy. Despite the complexities of its operation the European Community represents a forum where the problems of trans-national pollution law could eventually be resolved. The development of policy in North America, with Canada being keen to act and the US reluctant is covered by Ref. 158, while international co-operation is discussed by Refs 122 and 73.

8.7 A CONCLUDING THOUGHT

Although one could not argue that the acid rain problem has been solved, the situation in the 1990s will clearly be different from that in the 1980s. Public and media attention has shifted away from acid rain towards issues such as the ozone hole or the greenhouse effect. In some senses the focus of scientific interest may have shifted also. This shift in attention could result in future problems relating to atmospheric acidity. Inadequate research on the likely difficulties in developing and tropical countries, failure to monitor the implementation of control strategies in Europe and North America and a lack of appreciation of the long time-scale for ecosystems to recover from acid inputs may take their toll. Atmospheric acidity will continue to be worthy of study long after popular interest fades.

REFERENCES

1. MoA, *Acidification Today and Tomorrow.* Swedish Ministry of Agriculture, Stockholm, 1982.
2. Glenn, W., Why scientists should do history. *EOS,* **69** (1988) 1572.
3. Brimblecombe, P. & Pfister, C., eds, *The Silent Countdown.* Springer Verlag, Heidelberg, Germany, 1990.
4. Brimblecombe, P., Historical changes in atmospheric trace components. In *Evolution des Atmospheres Planetaires et Climatologie de la Terre.* CNES, France, 1978, pp. 533–40.
5. Smith, A. R., *Air and Rain.* London, UK, 1872.
6. Ducros, M., Observation d'une pluie acide. *J. Pharm. Chim.,* **3** (1845) 273–7.
7. Agnew, C. T., The Big Smoke. *Geog. Rev.,* April (1989) 246–7.
8. Ehrenberg, C. G., Passat-Staub und Blut-Regen. *Phys. Abhand. Kon. Akad. Wiss.* (Berlin) (1847).
9. Loye-Pilot, M. D., Martin, J. M. & Morelli, J., Influence of Saharan dust on the rain acidity and atmospheric input to the Mediterranean. *Nature,* **321,** (1986) 427–8.
10. Adebayo, S. I., Trajectories of advected Saharan dust in Nigeria. *Atmos. Environ.,* **23** (1989) 1581–9.
11. Britton, C. E., *A Meteorological Chronology to AD 1450,* Geophysical Memoirs No. 70. Met. Office, London, UK, 1937.
12. Derham, W., Observations of the late great storm. *Phil. Trans. Royal Soc.* (Lond.), **V** (1704) 60.
13. Thorndyke, L., *History of Magic and Experimental Science,* Columbia University Press, New York, 1923.
14. Beer, E. S. de, *The Diary of John Evelyn.* OUP, Oxford, UK, 1956, entry for 7 Nov 1651.
15. Lawrence, E. N., Raining fish. *Weather,* **10,** (1955) 345.
16. Douglas, D., 16th *De Naturae Mirabilius Opusculum Cuius Catalogum Veras Pagella Docet,* Brit. Mus. 538.e.27.
17. Evelyn, J., *Fumifugium.* London, UK, 1661.
18. Stirling, J., An account of a remarkable darkness, *Phil. Trans. Royal Soc.,* **53,** (1763) 63–4.
19. Brimblecombe, P. & Wigley, T. M. L., Early observations of London's urban plume. *Weather,* **33** (1978) 215–20.
20. Gervat, G. P., *Clouds at Ground Level: Samples from the Southern Pennines,* CEGB TPRD/L/2700/N84, 1984.
21. Brøgger, W. C., Mindre meddelelser. *Nature,* (1881) 47.
22. Frankland, E., *Sixth Report Rivers Pollution Commission.* HMSO, London, UK 1874.
23. Cowling, E. B., Acid precipitation in historical perspective. *Environ. Sci. Technol.,* **16** (1982) 110A–21A.
24. Brimblecombe, P., *The Big Smoke.* Methuen, London, UK, 1987.
25. Eriksson, E., Composition of atmospheric precipitation. I. Nitrogen compounds. *Tellus,* **4** (1952) 215.

26. Rigby, M. & Sinha, E. Z., Annotated bibliography on precipitation. *Meteorol. Astrophys. Abstr.*, **12** (1961) 1430–95.
27. Brimblecombe, P., Historical changes in the atmospheric trace components. In *Evolution des Atmospheres Planetaires et Climatologie de la Terre*, CNES, France, 1978, pp. 533–40
28. Rust, Rev. J., *The Black Scottish Rain Showers and Pummice Shoals of the Years 1862 and 1863*. Blackwood & Sons, Edinburgh, 1864.
29. Brimblecombe, P., Davies, T. D. & Tranter, M., Nineteenth century black Scottish showers. *Atmos. Environ.*, **20** (1986) 1053–7.
30. Bowler, C., Brimblecombe, P. & Wells, A., *Pollution History of Westminster*, PECD 7/12/07. Report to the UK Dept. of Environment, 1989.
31. Camuffo, D., Reconstructing the climate and air pollution of ancient Rome during the life of the Trajan Column, 1988.
32. Bowler, C. & Brimblecombe, P., *Air Pollution History of Westminster Abbey, Houses of Parliament and Hampton Court*. PECD 7/12/07 Report to the UK Dept of Environment, 1991.
33. Leach, F., Beverley town documents. *Seldon Soc.*, **14** (1900).
34. Platt, H., *A New Cheape and Delicate Fire of Cole-Balles*. . ., London, 1603.
35. Hawksworth, D. L., Rose, F. & Coppins, B. J., Changes in the lichen flora of England and Wales attributable to pollution of the air by sulphur dioxide. In *Air Pollution and Lichens*, ed. B. W. Ferry, M. S. Baddeley & D. L. Hawksworth. Althone Press, London, UK, 1973, pp. 330–67.
36. Helland, A., *Lister og Mandals Amt*. Norges Land og Folk Kristiania, 1903.
37. Strøm, K. M., pH values in Norwegian mountain lakes. *Nytt Magaz. F. Nature*, **IX** (1925) II.
38. Dahl, K., *Norsk Jaeger-Fiskeforenings*, **49** (1921) 249–67.
39. Dahl, K., *Salmon and Trout Magazine*, **46** (1927) 35–43.
40. Torgersen, H., Forsøk med oparbeidelse av sure fisketomme van, *Stangfiskeren*. Oslo, Norway, 1934.
41. Thunberg, B., Anglers first to discover acid rain. *Acid*, **7** (1988) 2–4.
42. Hjarne, *Acta et Testamenina Chemica Holmiensia* ii. Stockholm, Sweden, 1753, p. 28.
43. Marggraf, A. S., *Chym. Schriften*. Berlin (1761/7).
44. Bergman, T., *Opuscula Physica et Chemica*. Leipsig, Germany, 1788/90.
45. Hermbstadt, S., Beobachtungen uber die atmospare und das wasser der ostee. *Sweigger's J. Chem. Phys.*, **32** (1821) 281–91.
46. Mulder, G. J., Ueber den ammoniakgehalt der atmospharischen luft. *J. Prakt. Chem.*, **96** (1866) 339.
47. Zimmerman, W., Beitrage zur naheren Kenntniss der wassrigen. *Arch. Ges. Naturlehrs*, **1** (1824) 257–92.
48. Dalton, J., Saline impregnation of rain. *Edinburgh J. Sci.*, **2** (1925) 176.
49. Symons, G. J., Detection of sea spray when mingled with rain. *Brit. Rainfall*, **1–2** (1886).
50. Madden, H. R., On the advantages of extended chemical analysis to agriculture. *Trans. Highland Agric. Soc.*, **14** (1843) 568–86.
51. Church, A. H., Salt in rainwater. *Agric. Student's Gazette*, **2** (1877) 14.
52. Evelyn, J., *Philosophical Discourse of Earth*. London, UK, 1676.

53. Liebig, J. von, *Chemistry and its Application to Agriculture and Physiology*. Taylor & Walton, London, UK, 1843.
54. Asman, A. H. & Drukker, B., Historical concentrations and depositions of ammonia and ammonium in Europe. *Atmos. Environ.*, **22** (1988) 725–35.
55. Horvarth, L., Trend in nitrate and ammonium content of precipitation water in Hungary for the last 80 years. *Tellus*, **35B** (1983) 304–8.
56. Failyer, G. H. & Willard, J. T., Ammonia and nitric acid in atmospheric waters. In *Second Annual Report of the Kansas Agricultural Station*, 123–32, 1889.
57. MacIntire, W. H., Willis, L. G. & Holding, W. A., Effects of lime and magnesium upon soil sulphur. *Soil Sci.*, **4** (1917) 231–5.
58. Wilson, B. D., Nitrogen and sulfur in rainwater in New York. *J. Amer. Soc. Agron.*, **18** (1926) 1208–12.
59. Mooers, C. A., MacIntire, W. H. & Young, J. B., The recovery of soil nitrogen under various conditions. *Univ. Tennessee Agric. Exper. State. Bull.*, **138** (1927).
60. Collinson, R. C. & Mensching, J. E., Lysimeter investigations II. *New York State Agricultural Experimental Station Tech. Bull.*, **193** (1932).
61. Eriksson, E., Composition of atmospheric precipitation. II. Sulphur chloride iodide compounds. *Tellus*, **4** (1952) 280.
62. Russell, W. J., On London rain. *Monthly Weather Report*, April (1884) Appendix 1.
63. Crowther, C. & Ruston, A. G., The nature, distribution and effects upon vegetation of atmospheric impurities in and near an industrial town. *J. Agric. Sci.*, **4** (1911) 25–55.
64. Shaw, N. & Owens, J. S., Causes of acidification of four streams on Laurel Hill in Southwestern Pennsylvania. *The Smoke Problem of Great Cities* Constable, London, UK, 1925.
65. Swain, R. E., Smoke and fume investigations, a historical review. *Indust. Engng. Chem.*, **41** (1949) 2384–8.
66. Quinn, M.-L., Early smelter sites: a neglected chapter in the history and geography of acid rain in the United States. *Atmos. Environ.*, **23** (1989) 1281–92.
67. Gorham, E., On some factors affecting the chemical composition of Swedish fresh waters. *Geochim. Cosmochim. Acta*, **7** (1955) 129–50.
68. Gorham, E., The influence and importance of daily weather consititions in the supply of chloride, sulphate and other ions to fresh waters from precipitation. *Phil. Trans. Royal Soc.* (Lond.), **B241** (1958) 147–78.
69. Junge, C. E., *Air Chemistry and Radioactivity*. Academic Press, New York, 1963.
70. Oden, S., The Acidification of the Atmosphere and Precipitation and its Consequences in the Natural Environment. SNSF, Stockholm, 1968.
71. Popovics, M. M., Szepesi, D. J. & Hakkarinen, C. S., Historical data of atmospheric sulphur in three monitoring periods for Europe. *Water Air Soil Pollut.*, **36** (1987) 47–59.
72. Anon., Money refused. *Acid*, Winter (1984/5) p.26.
73. Park, C. C., *Acid Rain: Rhetoric or Reality*. Methuen, London, UK, 1987.

74. Walker, J. C. G., Possible limits on the composition of the Archaen ocean. *Nature,* **302** (1983) 518–20.
75. Prinn, R. G. & Fegley Jr, B., Bolide impacts, acid rain and biospheric traumas at the Cretaceous–Tertiary boundary. *Earth Plan. Sci. Lett.,* **83** (1987) 1–15.
76. Macdougal, J. D., Seawater strontium isotopes, acid rain and the Cretaceous-Tertiary Boundary, *Science,* **239** (1988) 485–6.
77. Oeschger, H. & Langway Jr, C. C., *The Environmental Record in Glaciers and Ice Sheets.* John Wiley, Chichester, UK 1989.
78. Saigne, C. & Legrand, M., Measurements of methane sulphonic acid in Antarctic ice. *Nature,* **330** (1987) 240–2.
79. Brimblecombe, P., Hannmer, C., Rodhe, H., Ryaboshapko, A. & Batron, C. F., Human influence on sulphur cycle. In *Evolution of the Biochemical Sulphur Cycle,* ed. P. Brimblecombe & A. Yu. Lein, John Wiley, Chichester, UK, 1989, pp. 77–121.
80. Ayers, G. P. & Gillett, R. W., Acidification in Australia. In Acidification in Tropical Countries, ed. H. Rodhe & R. Herrera. John Wiley, Chichester, UK, 1988, pp. 347–402.
81. Bettelheim, I. & Littler, A., *Historical Trends of Sulphur Dioxide Emission in Europe since 1865,* UK Central Electricity Generating Board, Report PL-GS/E/1/79, 1979.
82. Brimblecombe, P., Balans emissii seri v evrope do 1900g. *Trudi Ordena Trydogo Kracnogo Znameni Instituta Prikaladnoi Geofiziki,* **71** (1988) 52–8.
83. Afinogenova, O. G. & Ryaboshapko, A. G., Dinamika antroprogenoi emissii dvookici seri v atmosferoo na territorii CCCP c 1860 po 1970g. *Trudi Ordena Trydogo Kracnogo Znameni Instituta Prikladnoi Geofiziki,* **71** (1988) 38–52.
84. Eliassen, A., Kartina vipadenii antropogennoi seri v evrope v proshlom, nastoyashchem i boodooshchem. *Trudi Ordena Trydogo Kracnogo Znameni Instituta Prikladnoi Geofiziki,* **71** (1988) 59–64.
85. Brimblecombe, P., Is acid rain a new phenomenon? In *The Report of the Acid Rain Inquiry.* Scottish Wildlife Trust, Edinburgh, UK, 1985, pp. 5–8.
86. Gschwandtner, G., Gschwandtner, K. C., Eldridge, K., Mann, C. & Mobley, D., Historic emissions of sulphur and nitrogen oxides in the United States from 1900 to 1980. *J. Air. Pollut. Control Assoc.,* **36** (1986) 139–49.
87. Rodhe, H. & Rood, M. J., Temporal evolution of nitrogen compounds in Swedish precipitation. *Nature,* **321** (1986) 762–4.
88. Brimblecombe, P. & Stedman, D. H., Historical evidence for a dramatic increase in the nitrate component of acid rain. *Nature,* **298** (1982) 460–2.
89. Asman, A. H. & Diederen, S. M. A., *Ammonia and Acidification.* National Institute of Public Health and Environmental Hygeine,Bilthoven, The Netherlands, 1987.
90. Brimblecombe, P. & Pitman, J. I., Long term deposit at Rothamsted, England. *Tellus,* **32** (1980) 261–7.
91. UKPORG, *Ozone in the United Kingdom.* UK Department of the Environment, London, UK, 1987.
92. Paterson, M. P. & Scorer, R. S., Data quality of the European Air Chemistry Network. *Atmos. Environ.,* **7** (1973) 1163–71.

93. Davies, T. D., Barthelmie, R. J., Varley, M., Dorling, S., Framer, G. & Schaug, J., *European Precipitation Chemistry Atlas,* Vol. 2, EMEP/CCC Report 6/90, 1990.
94. Rodhe, H. & Granat, L., An evaluation of sulfate in European precipitation 1955–1982. *Atmos. Environ.*, **18** (1984) 2627–39.
95. Likens, G. E., Borman, F. H., Pierce, R. S., Eaton, J. S. & Munn, R. E., Long-term trends in precipitation chemistry at Hubbard Brook, New Hampshire. *Atmos. Environ.*, **18** (1984) 2641–7.
96. Burton, M. A. S., *Biological Monitoring.* MARC, London, UK, 1986.
97. Batterbee, R. W. & Charles, D. F., Diatom-based pH reconstruction studies of acid lakes in Europe and North America: a synthesis. *Water Air Soil Pollut.*, **30** (1986) 347–54.
98. Haines, T. A. & Baker, J. P., Evidence of fish population responses to acidification in the eastern United States. *Water Air Soil Pollut.*, **31** (1986) 605–29.
99. Sharpe, W., DeWalle, D., Leibfried, R., Dinicola, R., Kimmel, W. & Sherin, L., *J. Environ. Quality,* **13** (1984) 619.
100. Bauch, J., Biological alterations in the stem and root of fir and spruce due to pollution influence. In *Effects of Accumulation of Air Pollutants in Forest Ecosystems,* ed. B. Ulrich & J. Pankrath. D. Reidel, Dordrecht, The Netherlands, 1983, pp. 377–86.
101. Schroter, H., Is the West German experience of acid rain different. In *The Report of the Acid Rain Inquiry.* Scottish Wildlife Trust, Edinburgh, UK, 1985, pp. 27–9.
102. Davis, R. B. & Stokes, P. M., Overview of historical and paleoecological studies of acidic air pollution and its effects. *Water Air Soil Pollut.*, **31** (1986) 311–18.
103. Johnson, A. H., Friedland, A. J. & Dushoff, J. G., Recent and historic red spruce mortality: evidence of climatic influence. *Water Air Soil Pollut.*, **30** (1986) 319–30.
104. Tamm, C. O. & Hallbacken, L., Changes in soil pH over a 50-year period under different forest canopies in SW Sweden. *Water Air Soil Pollut.*, **31** (1986) 337–41.
105. Stohr, D., *Waldbodenversauerung in Osterreich.* Institute fur Forstokologie, Wein, Germany, 1984.
106. Berden, M., Nilsson, S. I., Rosen, K. & Tyler, G., *Soil Acidification, Extent, Causes and Consequences.* National Swedish Environment Protection Board, Solna; Sweden, 1987.
107. Semb, A. & Dovland, H., Atmospheric deposition in Fenno-Scandia: characteristics and trends. *Water Air Soil Pollut.*, **30** (1986) 5–16.
108. Mylona, S. N., *Detection of sulphur emission reductions in Europe during the period 1979–1986.* EMEP MSC-W Report 1/89, 1989.
109. Fowler, D. & Cape J. N., How is acid rain formed and where does it fall? In *The Report of the Acid Rain Inquiry.* Scottish Wildlife Trust, Edinburgh, UK, 1985, pp. 30–45.
110. Buishand, T. A., Kempen, G. T., Frantzen, A. J., Reijnders, H. F. R. & Eshof, A. J. van den, Trend and seasonal variation of precipitation chemistry data in the Netherlands. *Atmos. Environ.*, **22** (1988) 339–48.

111. Leck, C. & Rodhe, H., On the relation between anthropogenic SO_2 emissions and concentration on sulphate in air and precipitation. *Atmos. Environ.*, **23** (1989) 959–66.
112. Berge, E., Time trends of sulfate in precipitation in Norway (1972–1982). *Atmos. Environ.*, **22** (1988) 333–8.
113. Lins, H. F., Trend analysis of monthly sulfur dioxide emissions in the conterminous United States. *Atmos. Environ.*, **21** (1987) 2297–309.
114. Hedin, L. O., Likens, G. E. & Bormann, F. H., Decrease in precipitation acidity resulting from decreased SO4 concentration. *Nature,* **325** (1987) 244–6.
115. Dillon, P. J., Lusis, M., Reid, R. A. & Yap, D., Ten year trends in sulphate, nitrate, and hydrogen deposition in central Ontario. *Atmos. Environ.*, **22** (1988) 901–5.
116. DoE, *Acid Deposition in the United Kingdom 1986–1988.* Department of Environment, London, UK, 1990.
117. Valenta, P., Nguyen, V. D. & Nurnberg, H. W., Acid and heavy metal pollution by wet deposition. *Sci. Total Environ.*, **55** (1987) 311–20.
118. Henriksen, A., Lien, L., Traan, T. S. & Sevaldrud, I. H., 1000 sjøer's undersøkelsen 1986. Statlig program for forurensningsovervaking, Rapport 287/87, SFT/NIVA, (SFT, PO Box 8100 Dep., 0032 Oslo 1, Norway), 1987.
119. Dillon, P. J., Reid, R. A. & Girard, R., Changes in the chemistry of lakes near Sudbury Ontario following reductions in SO_2 emissions. *Water Air Soil Pollut.*, **31** (1986) 59–65.
120. MacIsaac, H. J., Keller, W., Hutchinson, T. C. & Yan, N. D., Natural changes in the planktonic rotifera of a small acid lake near Sudbury Ontario following water quality improvement. *Water Air Soil Pollut.*, **31,** (1986) 791–7.
121. Rodhe, H. & Herrera, R., eds, *Acidification in Tropical Countries.* John Wiley, Chichester, UK, 1988.
122. Elsom, D., *Atmospheric Pollution.* Blackwell, Oxford, UK, 1987.
123. Varma, G. S., Background trends of pH of precipitation over India. *Atmos. Environ.*, **23** (1989) 747–51.
124. Khemani, L. T., Monin, G. A., Prakasa Rao, P. S., Safai, P. D., Singh, G., Chatterjee, R. N. & Prakash, P., Long term effects of pollutants on pH of rain water in North India. *Atmos. Environ.*, **23** (1989) 753–6.
125. Galloway, J. N., Dianwu, Z., Jiling, X. & Likens, G. E., Acid rain: China, United States and a remote area. *Science,* **236** (1987) 1559–62.
126. Mamane, Y., Air pollution control in Israel during the first and second century. *Atmos. Environ.*, **21** (1987) 1861–3.
127. Brimblecombe, P., The antiquity of smokeless zones. *Atmos. Environ.*, **21** (1987) 2485.
128. Chagnon, S. A., Historical atmospheric transmission changes in midwestern air pollution. *Bull. Amer. Met. Soc.*, **68,** (1987) 477–80.
129. De Koning, H. W., Kretzschemar, J. G., Akland, G. G. & Bennett, B. G., Air pollution in different cities around the world. *Atmos. Environ.*, **20** (1986) 101–13.
130. Brimblecombe, P., Long term trends in London Fog. *Sci. Total Eviron.*, **22** (1982) 19–29.

131. Brimblecombe, P., Trends in the deposition of sulphate and total solids in London. *Sci. Total Environ.*, **22** (1982) 97–103.
132. Krier, J. E. & Ursin, E., *Pollution and Policy*. University of California Press, Berkeley, CA, USA, 1977.
133. Stephens, E. R., Smog studies of the 1950s *EOS*, **68** (1987) 89–93.
134. Haagen-Schmidt, A., Chemistry and physiology of Los Angeles smog. *Indust. Engng. Chem.*, **44** (1952) 1342.
135. Digby, K., *A late discourse*. London, UK, 1658.
136. Keepe, H., *Monumenta Westmonasteriensia*. London, UK, 1683.
137. Malcolm, J. P., *Anecdotes from the Manners and Customs of London During the Eighteenth Century*, 1810.
138. Witz, M., Sur la presence de l'acide sulfureux dans l'atmosphere des villes. *C.R. Acad. Sci.*, **100** (1885) 1385–9.
139. Brimblecombe, P., Trends in air pollution and the decay of building materials. *Euro. Cult. Heritage Newslett. Res.*, **2**(2) (1988*a*), 13–14.
140. BERG, *The Effects of Acid Deposition on Buildings and Building Materials in the United Kingdom*. Building Effects Review Group Report, HMSO, London, UK, 1989.
141. Graunt, J., Natural and Political Observations . . . Made upon the Bills of Mortality. In *The Earliest Classics*, eds. J. —, King & G. Gregg Int. Pub. Ltd., 1973.
142. Lipfert, F. W., Morris, S. C. & Wyzga, R. E., Acid aerosols: the next criteria air pollutant. *Environ. Sci. Technol.*, **23** (1989) 1316–22.
143. Frend, W., Is it possible to free the atmosphere of London from smoke? *The Pamphleteer*, **15** (1819).
144. Louis, H., *The Preparation of Coal for the Market*. Methuen, London, UK, 1928.
145. Bretschneider, B. & Kurfurst, J., *Air Pollution Control Technology*. Elsevier, Amsterdam, The Netherlands, 1987.
146. Marsh, A., *Smoke*. Faber & Faber, London, UK, 1947.
147. Williams, C. W., *The Combustion of Coal and the Prevention of Smoke*. London, UK, 1843.
148. Booth, W. M. & Kershaw, J. B. C., *Smoke Prevention and Fuel Economy*. Constable, London, UK, 1906.
149. Lessing, R., The economic value of coal cleaning in relation to the problem of dust and sulphur emission from chimneys. In *Second World Power Conference*. Berlin, 1930, pp.175–95.
150. White, H. J., Fifty years of electrostatic precipitation. *J. Air Pollut. Control Assoc.*, **7** (1957) 166–77.
151. Martin, J. C., A history of flue gas desulphurization since 1850. *J. Air Pollut. Control Assoc.*, **27** (1957) 948–61.
152. Bowler, C. & Brimblecombe P., Battersea Power Station and environmental issues 1929–1989. *Atmos. Environ.*, **25A,** 143–51.
153. Elsworth, S., *Acid Rain*. Pluto Press, London, UK, 1984.
154. Gilpin, A., *Control of Air Pollution*. Butterworths, London, UK, 1963.
155. Halliday, E. C., A historical review of atmospheric pollution. In *Air Pollution*. WHO, Geneva, Switzerland, 1961, pp. 9–37.
156. Koolmala, A., *Atmosfernii Vklad v Zagryaznenie Regiona Baltiiskogo*

Morya. Komissia po Naoochno-technicheskomoo Sotroodnichestboo Meshdoo Finlyandiei i CCCP, 1985.

157. O'Riordan, T., Air pollution legislation and acid regulation in the European Community: a review essay. *Atmos. Environ.*, **23** (1989) 293–306.
158. Schmandt, J., Clarkson, J. & Roderick, H., *Acid Rain and Friendly Neighbours.* Duke University Press, 1988.
159. Wilson, R., Colome, S. D., Spengler, J. D. & Wislon, D. G., *Health Effects of Fossil Fuel Burning.* Ballinger, Cambridge, MA, USA, 1980.

Chapter 9

Health Effects of Atmospheric Acidity

MORTON LIPPMANN
New York University Medical Center, Nelson Institute of Environmental Medicine, Long Meadow Road, Tuxedo, NY 10987, USA

9.1 INTRODUCTION

A variety of gaseous and particulate chemicals in the ambient air are acidic, and some of them are known to have produced health effects in exposed populations. The best documented of the effects are attributable to strong acids in aerosol form, i.e. sulfuric acid (H_2SO_4) and ammonium bisulfate (NH_4HSO_4). Strong acids which are present in the atmosphere as vapors, such as anhydrous nitric acid (HNO_3) and hydrochloric acid (HCl) have not been shown to produce measurable health effects at commonly encountered ambient air levels, although one controlled laboratory study did report a broncho-constrictive response in adolescent asthmatics at 50 parts per billion (10^9) by volume (ppb) HNO_3.[1] Among the weak acids, sulfur dioxide (SO_2) and its hydrolysis products have been associated with both acute broncho-constriction and elevated morbidity and mortality rates. With respect to these elevated rates, SO_2 may have been a surrogate for its oxidation products and other particulate pollutants which often coexist with SO_2. Nitrogen dioxide (NO_2) is both an oxidant and a source of hydrogen ions (H^+) upon hydrolysis on airway surfaces. Some of the health effects associated with NO_2 may therefore be due to its acidity.[2] In any case, the health effects associated with ambient exposures to nitrogen oxides (NO_x) are much harder to document and appear to be of lesser public health concern than those associated with exposures to the sulfur oxides (SO_x).

Other airborne acids include hydroxymethanesulfonate, a reaction product of SO_2 and formaldehyde (HCHO), as well as a variety of organic acids. However, these acids have not, to date, been associated with measurable health effects.

This chapter summarizes and discusses the health effects attributable to SO_2 and acidic aerosols at concentrations which have been monitored or measured in community and industrial atmospheres. It will not discuss effects associated with exposures to these or other acidic pollutants at much higher concentrations because of their lack of relevance to the subject at issue, i.e. the health effects of atmospheric acidity.

9.2 SOURCES AND EXPOSURES

9.2.1 Sources of SO_x

Most of the sulfur in fossil fuel is converted into SO_2 in the combustion zone and it is vented to the atmosphere with the other products of combustion. A small fraction of the sulfur, generally less than 10%, is emitted as H_2SO_4, with some of it being a surface film on ultrafine-sized mineral ash particles. When the discharge point is a tall stack, most of the SO_2 escapes local deposition on terrestrial surfaces, and is gradually (1–10% h^{-1}) converted into sulfur trioxide (SO_3), a highly hygroscopic vapor. The SO_3 rapidly combines with water vapor to produce an ultrafine droplet aerosol of H_2SO_4. The H_2SO_4 is then gradually neutralized by ammonia (NH_3); first to the strong acid, NH_4HSO_4, and then to ammonium sulfate ($(NH_4)_2SO_4$), a nearly neutral salt. Rates of NH_3 neutralization vary widely, depending on emission rates from ground-based sources. Rates are high over cities and agricultural areas, low over forests, and virtually nil over deep water bodies.

The ratios between SO_2, H_2SO_4, and total particulate sulfate (SO_4^{2-}) in the atmosphere are highly variable in space and time. While ambient concentration data are relatively plentiful for SO_2 and SO_4^{2-}, they are, unfortunately, very sparse for H_2SO_4 and NH_4HSO_4, the strong acid aerosols which are believed to account for most of the mortality and morbidity historically associated with mixtures of SO_2 and particulate matter (PM).

Both SO_2 and SO_4^{2-} are very poor surrogate indices for ambient concentrations of acid aerosols.

9.2.2 Exposures to SO_x

Current US ambient air levels of SO_2 are generally well within the current primary National Ambient Air Quality Standard (NAAQS) of 80 $\mu g\ m^{-3}$ for an annual average and 360 $\mu g\ m^{-3}$ for a 24 h maximum. There is an additional special concern for asthmatics' peak exposure to SO_2 while performing outdoor exercise. It has been estimated that the size of the asthmatic population with peak 5–10 min exposures at concentrations $> 0{\cdot}2$ ppm (520 $\mu g\ m^{-3}$) during light to moderate exercise, who may exhibit a broncho-constrictive response, varies from 5000 to 50 000.

For acidic aerosols, there is a very limited ambient concentration database. Reference 3 summarizes the available data from research investigations conducted over limited time periods, generally during the summer. Levels of acidic aerosol in excess of 20–40 $\mu g\ m^{-3}$ (as H_2SO_4) have been observed for time durations ranging from 1 to 12 h. These were associated with high, but not necessarily the highest, atmospheric SO_4^{2-} levels. Exposures (concentration–time product) of 100–900 $\mu g\ m^{-3}$ h were calculated for the acid events which were monitored. In contrast, earlier London studies indicated that acidity exposures in excess of 100 $\mu g\ m^{-3}$ h (as H_2SO_4) were present in the atmosphere, and exposures > 2000 $\mu g\ m^{-3}$ h were possible.

The first data on annual average acidic aerosol concentrations in contemporary US communities were recently reported.[4] In the four eastern US communities studies, the annual average ranged up to $1{\cdot}8\ \mu g\ m^{-3}$ (as H_2SO_4).

9.3 HEALTH EFFECTS

9.3.1 Health effects of SO_2

9.3.1.1 Dosimetry of inhaled SO_2

Highly water-soluble acidic vapors such as SO_2, HNO_3 and HCl are highly efficiently captured in the upper respiratory tract during inhalation, and virtually none of the vapor penetrates to the lungs during normal, quiescent breathing. However, during vigorous physical activity there is less residence time in the upper airways and a shift to oronasal breathing involving partial flow through the less efficient oral passages. Under these conditions, a small fraction of inhaled SO_2 can penetrate to the larger conductive airways of the lungs.

9.3.1.2 Acute broncho-constrictive effects of SO_2 in humans

When asthmatics and others with hyper-reactive airways are exposed to SO_2 at 0·25–0·50 ppm and higher while exercising, the most striking acute response is rapid broncho-constriction (airway narrowing), usually evidenced in increased airway resistance, decreased expiratory flow rates, and the occurrence of symptoms such as wheezing and shortness of breath. Similar responses can be produced in healthy persons, but require exposure concentrations about an order of magnitude higher, and outside the range of ambient levels.

The penetration of SO_2 to sensitive portions of the respiratory tract is largely determined by the efficiency of the oral or nasal mucosa in absorbing SO_2, which in turn depends on the mode of breathing (nasal, oral, or oronasal) and the rate of airflow. Controlled SO_2 exposure studies on asthmatics show that at comparable SO_2 concentrations, broncho-constrictive effects increase with increased ventilation rates and with the relative contribution of oral ventilation to total ventilation.[5–7] Increased oral ventilation not only allows more direct penetration of SO_2, but may also result in airway drying and changes in surface liquids, affecting SO_2 absorption and penetration. Evaporation of airway surface liquid and perhaps convective cooling of the airways caused by cold, dry air can act as direct broncho-constrictive stimuli in asthmatics.[8–10] The combined effect of SO_2 and cold, dry air further exacerbates the asthmatic response.[11–13] The broncho-constrictive effects of SO_2 are reduced under warm, humid conditions.[13].

The time required for SO_2 exposure to elicit significant broncho-constriction in exercising asthmatics is brief. Exposure durations as short as 2 min at 1·0 ppm have produced significant responses.[14] Little enhancement of response is apparent on prolonged exposure beyond 5 min, although some suggestion of an increase is seen with continuous excerise between 10 and 30 min.[15] Following a single SO_2 exposure during exercise, airway resistance in asthmatics appears to require a recovery period of 1–2 h.[16]

The magnitude of response induced by any given SO_2 concentration is variable among individual asthmatics. Exposure to SO_2 concentrations of 0·25 ppm or less, which do not induce significant group mean increases in airway resistance, also do not cause symptomatic broncho-constriction. On the other hand, exposure to 0·40 ppm SO_2 or greater (combined with moderate to heavy exercise) which induce significant group mean increases in airway resistance, also cause substantial broncho-constriction in some individual asthmatics. This broncho-constriction is often associated with wheezing and the perception of respiratory distress, sometimes necessitating the discontinuance of the exposure and the provision of medication. The significance of these observations is that some SO_2-sensitive asthmatics are at risk of experiencing symptomatic broncho-constriction requiring termination of activity and/or medical intervention when exposed to SO_2 concentrations of 0·40–0·50 ppm (1040–1300 μg m^{-3}) or greater when this exposure is accompanied by at least moderate activity. These concentrations can occur downwind of point sources at 10 min averages.

Various studies have examined exposure–response relationships over various concentration and ventilation ranges. Some examined the influence of various subject-related and environmental factors. Since the individual studies used different conditions of airway entry, ventilation rate, concentration, etc., it is difficult to compare directly the results from different investigations. One approach normalizes studies according to effective oral dose rate.[17,18] They showed that reasonably consistent results are derived from the various controlled SO_2 asthmatic studies when adjustments are made for differences in ventilation rates and oral–nasal breathing patterns.

9.3.1.3 Associations between ambient SO_2 and rates of mortality and morbidity

Evidence for effects of SO_2 other than short-term broncho-constriction is less direct. There is a considerable body of epidemiological evidence demonstrating statistically significant associations between SO_2 and rates of mortality and morbidity. However, it is less likely that SO_2 was a causal factor than that it was serving as a surrogate exposure index for other pollutants in the SO_x–particulate complex deriving from fossil-fuel combustion.

Mortality. After exhaustive review of the mortality data from London in the period 1958–1972, the US EPA conclusions were as follows.[19]

(1) Markedly increased mortality occurred, mainly among the elderly and chronically ill, in association with black smoke (BS) and SO_2 concentrations above 1000 μg m^{-3}, especially during episodes

when such pollutant elevations occurred for several consecutive days.

(2) The relative contributions of BS and SO_2 cannot be clearly distinguished from those of each other, nor can the effects of other factors be clearly delineated, although it appears likely that coincident high humidity (fog) was also important (possibly in providing conditions leading to formation of H_2SO_4 or other acidic aerosols).

(3) Increased risk of mortality is associated with exposure to BS and SO_2 levels in the range of 500–1000 $\mu g\ m^{-3}$, clearly at concentrations in excess of 700–750 $\mu g\ m^{-3}$.

(4) Less certain evidence suggests possible slight increases in the risk of mortality of BS levels below 500 $\mu g\ m^{-3}$, with no specific threshold levels having yet been demonstrated or ruled out at lower concentrations of BS (e.g. at 150 $\mu g\ m^{-3}$) nor potential contribution of other plausibly confounding variables having yet been fully evaluated.

Other studies examined pollutant–mortality relationships in New York City, Pittsburgh, and Athens, Greece. Reanalysis of 14 years of New York City data (1963–1976) found significant associations between excess daily mortality and airborne PM, SO_2 and temperature.[20] Differences in the rate of change of SO_2 and PM indicators during the study period allowed estimation of their separate effects. In joint regression analysis across all years, PM indicators (coefficient of haze and visibility extinction coefficient) together accounted for significantly greater excess mortality than did SO_2.

For Pittsburgh, a significant association between PM and excess deaths was found, but no effect of SO_2.[21] On the other hand, an association with SO_2 but not with smoke measurements was found in Athens, Greece.[22]

Morbidity. Associations have been found between 24-h average concentrations of ~ 0·18 ppm (500 $\mu g\ m^{-3}$), in association with BS of ~ 0·10 ppm (250 $\mu g\ m^{-3}$), and a worsening of health status among chronic bronchitis patients in London in the 1950s and 1960s.[23,24]

It has been reported that wheeze was more prevalent in non-smoking women living downwind from mine-mouth coal-burning electric utility plants than among women in control communities with lesser exposures to the effluents.[25] There was a significant association with SO_2, the only effluent measured, and the highest exposure group had 24 h and annual average SO_2 levels which were between 100 and 125% of the US standards.

It has also been reported that daily rates of hospital admissions for respiratory disease in southern Ontario during the summer were associated with ambient levels of SO_2 with 24 and 48 h lags.[26] However, subsequent analyses of data from this population showed that there were stronger associations with the sulfate content of the aerosol than with gaseous pollutants such as SO_2 or ozone (O_3).[27]

For an acidic pollution episode in January 1985 in the Ruhr district in Germany in which average concentrations of SO_2 and suspended particles were 800 and 600 $\mu g\ m^{-3}$, significant increases in deaths, hospital admissions, outpatient visits, and ambulance deliveries to hospitals in comparison to those in a less polluted control area were reported.[28]

Other short-term responses to PM–SO_2 mixtures have been seen in children. Repeated measurements of lung function in school children in Steubenville, OH in 1978–1980 showed that statistically significant, but physiologically small and apparently reversible, declines of forced vital capacity (FVC) and forced expiratory volume ($FEV_{0 \cdot 75}$, i.e. in first 0·75 second of a forced vital capacity manoeuver (in liters)) levels were associated with short-term increases in PM and SO_2.[29] The highest 24-h average PM and SO_2 concentrations were 422 and 455 $\mu g\ m^{-3}$, respectively. The small, reversible decrements persisted for up to 3–4 weeks after episodic exposures.

A study of the association between episodic exposures to particulate matter and SO_2 and pulmonary function in children was conducted in the Netherlands,[30] producing results similar to those of Ref. 29. Pulmonary function values measured during an air pollution episode in which both 24-h average PM and SO_2 levels reached 200–250 $\mu g\ m^{-3}$ were significantly lower (3–5%) than baseline values measured 1–2 months earlier in the same group of Dutch school children. Lung function parameters that showed significant declines included FVC and FEV, as well as measures of small airway function. Declines from baseline were observed 2 weeks after the episode in a different subset of children, but not after 3·5 weeks in a third subgroup.

Studies of associations between chronic exposure to SO_2 and PM and long-term changes in respiratory function in children have also been performed. Changes in baseline lung function between 1981 and 1983 in 1880 schoolchildren living in or near Turin, Italy, have been reported.[31] During that interval, annual average SO_2 in central Turin decreased from ~ 200 to ~ 110 $\mu g\ m^{-3}$, and total suspended particulate matter dropped from ~ 150 to ~ 100 $\mu g\ m^{-3}$. During the same period SO_2 in a suburban area declined from ~ 70 to ~ 50 $\mu g\ m^{-3}$. A group of 162 children from the suburban area served as controls. In the first survey, $FEV_{1 \cdot 0}$, forced expiratory flow rate (FEF_{25-75}), and maximal expiratory flow rate (MEF_{50}) of children from urban areas were significantly lower, while in the second survey they were not significantly different from those of the controls. The slopes over time of $FEV_{1\cdot0}$, FEF_{25-75}, and MEF_{50}, adjusted for sex and anthropometric variables, were closely related to the decrease of pollutant concentrations, suggesting that the decrease of air pollution produced an improvement of baseline lung function.

9.3.1.4 Summary of health effects of SO_2

In summary, the more quantitative epidemiological evidence from London suggests that effects may occur at SO_2 levels at or above 0·19 ppm (500 μg m^{-3}), 24-h average, in combination with elevated particle levels. Additional evidence suggests the possibility of short-term, reversible decline in lung function at SO_2 levels above 250–450 μg m^{-3} (0·10–0.18 ppm). These effects could be due to SO_2 alone, formation of H_2SO_4 or other irritant aerosols, other particles, or peak SO_2 values well above the daily mean, but the relative roles of these factors cannot be determined at this time. It is known that the capacity of fog particles to 'carry' untransformed SO_2 is limited. Thus, it appears more likely that the role of SO_2, in the presence of smoke, involved transformation products such as acidic fine particles.

The statistically significant associations between annual average levels of SO_2 and chronic disease end-points do not have any credible mechanistic basis and will not be reviewed here. To an even greater extent than the more acute response associations, they are likely to be artifacts of co-linear associations between SO_2 and fine particles from combustion processes.

9.3.2 Health effects of acidic aerosols

9.3.2.1 Deposition, growth, and neutralization within the respiratory tract

The deposition pattern within the respiratory tract is dependent on the size distribution of the droplets. Acidic ambient aerosol typically has a mass median aerodynamic diameter (MMAD) of 0·3–0·6 μm, while industrial aerosols can have an MMAD as large as 124 μm.[32] With hygroscopic growth in the airways, submicrometer-sized droplets can increase in diameter by a factor of two to four, and still remain within the fine particle range which deposits preferentially in the distal lung airways and airspaces. As droplet sizes increase above the 3 μm MMAD, deposition efficiency within the airways increases, with more of the deposition taking place within the upper respiratory tract, trachea, and larger bronchi.[33] For larger droplets, the residence time in the airways is too short for a large growth factor.

Some neutralization of inhaled acidic droplets can occur before deposition, due to the normal excretion of endogenous NH_3 into the airways.[34] Once deposited, free H^+ reacts with components of the mucus of the respiratory tract, changing its viscosity.[34] Unreacted H^+ diffuses into surrounding tissues. The capacity of the mucus to react with H^+ is dependent on the H^+ absorption capacity, which is reduced in acidic saturated mucus as found in certain disease states, e.g. asthma.[35]

9.3.2.2 *Effects on experimental animals*

Acute exposures

RESPIRATORY MECHANICAL FUNCTION. Alterations of pulmonary function, particularly increases in pulmonary flow resistance, occur after acute exposure. Reports of the irritant potency of various sulfate species are variable, due in part to differences in animal species and strains, and also to differences in particle sizes; pH, composition and solubility.[19] H_2SO_4 is more irritating than any sulfate salt in terms of increasing airway resistance. For short-term (1 h) exposures, the lowest concentration of H_2SO_4 shown to increase airway resistance was 100 $\mu g\ m^{-3}$ (in guinea pigs). The irritant potency of H_2SO_4 depends in part on droplet size, with smaller droplets having more effect.[36] Some recent animal inhalation studies are of interest to this discussion because they demonstrate that effects produced by single exposures at very low acid concentrations can be persistent.[37] They exposed guinea pigs by inhalation for 3 h to the diluted effluent from a furnace which simulates a model coal combuster. Pulverized coal yields large particle mineral ash particles and an ultrafine ($< 0{\cdot}1 \mu m$) condensation aerosol. The core of the ultrafine particles consists of oxides of Fe, Ca, and Mg, covered by a layer containing Na, As, Sb, and Zn. The Zn is important because it generally has the highest concentration on the surface of the solidified particle. As the particles cool further, there is surface formation and/or condensation of a layer of H_2SO_4. In the initial experiments, the model aerosol was a mixture containing SO_2, zinc oxide (ZnO), and water vapor, and there was a single 3-h exposure to a mixture containing 1 ppm ($2{\cdot}5$ mg m^{-3}) SO_2 and 5 mg m^{-3} of ZnO passed through a humid furnace. The amount of H_2SO_4 on the surface of the ZnO particles was ~ 40 μg m^{-3}. In control studies, neither 1 ppm of SO_2 nor 5 mg m^{-3} of ZnO alone produced any significant responses. There were also no significant responses to the mixture in the absence of water vapor and passage through the furnace. However, the humid mixture, passed through the furnace, where it acquired a surface coating of H_2SO_4, produced significant decrements in lung diffusing capacity (DL_{CO}). At 1 h after exposure, there was an increase in lung permeability. At 12 h after exposure, there was distention of perivascular and peribronchial connective tissues, and an increase in lung weight. The alveolar interstitium also appeared distended. At 72 h after exposure, total lung capacity (TLC), vital capacity (VC) and functional residual capacity (FRC) had returned to baseline levels, but DL_{CO} was still significantly depressed. Based upon prior experience with pure SO_2 and pure H_2SO_4 exposures in the guinea pig model, it was concluded that the humid furnace effluent effect was an acid aerosol effect because of its persistence.[37] In subsequent tests, 3 h exposures to the acid-coated ZnO aerosol were given on five successive days. Significant depressions of DL_{CO} were produced on the second and subsequent days for 30 μg m^{-3} of H_2SO_4, while 20 μg m^{-3} produced significant depressions on the fourth and fifth days. The most sensitive response was a change in airway reactivity, where a significant response was produced by a single 1 h exposure to 20 μg m^{-3} H_2SO_4 as a surface coating on the ZnO.[38] The persistant changes in function and morphological changes following exposure to very low levels of acidic aerosol suggest that repetitive exposures could lead to chronic lung disease. However, the implications of these changes in guinea pigs to human disease remain highly speculative.

PARTICLE CLEARANCE FUNCTION. Donkeys exposed by inhalation for 1 h to 0·3–0·6 μm H_2SO_4 at concentrations ranging from 100 to 1000 μg m^{-3} exhibited slowed bronchial mucociliary clearance function at concentrations ≥ 200 μg m^{-3}, while, as shown in Fig. 9.1, rabbits undergoing similar exposures exhibited an acceleration of clearance at concentrations between 100 and 300 μg m^{-3}, and a progressive slowing of clearance at ≥ 500 μg m^{-3}.[40]

Sub-chronic exposures

PARTICLE CLEARANCE FUNCTION. Donkeys exposed for 1 h day^{-1} (5 days week^{-1}) for 6 months to an aerosol (0·3–0·6 μm) or H_2SO_4 at a concen-

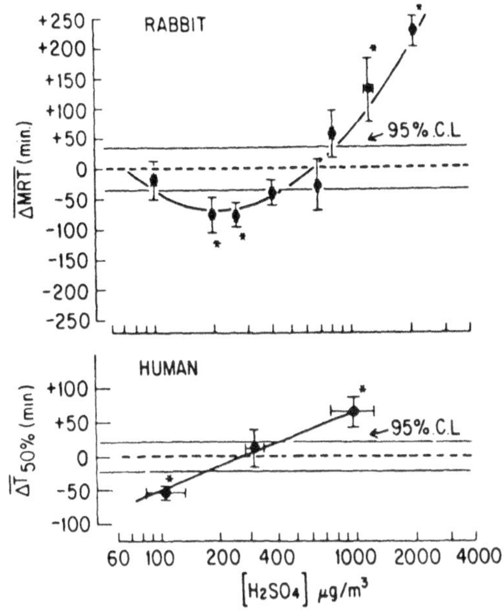

Figure 9.1. Exposure-dependent changes in characteristic bronchial mucociliary clearance times in rabbits (mean residence time) and humans (clearance half-time) due to 1 h exposures to submicrometer sized H_2SO_4 aerosols. Each point represents the average for the group, with the vertical bars indicating ± 1 SD. The horizontal bands represent the mean ± 1 SD of the measurements for the sham exposure controls. Asterisks indicate a significant change ($p < 0{\cdot}05$) at an individual concentration (paired *t* test, one-tailed). (Reproduced with permission from Ref. 39.)

tration of 100 μg m^{-3} developed highly variable clearance rates, and a persistent shift from baseline rate of bronchial mucociliary clearance during the exposures and for 3 months after the last exposure. Two animals had much slower clearance than their baseline during the 3 months of follow-up while two had faster than baseline rates.[1] Rabbits exposed for 1 h day^{-1} (5 days $week^{-1}$) for 4 weeks to 0·3 μm H_2SO_4 at 250 μg m^{-3} developed variable mucociliary clearance rates during the exposure period, and their clearance during a 2-week period following the exposures was substantially faster than their baseline rates.[42] For a group of rabbits undergoing daily exposures via the nose at 250 μg m^{-3} for 1 year, Fig. 9.2 shows that clearance was consistently slowed after the first few weeks, and became even slower during a 3 month period following the end of acid

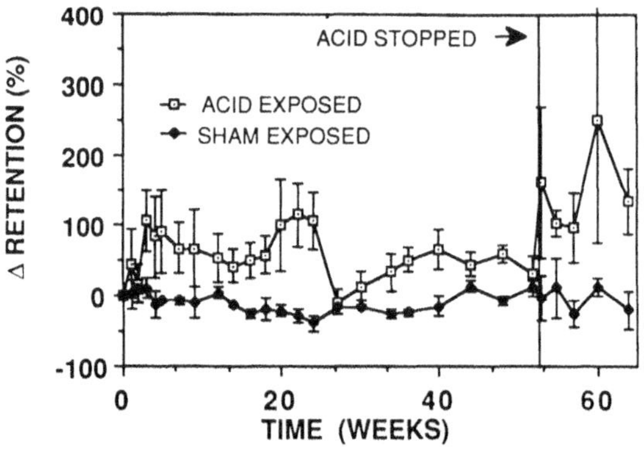

Figure 9.2. Mean change in percentage retention of tracer particles (±SD) during intermittent exposure to H_2SO_4, in acid and sham exposed animals from that established in pre-exposure control tests. (Reproduced with permission from Ref. 43.)

exposure.[44] During the course of a 1 year series of 1 h day^{-1}, 5 days $week^{-1}$ nasal exposure to submicrometer H_2SO_4 at 250 $\mu g\ m^{-3}$, groups of rabbits were exposed on three occasions to ^{85}Sr-tagged latex aerosols for determination of the rates of clearance from the non-ciliated alveolar region.[45] The latex aerosols were inhaled on days 1, 57 and 240 following the start of the H_2SO_4 exposures, and particle retention was followed for 14 days after each latex administration. As compared to baseline rates of clearance in control animals, early alveolar clearance was accelerated to a similar degree in all three tests performed during the chronic H_2SO_4 exposures.

AIRWAY HYPER-RESPONSIVENESS. The effects of daily, 1 h exposures of rabbits to 250 $\mu g\ m^{-3}$ of H_2SO_4 on bronchial responsiveness was assessed at the end of 4, 8 and 12 months by administration (IV) of doubling doses of acetylcholine and measurement of pulmonary resistance (R_L), as shown in Fig. 9.3.[46] Dynamic compliance (C_{dyn}) and respiratory rate (f) were also measured following agonist challenge. Those animals exposed for 4 months showed increased sensitivity to acetylcholine (i.e. the dose required to produce a 150% increase in R_L), and there was an increase in reactivity (i.e. the slope of dose versus change in R_L), by 8 months, with leveling-off of the response after this time. No changes in C_{dyn} or f were noted at any time. Thus, repeated exposures to H_2SO_4 resulted in the production of hyper-responsive airways in previously healthy animals. This

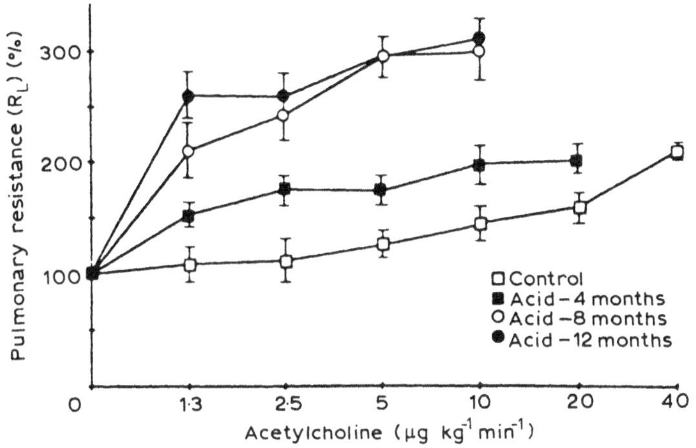

Figure 9.3. Effect of serial broncho-provocation challenges on pulmonary resistance (R_L). The abscissa is expressed in terms of doubling doses of acetylocholine. Data are expressed as the group mean (±SD) percentage of baseline R_L at each dose (n = 12 for control; n = 4 for each acid group). (Reproduced with permission from Ref. 46.)

has implications for the role of non-specific irritants in the pathogenesis of airway disease.

HISTOLOGY. In the study of Schlesinger *et al.* in which rabbits were exposed to 250 μg m^{-3} for 4 weeks and sacrificed 2 weeks later, histological examination showed increased numbers of secretory cells in distal airways and thickened epithelium in airways extending from mid-sized bronchi to terminal bronchioles.[42] There were no corresponding changes in the trachea or other larger airways. In the follow-up study, in which rabbits received daily exposures for 1 year via the nose at 250 μg m^{-3}, the secretory cell density was elevated in some lung airways at 4 months and in all lung airways at 9 months.[44,47] At 12 months the increased density remained in small and mid-sized airways, but not large airways. Partial recovery was observed at 3 months after the last exposure. In a study in which dogs were exposed daily for 5 years to 1100 μg m^{-3} SO_2 plus 90 μg m^{-3} H_2SO_4 and then allowed to remain in unpolluted air for 2 years, there were small changes in pulmonary functions during the exposures, which continued following the termination of exposure. Morphometric lung measurements made at the end of a 2 years' post-exposure period showed changes analogous to an incipient stage of human centrilobular emphysema.[48]

9.3.2.3 *Effects on humans*

Acute effects—controlled exposures

RESPIRATORY MECHANICAL FUNCTION. H_2SO_4 and other sulfates have been found to affect both sensory and respiratory function in humans. Respiratory effects from exposure to H_2SO_4 (350–550 μg m^{-3}) have been reported to include increased respiratory rates and tidal volumes.[49,50] However, other studies of pulmonary function in non-sensitive healthy adult subjects indicated little effect on pulmonary mechanical function when subjects were exposed to submicrometer H_2SO_4 at 10–1000 μg m^{-3} for 10–120 min. In one study, the broncho-constrictive action of carbachol was potentiated by 0·8 μm H_2SO_4 and other sulfate aerosols, more or less in relation to their acidity.[51] Asthmatics are substantially more sensitive in terms of changes in pulmonary mechanics than healthy people, and vigorous exercise potentiates the effects at a given concentration. The lowest-demonstrated-effect level was 68 μg m^{-3} of 0·6 μm H_2SO_4 via mouthpiece inhalation in exercising adolescent asthmatics,[1] with somewhat greater responses at 100 μg m^{-3}.[52] The effects disappeared within about 15 min. In adult asthmatics undergoing exposure to 0·8 μm H_2SO_4 for 2 h, the lowest-observed-effect level was 75 μg m^{-3}.[53] It has been concluded that these results are consistent when the exposure metric is total amount of H_2SO_4 inhaled rather than the concentration of H_2SO_4.[4] The effects of acid fog droplets on respiratory function and symptoms have been studied.[54] The study exposed both normal and mild asthmatic adult volunteers for 60 min to 8 μm MMAD fog droplets containing 0, 150 and 680 μg m^{-3} of H_2SO_4, with alternating 10 min periods of rest and heavy exercise. Both normals and asthmatics reported more symptoms with increasing concentration, and the asthmatics showed an increase in airway resistance at the higher acid concentration. There were no significant differences in either forced expiratory function or airway reactivity to methacholine between the sham and acid exposures.

PARTICLE CLEARANCE FUNCTION. In healthy non-smoking adult volunteers exposed to 0·5 μm H_2SO_4 at rest at 100 μg m^{-3} for 1 h there was an acceleration of bronchial mucociliary clearance of tracer particles which deposited primarily in the larger bronchial airways, and a slowing of

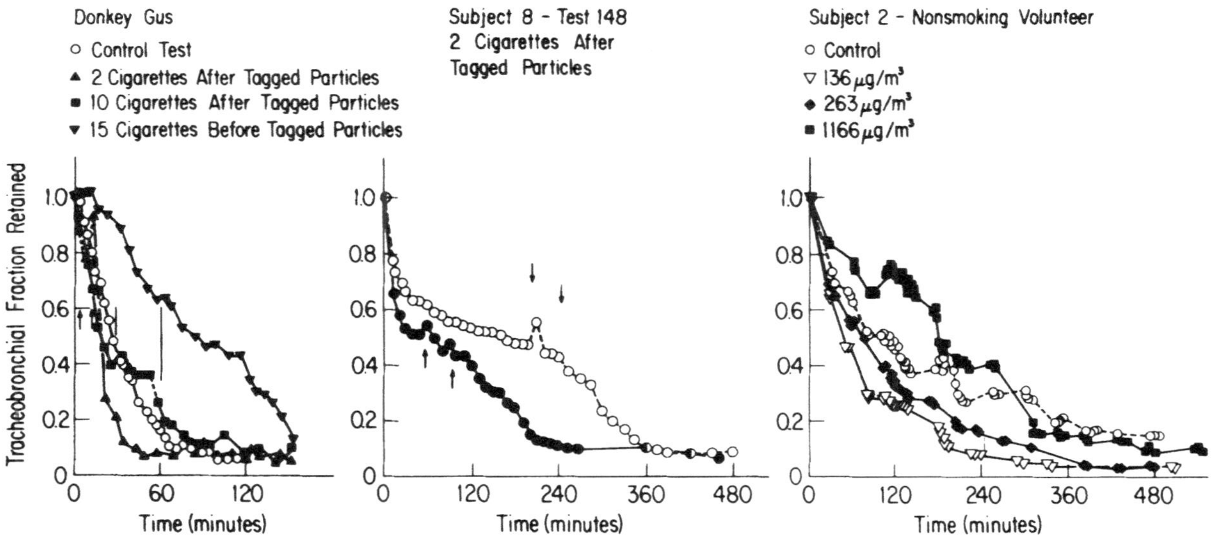

Figure 9.4. Tracheo-bronchial particle retention versus time (a) for donkey Gus in a control test, and in tests involving exposure to whole fresh cigarette smoke from the indicated number of cigarettes; (b) for a 38-year-old non-smoking man for two tagged aerosols inhaled 2·5 h apart on the same day (smoke from two cigarettes, which was inhaled beginning ~ 1 h after the inhalation of the second tagged aerosol, accelerated the clearance of both); (c) for the same man as in (B), but 7 years later. In (c), the man was exposed during the first hour to 0·5 μm H_2SO_4 via nasal mask at four different concentrations, i.e. 0 (sham or control), 136, 263, and 1166 μg m^{-3}. (Reproduced with permission from Ref. 55.)

clearance when the exposure was raised to 1000 μg m^{-3} (Figs 9.1 and 9.4).[56] For tracer particles which deposited primarily in mid-sized to small conducting airways, there was a small but significant slowing of clearance at 100 μg m^{-3} H_2SO_4 and a greater slowing at 1000 μg m^{-3}.[57] These changes are consistent with the greater deposition of acid in mid-sized to smaller airways. Exposures to 100 μg m^{-3} for 2 h produced slower clearance than the same exposure for 1 h, indicating a cumulative relationship to dose.[58] The results of these studies have been used to construct a model for the effects of surface deposition of acidic droplets on mucus transport velocity along the tracheo-bronchial airways.[59] Based on this model, mucus velocities are increased when less than about 10^{-7} g cm^{-2} of H_2SO_4 is deposited, while clearance is retarded when the acid deposition exceeds this limit. The effects of a 1 h inhalation of submicrometer H_2SO_4 aerosols via nasal mask on tracheo-bronchial mucociliary particle clearance and respiratory mechanics were studied in subjects with histories of asthma.[60] A brief inhalation of tagged aerosol preceded the 1 h H_2SO_4 or a sham exposure. Respiratory function was measured before, and 15 min and 3 h after the H_2SO_4 or sham exposure. After exposure to 1000 μg m^{-3} of H_2SO_4, the six subjects not on routine medication exhibited a transient slowing of mucociliary clearance and also decrements in specific airway conductance (SG_{aw}), FEV in 1 s (FEV_1), mid-MEF, and flow rate at 25% of total lung capacity (V_{25}) ($p < 0{\cdot}05$) in both sets of measurements. The four asthmatics on daily medication exhibited stepwise mucociliary clearance that was too variable to allow detection of any H_2SO_4 effect on clearance. Mucociliary clearance rates in both groups in the sham exposure tests were significantly slower than those of healthy non-smokers studied previously using the same protocols.[54] The extent of mucociliary clearance slowing following the 1000 μg m^{-3} exposure in the non-medicated subjects was similar to that in the healthy non-smokers. This similar change, from a reduced baseline rate of clearance, together with the significant change in respiratory function, indicates that asymptomatic asthmatics may respond to H_2SO_4 exposures with functional changes of greater potential health significance than do healthy non-smokers.

EFFECTS—AMBIENT AIR EXPOSURES. There are numerous studies of associations between SO_x pollutants such as SO_2 and SO_4^{2-} and various health effect indices in polluted communities.[20–30, 61–74] As discussed earlier in the section on SO_2, the associations have generally been stronger with the concentrations of indices of particulate pollution such as BS and suspended particulate matter (PM) than with SO_2 in those cases where attempts were made to determine the separate contributions. While strong suspicion that the strong acid component of the particulate matter was the most likely causal factor for the effects was expressed by early investigators,[75–77] very few attempts were made to collect the relevant exposure data, and epidemiologists have had to make do with surrogate exposure indices such as BS, total suspended particulate (TSP), so-called inhalable particulate (IP), respirable particulate (RP), fine particulate (FP), and SO_4^{2-}. IP, RP, and FP are the concentrations below specific aerodynamic size-cuts, i.e. 15, 3·5, and 2·5 μm, respectively.

As a result of the almost total reliance on surrogate measures of exposures, evidence of links between high concentrations of ambient H_2SO_4 and human health effects is largely circumstantial. For example, H_2SO_4 concentrations in the ambient air were certainly much higher than current levels during the classic episodes in London, the Meuse Valley, and Donora, but so were those of many other pollutants. Similarly, the decline in the prevalence of chronic bronchitis in the UK over the past 3 decades could have been due to the decline in any of several pollutants. However, on mechanistic grounds and known exposure–response relationships, H_2SO_4 is a more plausible candidate as an active agent than SO_2, carbonaceous particles, or other known constituents.[78]

In a re-examination of London mortality in relation to exposure to acidic aerosols during 1963–1972 winters,[79] it was shown that acid aerosol concentrations were more closely correlated with daily mortality than were either Black smoke or SO_2. The results were similar for all years and for each winter separately. A typical winter was 1956–1966, with an average concentration of H_2SO_4 of 7·5 $\mu g\ m^{-3}$, and a maximum daily value of 24·3 $\mu g\ m^{-3}$.

Associations between 1980 US mortality rates in 98 Standard Metropolitan Statistical Areas (SMSAs) and four measures of particulate air pollution have been reported.[72] These were TSP, IP, FP, and SO_4^{2-}, a major component of FP. Annual mean SO_4^{2-} concentrations in this data set varied from 3 to 20 $\mu g\ m^{-3}$. They found that FP and SO_4^{2-} were most consistently and significantly associated with reported SMSA-specific total annual mortality rates, while TSP and IP were often non-significant predictors of mortality. Figure 9.5 is an adaptation of the results of the SMSA mortality analyses, showing a hypothetical extrapolation of the analytical results based upon the hypothesis that the acid component of the particulate, a greater fraction as one goes from TSP to IP to FP to SO_4^{2-}, is the causal factor.[80]

In the US, the largest and most comprehensive study of the associa-

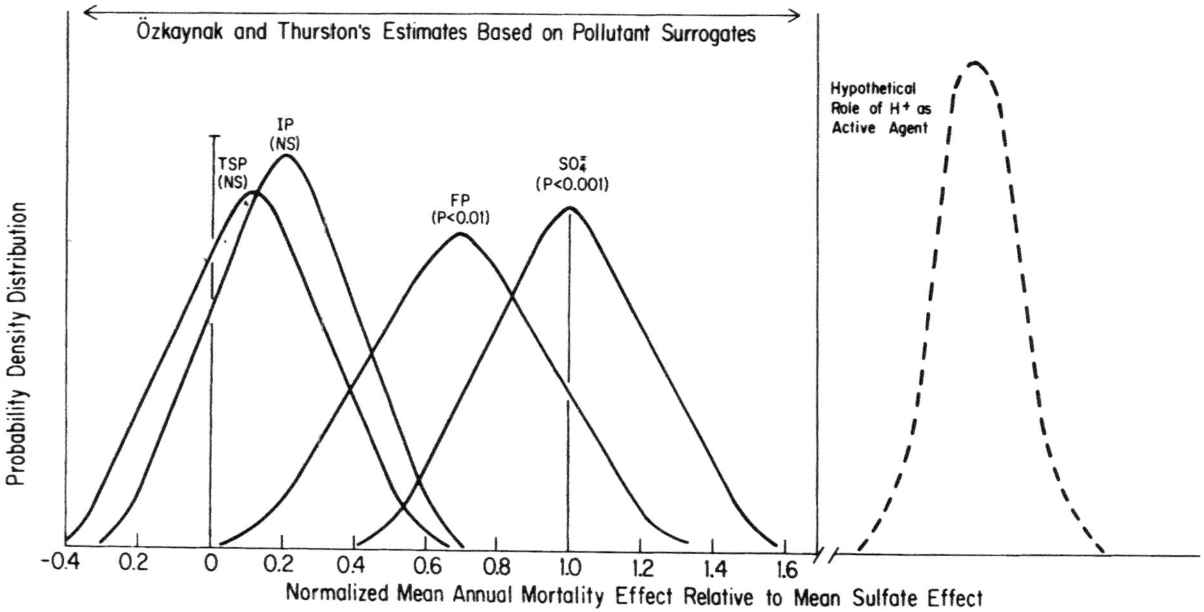

Figure 9.5. Comparative estimates of mean mortality effect of chronic exposures to TSP, IP, FP and SO_4^{2-} particle concentration measures for 98 SMSA model[72] and hypothetical effect of aerosol H^+ concentration. (Reproduced with permission from Ref. 80.)

tions between SO_x and PM pollution and human health is known as the Harvard Six-Cities study, which has been underway since 1974. Unfortunately, direct measurements of aerosol acidity in the six US cities—Kingston-Harriman, TN; Portage, WI; St Louis, MO; Steubenville, OH; Topeka, KS; and Watertown, MA—were not begun until 1985. For the health data gathered between 1974 and 1980 on over 5000 children, a strong association between frequencies of chronic cough, bronchitis, and chest illness in pre-adolescent school children and TSP, SO_4^{2-}, and SO_2 was reported.[70] Illness and symptom rates were higher by approximately a factor of two in the community with the highest air pollution concentrations compared with the community with the lowest concentrations. No association was found, however, between air pollution concentrations and two measures of pulmonary function, FVC and FEV_1. Analysis of limited data on spatial and temporal variability of air pollution concentrations and respiratory health within the six cities found an association between total sulfate concentrations and respiratory illness and symptom rates, but not with TSP or SO_2.

Results from a second cross-sectional assessment of children in the Six-Cities Study have been presented.[71] Air pollution measurements included TSP, IP, FP, SO_4^{2-}, SO_2, O_3, and NO_2. Reported rates of chronic cough, bronchitis, and chest illness during the 1980–1981 school year were positively associated with all measures of particulate pollution (TSP, IP, FP, and SO_4^{2-}) and positively but less strongly associated with concentrations of two of the gases (SO_2 and NO_2). No associations with PM indices were found with asthma, persistent wheeze, hay fever, or non-respiratory illness. No associations were found between pollutant concentrations and any of the pulmonary function measures considered (FVC, $FEV_{1\cdot0}$, $FEV_{0\cdot75}$, and FEF_{25-75}). Children with a history of wheeze or asthma had a much higher prevalence of respiratory symptoms, and there was some evidence that the association between air pollutant concentrations and symptom rates was stronger among children with these markers for hyper-reactive airways. These data provide further evidence that rates of respiratory illnesses and symptoms are elevated among children living in cities with high particulate pollution. They also suggest that children with hyper-reactive airways may be particularly susceptible to other respiratory symptoms when exposed to these pollutants. The lack of association between pollutant concentrations and measures of both pulmonary flow and volume suggests, however, that these increased rates of illness are not associated with permanent loss of pulmonary function, at least during the pre-adolescent years.

The earliest direct association between measured acidity and human health was the highly significant ($p < 0{\cdot}01$) correlation between mortality rates for bronchitis in 53 UK metropolitan areas in the period 1950–1954 and the pH of winter precipitation in these areas.[76] There was also a correlation with SO_4^{2-} at the 5% level of significance. When multiple regression analyses were performed, pH remained significant at the 1% levels, but SO_4^{2-} lost its statistical significance.

Results from the first, preliminary comparison of associations between respiratory symptoms and aerosol acidity from the Six-Cities Study are illustrated in Fig. 9.6. It was shown that the incidence of reported symptoms of bronchitis among 10–12-year-old children in the first four cities studied varied in proportion to the annual average acid aerosol concentration, which ranged from 8 to 35 nmols m^{-3} (0·4–1·8 $\mu g\ m^{-3}$ of H_2SO_4 equivalent).[81] The acid concentrations were for a different year than the symptom reports, but the results provide some support for the causal role of acid aerosols in that both the relative ranking of symptom scores in the Six-Cities Study and the relative ranking of their SO_x pollution levels have remained steady over the years.

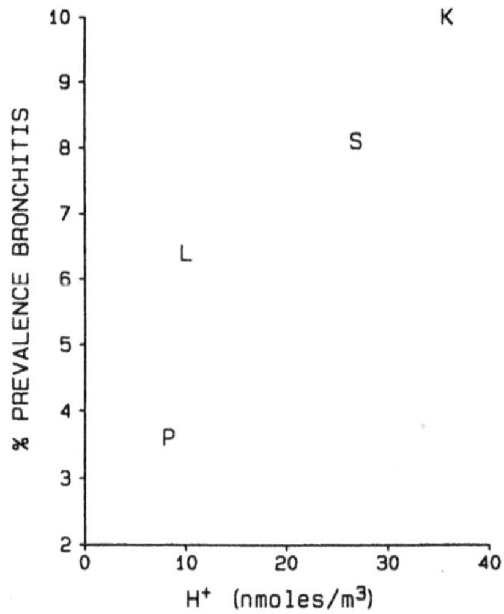

Figure 9.6. Bronchitis in the last year in children 10–12 years of age in four US cities by annual average concentration of H^+ (P—Portage, WI; L—St Louis, MO; S—Steubenville, OH; K—Kingston, TN). (Reproduced with permission from Ref. 81.)

An association based on some limited, but direct measurements was reported in Ref. 73, which identified H_2SO_4 as the probable causal agent for approximately 600 cases of acute respiratory disease in the Yokkaichi area in central Japan between 1960 and 1969. The patients' residences were cencentrated within 5 km of a titanium dioxide (TiO_2) plant with a 14 m stack which emitted from 100 000 to 300 000 kg month^{-1} of H_2SO_4 in the period 1961–1967. The average concentration of SO_2 in February 1965 in Isozu, a village 1–2 km from the plant, was 130 μg m^{-3}, equivalent to 159 μg m^{-3} of H_2SO_4. It was estimated that the peak concentrations might be up to 100-times as high with a north wind. Electrostatic precipitators were installed to control aerosol emissions in 1967, and after 1968 the number of newly found patients with 'allergic asthmatic bronchitis' or 'Yokkaichi asthma' gradually decreased. Although the quantitative estimates of exposure to H_2SO_4 and the criteria used to describe cases of respiratory disease may have differed from current methods, the unique aspect of this report is the identification of H_2SO_4 as the likely causal agent for an excess in morbidity.

In an independent analysis of mortality from asthma and chronic bronchitis associated with changes in SO_x air pollution in Yokkaichi from 1963 to 1983, mortality was correlated with sulfation index (lead peroxide candle measurements), and reduction in SO_x emissions from a petroleum refinery in the harbor area in 1972 were focused on.[74] Thus, it is not clear, from the analysis, what the SO_2 or H_2SO_4 exposures to the population from these emissions were. In any case, mortality rates for bronchial asthma were significantly elevated in Yokkaichi in the period 1967–1970, and the mortality rates due to chronic bronchitis were significantly elevated for the periods 1967–1970 and 1971–1972. There was a greater lag between the reduction in SO_x pollution and reduction in mortality rate for chronic bronchitis than for bronchial asthma.

The most recent report of a direct association between measured levels of ambient acidic aerosol concentration and human health effects was that of Ref. 82. Data were correlated on aerosol H^+, SO_4^{2-}, NO_3^-, and FP, as well as gaseous SO_2 and carbon dioxide (CO_2) with daily symptom, medication usage, and other variables for a panel of about 200 asthmatics in Denver, CO, between November 1987 and March 1988. The H^+ concentrations ranged from 2 to 41 neq m^{-3} (0·04 to 0·84 μg m^{-3} of H_2SO_4 equivalent), and were significantly related to both the proportion of the survey respondents reporting a moderate or worse overall asthma condition, and the proportion reporting a moderate or worse cough. Of all the pollutants considered, H^+ displayed the strongest association with asthma and cough. The magnitudes of the effects were compared by computing elasticities, or the percentage change in the health effect due to a given percentage change in the pollutant. Using asthma as an example, the results indicate elasticities with respect to SO_4^{2-}, FP, and H^+ of 0·060, 0·055 and 0·096, respectively. This means that 10% changes in the concentrations of these pollutants increase the proportion reporting a moderate or worse asthma condition by 0·6, 0·55, and 0·96%, respectively. Thus, of the pollutants considered in this analysis, the results indicate that H^+ had the largest impact on reported asthma status.

9.3.2.4 Implications on the effects of acidic aerosols on respiratory function in the exacerbation of asthma and chronic bronchitis

It has been demonstrated that brief exposures to acidic aerosols reduce airway conductance in healthy humans, and that asthmatic subjects are more sensitive than healthy individuals.[51,83] The lowest concentration that produced a significant response in the healthy group as a whole was 450 $\mu g\ m^{-3}$. A 40% increase has been reported in total airway resistance in a group of exercising asthmatic adolescents when they were exposed to 100 $\mu g\ m^{-3}$ of H_2SO_4,[52] and lesser but still significant effects were reported at 68 $\mu g\ m^{-3}$.[1] The responses were similar to those reported for the same protocols and kinds of subjects for exposure of 0·5 ppm of SO_2 (1·3 $\mu g\ m^{-3}$).[84,85] Thus, when SO_2 is oxidized and hydrolyzed, the resulting H_2SO_4 is 10 to 20 times more potent. While the average increases in airway resistance were small, the populations studied were small, and some subjects had much greater responses than the average. Also, the populations were carefully selected and did not include the more unstable and potentially more reactive asthmatics in the population.

Moderate exercise appears to enhance the response, by increasing the dose of irritant delivered to epithelial surfaces. With increasing exercise, more pollutant is inhaled. The greater inspiratory flow rates also act to increase the percentage of the highly soluble SO_2 vapor which can penetrate beyond the upper airways into those bronchial airways where reflex responses are most likely to be initiated. The greater flow rate also produces a thinner boundary layer around the airway bifurcations, enhancing 'hot spots' of deposition of particles and vapors from the air-stream. Thus, exercise results in increased deposition in this region for the submicrometer-sized H_2SO_4 droplets which would have minimal deposition in such airways at lower flow rates.

The irritant dose delivered to the larger bronchial airways is greater in asthmatics and bronchitics than in healthy individuals because the former groups have airways with smaller diameters. This may account for some, or perhaps all, of the greater responsiveness of asthmatics to inhaled irritants. They may also have a greater responsiveness at the sites of deposition to the delivered dose, but this has not been clearly established in in-vivo tests. In any case, an irritant-induced narrowing of the conducting airways of the lung can increase the surface deposition of subsequently inhaled irritant, resulting in further airway constriction.

The subjective responses to inhaled acid aerosols may include a feeling of chest tightness, and the work of breathing is increased. For individuals with chronic respiratory disease, any increment of work in breathing may be considered an adverse effect. For asthmatic individuals, the major concern is the induction of broncho-spasm. The few clinical laboratory studies on carefully selected asthmatics cannot be expected to generate data on the exact conditions that provoke broncho-spasm and acute respiratory insufficiency. It would be highly desirable to have an animal model for bronchial asthma so that this important issue could be systematically studied.

9.3.2.5 Implications of the effects of acidic aerosols on mucociliary clearance in the pathogenesis of chronic bronchitis

The effects of concentrations and duration of exposure to submicrometer-sized H_2SO_4 aerosols on tracheo-bronchial and alveolar rates of particle clearance in rabbits have been studied, and it was found that the effects increased with duration and concentration.[42] A similar finding for tracheobronchial particle clearance in humans has been reported.[86]

The altered clearance rates during and after the exposure period may be an adaptive response of the mucociliary system to acid exposures. On the other hand, they may be early stages in the progression towards more serious dysfunctions, e.g. those found in chronic bronchitis, which may result from continued irritant exposures.

A mechanistic basis for linkage between chronic exposure to H_2SO_4 and the pathogenesis of chronic bronchitis lies in the series of studies involving chronic exposures of animals and persistent histological alterations in lung structure. As noted previously,[43] these structural changes in the rabbit model have correlates in terms of clearance function changes. These, in turn, are indicative of changes in mucus secretion leading to mucus stasis, a hallmark of bronchitic disease.

The animal studies can be related to human responses in two ways. One is the concordance in functional and morphometric responses of animals to H_2SO_4 and cigarette smoke, a known causal factor for human chronic bronchitis. The other is that humans, rabbits, and donkeys all have essentially the same transient mucociliary clearance function response to single 1-h exposures to H_2SO_4. The fact that daily 1-h H_2SO_4 exposures in rabbits and donkeys produce persistent changes in clearance function makes it highly likely that humans would also if similarly exposed. Furthermore, a comparison of the human and rabbit responses to single exposures indicates that humans respond at lower concentrations than do rabbits.[87]

The effects of H_2SO_4 on the airways are very likely to be cumulative, during each exposure day, at least in part. Thus, the daily 1-h exposures at 250 $\mu g\ m^{-3}$ in the rabbits may be equivalent to $< 50\ \mu g\ m^{-3}$ for a 7–8-h day and to a still lower concentration for equivalent effects in humans. On the other hand, the effects produced by the 1 year series of exposures in the rabbits were less severe than the condition corresponding to a clinical diagnosis of chronic bronchitis in humans.

Unfortunately, there are few data concerning the response of the human mucociliary clearance system under prolonged insult by potentially harmful pollutants such as H_2SO_4. The most direct evidence for an association between chronic bronchitis and the exposure to H_2SO_4 comes from occupational exposures, but these were at high levels. An excess incidence of chronic bronchitis in workers occupationally exposed to H_2SO_4 levels above 1 $mg\ m^{-3}$ (with a probable diameter of 14 μm) has been observed; however, the excess was actually in increased incidence of episodes in affected workers, rather than an increase in the number of workers affected.[32]

While available evidence suggests that exposure to H_2SO_4 may exacerbate disease, it has not been clearly established that it can initiate it. Some limited evidence indicates that it can. For example, in two previously healthy human subjects, what appeared to be long-lasting symptoms of bronchitis have been found to develop due to repeated exposure to H_2SO_4 lasting 1 h and given no more than twice a week, with at least 24 h between exposures.[88] Concentrations were, however, high, ranging from 3 to 39 $\mu g\ m^{-3}$.

The suggestion for a role of H_2SO_4 in the development of chronic bronchitis is given added strength when results of studies of submicrometer H_2SO_4 or whole fresh cigarette smoke exposures, both conducted at New York University Medical Center with laboratory animals and humans, are compared.[55] Cigarette smoke is an agent known to be involved in the etiology of human chronic bronchitis. As shown in Fig. 9.4, the effects of both agents on the mucociliary clearance of tracer particles are essentially the same in terms of (i) transient acceleration of clearance in single low dose exposures; and (ii) transient slowing following single high dose exposures. Furthermore, alterations in clearance rates persisting for several months followed multiple exposures to both agents (Figs 9.2 and 9.7). Thus, although direct evidence for an association between intermittent low level exposures to H_2SO_4 and chronic bronchitis is lacking, the similarity in response between H_2SO_4 and cigarette smoke exposures suggests that such an association is likely.

Human chronic bronchitis is a clinically diagnosed disease, but one which is characterized by certain morphological changes associated with these clinical symptoms.[89–93] One of the basic manifestations is an increase in the number and/or site of epithelial mucus secretory cells in both proximal bronchi, as well as in peripheral airways where such cells are normally absent or few in number; this change is accompanied by an increase in the volume of secretion.[90] In some studies, an increase in epithelial secretory cell proportions in small airways was noted (42, 47).

The appearance of persistently increased secretory cell number in peripheral airways due to H_2SO_4 is a finding of major importance, since excessive mucus production in small airways, which is consistent with an increase in the propagation of secretory cells, may be an early feature in the pathogenesis of bronchitis.[94] Furthermore, it demonstrates an underlying histological change consistent with the observed physiological effects of the H_2SO_4, i.e. altered mucociliary clearance.

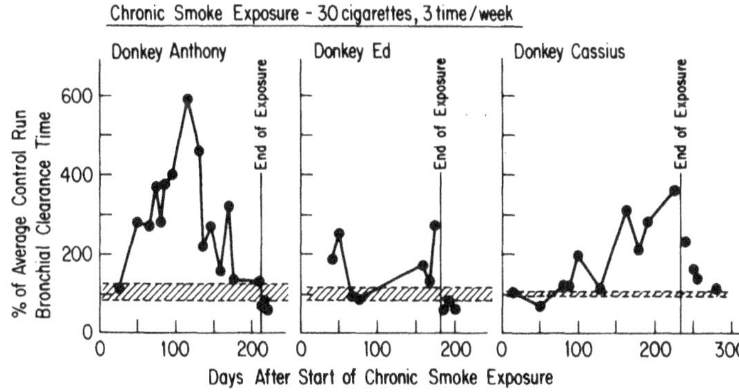

Figure 9.7. Effect of exposures to the whole fresh smoke from 30 cigarettes, 3 times week^{-1}, on the mean residence time for tagged particles on the tracheobronchial airways in three donkeys. The dashed lines indicate the range of three control tests for each animal which preceded the smoke exposures. (Reproduced with permission from Ref. 55.)

In addition to a change in the relative number of secretory cells in different airway levels of acid exposed animals, two other changes were noted after H_2SO_4 exposures.[36] There was an increase in epithelial thickness and a decrease in airway diameter. A significant increase in epithelial thickness of small bronchi and bronchioles occurred in rabbits exposed orally at approximately 250 $\mu g\ m^{-3}$, and nasally at approximately 500 $\mu g\ m^{-3}$. In addition, in the oral exposure series, the lumen diameter of smallest airways was significantly less than in the sham controls.

In human chronic bronchitis and in experimental bronchitis in laboratory animals, an initial change in secretory cell number or size is followed by intrabronchial narrowing, especially in small bronchi and bronchioles; in part, due to a thickening of the bronchial wall.[95,96]

In summary, the first stage of effect of acid exposure may be a change in secretory cell proportions in the airways, and thickening of the epithelium may occur later. Thus, Refs 42 and 47 provide further support for the role of H_2SO_4 in the pathogenesis of chronic bronchitis, via effects on the mucociliary clearance systems. However, the progression of clearance dysfunction in the pathogenesis of chronic bronchitis is not known.

A schema describing the pathogenesis of chronic bronchitis in man due to cigarette smoking is illustrated in Fig. 9.8.[97] It may also apply to repeated exposures to other irritants, such as H_2SO_4. According to this schema, irritant inhalation initially results in a tendency towards some acceleration of clearance, as excess mucus is produced but mucosal damage has not occurred. The H_2SO_4 dose delivered to the rabbit in nasal breathing at 250 $\mu g\ m^{-3}$ may have been enough to initiate this first stage. An increase in the number of airways containing epithelial secretory cells is consistent with increased mucus production. However, the degree of clearance rate change could vary with the individual rabbit and with the time after exposure at which the clearance was measured. This may account for the fact that a significant acceleration was often observed when clearance was measured immediately after H_2SO_4 exposure, while retardation was more commonly observed when clearance was measured 1 day after the last H_2SO_4 exposure.[44]

In the next stage of Fig. 9.8, a further increase in the level of secretion, coupled with some mucosal damage, results in an overloading of transport mechanism; the result is a retardation of clearance. Such a retardation was observed in a study involving 6 months of daily 1 h H_2SO_4 exposures at 100 $\mu g\ m^{-3}$ in donkeys and a study involving daily 1-h H_2SO_4 exposures in rabbits for 1 year.[41,44]

Since H_2SO_4 produces essentially the same sequence of effects on

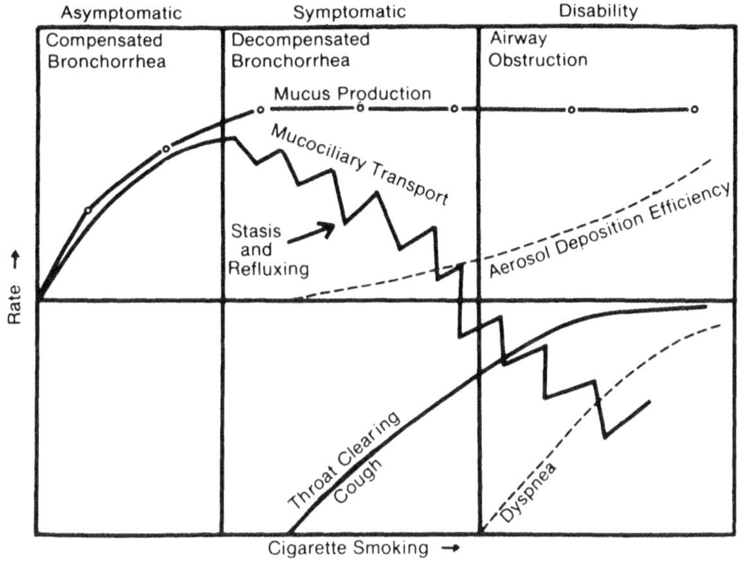

Figure 9.8. Tentative schema for the pathogenesis of obstructive lung diseases resulting from exposure to inhaled airborne irritants. (Reproduced with permission from Ref. 97.)

mucociliary bronchial clearance as cigarette smoke, both following short-term and chronic exposures, it may be capable of contributing to the development of bronchitis. But the question still remains whether variable clearance rates, and persistent clearance rate changes merely predispose to chronic bronchitis, or are the actual initiating events in a pathogenic sequence leading to its development. Furthermore, the response of the mucociliary clearance system observed in the rabbits may be adaptive, rather than pathological. Many irritants may stimulate clearance at low doses or after exposure for a short time, and then retard it at higher doses, or with prolonged exposures.[98] An increase in secretory cell proportions is consistent with hypersecretion. Thus, low-level exposures may initially increase secretion, which can be coped with, and may even be protective. However, pathological changes appear when adaptive capacity is overloaded. Thus, with increasing exposure time or dose, the degree of enhanced secretion may be too great, resulting in overwhelming of clearance, leading to retardation and, eventually, bronchitis.[42]

Airborne acidity and cancer. There is a possibility that exposure to acidic aerosols may play a role in carcinogenesis. A review paper has recently been published that examined a broad array of epidemological and toxicological literature, which, it was concluded, supports the hypothesis that acidic pollutants contribute to carcinogenesis in humans.[99] Possible biological mechanisms for such a contribution were examined, including pH modulation of toxicity of xenobiotics, and pH-induced changes of cells involving mitotic and enzyme regulation. There is, however, no conclusive evidence at this time linking ambient air exposures to carcinogenesis, and public health authorities should consider the hypothesis as a margin-of-safety factor in the establishment of air quality guidelines and/or standards based on the more clearly established health effects associated with exposure to acidic aerosols.

9.3.2.6 Summary of health effects of acidic aerosols

The human health effects of major concern with respect to the inhalation of acidic aerosol are broncho-spasm in asthmatics and chronic bronchitis in all exposed persons. The former relates to acute exposure, while the latter can be related more closely to chronic or cumulative exposures. In either cases, the effects are produced by droplets depositing on the surface of the conductive airways of the lungs.

The studies related to the provocation of broncho-spasm show evidence for increased airway resistance in exercising mild to moderate asthmatics following H_2SO_4 exposures for 30 min at rest and 10 min of exercise for adolescents at 68 or 100 $\mu g\ m^{-3}$ (0·6 μm diameter droplets).[1,52] Similar effects have been seen in adults exposed for 2 h at 75 $\mu g\ m^{-3}$ while exercising, or for 16 min at 450 $\mu g\ m^{-3}$ while at rest (0·8 μm).[53] The effects previously seen at 68 $\mu g\ m^{-3}$ have also been shown to increase when there was co-exposure to 0·1 ppm (260 $\mu g\ m^{-3}$) of SO_2.[1]

Evidence for synergism is also to be found elsewhere,[100] showing that H_2SO_4 (1 h at 20 $\mu g\ m^{-3}$) on ultrafine ZnO particles which simulate coal-combustion effluent, when present in a mixture with SO_2, produce increased lung reactivity response about 10-fold greater than those produced by pure droplets of H_2SO_4 of comparable size.

A causal basis for the historic association between acidic aerosols and the prevalence of bronchitis in humans has been established, at least in part, by the acute and chronic exposure studies in which rates of particle clearance from lungs have been measured. The short-term effects are cumulative over at least several hours, and total exposures at the equivalent of current peak ambient levels produce similar transient changes in mucociliary clearance rates in rabbits and donkeys, as well as in humans. Repetitive daily exposures of rabbits and donkeys, at comparable rates of exposure, produce both transient and persistent clearance abnormalities which are essentially the same as those produced by whole fresh cigarette smoke, a known causal factor for chronic bronchitis in humans.

Conclusive evidence for human health responses to ambient acidic aerosols is lacking, and inferences have had to be drawn from associations between health and likely surrogate indices of acidic aerosol pollution such as SO_2, BS, TSP, IP, FP, and SO_4^{2-}. Of particular interest are the studies reported in Refs 29 and 30 showing the effects of episodic overexposures to ambient mixtures containing concentrations of SO_2 and PM close to current US standards. These exposures produced an apparent continuum of response, with a substantial fraction (perhaps 25% or more) of children having at least a small loss of lung function persisting for at least several weeks. A smaller percentage may have persistent functional decrements exceeding 10%. At concentrations which are about twice the current US standards an episode in Germany in 1985 produced increases in deaths, hospital admissions, outpatient visits and ambulance deliveries to hospitals.[28]

More direct semi-quantitative evidence for a causal role for acidic aerosols comes from the studies reported in Refs 73 and 79. The former study showed an association between acid aerosols and morbidity in Japan, while the latter showed that total daily mortality in London in the period 1963–1972 was more closely associated with daily measured acid aerosol than with $BS+SO_2$.

Epidemiological studies which directly address the role of acidic aerosols on human health are finally beginning to produce results. The preliminary reports of Refs 81 and 82 are both encouraging and dismaying. They are encouraging in that they directly address serious public health concerns. The initial findings are, however, quite disturbing. Reference 81 shows that the prevalence of bronchitic symptoms in school children varies from about 3·5 to 10% as annual average H^+ concentrations (expressed as H_2SO_4 equivalent) varied from 0·4 to 1·8 $\mu g\ m^{-3}$. Reference 82 reports that responses among adult asthmatics were more closely associated with H^+ than with any other pollution variable for concentrations (expressed as H_2SO_4 equivalent) ranging from 0·04 to 0·84 $\mu g\ m^{-3}$. These are commonly encountered levels in the US, and well below historic ambient levels in the US and Europe. More direct measurement data need to be made and coupled to health effects studies. The implications of the preliminary data to public health and ultimately, to the health costs of fossil-fuel consumption need to be much better documented.

ACKNOWLEDGEMENTS

This research was supported by Cooperative Agreement No CR 811563 between the US Environmental Protection Agency and NYU Medical Center, and is part of a Center Program supported by Grant ES 00260 from the National Institute of Environmental Health Sciences.

REFERENCES

1. Koenig, J. Q., Covert, D. S. & Pierson, W. E., Effects of inhalation of acidic compounds on pulmonary function in allergic adolescent subjects. *Environ. Health Perspect.*, **79** (1989) 173–8.
2. US EPA, *Review of the National Ambient Air Quality Standard for Nitrogen Oxides: Assessment of Scientific and Technical Information,* QAQPS Staff Paper, EPA-450/5-82-002. Office of Air Quality Planning and Standards, Research Triangle Park, NC, USA, August 1982.
3. Lioy, P. J. & Waldman, J. M., Acidic sulfate aerosols: characterization and exposure. *Environ. Health Perspect.*, **79** (1989) 15–34.
4. Spengler, J. D., Keeler, G. J., Koutrakis, P., Ryan, P. B., Raizenne, M. & Franklin, C. A., Exposures to acidic aerosols. *Environ. Health Perspect.*, **79** (1989) 43–51.
5. Bethel, R. A., Erle, D. J., Epstein, J., Sheppard, D., Nadel, J. A. & Boushey, H. A., Effect of exercise rate and route of inhalation on sulfur-dioxide-induced bronchoconstriction in asthmatic subjects. *Amer. Rev. Respir. Dis.*, **128** (1983) 592–6.
6. Roger, L. J., Kehrl, H. R., Hazucha, M. & Horstman, D. H., Bronchoconstriction in asthmatics exposed to sulphur dioxide during repeated exercise. *J. Appl. Physiol.*, **59,** (1985) 784–91.
7. Koenig, J. Q., Morgan, M. S., Horike, M. & Pierson, W. E., The effects of sulfur oxides on nasal and lung function in adolescents with extrinsic asthma. *J. Allergy Clin. Immunol.*, **76** (1985) 813–18.
8. Deal, E. C., McFadden, E. R., Ingram, R. H. & Jaeger, J. J., Hyperpnea and heat flux: initial reaction sequence in exercise-induced asthma. *J. Appl. Physiol.*, **46** (1979) 476–83.
9. Strauss, R. H., McFadden Jr, E. R., Ingram Jr, R. H. & Jaeger, J. J., Enhancement of excerise-induced asthma by cold air. *New Engl. J. Med.*, **297** (1977) 743–7.
10. Anderson, S. D., Issues in exercise-induced asthma. *J. Allergy Clin. Immunol.*, **76** (1985) 763–72.

11. Bethel, R. A., Sheppard, D., Epstein, J., Tam, E., Nadel, J. A. & Boushey, H. A., Interaction of sulfur dioxide and cold dry air in causing bronchoconstriction in asthmatic subjects. *J. Appl. Physiol.: Respir. Environ. Exercise Physiol.*, **57** (1984) 419–23.
12. Sheppard, D., Eschenbacher, W. L., Boushey, H. A. & Bethel, R. A., Magnitude of the interaction between the bronchomotor effects of sulfur dioxide and those of dry (cold) air. *Amer. Rev. Respir. Dis.*, **130** (1984) 52–5.
13. Linn, W. S., Shamoo, D. A., Anderson, K. R., Whynot, J. D., Avol, E. L. & Hackney, J. D., Effects of heat and humidity on the responses of exercising asthmatics to sulfur dioxide exposure. *Amer. Rev. Respir. Dis.*, **131** (1985) 221–5.
14. Horstman, D. H., Seal, Jr, E., Folinsbee, L. J., Ives, P. & Roger, L. J., The relationship between exposure duration and sulfur dioxide-induced bronchoconstriction in asthmatic subjects. *Amer. Indust. Hyg. Assoc. J.*, **49**, (1988) 38–47.
15. Kehrl, H. R., Roger, L. J., Hazucha, M. J. & Horstman, D. H., Differing response of asthmatics to SO_2 exposure with continuous and intermittent exercise. *Amer. Rev. Respir. Dis.*, **135** (1987) 350–5.
16. Hackney, J. D., Linn, W. S., Bailey, R. M., Spier, C. E. & Valencia, L. M., Time course of exercise-induced bronchoconstriction in asthmatics exposed to sulfur dioxide. *Environ. Res.*, **34** (1984) 321–7.
17. Kleinman, M. T., Sulfur dioxide and exercise: relationships between response and absorption in upper airways. *J. Air Pollut. Control Assoc.*, **34** (1984) 32–7.
18. Linn, W. S., Venet, T. G., Shamoo, D. A., Valencia, L. M., Anzar, U. T., Spier, C. E. & Hackney, J. D., Respiratory effects of sulfur dioxide in heavily exercising asthmatics: A dose-response study. *Am. Rev. Respir. Dis.*, **127** (1983) 278–83.
19. US EPA, *Second Addendum to Air Quality Criteria for Particulate Matter and Sulfur Oxides (1982)*, EPA/600/8-86/020F. Office of Health and Environmental Assessment, Washington, DC, USA, December 1986.
20. Ozkaynak, H. & Spengler, J. D., Analysis of health effects resulting from population exposures to acid precipitation precursors. *Envion. Health Perspect.*, **63** (1985) 45–55.
21. Mazumdar, S. & Sussman, N., Relationships of air pollution to health: results from the Pittsburgh Study. *Arch. Environ. Health*, **38** (1983) 17–24.
22. Hatzakis, A., Katsouyanni, K., Kalandidi, A., Day, N. & Trichopoulos, D., Short-term effects of air pollution on mortality in Athens. *Int. J. Epidemiol.*, **15** (1986) 73–81.
23. Lawther, P. J., Climate, air pollution and chronic bronchitis. *Proc. Royal Soc. Med.*, **51** (1958) 262–4.
24. Lawther, P. J., Waller, R. E. & Henderson, M., Air pollution and exacerbations of bronchitis. *Thorax*, **25** (1970) 525–39.
25. Schenker, M. B., Samet, J. M., Speizer, F. E., Gruhl, J. & Batterman, S., Health effects of the air pollution due to coal combustion in the Chestnut Ridge region of Pennsylvania: results of cross-sectional analysis in adults. *Arch. Environ. Health*, **38** (1983) 325–30.

26. Bates, D. V. & Sizto, R., Relationships between air pollutant levels and hospital admissions in Southern Ontario. *Can. J. Public Health,* **74** (1983) 117–22.
27. Bates, D. V. & Sizto, R., Air pollution and hospital admissions in Southern Ontario: the acid summer haze effect. *Environ. Res.,* **43** (1987) 317–31.
28. Wichmann, H. E., Mueller, W., Allhoff, P., Beckmann, M., Bochter, N., Csicsaky, M. J., Jung, M., Molik, B. & Schoeneberg, G., Health effects during a smog in West Germany in 1985. *Environ. Health Perspect.,* **79** (1989) 89–99.
29. Dockery, D. W., Ware, J. H., Ferris Jr, B. G., Speizer, F. E., Cook, N. R. & Herman, S. M., Change in pulmonary function in children associated with air pollution episodes. *J. Air Pollut. Control Assoc.,* **32** (1982) 937–42.
30. Dassen, W., Brunekreef, B., Hoek, G., Hofschreuder, P., Staatsen, B., de Groot, H., Schouten, E. & Biersteker, K., Decline in children's pulmonary function during an air pollution episode. *J. Air Pollut. Control Assoc.,* **36** (1986) 1223–7.
31. Arossa, W., Spinaci, S., Bugiani, M., Natale, P., Bucca, C. & de Candussio, G., Changes in lung function of children after an air pollution decrease. *Arch. Environ. Health,* **42** (1987) 170–4.
32. Williams, M. K., Sickness, absence and ventilatory capacity of workers exposed to sulphuric acid mist. *Brit. J. Indust. Med.,* **27** (1970) 61–6.
33. Lippmann, M., Yeates, D. B. & Albert, R. E., Deposition, retention and clearance of inhaled particles. *Brit. J. Indust. Med.,* **37** (1980) 337–62.
34. Larson, T. V., Covert, D. S., Frank, R., & Charlson, R. J., Ammonia in the human airways: neutralization of inspired acid sulfate aerosols. *Science,* **197** (1977) 161–3.
35. Holma, B., Influence of buffer capacity and pH-dependent rheological properties of respiratory mucus on health effects due to acidic pollution. *Science Total Environ.,* **41** (1985) 101–23.
36. Amdur, M. O., Dubriel, M. & Creasia, D., Respiratory response of guinea pigs to low levels of sulfuric acid. *Environ. Res.,* **15** (1978) 418–23.
37. Amdur, M. O., Sarofim, A. F., Neville, M., Quann, R. J., McCarthy, J. F., Elliot, J. F., Lam, H. F., Rogers, A. E. & Connor, M. W., Coal combustion aerosols and SO_2: an interdisciplinary analysis. *Environ. Sci. Technol.,* **20** (1986) 138–45.
38. Amdur, M. O., Sulfuric acid: the animals tried to tell us. *Appl. Indust. Hyg.,* **4** (1989) 189–97.
39. Lippmann, M., Respiratory tract deposition and clearance of aerosols. In *Aerosols,* ed. S. D. Lee, T. Schneider, L. D. Grant & P. J. Verkerk. Lewis Publishers, Chelsea, MI, USA, 1986, pp. 43–57.
40. Schlesinger, R. B., Effects of inhaled acids on respiratory tract defense mechanisms. *Environ. Health Perspect.,* **63** (1985) 25–38.
41. Schlesinger, R. B., Halpern, M., Albert, R. E. & Lippmann, M., Effect of chronic inhalation of sulfuric acid mist upon mucociliary clearness from the lungs of donkeys. *J. Environ. Pathol. Toxicol.,* **2** (1979) 1351–67.
42. Schlesinger, R. B., Naumann, B. D. & Chen, L. C., Physiological and histological alterations in the bronchial mucociliary clearance system of rabbits following intermittent oral or nasal inhalation of sulfuric acid mist. *J. Toxicol. Environ. Health,* **12** (1983) 441–65.

43. Lippmann, M., Gearhart, J. M. & Schlesinger, R. B., Basis for a particle size-selective TLV for sulfuric acid aerosols. *Appl. Indust. Hyg.*, **2** (1987) 188–99.
44. Gearhart, J. M. & Schlesinger, R. B., Response of the tracheobronchial mucociliary clearance system to repeated irritant exposure: effect of sulfuric acid mist on function and structure. *Exp. Lung. Res.*, **14** (1988) 587–605.
45. Schlesinger, R. B. & Gearhart, J. M., Early alveolar clearance in rabbits intermittently exposed to sulfuric acid mist. *J. Toxicol. Environ. Health*, **17** (1986) 213–20.
46. Gearhart, J. M. & Schlesinger, R. B., Sulfuric acid-induced airway hyperresponsiveness. *Fund. Appl. Toxicol.*, **7** (1986) 681–9.
47. Gearhart, J. M. & Schlesinger, R. B., Sulfuric acid-induced changes in the physiology and structure of the tracheobronchial airways. *Environ. Health Perspect.*, **79** (1989) 127–37.
48. Stara, J. F., Dungworth, D. L., Orthoefer, J. G. & Tyler, W. S., *Long-Term Effects of Air Pollutants in Canine Species*. EPA-600/8-80-014. US Environmental Protection Agency, Research Triangle Park, NC, USA, 1980.
49. Amdur, M. O., Silverman, L. & Drinker, P., Inhalation of H_2SO_4 mist by human subjects. *Arch. Indust. Hyg. Occup. Med.*, **6** (1952) 305–13.
50. Ericsson, G. & Camner, P., Health effects of sulfur oxides and particulate matter in ambient air. *Scand. J. Work Environ. Health*, **9** (1983) 52.
51. Utell, M. J., Morrow, P. E. & Hyde, R. W., Airway reactivity to sulfate and sulfuric acid aerosols in normal and asthmatic subjects. *J. Air Pollut. Control Assoc.*, **34** (1984) 931–5.
52. Koenig, J. Q., Pierson, W. E. & Horike, M., The effects of inhaled sulfuric acid on pulmonary function in adolescent asthmatics. *Amer. Rev. Respir. Dis.*, **128** (1983) 221–5.
53. Bauer, M. A., Utell, M. J., Speers, D. M., Gibb, F. R. & Morrow, P. E., Effects of near ambient levels of sulfuric acid aersol on lung function in exercising subjects with asthma and COPD. *Amer. Rev. Resp. Dis.*, **137**(4) (1988) A167.
54. Avol, E. L., Linn, W. S.,.Wightman, L. H., Whynot, J. D., Anderson, K. R. & Hackney, J. D., Short-term respiratory effects of sulfuric acid in fog: a laboratory study of healthy and asthmatic volunteers. *J. Air Pollut. Control Assoc.*, **38** (1988) 258–63.
55. Lippmann, M., Schlesinger, R. B., Leikauf, G., Spektor, D. & Albert, R. E., Effects of sulphuric acid aerosols on respiratory tract airways. *Ann. Occup. Hyg.*, **26** (1982) 677–90.
56. Leikauf, G., Yeates, D. B., Wales, K. A., Albert, R. E. & Lippmann, M., Effects of sulfuric acid aerosol on respiratory mechanics and mucociliary particle clearance in healthy nonsmoking adults. *Am. Ind. Hyg. Assoc. J.*, **42** (1981) 273–82.
57. Leikauf, G. D., Spektor, D. M., Albert, R. E. & Lippmann, M., Dose depedent effects of submicrometer sulfuric acid aerosol on particle clearance from ciliated human lung airways. *Amer. Indust. Hyg. Assoc. J.*, **45** (1984) 285–92.
58. Schlesinger, R. B., The effects of inhaled acids on lung defenses. In *Aerosols*, eds S. D. Lee, T. Schneider, L. D. Grant & P. J. Verkerk. Lewis, Chelsea, MI, USA, 1986, pp. 617–35.

59. Yu, C. P., Hu, J. P., Yen, B. M., Spektor, D. M. & Lippmann, M., Models for mucociliary particle clearance in lung airways. In *Aerosols,* ed. S. D. Lee, T. Schneider, L. D. Grant & P. J. Verkerk, Lewis, Chelsea, MI, USA, 1986, pp. 569–78.
60. Spektor, D. M., Leikauf, G. D., Albert, R. E. & Lippmann, M., Effects of submicrometer sulfuric acid aerosol on mucociliary transport and respiratory mechanics in asymptomatic asthmatics. *Environ. Res.,* **37** (1985) 174–91.
61. Chapman, R. S., Calafiore, D. C. & Hasselblad, V., Prevalence of persistent cough and phlegm in young adults in relation to long-term ambient sulfur oxide exposure. *Amer. Rev. Respir. Dis.,* **132** (1985) 261–7.
62. Dodge, R., Solomon, P., Moyers, J. & Hayes, C., A longitudinal study of children exposed to sulfur oxides. *Amer. J. Epidemiol.,* **121** (1985) 720–36.
63. Charpin, D., Kleisbauer, J. P., Fondarai, J., Graland, B., Viola, A. & Gouezo, F., Respiratory symptoms and air pollution changes in children: the Gardann coal-basin study. *Arch. Environ. Health,* **43** (1988) 22–7.
64. Martin, A. E., Mortality and morbidity statistics and air pollution. *Proc. Royal Soc. Med.,* **57** (1964) 969–75.
65. Martin, A. E. & Bradley, W. H., Mortality, fog and atmospheric pollution—an investigation during the winter of 1958–59. *Mon. Bull. Minist. Health Public Health Lab. Serv.,* **19** (1960) 56–72.
66. Mazumdar, S., Schimmel, H. & Higgins, I. T. T., Relation of daily mortality to air pollution: an analysis of 14 London winters, 1958/59–1971/72. *Arch. Environ. Health,* **37** (1982) 213–20.
67. PAARC Cooperative Group, Atmospheric pollution and chronic or recurrent respiratory disease. II. Results and discussion. *Bull. Euro. Physiopathol. Respir.,* **18** (1982) 101–16.
68. Dohan, F. C., Everts, G. S. & Smith, R., Variations in air pollution and the incidence of respiratory disease. *J. Air Pollut. Control Assoc.,* **12** (1962) 418–36.
69. Saric, M., Fugas, M. & Hrustic, O., Effects of urban air pollution on school-age children. *Arch. Environ. Health,* **36** (1981) 101–8.
70. Ware, J. H., Ferris Jr, B. G., Dockery, D. W., Spengler, J. D., Stram, D. O. & Speizer, F. E., Effects on ambient sulfur oxides suspended particles on respiratory health of preadolescent children. *Amer. Rev. Respir. Dis.,* **133** (1986) 834–42.
71. Dockery, D. W., Speizer, F. E., Strame, D. O., Ware, J. H., Spengler, J. D. & Ferris Jr, B. G., Effects on inhalable particles on respiratory health of children. *Amer. Rev. Respir. Dis.,* **139** (1989) 587–94.
72. Ozkaynak, H. & Thurston, G. D., Association between 1980 US mortality rates and alternative measures of airborne particle concentration. *Risk Anal.,* **7** (1987) 449–60
73. Kitagawa, T., Cause analysis of the Yokkaichi asthma episode in Japan. *J. Air Pollut. Control Assoc.,* **34** (1984) 743–6.
74. Imai, M., Yoshida, K. & Kitabatake, M., Mortality from asthma and chronic bronchitis associated with changes in sulfur oxides air pollution. *Arch. Environ. Health,* **41** (1986) 29–35.

75. Waller, R. E. & Lawther, P. J., Further observation on London fog. *Brit. Med. J.,* **21 Dec.** (1957) 1473–5.
76. Gorham, E., Bronchitis and the acidity of urban precipitation. *Lancet,* **27 Sept.** (1958) 691–3.
77. Hudson, J. R., The effect of the fog on cattle as the Smithfield Show. In *Reports on Public Health and Medical Subjects,* No. 95—*Mortality and Morbidity during the London Fog of December 1952.* HMSO, London, UK, 1954, Appendix A.
78. Lippmann, M., Airborne acidity: estimates of exposure and health effects. *Environ. Health Perspect.,* **63** (1985) 62–70.
79. Thurston, G. D., Ito, K., Lippmann, M. & Hayes, C., Re-examination of London mortality in relation to exposure to acidic aerosols during 1963–1972 winters. *Environ. Health Perspect.,* **79** (1989) 73–82.
80. Lippmann, M., Background on health effects of acid aerosols. *Environ. Health Perspect.,* **79** (1989) 3–6.
81. Speizer, F. E., Studies of acid aerosols in six cities and in a new multicity investigation: design issues. *Environ. Health Perspect.,* **79** (1989) 61–7.
82. Ostro, B. D., Lipsett, M. J., Wiener, M. B. & Selner, J. C., Asthmatic response to airbourne acid aerosols. *An. J. Public Health,* **81** (1991) 694–702.
83. Utell, M. J., Morrow, P. E. & Hyde, R. W., Comparison of normal and asthmatic subjects' response to sulfate pollutant aerosols. *Ann. Occup. Hyg.,* **26** (1982) 691–7.
84. Koenig, J. Q., Pierson, W. E., Horike, M. & Frank, R., Effects of inhaled SO_2 plus NaCl aerosol combined with moderate exercise on pulmonary function in asthmatic adolescents. *Environ. Res.,* **25** (1981) 340–8.
85. Koenig, J. Q., Pierson, W. E., Horike, M. & Frank, R., A comparison of the pulmonary effects of 0·5 ppm versus 1·0 ppm sulfur dioxide plus sodium chloride droplet in asthmatic adolescents. *J. Toxicol. Environ. Health,* **11** (1983) 129–39.
86. Spektor, D. M., Yen, B. M. & Lippmann, M., Effect of concentration and cumulative dose of inhaled sulfuric acid on lung clearance in humans. *Environ. Health Perspect.,* **79** (1989) 167–72.
87. Schlesinger, R. B., The effects of inhaled acids on lung defenses. In *Aerosols,* ed. S. D. Less, T. Schneider, L. D. Grant & P. J. Verkerk. Lewis, Chelsea, MI, USA, 1986, pp. 617–35
88. Sim, V. M. & Pattle, R. E., Effect of possible smog irritants on human subjects. *J. Amer. Med. Assoc.,* **165** (1957) 1908–13.
89. Lourenco, R. V., Distribution and clearance of aerosols. *Amer. Rev. Respir. Dis.,* **101** (1969) 460–1.
90. Reid, L., An experimental study of the hypersecretion of mucus in the bronchial tree. *Brit. J. Exper. Pathol.,* **44** (1963) 437–45.
91. Mitchell, R. S., Clinical and morphologic correlations in chronic airway obstruction. *Amer. Rev. Respir. Dis.,* **97** (1967) 54–62.
92. Suhs, R. H., Lumeng, J. L. & Leppe, M. H., An experimental immunologic approach to the induction and perpetuation of chronic bronchitis. *Arch. Environ. Health,* **18** (1969) 564–73.
93. Jefferey, P. K., The effects of irritation on the structure of bronchial

epithelium. In *The Lung in the Environment,* ed. G. Bonsignore & G. Cummings. Plenum Press, New York, USA, 1982, pp. 303–13.

94. Hogg, J. C., Macklem, P. T. & Thurlbeck, W. M., Site and nature airway obstruction in obstructive lung disease. *New Engl. J Med.,* **278** (1968) 1355–60.
95. Matsuba, K. & Thurlbeck, W. M., Disease of the small airways in chronic bronchitis. *Amer. Rev. Respir. Dis.,* **107** (1973) 552–8.
96. McKenzie, H. I., Glick, M. & Outhred, K. G., Chronic bronchitis in coal miners: antemortem/post-mortem correlations. *Thorax,* **24** (1969) 527–35.
97. Albert, R. E., Lippmann, M., Peterson Jr, H. T., Berger, J. M., Sanborn, K. & Bohning, D. E., Deposition and clearance of aerosols. *Arch. Intern. Med.,* **131** (1973) 115–27.
98. Wolff, R. K., Mauderly, J. L. & Pickrell, J. A., Chronic bronchitis and asthma: biochemistry, rheology and mucociliary clearance. In *Lung Connective Tissue: Location, Metabolism and Response to Injury,* ed. J. A. Pickrell. CRC Press, Boca Raton, FL, USA, 1981, pp. 169–83.
99. Soskolne, C. L., Pagano, G., Cipollaro, M., Beaumont, J. J. & Giordano, G. G., Epidemiologic and toxicologic evidence for chronic health effects and the underlying biologic mechanisms involved in sublethal exposures to acidic pollutants. *Arch. Environ. Health,* **44** (1989) 180–91.
100. Amdur, M. O. & Chen, L. C., Furnace generated acid aerosols: speciation and pulmonary effects. *Environ. Health Perspect.,* **79** (1989) 147–50.

Chapter 10

Effects of Acidic Pollutants on Terrestrial Ecosystems

DAVID FOWLER
Institute of Terrestrial Ecology, Bush Estate, Penicuik, Midlothian, EH26 0QB, UK

10.1 INTRODUCTION

The pollutants associated with acidic deposition include the emitted gaseous pollutants sulphur dioxide (SO_2), nitric oxide (NO), nitrogen dioxide (NO_2) and their oxidation products which, with the exception of gaseous nitric acid (HNO_3), are in the particulate phase as sulphate (SO_4^{2-}) and nitrate (NO_3^-). Acidity per se as hydrogen ions (H^+) is an important component of the pollutant mixture but few of the terrestrial effects are thought to result directly from acidity. The consideration of acidic deposition has been widened considerably in recent years by the role of ammonia (NH_3) and ammonium (NH_4^+) deposition in the Netherlands. The biological fate of NH_4^+ taken up by plants results in the release of H^+ into the soil and net acidification. In regions of the Netherlands where inputs of NH_3 and NH_4^+ are very large (30–100 kg N ha^{-1} $year^{-1}$), this acidification is the dominant term. The Dutch acidic deposition research programme therefore regards NH_3 and NH_4^+ as important components of the overall problem of terrestrial effects of acidic deposition, or in this case more correctly 'acidifying deposition'. As nitrogen deposited as NH_4^+ is large elsewhere in Europe, other countries are adopting a similar approach. Other major research programmes designed to quantify the effects of acidic pollutants on terrestrial ecosystems have also expanded the literal definition of acidic deposition to include other compounds which are not acidic and are not directly linked with acidity in the atmosphere or soil. The effects of photochemical oxidants and ozone (O_3) in particular on terrestrial ecosystems was addressed by the US National Acidic Precipitation Assessment Program (NAPAP).

The specific reasons for expanding the remit of NAPAP to include O_3 are not known to this reviewer but the following should be noted.

(1) The emitted pollutants NO and NO_2 which, following oxidation ultimately contribute to acidic deposition also play a central role in the atmospheric photochemical oxidant cycle, through which O_3 is produced.
(2) The effects of acidic deposition may at many locations be the consequence of multiple pollutant stresses on plants and contribute to the observed effects.
(3) The most widespread terrestrial effects which have been associated, although not necessarily correctly, with acidic deposition are the forest declines observed in the eastern US and in parts of Continental Europe. Large inputs of deposited acidity *and* prolonged exposure to phytotoxic concentrations (> 60 ppbv) are common characteristics, among the sites at high-altitude locations in Central Europe and on the Appalachians of the eastern US showing a marked decline.

In this review therefore, to provide a balanced interpretation of the most likely causal agents associated with observed effects in terrestrial ecosystems, the following pollutants will be considered under the inappropriate but widely applied title of 'acidic deposition': acidity (H^+), SO_2, SO_4^{2-}, NO, NO_2, NO_3^-, HNO_3, NH_3, NH_4^+ and O_3.

Rather than attempt a comprehensive treatment of the effects of these compounds or ions on terrestrial ecosystems, the review will be restricted to the major problems of long-range transport and deposition of the major pollutants listed above. The major acute injury effects of gaseous pollutants such as SO_2 in the proximity of large point sources such as metal smelters are well documented elsewhere. While these effects may still be observed, there is no longer any scientific doubt about the cause or the concentration thresholds for visible injury and long-term survival of many common plant species in the presence of large ambient concentrations (> 50 ppbv) of SO_2 and to a lesser extent NO_2.

The review does, therefore, concentrate on regional (> 10^4 km^2) effects of dry and wet deposited acidity, sulphur and nitrogen compounds on natural ecosystems, in particular, heathland, moorland and forests. A short section on agricultural crops and O_3 is included to provide the necessary perspective for effects on forests and unmanaged ecosystems.

10.2 AGRICULTURAL CROPS

10.2.1 Gases, SO_2, NO_2

The effects of gaseous air pollutants on agricultural crops have been reviewed previously, and the effects of SO_2 have been described.[1,2] Although these detailed reviews describe important physiological responses by major crop plants to SO_2 and NO_2 at concentrations in the range 20–100 ppbv, there are no studies which show that either of these two pollutants cause crop losses greater than 10% on a regional scale (> 10^4 km^2). The ambient concentrations of SO_2 and NO_2 in the major crop growing regions of North America and Western Europe are in the range 2–20 ppbv, with comparatively few areas in which annual mean concentrations of either of these pollutants exceed 15 ppbv. At concentrations smaller than 15 ppbv the effects on crop yield are very sensitive to other stresses and environmental conditions in general. For example, in studying cereal yield responses to small ambient concentrations of SO_2 and NO_2 (both in the range 5–10 ppbv) the grain and total dry matter yield in polluted air treatments of barley have been shown to exceed those in clean air.[3,4] The stimulation in dry matter yield of grasses at small SO_2 concentrations have been noted by other groups (Fig. 10.1).[5,6] The strong interaction between pollutant stress and the physical and biological stresses to which field crops are commonly subjected have made the prediction of regional crop losses due to SO_2 and NO_2 difficult and the results very uncertain. In the proximity of major sources of SO_2 the acute injury and resulting crop losses have been quantified.[7]

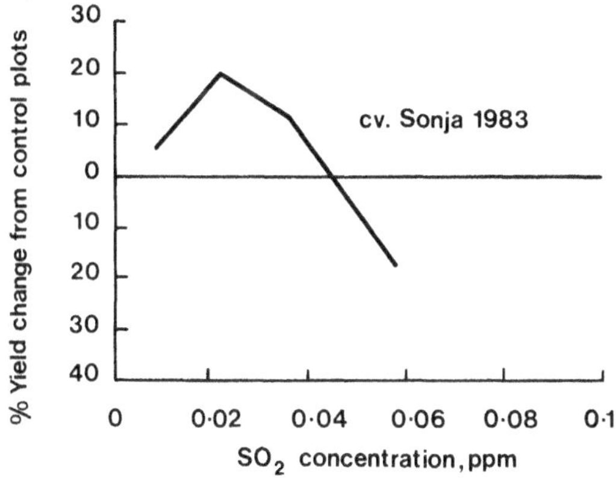

Figure 10.1. The yield response of winter barley (cv. Sonja) to SO_2 in the field exposure experiment at Littlehampton showing stimulation of yield at small SO_2 concentrations. (Adapted from Ref. 4.)

The interactive effects of insect pests and ambient gaseous pollutants on crop plants has become an important aspect of air pollution effects research during the last 5 years. The modification of plant nutritional status or chemistry of the intercellular and intracellular fluids by pollutant uptake influences the feeding behaviour of insects and thus their damage to the crops.[8,9,10] These interactions may well induce much greater crop loss as a con-sequence of the insect pest rather than as a direct effect of the pollutant, but have not so far been quantified and remain very uncertain.

10.2.2 O_3

The occurrence of photochemical O_3 episodes during which ambient concentrations exceed 60 ppbv is now well documented for most agricultural regions of North America and Western Europe.[11,12] The areas exposed to frequent summer O_3 episodes (times during which concentrations exceed 60 ppbv) are large and are regional in scale (10^4 km upwards) unlike the fumigation of primary pollutants in the vicinity of major sources (10–100 km^2). Furthermore, the National Crop Loss Assessment Program (NCLAP) in the US showed that crop losses due to O_3 were large and exceed the likely losses as a consequence of direct effects of SO_2 and/or NO_2. In particular, the form of the yield, and response function for several major North American crops to daytime seasonal mean O_3 concentrations have been shown.[13]

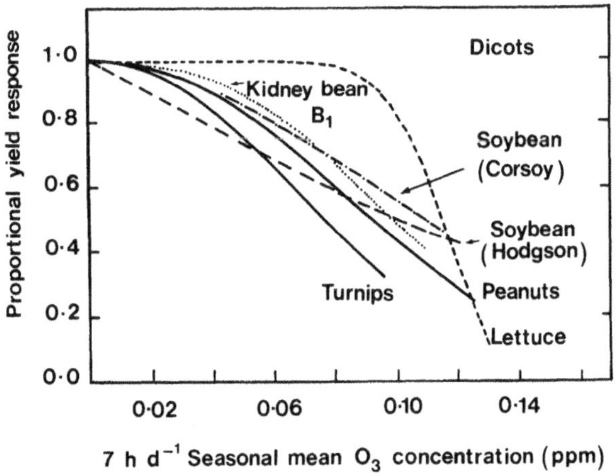

Figure 10.2. Proportional response of crop yield to ozone in open-top chamber studies by the US NCLAP. (Adapted from Ref 13.)

An exampleof the results from the NCLAP studies is shown in Fig. 10.2, in which 20% yield losses are found for sensitive varieties of soybean subjected to seasonal daytime (7 h) mean O_3 concentrations of 50 ppbv. Much discussion on the form of the yield loss–O_3 concentration relationship has been published.[11,14] The response of crop plants to O_3 in the range 10–50 ppbv remains uncertain and it has yet to be demonstrated that the Weibull function as a dose–response model is appropriate at concentrations in this range. More progress may be made in the prediction of crop losses from process rather than statistical models,[15] though as yet little effort has been made to apply such approaches to the problems of air pollution effects.

It is clear that on regional scales O_3 is the major gaseous pollutant causing crop loss in North America. For Western Europe a programme of research analagous to that in NCLAN is in progress under the auspices of the Commission of the European Communities. The programme is at an earlier stage and has a broader focus, including as many studies on tree species and natural vegetation as on agricultural crops.[16,17] The European areas experiencing O_3 concentrations similar to those of eastern North America are smaller, but still occupy substantial areas of the countries on the north Mediterranean coast. Given that many of the crops of southern Europe are similar to those of the eastern US, it would be surprising if a similar scale of agricultural crop loss in Europe as a consequence of photochemical oxidants is not occurring.

10.2.3 Direct effects of acid deposition on crops

The acidity deposited in precipitation and the major anions associated with the acidity (SO_4^{2-} and NO_3^-) have been applied to agricultural crops as spray and mist treatments in controlled conditions. In general, few significant effects have been observed at treatment pH values above 3·0. There are data which show effects on soybean yield at rather large acidities for typical ambient rainfall (pH 2·0–3·0), but such results are the exception rather than the rule. The effects appear to be mediated through damage to the processes of flowering and development of seeds and not to carbon assimilation or allocation.[18]

10.3 AGRICULTURAL SOILS

The effects of deposited acidity on agricultural soils represent a potential source of base cation leaching and thus soil acidification. In practice, however, the quantity of acidity deposited onto agricultural soils from the atmosphere is small relative to the natural production of acidity in soil as a consequence of crop production. The underlying chemical processes of acidification in agricultural soil have been described and quantified in detail (e.g. Ref. 19) and the pH values of agricultural soils are actively managed by the regular application of lime.

The focus of interest in terrestrial effects of 'atmospheric acidity' during the last 10 years has been on natural ecosystems, in particular heathland, and forests (including managed forests). This is in part the result of observed changes in either species composition or health of individual species in these ecosystems, but it also follows the recognition that much of the heathland, and forests grow on poorer soils which are more susceptible to soil acidification and the acidity of the soil is not 'managed' actively.

10.4 EFFECTS OF ACIDIC DEPOSITION ON HEATHLAND

Heathland plant communities have developed on poorly buffered soils which are naturally quite acidic with a limited supply of the major plant nutrients. Typical soil pH values (in water) of 4·0–5·0 are observed, for example, in Dutch heathland. At the lower end of this range, pH 4·0–4·3 the acidity of the soil is of the same order as that of incoming precipitation. However, the local variability in soil pH in some of the Dutch heathlands, which have enabled species such as *Thymus serpyllum* (L.) and *Pedicularis sylvatica* (L.) to occur on small areas of soil with pH values in the range 4·7–5·0 has been reduced by further acidification. The acidification is believed to result mainly from deposition of NH_3 and NH_4^+ which is rapidly nitrified, causing acidification of the soil to values of typically pH 4·1 at which nitrification is strongly inhibited but at which the above plant species cannot survive.[20] The loss of the slightly buffered sites within soils in some Dutch heathlands has led to the loss or marked decline of such species as *T. serpyllum* (L.) in areas of large NH_3 and NH_4^+ deposition. A much more dramatic change in species composition in the same heathlands as a result of a decline in heather species such as *Erica tetralix* (L.) and *Calluna vulgaris* (L.) and the rapid growth of grasses such as *Molinia caerulea* (L.) and *Agrostis canina* (L.) has occurred.[21] The changes result from competition between the grasses and heather species in which the grasses profit from the larger nitrogen deposition.[22]

The hypothesis that atmospheric inputs of nitrogen are the cause of these field observations has been subjected to experimental investigation in which small heathland plant communities have been subjected to simulated atmospheric inputs of the major pollutants NH_4^+ and SO_4^{2-}. The results of the experiments with NH_4^+ addition comparable with current atmospheric inputs of 20–60 kg N ha^{-1} $year^{-1}$ demonstrate a clear increase in biomass of the two grasses relative to the heather species. These experiments and related field studies show that NH_4^+ is the primary cause of the growth stimulation in the grasses and that the acidity of the treatment is not a major factor (Fig. 10.3).

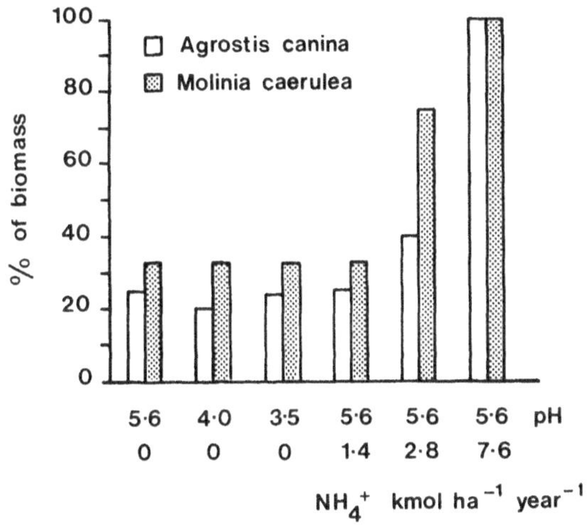

Figure 10.3. The relative increase in biomass of *Agrostis canina* and *Molina caerulea* nitrogen as NH_4^+ applied to experimental plots of heathland plant communities. (Adapted from Ref. 22.)

The rate of change of species composition in Dutch heathlands has been rapid. Of the 4×10^4 ha of heathland in 1980, 55% was considered to be 'pure heather heathland' while 25% comprised a mixture of grasses and heather and a further 20% was almost pure grassland. By 1987 the pure heathland area had decreased to 36% of the total while the mixed heather–grass community was still 25% of the total but the grass dominated heathland had increased from 20 to 34%.

Many questions remain about the processes controlling the interspecies competition and the role of pests, pathogens and other stresses, but the dramatic species composition changes observed are a clear indication of the way in which plant species (and the fauna on which they depend) may change in sensitive ecosystems in response to atmospheric inputs.

The very large nitrogen deposition in the Netherlands is dominated by NH_3 and NH_4^+, with the total input in the range 30–100 kg N ha^{-1} $year^{-1}$. Elsewhere in Europe there are sites with similar inputs. For example, in the border counties of England and Wales and in parts of East Anglia where ambient NH_3 concentrations are typically 10 ppbv inputs of nitrogen on short vegetation are between 30 and 40 kg N ha^{-1} $year^{-1}$.[23]

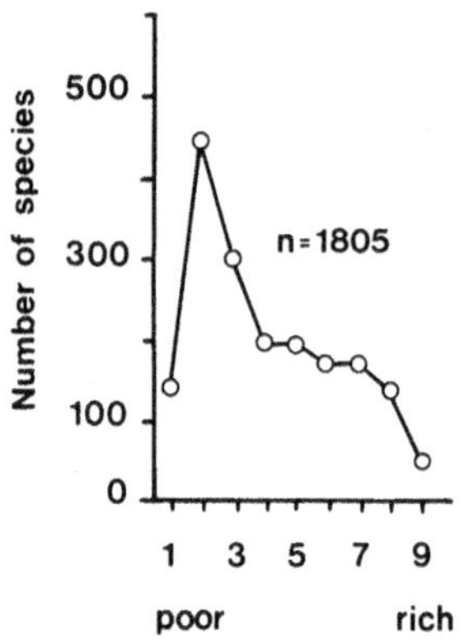

Figure 10.4. The distribution of 2164 Central European plant species arranged along a gradient of increasing nitrogen availability. The scale in the ordinate is arbitrary and corresponds to the following: 1 — pronounced nitrogen deficiency ('background' 1–3 kg N ha^{-1} $year^{-1}$); 5—'just sufficient' nitrogen; and 9—large inputs from the atmosphere (> 30 kg N ha^{-1} $year^{-1}$). (Adapted from Ref. 24.)

The extreme conditions of very large atmospheric inputs of nitrogen in areas polluted by large ambient NH_3 concentrations contrasts with the majority of the natural and afforested ecosystems of Europe where pollutant inputs of nitrogen are smaller and in the range 10–40 kg N ha^{-1} $year^{-1}$ but are much larger than in areas remote from sources of pollution.[25] Large numbers of plant species in Europe have adapted to survive at sites with a very limited nitrogen supply. Such plants have been estimated to constitute 50% of the plant species in Central Europe. It is suggested in Ref. 24 that the species diversity in many plant communities is strongly linked to the nitrogen supply. The data in Fig. 10.4 show the decline in species numbers in wetland, and heathland plant communities with increasing nitrogen supply.

10.5 EFFECTS ON WETLANDS

Ombrotrophic blanket mires in the uplands of western Britain are examples of another ecosystem which is sensitive to acidic deposition. These ecosystems like the heathland plant communities are naturally acidic, as a result of H^+ release into the bogwater following cation exchange with polyuronic acids produced by species of *Sphagnum*.[26] However, these species are very sensitive to the supply of nitrogen as NO_3^- and NH_4^+ and to HSO_3^-, the latter resulting from SO_2 uptake.

The blanket mires of the southern areas of the Pennine hills in northern England have been widely affected by air pollutants during the last 100 years, and in many areas *Sphagnum* species. have been eliminated. The role of air pollutants in the decline of *Sphagnum* was originally proposed by natur-alists in the 19th century (see, for example, Ref. 27), but more recent work has shown a clear association between indicators of industrial air pollutants (soot particles) and the loss of *Sphagnum* species.[28] In particular, the important species of *Sphagnum* have been shown to be sensitive to HSO_3^- which was present in the Pennine region as a consequence of the large annual average concentrations of SO_2, in excess of 30 ppbv.[29] It is more likely that the phytotoxic agent was dissolved SO_2 rather than HSO_3^- in the early decades of this century.[30] More recently, ambient SO_2 concentrations have declined sharply

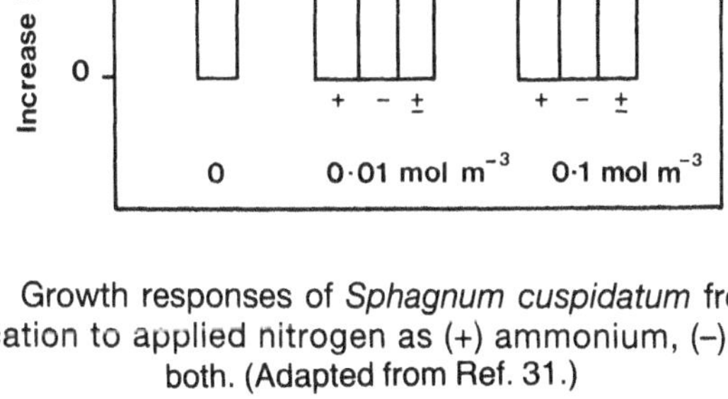

Figure 10.5. Growth responses of *Sphagnum cuspidatum* from a relatively clean air location to applied nitrogen as (+) ammonium, (–) nitrate, or (±) both. (Adapted from Ref. 31.)

and are now in the range of 5–10 ppbv in these areas. However, the steady increase in emission of the oxides of nitrogen (NO_x) has presented a different pollutant stress for these mire ecosystems. The large concentrations of nitrate in precipitation and in deposited cloudwater have been shown[32] to be associated with poor growth of *Sphagnum* transplanted into polluted Pennine sites from areas with much smaller concentrations of NO_3^- in precipitation. The data in Fig. 10.5 show the effects of both NH_3 and NO_3^- concentrations in surface water on the growth of *Sphagnum cuspidatum*.[7]

One of the most important features of the current effects of acidic deposition on these natural ecosystems is the large area in Europe and North America which is influenced. Unlike the effects due to acute injury to vegetation in the vicinity of point sources the changes to wetland plant communities are occurring over much larger scales as a consequence of the long-range transport of pollutants in the air over Europe, and over eastern North America. The changes are more subtle and occur slowly relative to changes induced by acute injury. Thus, many changes may take place unseen. The gradual nature of these changes also provides opportunity for other biotic and physical stresses to play a part.[9] For example, the ability of plants to survive extreme drought or early frost has been shown to be influenced by pre-treatment with the pollutants SO_2 and NO_2.[33]

10.6 FORESTS AND AIR POLLUTANTS

The effects of acidic deposition on forests has been an important focus of air pollution effects research during the last decade. Unlike the problems of SO_2 injury to forests close to major point sources (e.g. Ref. 34), there have been considerable problems in linking observed cases of forest decline with individual or even combinations of air pollutants. The widely separated areas of Europe and eastern North America in which forest decline has been observed have led many to seek a common phytotoxic agent responsible for the range of symptoms.[35, 36] There have also been great swings of scientific opinion over the cause and the mechanisms leading to decline during this period. The studies of forest decline during the early 1980s were greatly stimulated by Ref. 37, which proposed a hypothesis in which the large atmospheric inputs of the anions SO_4^{2-} and NO_3^- to sensitive forest soils led to base-cation leaching, soil acidification and ultimately to a nutritional imbalance in the trees. While attractive as a hypothesis and consistent with observations of soil solution chemistry, root damage and tree health at some locations,[31] there were many sites showing clear evidence of a decline in tree health on well-buffered calcareous soils. Other workers also studying the particular problems of forest decline in Germany observed a marked increase in symptoms of forest decline with increases in altitude.[38] The changes in chemical and physical characteristics of the climate of upland sites led them to suspect O_3 as the causal agent.

These represented two of the leading hypotheses for forest decline during the 1980s, but many others including changes in climate, the presence of acid cloudwater and 'excessive' nitrogen deposition have also been proposed. As more detailed measurements of atmospheric composition, forest condition and physiological responses to pollutants have become available, clear links between air pollutants and forest decline are being established.[39, 40] However, the mechanisms by which the effects are being mediated are in-direct and there are important areas of declining forest in which the role of atmospheric pollutants has been difficult to quantify. In particular the 'type 1' decline of Norway spruce (*Picea abies* (L.) Karst) and 'silver fir' (*Abies alba* Mill) in mountainous areas of Central Europe results directly from foliar magnesium deficiency,[41] but the role of acidic deposition in leaching exchangeable reserves of magnesium from the soils at sites with decline relative to other predisposing factors, has yet to be quantified.

In this section the particular characteristics of trees which enhance the inputs of acidifying substances are briefly described. The interactive effects of forests in the acidification of freshwaters are briefly outlined to provide an example of the way in which land use interacts with pollutants in the atmosphere and predisposes either the vegetation or freshwaters to subsequent effects. Lastly, two examples of clear links between acidic pollutants and forest health problems are described.

10.6.1 Pollutant deposition on forests

The tall, aerodynamically rough nature of forest canopies provides for very efficient exchange of heat, mass and momentum with the atmosphere[42] and, therefore, to potentially large rates of deposition of atmospheric pollutants. The large rates of deposition on forests occur for any gaseous pollutant which is readily absorbed on contact with foliar surfaces, such as gaseous HNO_3. However, for any gas which is primarily absorbed through stomata (Fig. 10.6) and which is not taken up by cuticular surfaces at significant rates (such as O_3 or NO_2) large rates of deposition on forests do not occur. For forests the bulk canopy stomatal resistances are generally larger than those for shorter vegetation (grass, cereals, etc.)[43] and limit deposition velocities to about 8 mm s^{-1}. For particles, the rates of deposition of submicrometer aerosols on forests are small relative to the maximum rates of turbulent deposition as a consequence of inefficient impaction efficiency in this size range.[44] However, cloud droplets commonly encountered in orographic cloud which are generally in the size range 3—15 μm (radius) impact efficiently on foliar surfaces, especially the small rigid needles of conifers. For wind speeds typical of high elevation forests (≥ 3 m s^{-1}) and droplets larger than 5 μm radius the cloud water is deposited onto the forests at rates similar to those of momentum,[45] and these are similar to those of HNO_3 vapour deposition.

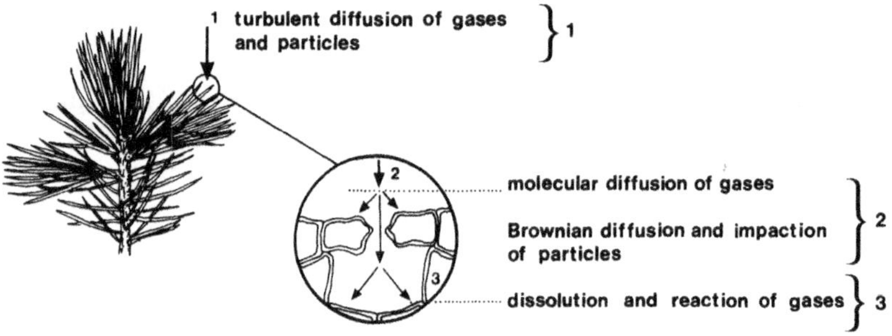

Figure 10.6. The mechanisms of transport of gases and small particles to foliar surfaces. (Adapted from Ref. 43.)

The net effect of these processes is that forests increase the total flux of acidic pollutants to the ground and the magnitude of the 'forest effect' varies with the 'pollution climate' and altitude.

An example of the effects of forests on inputs of pollutants is shown diagrammatically in Fig. 10.7.[46] The inputs of the major gaseous and particulate pollutants have been modelled from measured air concentrations close to the site and a knowledge of the mechanisms of deposition for each pollutant from the literature.

The effects of forests capturing pollutants from the atmosphere have been variously described as the filtering effect or scavenging effect and various attempts have been made to measure directly the net input of the gaseous and particulate pollutants or to deduce the inputs from a budget of the major ions deposited on soils beneath a forest canopy[47] compared with above canopy wet deposition (reviewed by Ref. 48). In general, there is no simple gauge which adequately simulates the chemical and physical 'collecting' properties of a forest canopy, and the mass balance methods are subject to considerable uncertainty in the case of ions which are readily taken up by or leached from foliage (notably SO_4^{2-}, NO_3^- and NH_4^+, K^+). However, there are indications that at some sites the leaching of soil-derived SO_4^{2-} from forest canopies is an insignificant fraction of the total input.[49]

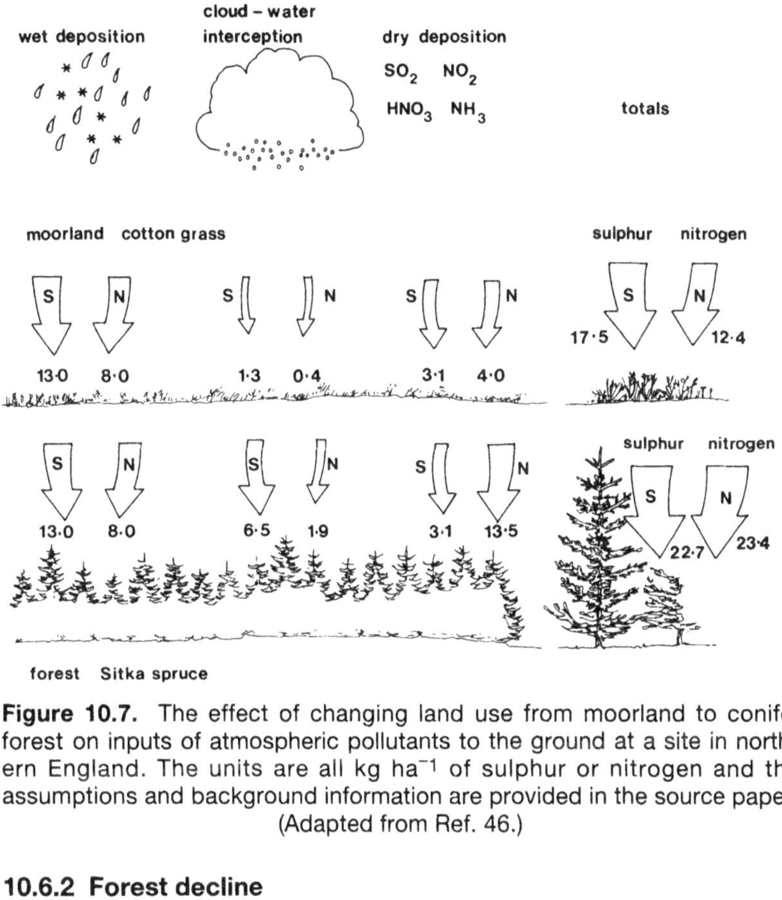

Figure 10.7. The effect of changing land use from moorland to conifer forest on inputs of atmospheric pollutants to the ground at a site in northern England. The units are all kg ha^{-1} of sulphur or nitrogen and the assumptions and background information are provided in the source paper. (Adapted from Ref. 46.)

10.6.2 Forest decline

Declines in the 'regional' health of forest species have been reported during the last century but the declines observed in Europe and eastern North America during the last decade appear to be unique in extending over a range of species and soil types as well as a wide geographic distribution. In the former FRG a rapid development of 'novel forest decline' was observed between 1975 and 1985. The surveys initiated during this period reported 8% of forest to be affected in 1982, 34% in 1983 and 50 and 54% in 1984 and 1987, respectively.[54] The absolute values for the extent of forest affected are subject to large uncertainties especially during a period of observer training,[50] but there is little doubt that during the early 1980s visible symptoms of forest decline increased over a wide area. The total area of forest in the former FRG showing reduced vitality has been estimated at 4×10^6 ha. The principal species involved are silver fir, *Abies alba* (Mill.), Norway spruce, *Picea abies* (L.) Karst, and beech, *Fagus sylvatica* (L.) (Fig. 10.8). While the initial European focus of scientific interest was on German forests, declines in forest health were reported elsewhere in Europe, notably Switzerland, France, the Netherlands and Belgium,[17] but there were reports of forest health problems in almost all countries of Europe. In the industrial areas of Eastern Europe there were also reports of widespread acute injury in the vicinity of major sources of SO_2.

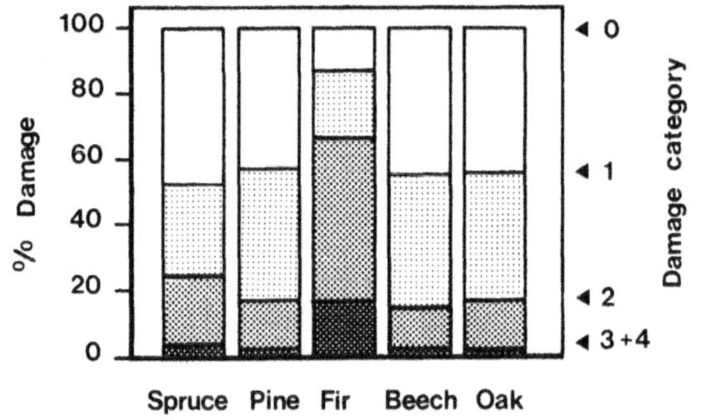

Figure 10.8. Indices of forest decline in the former FRG for 1985. The percentages refer to the damaged trees within each of the five categories. The damage categories are based on needle loss and are 0–10% needle loss, healthy; 1—11-25% needle loss; 2—26-60% needle loss; 3—> 60% needle loss; 4—dead. (Adapted from Ref. 50.)

In eastern North America a decline in red spruce (*Picea rubens* Sarg) in the Appalachians from the Great Smoky mountains of North Carolina, northwards to Vermont has been reported for the period 1960–1980.[51] The characteristics of the damaged trees include (i) extensive needle loss; (ii) dieback of leading and lateral branches at the tips of saplings and on mature trees;[52] and (iii) a decrease in radial growth at many sites beginning in the 1960s (Fig. 10.9).[53] The hypotheses proposed during the last decade are legion and some are briefly described in Ref. 54. In many of these hypotheses problems with underlying nutritional balance of the trees play a major part though it is not clear in many cases whether the initial problems occur in the foliage or the root systems.

Experimental studies however, have now established for some specific physiological effects that acidic deposition is the causal agent. To develop this point, two contrasting hypotheses are developed in more detail below.

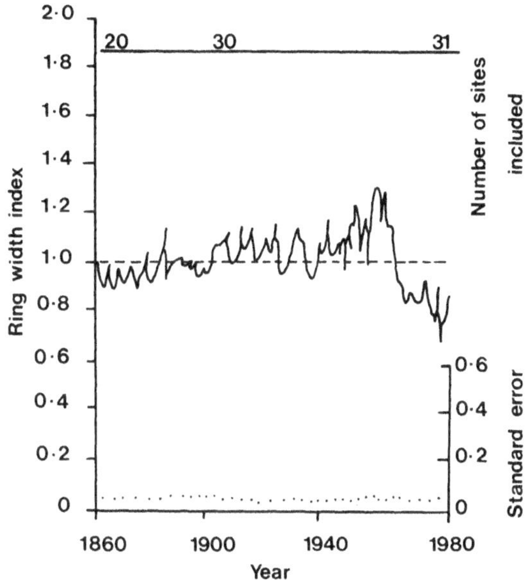

Figure 10.9. A combined tree ring width chronology for 508 red spruce trees collected from 31 stands predominantly high altitude sites in eastern North America showing a marked decline in ring width increment during the period 1960–1980. (Adapted from Ref. 18.)

10.6.3 Effects of acid mist on frost hardiness and winter injury

The decline in red spruce has been shown to be very altitude dependent. Damaged stands occur on mountains above 800 m with the greatest damage above 1200 m. These mountain-top locations are frequently enveloped in cloud. For example, at White Top mountain, a site with extensive red spruce decline, the summit is in cloud between 30 and 40% of the year. The cloud-water at such sites generally contains very large concentrations of the major ions SO_4^{2-}, NO_3^-, NH_4^+ and H^+, in the range 50–2000 μeq litre^{-1}. These are larger than the concentrations in rain by typically a factor of two to five and, as explained earlier, the cloud droplets are deposited onto conifer trees very efficiently. Simulated cloudwater composition in the range 0·01–3 mol m^{-3} ,

H^+, SO_4^{2-} and NO_3^- and 0·02–3·0 mol m^{-3} for NH_4^+ applied in open-top chambers were shown to reduce the frost hardiness of red spruce by up to 12°C during October and November (Fig. 10.10).[55] Subsequent work, in which the effects of pairs of ions were contrasted, showed that SO_4^{2-} is the ion mainly responsible for the frost hardiness effects.[56] The NO_3^- ion played no part on the frost hardiness damage and there was evidence that its presence mitigated the effects of SO_4^{2-}. The frost hardiness response to acid mist has also been detected,[57] and it represents an important mechanism by which 'acidic pollutants' transported over large distances may cause effects on forests.

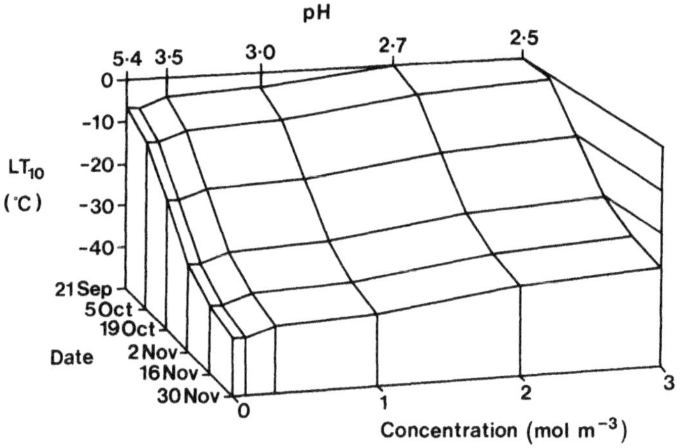

Figure 10.10. The frost hardiness response (LT_{10}) of seedling red spruce to simulated acid mist treatments at different times during the autumn frost hardening period. (LT_{10} is the freezing temperature required to kill 10% of the foliage following frost testing.) (Adapted from Ref. 55.)

Winter injury to red spruce has been a characteristic of the decline of this species at many of the high-altitude sites in the Appalachians.[58, 59] However, there is evidence that the nutritional status of the red spruce during the period of frost hardening strongly influences the response to acid mist treatment, with the plants with lower foliar nitrogen, phosphorus and potassium levels showing larger negative responses to acid mist treatments than plants with higher levels of the same nutrients. The same acid mist treatment caused a range of other physiological responses including an increase in cell wall elasticity deduced from pressure volume analysis[60] and reduced rates of photosynthesis expressed per unit chlorophyll and impaired stomatal function.[61] The large concentrations of major ions and SO_4^{2-} in particular can therefore be shown to be responsible for important cellular dysfunction in the species of interest.

10.6.4 Soil acidification and aluminium toxicity

The second of the specific hypotheses linking acidic deposition to tree damage was advanced during the early 1980s. In its original form, the hypothesis states that the natural acidification of forest soils is accelerated by additional cation leaching of soils by mobile anions.[37] These are notably the $SO_4^{2-}$and NO_3^- from wet and dry deposition, but also included are the NH_3 and NH_4^+, which following nitrification, provides additional NO_3^-. The very large acidifying inputs at sites in the Hartz mountains have led, in sensitive soils, to potentially toxic concentrations of the aluminium ion Al^{3+} (Fig. 10.11)

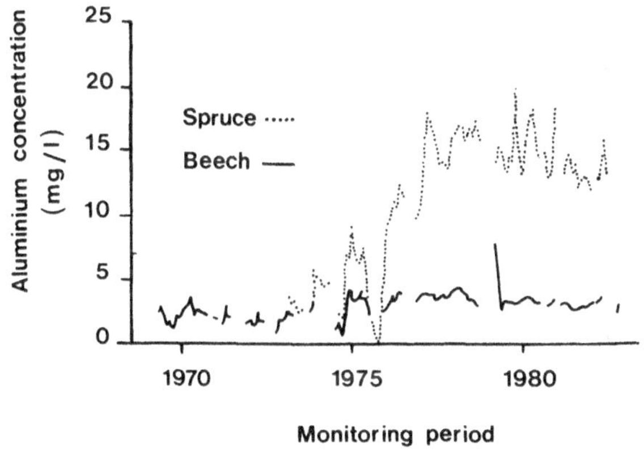

Figure 10.11. Concentrations of aluminium in soil water 0·9 m below spruce and beech canopies in the former FRG following the 1976 drought. (Adapted from Ref. 37.).

which have caused visible damage to the root systems. This is characterised by disrupted cell walls in the endodermis of mature roots and widespread cellular disruption in younger undifferentiated roots.[62] At the low soil pH in affected forest stands calcium and magnesium nutrition of the trees is poor and is believed to be inhibited by the presence of Al^{3+}. In particular, aluminium replaces calcium and magnesium at cell wall bonding locations. This in turn leads to structural deficiencies in the cell wall, reduced cell wall elasticity and impaired cell wall growth. The damaged root systems predispose the entire tree to nutrient and water stress. These predisposing factors may interact with warm dry summers, enhancing the natural stress of large soil water defects, and notable increases in forest decline follow such dry warm summers.

The two hypotheses very briefly described above show mechanisms by which acidic deposition may lead to forest decline. The mechanisms by which damage to the trees occurs is in both cases indirect and closely linked with tree nutrition. Physical stresses (e.g. freezing temperatures or hot dry summers) have also amplified the initial effects to produce the observed scale of injury.

The role of plant nutrition is central to many of the hypotheses advanced to explain forest decline. Leaching of nutrients from foliage damaged by photochemical oxidants, reduced uptake of nutrients from acidified soils or by damaged root systems, and nutrient imbalance as a consequence of excess nitrogen supply, are vital to three of the hypotheses and similar arguments are advanced in support of others. In recent work on high-elevation red spruce, the ratio of photosynthesis to dark respiration has been shown to be increased in the high-altitude trees along with reduced foliar calcium, magnesium and phosphorus and a marked elevation in aluminium.[63] The lower foliar calcium levels are regarded as particularly important, and it is speculated that this is the nutrient of primary concern for acid deposition effects in the forests of eastern North America. While mechanistically different from the hypotheses originated by Ref. 37 the similarity is striking.

Current research programmes on forest decline are therefore concentrating a major emphasis on nutrient cycling especially of the base cations and their relationship with inputs of NO_3^- and SO_4^{2-}.

The widespread interest in base cation nutrition of forests and its links with forest decline, and other physical and biological stresses, is leading slowly towards a consensus that the underlying cause is the total acidifying input. While this includes wet deposition of SO_4^{2-}, NO_3^-, NH_4^+ and dry deposited SO_2, NO_2, HNO_3 and NH_3, the role of O_3 should also be incorporated, in this case, since many of the effects of O_3 on physiological processes in controlled environment or open-top chamber studies are similar to those which may be induced using acid mist. It is probable therefore that a dose-equivalent of acid deposition (expressed in mol m^{-2} $year^{-1}$) may be substituted for the exposure to phytotoxic O_3 during a season (ppbv hours).

In this way the next few years of research activity should provide the policy makers with critical loads for forest decline in which the contribution of individual components of the total input may be quantified.

ACKNOWLEDGEMENTS

The part of this work which was carried out at ITE (Edinburgh) was supported by research contracts from the UK Department on the Environment and the Commission of the European Countries and the US Forest Service. The author is grateful to ITE colleagues and Dr S. B. McLaughlin for much helpful discussion.

Some research was supported by funds provided by the North-Eastern Forest Experiment Station's Spruce-Fir Research Cooperative within the joint US Environmental Protection Agency–USDS Forest Service Forest Response Program. The Forest Response Program is part of the National Acid Precipitation Assessment Program. This paper has not been subject to EPA or Forest Service peer review and should not be construed to represent the policies of either Agency.

REFERENCES

1. Unsworth, M. H. & Ormrod, D. P., ed., *Effects of Gaseous Air Pollution in Agriculture and Horticulture*. Butterworths, London, UK, 1982.
2. Winner, W. E., Mooney, H. A. & Goldstein, R. A., eds., *Sulphur Dioxide and Vegetation*. Stanford University Press, Stanford, CA, USA, 1985.
3. Fowler, D., Cape, J. N., Leith, I. D. & Nicholson, I. A., Effects of air filtration at small SO_2 and NO_2 concentrations on the yield of barley. *Environ. Pollut.*, **53** (1988) 135–49.
4. McLeod, A. R. & Baker, C. K., The use of open field systems to assess yield response to gaseous pollutants. In *Assessment of Crop Loss from Air Pollutants*, ed. W. W. Heck, O. C. Taylor & D. T. Tingey. Elsevier, London, UK, 1988, pp. 181–210.
5. Bell, J. N. B., Sulphur dioxide and the growth of grasses. In *Effects of Gaseous Air Pollution in Agriculture and Horticulture*, ed. M. H. Unsworth & D. P. Ormrad. Butterworths, London, UK, 1982, pp. 225–46.
6. Mansfield, T. A. & McCune, D. C., Problems of crop loss assessment when there is exposure to two or more gaseous pollutants. In *Assessment of Crop Loss from Air Pollutants*, ed. W. W. Heck, O. C. Taylor & D. T. Tingey. Elsevier, London, UK, 1988, pp. 317–44.
7. McLaughlin, S. B. & Taylor, G. E., SO_2 effects on dicot crops: some issues, mechanisms and indicators. In *Sulphur Dioxide and Vegetation*, ed. W. E. Winner, H. A. Mooney & R. A. Goldstein. Stanford University Press, Stanford, CA, USA, 1985, pp. 227–49.
8. Flückiger, W., Overtli, J. J. & Baltensweiler, W., Observations of an aphid infestation on hawthorn in the vicinity of a motorway. *Naturwissenschaften*, **65** (1978) 654–5.
9. Hughes, P. R. & Laurence, J. A., Relationship of biochemical effects of air pollutants in plants to environmental problems: insect and microbial interactions. In *Gaseous Air Pollutants and Plant Metabolism*, ed. M. K. Koziol & F. R. Whatley. Butterworths, London, UK, 1984, pp. 361–77.

10. McNeil, S. & Southwood, T. R. E., The role of nitrogen in the development of insect/plant relationships. In *Biochemical Aspects of Plant and Animal Coevolution*, ed. J. B. Harborne. Academic Press, London, UK, 1978, pp. 77–98.
11. Heck. W. W., Taylor, O. C. & Tingey, D. T., eds, *Assessment of Crop Loss from Air Pollutants*. Elsevier, London, UK, 1988.
12. Lefohn, A. S. & Mohnen, V. A., The characterization of ozone, sulphur dioxide and nitrogen dioxide for selected monitoring sites in the Federal Republic of Germany. *J. Air. Pollut. Control Assoc.*, **36** (1986) 1329–37.
13. Heagle, A. S., Kness, L. W., Temple, P. J., Kohut, R. J., Miller, J. E. & Heggestad, Factors influencing ozone close-yield response relationships in open-top field chamber studies. In *Assessment of Crop Loss from Air Pollutants*, ed. W. W. Heck, O. C. Taylor & D. T. Tingey. Elsevier, London, UK, 1988, pp. 141–79.
14. Rawlings, J. O., Lesser, V. M. & Dussel, K. A., Statistical approaches to assessing crop losses. In *Assessment of Crop Loss from Air Pollutants*, ed. W. W. Heck, O. C. Taylor & D. T. Tingey. Elsevier, London, UK, 1988, pp. 389–416.
15. Luxmoore, R. J., Assessing the mechanisms of crop loss from air pollutants with process models. In *Assessment of Crop Loss from Air Pollutants*, ed. W. W. Heck, O. C. Taylor & D. T. Tingey. Elsevier, London, UK, 1988, pp. 417–44.
16. Mathy, P., The European open-top chambers programme. In *Assessment of Crop Loss from Air Pollutants*, ed. W. W. Heck, O. C. Taylor & D. T. Tingey. Elsevier, London, UK, 1988, pp. 505–13.
17. Mathy, P., ed., *Air Pollution and Ecosystems* (Proceedings International Symposium). D. Reidel, Dordrecht, The Netherlands, 1988.
18. Evans, L. S., Lewin, K. F., Parri, M. J. & Cunningham, E. A., Productivity of field-grown soybeans exposed to simulated acidic rain. *New Phytol.*, **93** (1983) 377–88.
19. Bache, B. W., Soil–water interactions. *Phil. Trans. Royal Soc.* (Lond.), **B305** (1984) 393–407.
20. Van Breeman, N. & Van Dijk, H. F. G., Ecosystem effects of atmospheric deposition of Nitrogen in the Netherlands. *Environ. Pollut.*, **54** (1988) 249–74.
21. Scheckh, K. H., The effects of competition and nutrition on the interrelations of some wet-heath plants. *J. Ecol.*, **57** (1969) 87–99.
22. Heil, G. W. & Diemont, W. M., Raised nutrient levels change heathland into grassland. *Vegetatio*, **53** (1983) 113–20.
23. RGAR, *Review Group on Acid Deposition in the UK*, a report for the UK Department of the Environment. HMSO, London, UK, 1990.
24. Ellenberg, H., Floristic changes due to eutrophication. In *Ammonia and Acidification*, ed. W.A.H. Asman & H.S.M.A. Diederen. RIVM, Bilthoven, The Netherlands, 1987
25. Galloway, J. N. & Gaudry, A., The composition of precipitation on Amsterdam Island, Indian Ocean. *Atmos. Environ.*, **12** (1984) 2649–56.
26. Clymo, R.S., Sphagnum dominated peat bog: a naturally acid ecosystem. *Phil. Trans. Royal Soc.(Lond.)*, **B305** (1984) 487–99.
27. Press, U. C. & Lee, J. A., Nitrate reductase activity in Sphagnum species in the South Pennines. *New Phytol.*, **92** (1982) 487–94.

28. Tallis, J. H., Studies on the Southern Pennine peats. III. The behaviour of Sphagnum. *J. Ecol.*, **52** (1964) 345–53.
29. Ferguson, P. & Lee, J. A., Some effects of bisulphate and sulphate on the growth of sphagnum species in the field. *Environ. Pollut.* (Ser. A), **21** (1980) 59–71.
30. Cape, J. N., The importance of solutions equilibria in studying the effects of sulphate on plants. *Environ. Pollut.* (Ser. A), **34** (1984) 259–74.
31. Lee, J. A., Press, M. C., Woodin, S. & Ferguson, P., Responses to acidic deposition in ombrotrophic mires in the UK. In *Effects of Atmospheric Pollutants on Forests, Wetlands and Agricultural Ecosystems*, NATO ADSI Vol. 916, ed. T. C. Hutchinson & K. M. Meema. Springer-Verlag, Berlin, Germany, 1987, pp. 549–60.
32. Woodin, S. J., Press, M. C. & Lee, J. A., Nitrate reductase activity in Sphagnum fascum in relation to atmospheric nitrate deposition. *New Phytol.*, **99** (1985) 381–8.
33. Barnes, J. D., Reiling, K., Davison, A. W. & Renner, C. J., Interactions between air pollutants and winter stress. *Environ. Pollut.*, **53** (1988) 235–54.
34. Hutchinson, T. C. & Whitby, L. M., The effects of acid precipitation and heavy metal particulates on a boreal forest ecosystem near the Sudbury smelting region in Canada. In *Proceedings of 1st International Symposium on Acid Precipitation and the Forest Ecosystem*, USDA forest service technical report NE23 1976, pp. 745–760.
35. Cowling, E.B., Regional declines on forests in Europe and North America: the possible role of airborne chemicals. In *Aerosols, Research Risk Assessment and Control Strategies*. Lewis, Chelsea, MI, USA, 1986, pp. 855–64.
36. Rehfuess, K. E., On the impact of acid precipitation in forest ecosystems. *Forstwiss. Centralbl.*, **100** (1981) 363–70.
37. Ulrich, B. & Pankrath, J., ed., *Effects of Accumulation of Air Pollutants in Forest Ecosystems*. D. Reidel, Doordrecht, The Netherlands, 1983.
38. Prinz, B. Causes of forest damage in Europe. *Environment*, **29(9)** (1987) 11–15, 432–7.
39. Schulze, E. D., Air pollution and forest decline in a spruce (*Picea abies*) forest. *Science*, **244** (1989) 776–83.
40. NAPAP, National acidic precipitation assessment programme. Reports SOS/T 1–28 presented at the NAPAP International Conference, Hilton Head, SC, USA. Documentation published by NAPAP, 722 Jackson Place, Washington, DC, USA, 1990.
41. Roberts, T. M., Skeffington, R. A. & Blank, L. W., Causes of type 1 spruce decline in Europe. *Forestry*, **62(3)** (1989) 179–222.
42. Jarvis, P. G., James, G. B. & Landsberg, J. J., Coniferous forest. In *Vegetation and the Atmosphere*, Vol. 2, ed. J. L. Monteith. Academic Press, New York, USA, 1976 pp. 171–240.
43. Fowler, D., Transfer to terrestrial surfaces. *Phil. Trans. Royal Soc.* (Lond.), **B305** (1984) 281–97.
44. Chamberlain, A.C., Deposition of lycopudium spores and other small particles to rough surfaces. *Proc. Royal Soc (Lond.)* **A296** (1967) 45–70.
45. Beswick, K. M., Hargreaves, K., Gallagher, M. W., Choularton, T. W. & Fowler, D., Size resolved measurements of cloud droplet deposition velocity to a forest canopy using an eddy correlation technique. *Quart. J. Roy. Meteorol.*

Soc., **116.** In press.
46. Fowler, D., Cape, J. N. & Unsworth, M. H., Deposition of atmospheric pollutants in forests. *Phil. Trans. Royal Soc.* (Lond.), **B324** (1989) 247–65.
47. Miller, H. G., Deposition–plant–soil interactions. *Phil. Trans. Royal Soc.* (Lond.), **B305** (1984) 339–52.
48. Parker, G. G., *Throughfall and stemflow in the Forest Nutrient Cycle*, Advances in Ecological Research 13. Academic Press, London, UK, 1983, pp. 58–121.
49. Garten, C. T., Foliar leaching, translocation and biogenic emission of ^{35}S in radiolabelled loblolly pine. *Ecology*, **71** (1990) 239–51.
50. Innes, L. J., Forest health surveys: a critique. *Environ. Pollut.*, **54** (1988) 1–5.
51. Johnson, A. H. & Siccama, T. G., Acid deposition and forest decline. *Environ. Sci. Technol.*, **17** (1983) 294–305A.
52. Siccama, T. G., Bliss, M. J. & Vogelmann, H. W., Decline of red spruce in the green mountains of Vermont. *Bull. Torrey Bot. Club*, **109** (1982) 163–70.
53. Baes, C. F. & McLaughlin, S. B., Trace elements in tree rings: evidence of recent and historical air pollution. *Science*, **224** (1984) 494–7.
54. Krahl-Urban, B., Papke, H. E., Peters, K. & Schimansky, C., eds, *Forest Decline*. USEPA and K.F.A. Julich, 1988.
55. Fowler, D., Cape, J. N., Deans, J. D., Leith, I. D., Murray, M. B., Smith, R. I., Sheppard, L. J. & Unsworth, M. H., Effects of acid mist on the frost hardiness of red spruce seedlings. *New Phytol.*, **113** (1989) 321–35.
56. Cape, J. N., Leith, I. D., Fowler, D., Murray, M. B., Sheppard, L. J., Eamus, D. & Wilson, R. H. F., Sulphate and ammonium in mist impair the frost hardening of red spruce seedlings. *New Phytol.* In press.
57. DeHayes, D.H., Ingle, M.A. & Waite, C.E.,Nitrogen fertilization enhances cold tolerance of red spruce seedlings. *Can. J. For. Res.*, **19** (1989) 1037–943.
58. Friedland, A. J., Howley, G. I. & Gregory, R. A., Investigations of nitrogen as a possible contributor to red spruce decline. In *Air Pollutant Effects on Forest Ecosystems*. The Acid Rain Foundation, St Paul, MN, USA, 1985, pp. 95–106.
59. Johnson, A. H., Deterioration of red spruce in the Northern Appalachian mountains. In *Effects of Atmospheric Pollutants on Forest, Wetlands and Agricultural Ecosystems*, NATO ADSI Vol. 916, ed. T. C. Hutchinson & K. M. Meema. Springer-Verlag, Berlin, Germany, 1987, pp. 83–9.
60. Eamus, D. & Fowler, D., Photosynthetic and stomatal conductance responses to acid mist of red spruce seedlings. *Plant Cell Environ.* **13** (1990) 349–57.
61. Eamus, D., Leith, I.D. & Fowler, D., Water relations of red spruce seedlings treated with acid mist. *Tree Physiol.* **5** (1989) 387–97.
62. Huttermann, A. & Ulrich, B., Solid phase–solution–root interactions in soils subjected to acid deposition. *Phil. Trans. Royal Soc.* (Lond.), **B305** (1984) 353–68.
63. McLaughlin, S. B., Anderson, C. P., Edwards, N. T., Roy, W. K. & Layton, P. A., Seasonal patterns of photosynthesis and respiration of red spruce saplings from two elevations in declining southern Appalachian stands. Can. J. For. Res., 20 (1990) 485–95.

Chapter 11

Effects on Aquatic Ecosystems

S. J. Ormerod
Catchment Research Group, University of Wales College of Cardiff, PO Box 915, Cardiff, CF1 3TL, UK

11.1 INTRODUCTION

Of all the influences of acid deposition, the acidification of surface waters has been amongst the most pronounced, the most widely recognised, and of greatest scientific, political and managerial concern. Impacts involve complex effects by incident precipitation mediated through features of the catchment, in turn interacting with many components of the aquatic ecosystem through a wide variety of processes (Fig. 11.1). Moreover, acidification through these pathways has been recognised across large areas of the industrialised world including North America, Europe and the former USSR. Further, debates have been vehement regarding the relative effects of 'natural' versus anthropogenic acidification, and the relative contributions of acid deposition and changing land use. The result has been a vast research effort within and between disciplines, leading to an equally vast scientific literature, growing particularly in the last 10–20 years. Numerous major reviews are available at the general (e.g. Refs 1–3), national (e.g. Refs 4–6) or topic (e.g. Refs 7–12) level.

In general terms, surface water acidification has been seen widely as the result of acid deposition on sensitive catchments. The responses in affected surface waters include a reduction in pH, elevated metal concentrations (notably aluminium), and ultimately base cation depletion. Together, these chemical changes not only have pronounced consequences for freshwater organisms from all trophic levels, but also for the use of aquatic resources such as water, fisheries and features of conservation value. This chapter aims to provide a general outline of some of the causes, consequences and pathways of these changes.

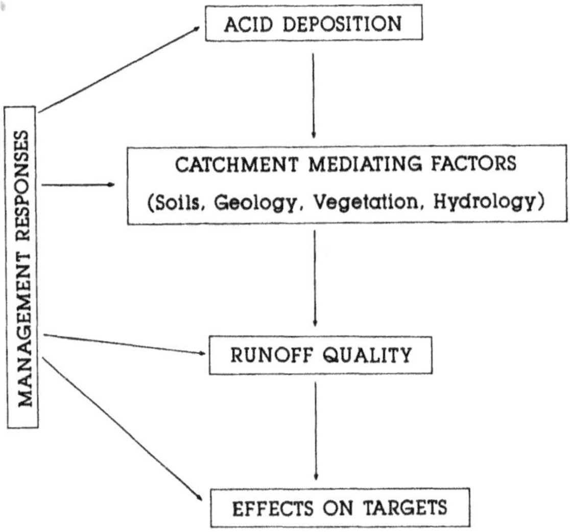

Figure 11.1. A schematic representation of pathways through which acid deposition influences freshwater ecosystems.

11.2 CATCHMENT MEDIATING FACTORS: FROM ATMOSPHERE TO RUNOFF

Several factors will influence the degree of sensitivity of a given catchment to acid deposition, but four features of great importance are geology, soils, hydrology and vegetation. Soils and rocks together determine buffering reactions, while vegetation systems can influence soil conditions and also influence deposition processes (see Chapter 10). Hydrological flow paths within a catchment will affect, for example, the generation of acid episodes, the time available for buffering reactions, and the extent to which different soil horizons contribute to runoff quantity and quality.

11.2.1 Geology

Two principal processes which assimilate hydrogen ions (H^+) and neutralise acidity as water percolates through rocks are the dissolution of carbonate minerals (e.g. calcite) and the hydrolysis of silicates (e.g. orthoclase):

$$CaCo_3 + H^+ = Ca^{2+} + HCO_3^-$$
(Calcite)

$$2KAlSi_3O_8 + 2H^+ + 9H_2O = Al_2Si_2O_5(OH)_4 + 2K^+ + 4H_4SiO_4$$
(Orthoclase)

The amount of carbonates and weatherable silicate minerals in the underlying geology of a catchment thus influence buffering capacity. Several schemes which predict acid sensitivity on the basis of rock classification have been proposed and used in the US[13], the UK[14] and Wales[15] (Table 11.1).

11.2.2 Soils

In some locations, soil characteristics can overide geology as an influence on runoff chemistry. For example, in Florida, acidic lakes overlie highly calcareous and phosphatic rocks.[13] They occur on acid soils from which carbonates have been heavily depleted by leaching. By contrast, some lakes in Maine are underlain by granite, but are not acidified because they receive drainage from calcareous tills and marine clays. Hydrological factors will also alter drainage routes to surface waters so that streams under some circumstances might not receive groundwaters (from bedrock geology), instead receiving surface flow or throughflow solely from soil. Water chemistry in these circumstances will reflect soil chemistry, thus soil characteristics have also been included prominently into schemes for predicting acid sensitivity.[13,15] As with geological systems, the buffering capacity of soils will be determined by the presence of carbonates and weatherable silicate minerals (see Table 11.2), although the array of chemical processes involved in soils is more complex.

11.2.3 Acid deposition and soil chemistry

In general terms, two schools of thought have governed the thinking on the chemistry of surface water acidification. On the one hand, observers of 'natural' soil processes point out that acid waters occur overwhelmingly in areas with acid soils. Here, acidic podzolic soils contain amounts of naturally produced acid, equivalent to thousands of years of atmospheric deposition, so that further additions of acidity from recent deposition are of little additional consequence.[16] Since most rain, even in the absence of air pollutants, is acidic, there has been large potential for the acidification of soils and surface

Table 11.1
The impact of acidic deposition on surface and groundwaters of areas underlain by rocks with different buffering capacities (Adapted from Ref. 15.)

Class	*Buffering capacity*	*Rock types*	*Impact of acidic precipitation on surface waters*	*Impact of acidic precipitation on groundwaters*	*Characteristics of first and second order streams*
1	Little or no buffering capacity	Granite and acid igneous rocks or metamorphic equivalents; granite gneisses, quartz sandstones and metamorphic equivalents; decalcified sandstones; most metasediments, including slates; non arkosic grits	Widespread impact expected	Most areas susceptible to acidification unless significant thicknesses of glacial drift present	Naturally acidic, low conductivity, poorly buffered
2	Low to medium buffering capacity	Sandstones, shales, conglomerates and metamorphic equivalents; coal measures; intermediate igneous rocks; high-grade metamorphic felsic to intermediate volcanic rocks	Impact restricted to first and second-order streams and small lakes	Many areas could be susceptible	Weakly acidic, low conductivity, poorly buffered
3	Medium to high buffering capacity	Slightly calcareous rocks, e.g. marlstones; basic and ultrabasic igneous rocks; Mesozoic mud-stones; low grade intermediate to mafic volcanic rocks	Impact improbable except for near surface drainage in areas of acid soils	Little general likelihood of acidification	Circum-neutral, well buffered
4	Infinite buffering capacity	Limestones, chalk, dolomitic limestones, highly fossiliferous sediments or metamorphic equivalents	No impact	No impact	Alkaline, high conductivity, highly buffered

Table 11.2
Soil buffering classes[a]

Soils with little or no neutralising capacity	Holocene podzols; deeply weathered palaeo-argillic soils, thin rankers on non-calcareous Palaeozoic rocks; brown earths on silicious gravels; gley soils on non-calcareous or pyritic clays and shales; raw bog, basin and blanket peat soils
Soils with moderate neutralising capacity	Non-calcareous pelosols; brown earths on base-rich materials; gley soils on non-calcareous clay; fen peats
Soils with large neutralising capacities	Little weathered soils on Holocene marine clays and calcareous sands; rendzinas; calcareous pelosols; brown calcareous earths; calcareous gleys formed during the Holocene or calcareous sediments

[a] Terminology based on the soil classification of Ref. 15.

waters over time-scales of millennia.[17] Natural acidification processes in these circumstances include those involved in soil formation and podsolization: the leaching of bases from exposed siliceous bedrock and the biological oxidation of carbon, nitrogen and sulphur in decomposing plant litter to produce humic acids.[15] Waterlogging and low temperatures may considerably retard litter decomposition so that gleying and peat formation occurs, resulting in highly acid soils with deep organic horizons and a large capacity to acidify runoff. *Sphagnum* species, growing at the soil surface, export further acidity in exchange for cations.[18] Under the acid conditions in strongly podsolized soils and peats, aluminium may be released by dissolution or ion exchange from alumno-siliceous clays to buffer runoff pH at around 5. Scandinavian scientists (see, for example, Ref. 19) also pointed out that acid episodes could occur from such soils because runoff during storms or snowmelt was in contact with the acidic upper soil horizons, while there would be little percolation or buffering contact with deeper mineral horizons. Both these features, aluminium release and the occurrence of acid episodes, were widely believed amongst more 'popular' science writers to be a consequence of acid deposition.[20]

The second school has largely comprised those who subscribe to the so-called 'mobile anion' hypothesis of surface water acidification. While mindful of the natural processes of acidification, proponents in this school point out that two factors are necessary to explain the distribution of acidified surface waters today: inherent sensitivity to acidification, and the presence of acidic atmospheric deposition.[11] Sensitivity in this context is defined as waters with low ionic strength, with poor buffering capacity, typically in regions dominated by granitic or highly siliceous bedrock (Fig. 11.2). The proponents point out that, in sensitive regions not receiving acid deposition, freshwaters are generally calcium–magnesium bicarbonate waters with pH levels above 5·5. Acidity in these waters is predominantly of organic and carbonic origins and arises from biological activity in their catchments. Waters of this type occur in northern Scandinavia, Labrador, north-western Ontario, the western US, Florida and parts of the Amazon basin.[11] By contrast, waters in sensitive areas which receive acid deposition are sulphate dominated, with strong mineral acidity, elevated aluminium concentrations, and pH below 5·5. They occur in southern Quebec, the Adirondacks of New York, southern Scandinavia and in parts of western Europe including Scotland and Wales.[11] The 'mobile anion' component of this view arises from a consideration of the soil processes which contribute to acidification or buffering: the adsorption and desorption of anions and cations on soil surfaces, the dissolution of minerals, and the natural formation of soil alkalinity.

Figure 11.2. Areas of the world with freshwaters potentially sensitive to acid deposition. (Adapted from Ref. 11.)

In simple terms, acid deposition increases the external input of strong acid anions (sulphate (SO_4^{2-}) and nitrate (NO_3^-)) accompanied by the cations H^+ and ammonium (NH_4^+). The latter may be oxidised to NO_3^- with further consequences for acidification. The fluxes from the soil of these two 'mobile' acid anions are then crucial in determining the extent to which acidity is transferred to surface waters. Most attention has been given to SO_4^{2-} since it was long thought that NO_3^- would be retained by biological fixation in the catchment, a view supported for some time by the ratio of $SO_4^{2-}:NO_3^-$ in acidified surface waters (5–10:1). By contrast, changing fluxes or concentrations of SO_4^{2-} in atmospheric deposition are often reflected in elevated concentrations in soils and surface waters (e.g. Fig. 11.3). Though there may be a delayed response to changing SO_4^{2-} loading

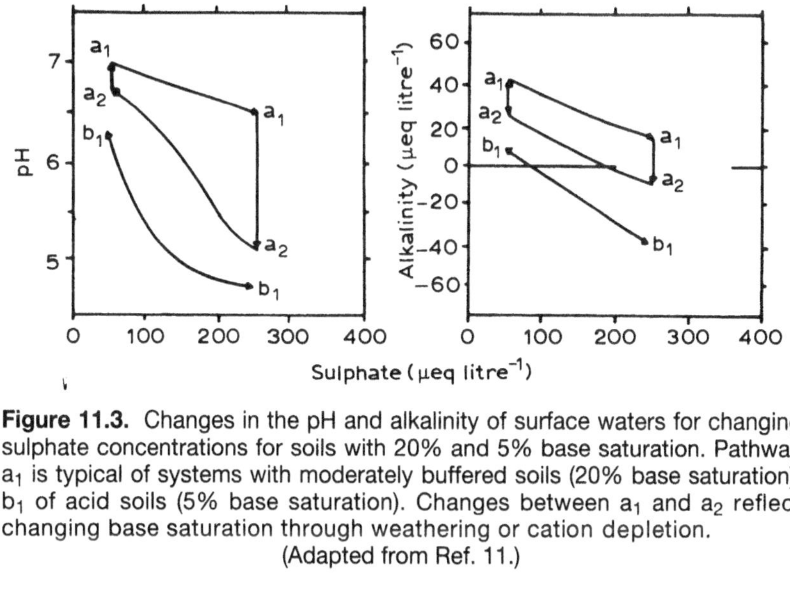

Figure 11.3. Changes in the pH and alkalinity of surface waters for changing sulphate concentrations for soils with 20% and 5% base saturation. Pathway a_1 is typical of systems with moderately buffered soils (20% base saturation), b_1 of acid soils (5% base saturation). Changes between a_1 and a_2 reflect changing base saturation through weathering or cation depletion. (Adapted from Ref. 11.)

due to strong adsorption in the case of highly weathered soils, most soils in North America and Europe appear to be in steady state with respect to SO_4^{2-} inputs and outputs. As a result of acid deposition, SO_4^{2-} adsorption sites in the soil are filled, so that increased SO_4^{2-} concentrations appear in soil solutions and runoff. To maintain a balanced charge, increased cation (calcium ion (Ca^{2+}), magnesium ion (Mg^{2+}), aluminium ion (Al^{3+})) concentrations must thus occur in the same solutions. An increase in the flux of SO_4^{2-} through the soil will thus increase the flux of bases, leading to base depletion in soils where flux losses are not matched by inputs from weathering or the atmosphere. Soil pH and alkalinity decline, and other cations, such as Al^{3+} will begin to occur in runoff in place of bases. At the same time, increased

SO_4^{2-} concentrations in soil solution can considerably reduce alkalinity because of the increased concentration of acid cations required to balance the SO_4^{2-} charge. In acid soils, with low base saturation, alkalinities can undergo a shift from positive to negative with major consequences for runoff pH (Fig. 11.4).

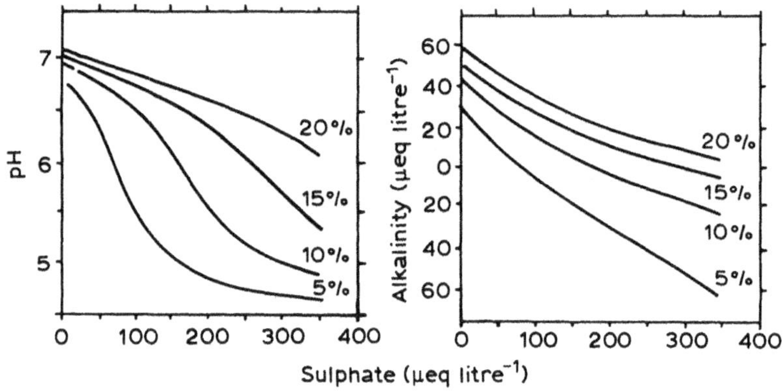

Figure 11.4. The effect of SO_4^{2-} on the pH and alkalinity of soils in which 5, 10, 15 and 20% of exchange sites are occupied by calcium. Calculations were made using the MAGIC model assuming P_{CO_2} in the soil and surface water of $10^{-1.5}$ and $10^{-3.5}$ atm, respectively.[11]

The inputs of H^+ to acid soils lead to key buffering processes: the dissolution of soil minerals to release inorganic aluminium, and cation exchange. Where they are available, base cations tend to be exchanged for Al^{3+}. However, where initial base saturation is low and aluminium solubility is high, as in acid soils, high concentrations of Al^{3+} may remain in solution.

Most researchers involved in acidification subscribe to the mobile anion hypothesis, and it has been responsible for the development of many of the conceptual and mathematical models now used extensively in acid rain research. Support for the hypothesis comes from experiments involving both artificial addition and removal of SO_4^{2-} in atmospheric deposition in Norway.[21] Evidence for a strong mobile anion effect and cation exchange also comes from natural and experimentally induced 'sea salt' episodes, where rainstorms laden with marine salt produce episodes of increased acidity in streams draining acid soils.[22] In these cases sodium ions (Na^+) in rain displaces H^+ from soil while the mobile anion is chloride (Cl^-).

Amongst the most recent developments in the understanding of the chemistry of acidification, it is now being increasingly recognised that NO_3^- in some circumstances acts as an acidifying mobile anion. That NO_3^- can contribute strongly to the acidification of runoff from acid soils has long been known from studies of catchment clearfelling.[23] Under these circumstances, NO_3^- sinks within the terrestrial vegetation are temporarily removed by vegetation clearance, so that it contributes to pH reduction in runoff. At the same time, experimental additions of NO_3^- fertiliser to upland catchments show how this anion can be highly effective in mobilising aluminium from sources within acid soils. The prerequisite for NO_3^- making a contribution to runoff acidity and aluminium concentration thus would appear to be that sinks of nitrogen within catchments are overtopped (i.e. the catchment is nitrogen saturated). There is evidence that this phenomenon may be occurring at sites in Europe,[24] but so far the exact pathways and sources are not clearly known and data are scarce and equivocal.[25]

It should be noted that soils within any given catchment are far from homogeneous, and variations in soil chemistry will arise both vertically and horizontally in the soil profile. Understanding such within-soil variability is of importance in understanding how acid soils can generate episodes of altered chemistry (see below). Taking as an example a Welsh stagnopodzol (a typical and natural soil under acid grasslands or heathlands), indications of changes in chemistry down the profile are shown in Table 11.3. Water from the organic (O) horizon is more acid than throughfall input, but pH increases with soil depth as aluminium is released probably in ion exchange for H^+.

The speciation of aluminium also changes down the soil profile, with organic forms dominant in the organic soil horizons, but toxic, inorganic monomeric aluminium is dominant lower in the profile.[26] Reflecting the importance of marine salts in deposition in Wales, Na^+ is the major cation in soil solution and contributes about 50% of the cationic charge. Cl^- is the dominant anion (*c.* 65% of the charge), with SO_4^{2-} accounting for a further 25%. This is greater than would be expected in an unpolluted area.

11.2.4 Hydrology

The chemistry of surface waters, even within a given catchment, is variable spatially and temporally. In general, acidity declines with downstream progression and during periods dominated by low flows. By contrast, heavy rainfall or snowmelt often leads to marked pH reduction. As a result, 'episodes' of reduced pH, often with elevated aluminium concentration but reduced base cation concentrations, occur in streams and in some lakes.[27,28]

Table 11.3
Annual mean solute concentrations in bulk precipitation, throughfall, soil waters and groundwaters from a *Nardus festuca* grassland on stagnopodzols over Ordovician mudstones at Plynlimon, mid-Wales (μeq litre^{-1} except pH, Si μmol litre^{-1})[a]

	pH	*H*	*Na*	*K*	*Ca*	*Mg*	*Al*	*Si*	NH_4	NO_3	SO_4	*Cl*	*DOC*[b]
Bulk precipitation	4.63	23	83	2	9	20	6	8	15	12	51	105	1
Throughfall	4.62	24	94	38	30	35	6	8	10	4	53	124	10
O horizon	4.11±0.02	77±4	100±8	8±2.0	14±1.0	39±4	29±4	18±1	n.d.	29±6	68±10	124±12	9.5±1.1
Soil E water	4.26±0.03	55±4	187±15	4±0.4	20±0.8	44±2	55±3	42±2	n.d.	39±5	87±5	209±18	6.0±0.7
Bs horizon	4.36±0.03	44±3	173±11	4±0.4	14±0.6	39±2	44±4	36±2	n.d.	28±4	71±5	185±15	3.7±0.4
C horizon	4.70±0.04	20±3	194±13	6±0.5	17±0.4	47±2	66±5	34±1	n.d.	44±6	77±5	202±5	2.3±0.4
Groundwater	6.2	1	152	5	144	126	6	15	n.d.	3	144	155	

[a] Arithmetic means for soil waters; otherwise volume-weighted means.
[b] DOC: dissolved organic carbon.

They expose organisms to extremes of water quality which may be important physiological stressors. Catchment hydrology is an important key to understanding such variations.

On the spatial scale, downstream increases in pH reflect an increase in the contribution by base-rich groundwater to stream runoff. The latter will have had prolonged contact with bedrock and other base-rich substrates. In-stream chemical processes also contribute, due, for example, to mineral dissolution from the stream bed. Opportunities for in-stream buffering will be reduced during periods of increased discharge, but it is now known that acid episodes are generated predominantly by the dilution of bases, and by a change in the pathways contributing to runoff. In upland regions, for example, five broad categories of runoff are recognisable:

(1) shallow throughflow (often partly via macropores in the soil);
(2) pipe flow (i.e. flow through large contiguous macropores);
(3) saturated overland flow;
(4) deep throughflow at the soil–rock or soil–till interface; and
(5) groundwater.

Changes in the relative contributions of each of these will affect runoff in ways which reflect their individual character within any given catchment. In the Norwegian Birkenes catchment, for example, much of the runoff during high flows originates as shallow throughflow from upper soil layers rich in acidity and aluminium. By contrast, at low flow most water arises as well-buffered base flow.[29] At the Svartberget Research Station in northern Sweden, contributions by alkaline groundwater to the stream continued before, during and after acid episodes. pH depressions occurred due to contributions of organic acidity from shallow organic soil horizons through which water flowed during storm conditions.[30] In other cases, mineral acidity could be added to runoff in analogous ways and there is some evidence that the severity of episodes can be determined in part by the antecedent conditions of air pollution.[31]

Although they are widely regarded as important,[32] the ecological effects of episodes are often difficult to separate, using field observations, from the long-term effects of chronic acid stress. This separation is not possible in many of the studies reviewed in this chapter, and no assumptions have been made over the respective importance of chronic and episodic influences, except where stated. Additionally, in areas such as Wales where episodes are generated predominantly by westerly, orographic rainstorms, the likelihood of acid episodes occurring across a range of sites can be predicted well by their mean chemistry (e.g. Fig. 11.5). Separating the relative biological effects of episodic and chronic chemical conditions, for example for the purposes of predicting the biological status of a site from its chemistry, may not always be necessary.

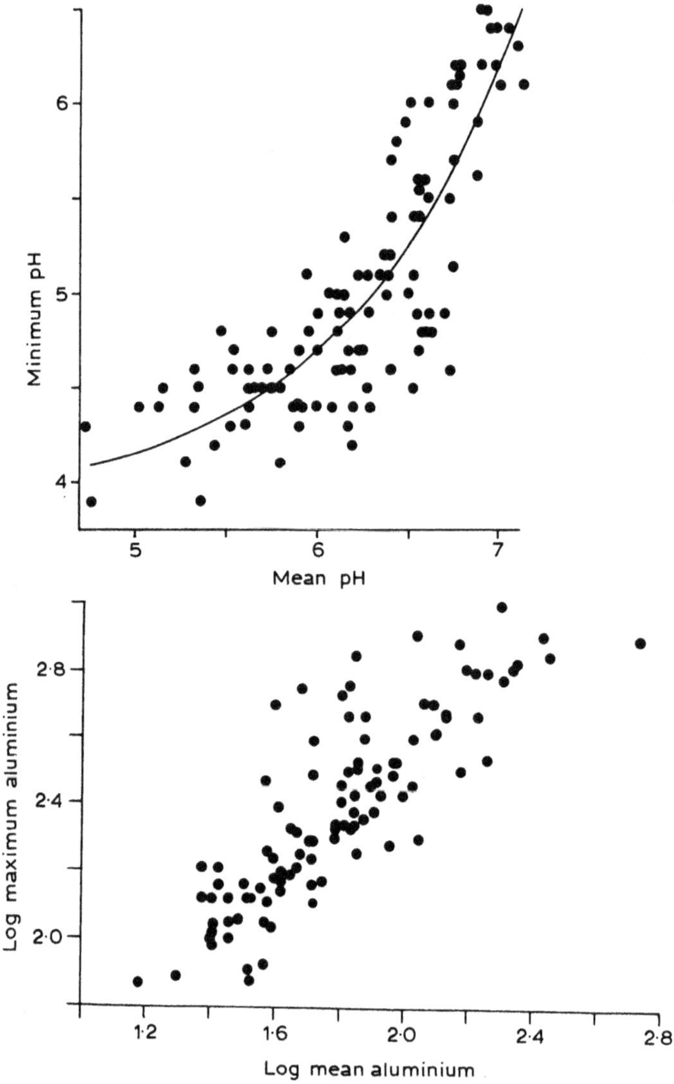

Figure 11.5. The relationship between maximum Al concentration and mean Al concentration, and mean and minimum pH, based on weekly samples, for streams in upland Wales. (Adapted from Ref. 28.)

11.2.5 Vegetation

By far the major part of precipitation deposited onto water catchments falls onto vegetated surfaces. Even before contact with soils and rocks, there is thus an opportunity for modification in the quality and quantity of catchment inputs as water falls through leaves (throughflow) or runs along plant stems (stemflow). This opportunity for plants to have major influences has led to pronounced and prolonged debate over the importance of vegetation and land use in surface water acidification.[19,33]

The influences of vegetation on deposition are reviewed elsewhere in this volume (Chapter 10), but four other principal processes modify the chemistry of precipitation as it passes through vegetation:

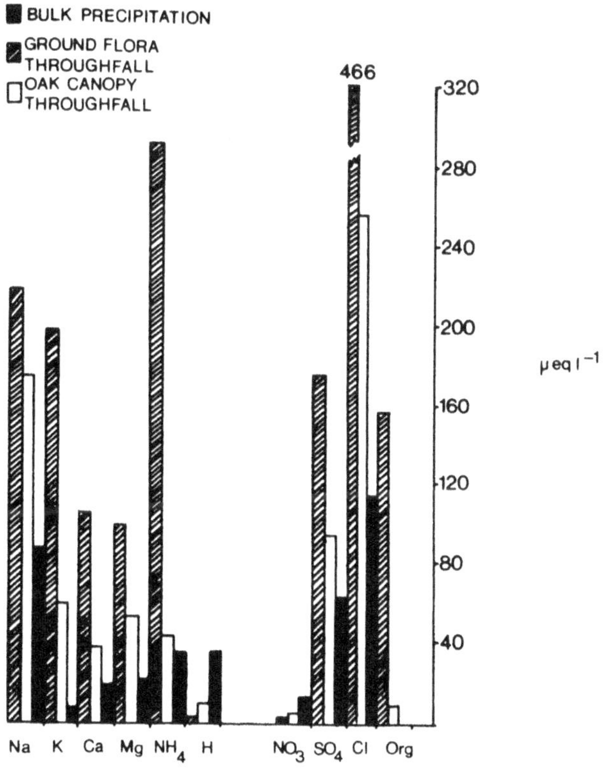

Figure 11.6. Solute concentrations relative to bulk precipitation beneath an oak and grass–bracken undersorey at Llyn Brianne, mid-Wales. (Adapted from Ref. 26.)

(1) concentration of solutes due to evaporation from the vegetation canopy;
(2) solution and wash-off of material previously deposited as aerosols, 'occult' or gaseous dry deposition;
(3) leaching of materials from vegetation as a result of ion exchange and the solution of plant products; and
(4) uptake of ions from precipitation by the foliage, or organisms living on it.

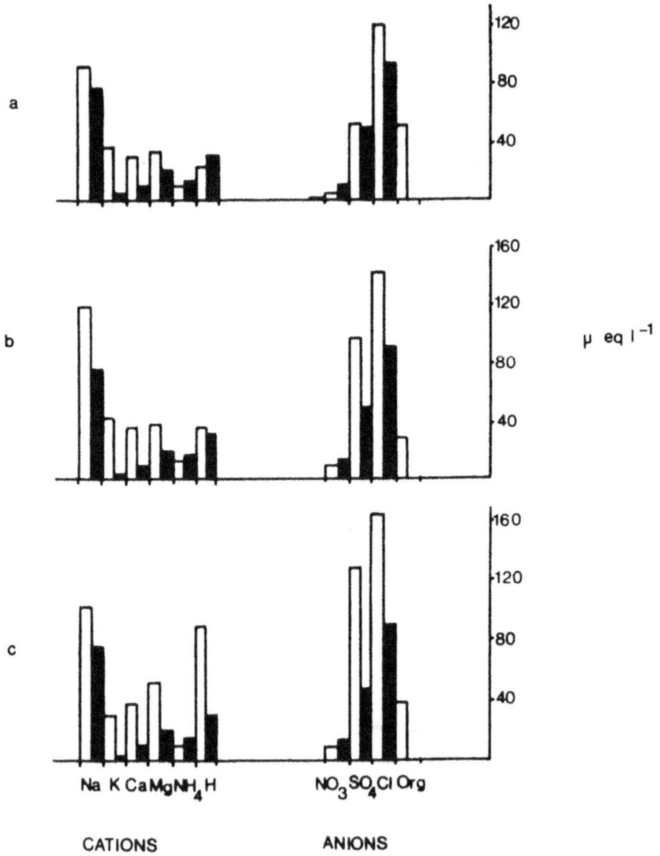

Figure 11.7. Mean solute concentrations in throughfall from (a) *Nardus festuca* grassland, (b) Sitka spruce, and (c) Japanese Larch at Plynlimon, mid-Wales. (□) Throughfall, and (■) bulk precipitation. (Adapted from Ref. 26.)

In Wales,[26] data from three locations (Beddgelert, Plynlimon and Llyn Brianne) show how the concentrations of most solutes increase in throughfall relative to bulk precipitation (Figs 11.6 and 11.7), although those increasing by the largest amounts are K^+, Na^+, Mg^{2+}, Cl^- and SO_4^{2-}. Moreover, these increases are more marked beneath spruce by comparison with the grasses *Nardus* and *Molinia*. Current estimates are that plantation forests in areas receiving acid deposition may increase deposited sulphur by 40–80% above values in adjacent moorland. The highest values in this range are likely at high elevations where mist and cloud are significant sources of precipitation. By contrast, oak forest at Llyn Brianne, while increasing SO_4^{2-} concentration relative to bulk deposition, also increases base cation concentrations. Here, contributed bases are from canopy leaching, including the ground flora, with the result that the acidity of incident precipitation is reduced (Figs 11.6 and 11.7).[26]

Because of the important role played by vegetation in influencing the quantity and quality of deposition, considerable attention has been paid to the acidifying role of land use change in acid sensitive areas. In Britain, for example, plantations of spruce and pine have expanded considerably over recent decades.[34] Catchment comparisons and regional surveys in areas of Wales and Scotland suggest that acidity and, particularly, Al concentrations may be increased in runoff from forest catchments relative to adjacent moorland on similar soil types.[33,35,36] These chemical changes are sufficient to result also in marked biological differences between forest and moorland streams in acid-sensitive areas.[36] Data on soil waters in Welsh catchment studies show several features consistent with this effect.[26] In stagnopodzols under conifer, for example, concentrations of Na^+, SO_4^{2-}, Cl^-, Al and silica are all elevated by comparison with the same soil types under moorland. Al concentrations are greater by two to three times, reaching over 3 mg Al $litre^{-1}$. Na^+ still accounts for a major part of the cationic charge (*c*. 45%), but Al accounts for 28% (*cf.* 15% in moorland). Cl^- remains the dominant anion, but SO_4^{2-} now accounts for more than 35% of the charge (*cf.* 25% in moorland) (Fig. 11.8).

At present, the major pathway thought to be responsible for the chemical effects of forestry is enhanced scavenging of atmospheric material (see Chapter 5). Other possible acidifying effects by forest include soil drying leading to the oxidation of organic sulphur and nitrogen, litter decomposition leading to the production of organic acids, base cation uptake by the growing trees, increased evapotranspiration and hence increased concentration of scavenged pollutants, and altered site hydrology. Some authors have suggested that the evidence for major effects by these other factors (with the exception of evaporation) is slight, because there are so far no indications that forestry exacerbates acidification in areas without polluted deposition (where these factors would still operate). It should be noted, also, that time-series data support acidifying effects by forestry only with some equivocation. On the one hand, clearfelling studies have reduced SO_4^{2-} and Cl^- concentrations in runoff, with Al subsequently declining after a brief period of NO_3^- release. On the other hand, temporal studies following ploughing and planting have not detected any acidification up to canopy closure, although few such studies have been carried out.[37] Longer-term assessments of acidification using fossil diatoms from lake sediments have shown in several cases that acidification occurred primarily due to acid deposition, prior to afforestation. Additional effects due to forestry were not detectable. In one case,[38] afforestation was thought to be a prime influence on lake acidification, though in two other cases (Llyn Cwn Mynach and Loch Doilet) there was no acidification despite afforestation occurring on sensitive catchments.

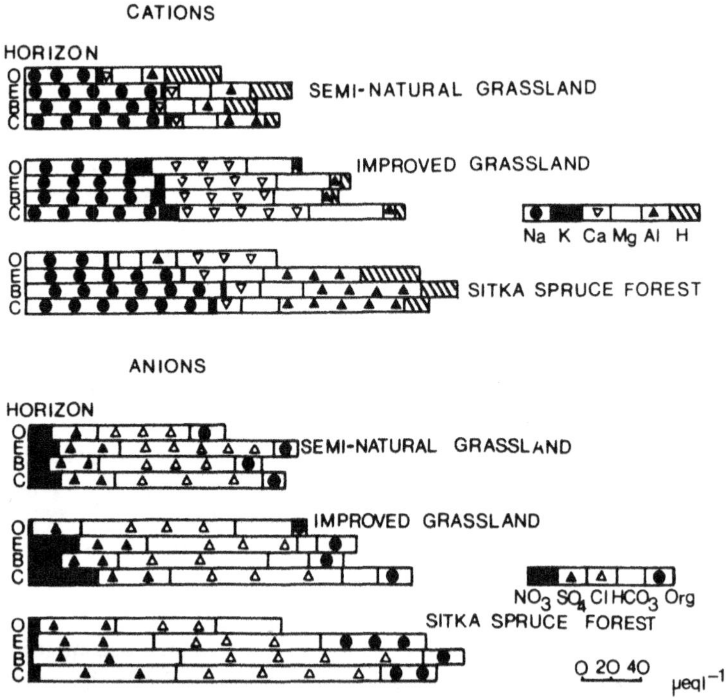

Figure 11.8. Mean solute concentrations in soil solutions from stagnopodzols in semi-natural grassland, improved grassland and Sitka spruce. (Adapted from Ref. 26.)

11.3 EVIDENCE FOR SURFACE WATER ACIDIFICATION

Some consideration of the circumstances under which, where and when acidification has occurred is germane to assessing the relative roles of the different natural and anthropogenic processes involved. There are many water bodies which today have low pH, low alkalinity and elevated metal concentrations,[1] indicating that acidification has occurred at some stage in the past. However, both natural[17,18] and man-induced phenomena could be respons-ible for such change, and the key question is not 'has acidification occurred?', but when, how much and by what processes? In many cases inferential techniques such as modelling[39] or palaeoecology[40] are relied on to provide the answers. These inferential techniques give important indications of the causes and timing of acidification, and long-term modelling and diatom stratigraphy are both techniques which have now been applied in many areas. pH records inferred from the diatom remains in the sediments of sensitive UK lakes show a pH reduction of 0·5–1·8 units from initial values around 1800 of pH 6–6·6. Pronounced chemical change in these systems occurred from 1850–1940 onwards, and can only be explained adequately by acid deposition since no other confounding changes occurred in the catchments concerned[40] (e.g. Fig. 11.9). Many similar examples exist from Norway, Sweden, Finland, Germany and Austria. In the lowlands of the Netherlands, Belgium and Denmark, diatom samples associated with old macrophyte collections have been used to trace trends in acidity, and show marked acidification.[40] At the same time, models linking deposition quality to runoff chemistry via simple chemical processes, show that historical trends in acid deposition can explain the pattern reconstructed from diatom remains in sensitive catchments.[41] Thus, the two techniques can be mutually corroborative in suggesting recent acidification induced by acid deposition. In North America, diatom data from the Adirondacks, New England, the Rockies and a small number of Canadian lakes have been reviewed,[42] showing that weakly buffered lakes in some regions became more acid from around the 1920s–1950s onwards. Declines in pH were strongest in the Adirondacks, but not apparent even in sensitive lakes from the southern Rockies. This is closely in line with what would be expected if acid deposition were a major cause of recent acidification, since only those sensitive areas receiving acid deposition have changed markedly.

In fewer instances have trends in acidification been described directly using historical data on chemical and biological change, such as the loss of fish populations. However, there are numerous instances where these two types of data have been mutually corroborative in indicating recent acidification due most likely to acid deposition.[7,8] Examples are in the Adirondacks region of New York state, and in the polluted areas of acid-sensitive Norway.

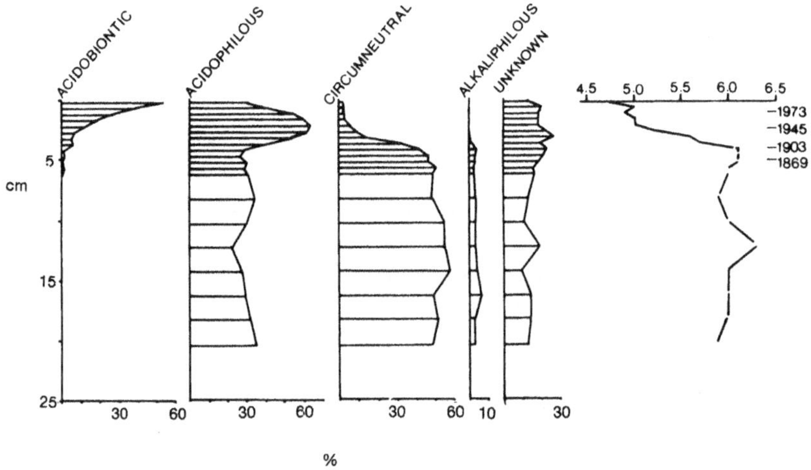

Figure 11.9. Changes in the percentage contribution by different diatom groups in Llyn-y-bi, Wales, and the reconstruction of pH trends. (Adapted from Ref. 4.)

Many recent studies of acidification have been concerned with the potential for reversibility as acid deposition declines. Both biological evidence from diatoms[41,43] and chemical data support the potential for such reversibility, although not so far in all locations. For example, in Scotland declines in SO_4^{2-} deposition at Loch Ard have been accompanied by reductions in stream SO_4^{2-} concentrations and increases in pH, even in forest streams (Harriman, R., unpublished data). No similar trends have been detected in Wales despite some reductions in the rate of SO_4^{2-} deposition. Another important example has been provided by Ref. 44, which describes the rate of acidification of Plastic Lake in southern Ontario. Here, there was a three-fold decrease in alkalinity and a 0·2 unit reduction in pH between 1979 and 1985 despite a reduction in the deposition of strong acids over the same time period. Acidification was accompanied by a decrease in base cation content in the lake, but no increase in strong acid anion content, suggesting that the supply of base cations to the lake may have been depleted. Clearly, much will depend on the characteristics of individual catchments affected by acid deposition.

11.4 EFFECTS ON TARGETS

The principal targets of effects by surface water acidification are the ecosystems in and around them. In order to appreciate the effects of acidification on such systems, it is helpful to have some understanding of their structure, organisation and functioning.

In both lakes and streams, primary producers are planktonic, epiphytic or epilithic algae, bryophytes and higher plants. In fast-flowing rivers, however, higher plants are restricted, whilst most of the algae present are benthic. Besides fixing solar energy into a form which can be used by herbivores (i.e. primary production), plants in aquatic ecosystems influence oxygen concentrations during photosynthesis and respiration, leach exudates rich in organic carbon, and act as a source of detritus after death and decomposition. In addition, macrophytes provide a surface area for epiphytic algae or bacteria, trap detritus from the water column, influence the characteristics of flow and sedimentation, and provide microhabitats in which animals avoid predators or turbulence. Some groups influence the flux or concentrations of metals or other ions through the ecosystem, either through fixation or by adsorption.[45]

In addition to primary production, a large part of the energy base of freshwaters arrives as organic material from terrestrial ecosystems. Indeed, leaves, woody debris and other organic matter from these 'allochthonous' sources can form the major energetic base of upland streams, particularly those which rise in areas of woodland.[46] Once in the aquatic system, an important part of the conversion of such plant litter to a form which can be used by invertebrates is degradation by fungi and bacteria. The fate of some aquatic primary production is also microbial decay. Leaves falling into freshwater leach their soluble constituents rapidly, and are quickly colonised by freshwater fungi, particularly hyphomycetes.[47] These fungi produce all the enzymes necessary to degrade leaf cell walls and some also degrade woody materials.[48] Bacteria become important as degradation progresses.[49,50] Bacteria and fungi, in combination with algae, are also important constituents of the epilithic layer on the stream or lake bed. Such layers are complex mixtures of these organisms in a slime matrix, and may be important in removing dissolved organic material or nutrients from the water column, in producing or holding enzymes for heterotrophic activity, in acting as a site for adsorption of metals and other ions, and providing a food source for grazing invertebrates.[51] Once sloughed from the epilithon and suspended in the water column of lakes or streams, such bacteria sometimes utilise and concentrate dissolved organic matter, or act as a food source to filter-feeding invertebrates.

Many micro- and macro-invertebrate groups are represented in lakes and streams, although flatworms, true worms, molluscs, crustaceans, and immature insects (dragonflies, mayflies, stoneflies, caddisflies and trueflies) are most numerous. Their roles in aquatic ecosystems are many and varied. Some convert primary production directly into animal tissue by eating plants. Others are part of the detrital processing system, 'shredding' coarse particulate organic matter, or 'collecting' fine organic matter by grazing bacteria or filtering suspended material from the water column. Others are predators, whilst all at some stage may provide food for vertebrates such as fish, amphibians, birds and mammals. Besides these energetic roles, some groups have other important functions, such as assisting the interchange of oxygen and nutrients between sediments and the water column.

Fish are represented at several trophic levels in aquatic systems, from detritivore and herbivore to carnivore. In waters vulnerable to acidification, most fish are carnivores from a range of families, which in Britain include salmonids, eels, and cottids, with perch (*Perca fluviatilis*), minnow (*Phoxinus phoxinus*) and coregonids less frequent. Besides their intrinsic value to man as subsistence, conservation and recreational resources, they are often a key part of energy transfer to other systems, for example via predation.

Amongst other vertebrate groups, amphibians occupy a unique niche in temperate regions, being dependent on freshwaters at the spawning and larval stages, but generally occurring at other times in terrestrial systems. Their ecological importance is often overlooked, but they can be an abundant component of aquatic systems, influencing algal or zooplankton composition and affecting energy or nutrient transfer to surrounding habitats. In the Hubbard Brook forest of the north-eastern US, amphibian biomass was twice that of birds and equal to that of small mammals.[52] Many species of birds are wholly dependent on production in freshwaters for food. Many are divers (e.g. loons), herons or waterfowl, and occur on and around lakes or wetlands. There are also several species of wading and perching birds particularly associated with rivers, including some sandpipers (*Tringinae*), American water-thrushes and flycatchers (*Parulidae* and *Tyrannidae*), pipits and wagtails (*Motacillidae*), and one specially adapted family, the dippers (*Cinclidae*). Fewer mammals feed exclusively from freshwaters, but those which often do, such as the otters (*Lutra* spp.) are of importance to conservation, or as predators in the freshwater system.

11.4.1 The influence of acidity

Three ions, H^+, Al^{3+} and Ca^{2+}, are particularly important, biologically, in acid waters. The first two may be toxic to many organisms, whilst Ca^{2+} may ameliorate acid or Al stress. All three are often closely correlated, and hence their individual effects are difficult to determine from field studies alone. Other chemical characteristics, such as the concentrations of fulvic and humic acids, may be important in 'brown water' acid systems. These organic acids are able to sorb metals such as Al thereby mitigating their toxic effects. Features which change the concentrations of these ions, such as acid deposition or recovery, are of great importance in the ecology of temperate freshwaters, affecting most trophic levels and many of the above groups of organism. Demonstrable effects on the distributions of algae, macrophytes, invertebrates and fish have been recorded and associated with the direct toxic action of factors accompanying low pH. Influences on ecosystem processes such as nutrient cycling, primary production and decomposition have also been suggested.[2,7]. Thus, in addition to having direct physiological effects, acidity and elevated metal concentrations can influence some organisms indirectly through functional links such as trophic pathways. These indirect effects also affect semi-aquatic organisms, such as mammals and birds, which depend on freshwaters for food.

11.4.2 Impacts on primary producers

Initial hypotheses concerning the effect of pH on primary productivity proposed that 'oligotrophication' could occur at reduced pH because of decreased nutrient release from sediments or because phosphorus co-precipitated with Al.[27,53]. In the few instances where measurements have been made, however, effects of low pH on primary productivity have not been pronounced.[54] Indeed, artificial acidification in both streams[55–57] and lakes[58] has often increased standing crop. Reasons proposed for the increase at low pH include reduced grazing pressure because of changes to the invertebrate fauna, reduced decomposition, and a succession to acid-tolerant species not readily usable by the grazers present. The biomass of macrophytes in acidic lakes can also be large,[54] though variable.

There is more information to indicate an influence of low pH on floral composition. Phytoplankton, periphyton and macrophytes usually show a reduced diversity with declining pH.[58–60] Acid-tolerant taxa include the dinoflagellates *Peridinium* and *Gymnodinium* amongst the phytoplankton, filamentous chlorophytes such as *Mougeotia* or blue–green algae amongst the periphyton, and *Sphagnum* and *Juncus bulbosus* amongst macrophytes.[54] Some diatoms are particularly sensitive algae, often being restricted to *Cyclotella* spp. in the plankton or *Eunotia* and *Tabellaria* in the benthos of acid streams and lakes. This group has provided an effective indicator system for assessing trends in lake pH.[43] Various physiological hypotheses have been proposed to explain why qualitative changes in floral composition accompany reduction in pH.[54,59]

Clearly, even in the absence of quantitative differences in the flora of water-bodies of different chemistry, qualitative differences could have consequences for animals which preferentially graze certain plants, or are dependent on them for other functions such as habitat provision.

Fewer data are available on the floral composition of streams. Macrophyte composition at 88 sites on Welsh streams (hardness < 25 mg $CaCO_3$ litre^{-1}) have been reported as part of a regional survey of acid waters in 1984.[61] Bryophytes, macroscopic algae and aquatic angiosperms were all recorded, although the latter were only present at 8% of the sites. Macrofloral composition differed markedly according to pH, with the alga *Lemanea*, the moss *Fontinalis squamosa* and the liverworts *Nardia compressa* and *Scapania undulata* characterising progressive increases in acidity and Al concentration (Fig. 11.10). Chlorophytic algae such as *Mougeotia, Ulothrix* and *Stigeoclonium* were also most frequent at low pH. TWINSPAN classification of the flora revealed four site-groups, which differed mostly in pH, Al and Ca^{2+} concentration. Similar patterns were found in streams around Llyn Brianne, where the liming of catchment 9 has since led to increased pH and a large reduction in the abundance of *N. compressa* (unpublished data). The addition of lime to the Welsh lakes Llyn Hir and Llyn Berwyn also led to reductions in the abundance of *Sphagnum* and *N. compressa*, whilst *F. squamosa* and *Lobelia dortmanna* increased. These experimental manipulations support an effect by water chemistry on the distribution and abundance of such plants.

11.4.3 Impacts on decomposition

Several studies have now revealed that the decomposition of plant litter is retarded at low pH in streams[62] and lakes,[63] although not consistently.[64] Reduced shredding activity because of invertebrate scarcity has sometimes been suggested as a reason for impaired decomposition. However, material such as wood, not normally shredded by invertebrates, also decays more slowly in acid waters.[63] Moreover, the abundance of shredding invertebrates sometimes increases when leaf decay is retarded.[62]

A more likely reason for reduced decomposition is through the effect of low pH on microbial activity. In streams in the English Lake District, slow decay of oak (*Quercus robur*) and alder (*Alnus glutinosa*) at pH 4·9 was associated with reduced counts of bacteria and fungi on leaves, and reduced species diversity of hyphomycetes.[49] Experimental acidification of a stream to pH 4 also reduced hyphomycete abundance.[55] Whilst such effects could be detrimental to invertebrates which favoured conditioned litter as a food source, it might benefit those which feed on coarse leaf particles.

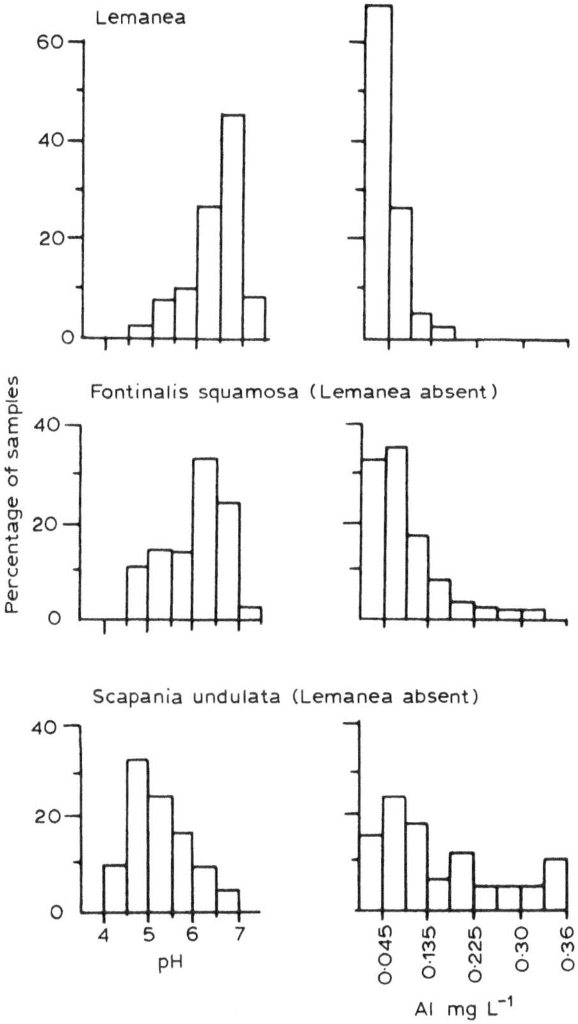

Figure 11.10. The stream chemistry at sites in upland Wales occupied by different macrofloral species. (Adapted from Ref. 4.)

11.4.4 Other microbial populations

In lakes, there appears to be no consistent relationship between pH and counts of bacteria in the water column or sediments.[63,65] Nevertheless, bacteria in the sediments of acidic lakes are sometimes important in generating alkalinity through the processes of denitrification and SO_4^{2-} reduction.[66]

Few data are available from streams, but the epilithon at an acidic site in southern England was lower in organic carbon content, and contained fewer living cells, than at adjacent sites which were less acidic.[67] Elsewhere, counts of epilithic bacteria were also lower in acidic headwaters than in adjacent circumneutral streams,[68] although the streams in this study were unusually acidic (pH < 4) and also had reduced chlorophyll a. Suspended bacteria in circumneutral streams in north-eastern England were only slightly more numerous than in acidic streams in Galloway, but those in the former were larger and more active.[69]

11.4.5 Impacts on macro-invertebrates

Relationships between acid–base status and the macro-invertebrate fauna of streams have been recognised for many years. Typically, taxon richness is reduced at low pH. Mayflies, philopotamid or hydropsychid caddis, molluscs and crustaceans are scarce under acidic conditions (< pH 5·7), despite widespread occurrence elsewhere. By contrast, stonefly nymphs are often abundant. Patterns of total abundance have been less clearly related to pH in soft waters, and invertebrate biomass is sometimes similar in acid and circumneutral streams.[35]

In lakes, molluscs and zooplankton show reduced diversity at low pH whilst acid-tolerant taxa include dragonfly nymphs, alder-fly larvae (*Sialis*), and corixid bugs. In some cases, acid lakes have increased diversity and abundance of nektonic organisms (e.g. Corixidae), a feature ascribed to decreased predation by fish.

Much of the debate concerning the impact of low pH on aquatic invertebrates has centred on whether effects are direct (i.e. physiological) or indirect (e.g. through trophic effects). Animals scarce at low pH are often grazing herbivores or filter-feeders,[70] whilst shredders of leaf litter are most abundant.[71] This might indicate either that acid tolerance varies systematically between feeding groups, or that the nature of the food base changes with pH reduction. Toxicological investigations in the field and laboratory have confirmed that invertebrates which are scarce in acid waters are often pH sensitive.[2] In general terms, physiological studies on invertebrates have shown toxic mechanisms similar to those for fish (see below), although Al can sometimes exacerbate acid stress, and sometimes ameliorate it. Some tolerant insect groups such as corixid bugs have evolved remarkable mechanisms for tolerating the ionoregulatory stress which would otherwise accompany exposure to low pH.[10]

11.4.6 Impacts on fish

Amongst fish, salmonids have received most attention in relation to acidification and there is now much evidence to show that they are directly affected by acid-related factors, sometimes acutely during acidic episodes of snowmelt or rainstorms. There is no indication that brown trout, at least, are limited by food scarcity in acid streams, partly because differences in invertebrate abundance between streams of different pH are not pronounced, but also because they can feed on insects entering the stream from the surrounding terrestrial habitats.

The toxic pathways responsible for the effects of acidification on fish have been very widely reviewed.[9,10,12,72] Initial assessments were that acute effects by low pH could involve disturbances to oxygen uptake and transport, effects on acid–base regulation, and on ion regulation. Of these, the case for disturbed oxygen (O_2) exchange is weakest, since trout stressed in laboratory experiments by low pH alone did not show pronounced accumulation of lactate; this would be expected if anoxia was occurring.[72] Nor were there strong indications of blood acidosis. Instead, both Na^+ efflux and influx across the gill surface are strongly affected by realistically low pH in soft waters so that Na^+ and Cl^- are lost progressively once pH falls to around 1 unit below circum-neutral. By pH 4, influx is virtually eliminated and efflux stimulated so that it accounts for major losses (Fig. 11.11). Death occurs once either plasma (Na^+) or (Cl^-), or both, fall to around 30% below normal, though this in itself is unlikely to be the cause of mortality since migratory salmonids experience similar dilution when moving from seawater to freshwater. Instead, it has been proposed that rapid ion losses also involve haematological and fluid volume disturbances which kill the fish through circulatory failure.[72]

While episodic changes in chemistry may stress fish acutely, chronic long-term effects of low pH may be important. In salmon these include retarded growth and development, reduced activity of branchial ATPase, impaired ionoregulation, and failure of smoltification.[72] Under these circumstances there may be some slight acclimation or recovery with continued exposure, though not to levels similar to those in unstressed fish. Long-term effects on Ca^{2+} metabolism, such as skeletal deformities, may also occur.

In many, and probably most, cases fish in acidic waters are not exposed to low pH alone, but also to elevated levels of Al. In surface waters, the speciation of this metal is dependent on pH in a complex way (Fig. 11.12), but one which is important to understanding impacts on fish. At pH 5, $Al(OH)_2^+$ and Al $(OH)^{2+}$ predominate, but are increasingly replaced by Al^{3+} as pH falls. Fish kills predominantly occur in the pH range 4·8–6, either because the hydroxides (part of the so-called 'inorganic monomeric fraction' of Al) are themselves toxic, or because their speciation is influenced further by the alkaline environment around the gill surface.[72] Al at around 0·1–0·5 mg litre^{-1} appears to cause greater stress in ionoregulation than low pH alone, with the effect increasing in severity from pH 4·4 to 5·2. However, respiratory stress is also apparent, with arterial O_2 concentrations falling markedly, carbon dioxide (CO_2) concentrations increasing, and hyperventilation occurring.

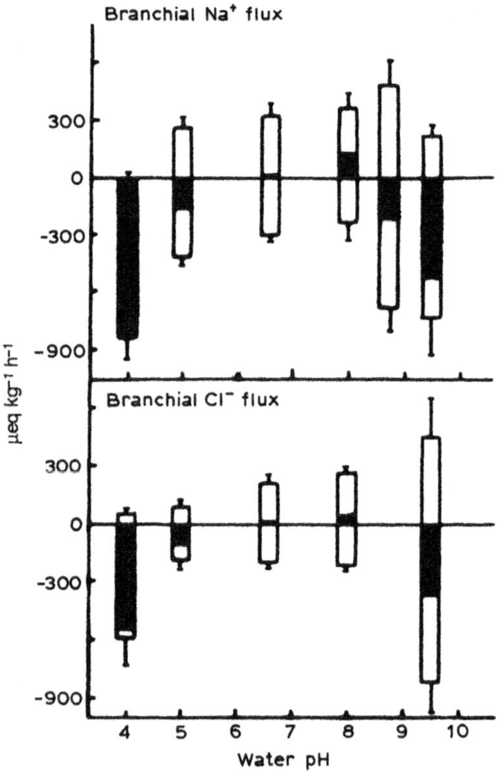

Figure 11.11. The influence of exposure for 3 h to different pH levels on sodium and chloride fluxes in rainbow trout. Upward bars are influx, downward are efflux, and shaded areas are net flux. (Adapted from Ref. 72.)

Al accumulates on the gills, and there is often excessive mucus production. In nature, changes in Al speciation are highly dynamic during acid episodes, over which time adsorption to particles and binding to ligands (e.g. organic acids, fluoride, SO_4^{2-}, silica) can also occur (Fig. 11.13). As a result it can be extremely difficult from field data to identify the key toxic components during a real acid–Al episode since fish are exposed to different species at different times.[73] Moreover, studies of Al speciation in the important, but microscopic, boundary environment around the fish gill have yet to be undertaken.

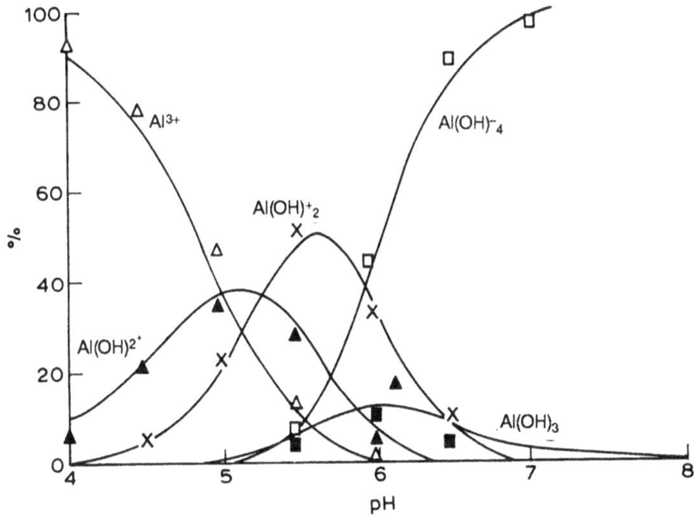

Figure 11.12. Aluminium speciation as a function of pH in a system considering only aluminium hydrolysis. (Adapted from Ref. 4.)

Fish population responses to acidification will be determined not only by toxicity to adult fish, but also by toxicity to different life stages. In general, young fry are thought most sensitive to low pH and elevated Al, so that fish populations being affected by acidification often show missing year classes or progressive recruitment failure. Loss of fish in itself can act as an ecological factor, and lakes without fish sometimes show a proliferation of certain insect groups normally controlled by fish predation. There is experimental evidence from river systems that fish sometimes exert a 'top-down' predatory effect also on the benthic fauna of streams. For example, the proliferation of large-bodied benthic organisms such as *Plectrocnemia conspersa*[74] in acid streams in southern England occurs either because of reduced predatory load, or because they replace fish as top predators in acid systems.

11.4.7 Impacts on amphibians

Most work on amphibians in relation to pH has centred on physiological effects. Many species show increasing mortality only below pH 4, but lethal and sub-lethal responses, including suppressed growth and developmental abnormality, vary widely between species.[75] The European common frog (*Rana temporaria*) shows poor development at metamorphosis when previously exposed to pH < 4·8 and Al concentration of 0·8 mg Al litre^{-1}.[76] The ecological significance of such physiological effects in this species have yet to be determined. Toxicity of low pH and elevated Al to amphibians probably involve similar pathways to those affecting fish.

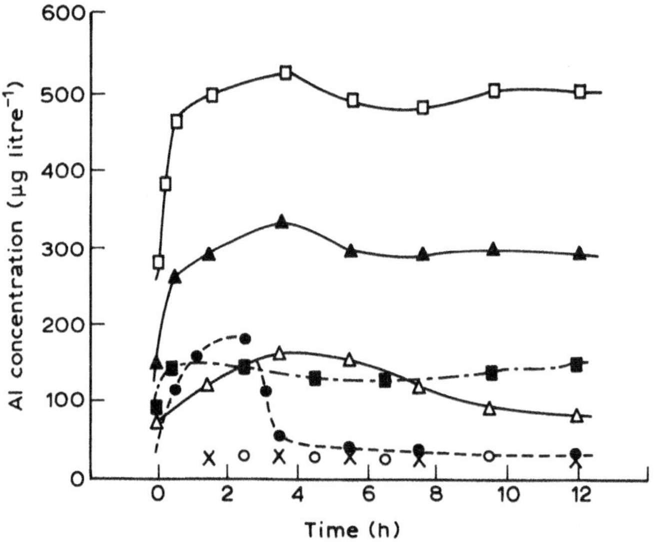

Figure 11.13. Changes in aluminium speciation during a storm episode in a stream at Llyn Brianne, mid-Wales. (□) Inorganic monomeric Al, (▲) Al^{3+}, (●) adsorbed Al, (Δ) organically complexed Al, (■), Al–OH hydroxo complexes, (×) Al–SO_4 complexes, (O) Al–F complexes.
(Adapted from Ref. 4.)

11.4.8 Impacts on birds and mammals

There are now a few instances where acidity has been demonstrated as an influence on the ecology of semi-aquatic vertebrates such as birds or mammals. Impaired fish recruitment in acidic lakes probably affects some lake-dwelling piscivorous birds, but insectivorous birds sometimes benefit from reduced competition in lakes where fish are scarce. For example, the osprey (*Pandion haliaetus*) is thought to show impaired reproductive performance around some Scandinavian lakes, and in North America divers (e.g. loons) *Gavia* spp. may be similarly affected.

In the UK, two bird species have been the subjects of particular attention, and provide a useful comparative study illustrating how acidification can affect animals through trophic pathways. The dipper (*Cinclus cinclus*) and grey wagtail (*Motacilla cinerea*) both breed in linear territories along rivers, the former being entirely dependent on aquatic insects and fish for food. In both Wales and Scotland, dippers are scarce along rivers which have low mean pH (< pH 5·7–6·0) and elevated concentrations of Aluminium (> 0·1 mg Al $litre^{-1}$), despite providing otherwise suitable habitat. In one documented case from the Welsh River Irfon, a sharp fall in the dipper population accompanied a pH reduction between the 1960s and 1980s when the catchment was heavily afforested.[77]

Considerable evidence has now been gathered to show that distribution and decline reflect cause–effect mechanisms by increased acidity. Prey which are important to dippers, particularly when feeding nestlings, include mayfly nymphs and caddis larvae. The time spent foraging by pre-breeding birds on acidic streams is strongly increased as a result, with body mass significantly reduced by comparison with birds on circum-neutral streams. Chick growth, and provisioning are reduced at low pH. Dippers also require prey rich in calcium, such as fish, molluscs or shrimps, for egg formation. All these groups are scarce in acid streams, with the result that blood chemistry in pre-breeding birds differs between acid and circum-neutral streams, with plasma calcium reduced at low pH. At the same time, skeletal sources of calcium are probably less than those required for egg formation. Consistent with calcium scarcity, egg laying is significantly delayed on acidic streams while clutch and brood sizes are significantly reduced. Additionally, eggs are lighter, and shells thinner at low pH.[78]

In complete contrast to the dipper, there is no strong evidence relating the breeding distribution of grey wagtails to stream acidity.[77] They are less reliant on aquatic prey than dippers, their diet and foraging strategies reflecting the use of resources other than the river channel. Prey abundance in the terrestrial riparian zones, where grey wagtails forage, is not affected by the acidity of adjacent streams, though it is by land use.[79] Time of laying, clutch size and brood size are unaffected by acidity, as is brood provisioning and chick growth. This generally negative result in grey wagtails is instructive in understanding why some species of birds should be affected by acidification, while others are not. Different responses are underpinned by differences in foraging ecology, with riparian species dependent on aquatic resources most likely to experience impacts.

11.5 MANAGEMENT RESPONSES

Together, the impacts of surface water acidification bring the need to appraise the need and scope for managerial action. Amongst the possibilities, four which are not mutually exclusive, have been most widely considered. They involve

(1) causative management, by reducing acidic deposition;
(2) symptomatic management, by chemical interventions in affected areas, such as liming or diverting base-rich groundwaters;
(3) preventive management, by controlling those features, such as land use change, which exacerbate the effects of acid deposition in sensitive areas; and
(4) taking no action.

Clearly, all these approaches are liable to have costs and benefits, in both economic and environmental terms, so that their implementation requires careful appraisal beforehand. Because there will be a long lead time between the implementation of management action and response in the affected systems, management strategies cannot be determined by trial and error. Instead, choice between strategies clearly requires prediction of their relative effectiveness in achieving desired objectives (e.g. restoration of a fishery, re-establishment of a viable population of some species of conservation importance). There is a need also to quantify the magnitude of change required to achieve the desired ends in a cost effective fashion. This problem is particularly difficult, since it involves forecasting the response of highly complex soil, water and biological systems to any actions taken. Moreover, because biological resources are amongst those most seriously affected by acidification, there is a need to consider how biological resources will respond to management action. Only a modelling approach permits quantitative forecasting over the time-scales of years to decades which are involved in the acidification or recovery of surface waters as a result of changing deposition.

In essence, the approach to modelling used most often has been to simulate acidification or recovery in rivers in the acid-sensitive regions, and to indicate their changing biological status as a result. Chemical changes have been reconstructed using a hydrochemical model, such as MAGIC[39] (the *M*odel of *A*cidification of *G*roundwaters *I*n *C*atchments), or other conceptually similar models.[40,41] Originally developed and tested for catchments in the US, the MAGIC model has now been adapted for Scotland, Wales, Norway and Sweden.[80] MAGIC uses data on soils and groundwater chemistry, together with information of rainfall chemistry, to simulate the changes in stream acidity which occur over time-scales of years to decades. Besides indicating changes which have occurred in the past, the model can be used to forecast the chemistry of streams under alternative future scenarios of acid deposition (e.g. 10, 20, 30% reduction, etc.).

The conceptual basis of MAGIC is that atmospheric deposition, mineral weathering and exchange processes in the soil and soil solutions are responsible for the chemistry in surface waters. Output from the model includes pH, alkalinity, the concentrations of strong acid anions (SO_4^{2-}, Cl^-, NO_3^- and F^-), base cations (Mg^{2+}, Ca^{2+}) and Al. Long-term trends in soil and stream chemistry occur partly because bases in the soil are used up, in turn affected by leaching and weathering due to the atmospheric deposition of strong acids. As a further, recent, extension to MAGIC, some of the acidifying influences of conifer forest have also been modelled by enhancing terms for sulphate deposition.[82] This simulates the effect of trees in increasing anion capture from the atmosphere. The principal requirements for the model are rainfall and stream chemistry data in the present day, some estimate of changes in atmospheric deposition over time, and information on soil characteristics (e.g. bulk density, depth, cation exchange capacity, sulphate adsorption rates, base saturation, temperature and the pCO_2, the partial pressure of this gas in the soil). Prediction has involved assessing the effect of different reductions in sulphate deposition (0 to 90%), usually phased in over a 20 year period.

Biological modelling has been carried out in a simple way as an adjunct to the hydrochemical models. Empirical relationships, from site-to-site, between biological phenomena and water chemistry are assumed to be causative, and are extrapolated to represent changes which would occur if chemical conditions changed with time. The resulting linear models from regression or discriminant analysis are essentially empirical, adopting a 'black box' approach which avoids detailed understanding of processes. In Wales, for example, biological models currently available predict the survival of first-year brown trout *Salmo trutta*, total trout density, macroinvertebrate community type, macro-floral community type, and chemical suitability of streams for the Dipper *Cinclus cinclus*. The development and use of these models is described elsewhere.[83–86] For any given year in the past or future, chemical output from MAGIC is used as input to the biological models. In all but one of them, Al concentration is a dominant component, reflecting its pronounced role in the biology of acid waters.

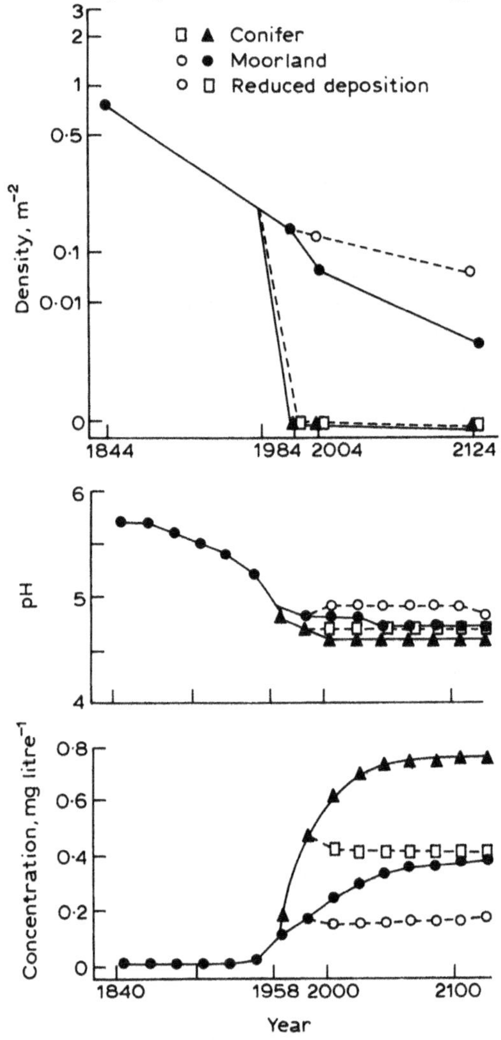

Figure 11.14. Changes in the chemistry and biology of a Welsh stream under different scenarios of land use and sulphate deposition according to the MAGIC model (adapted from Ref. 84). Open symbols involve deposition reduced by 50% from 1984 onwards.

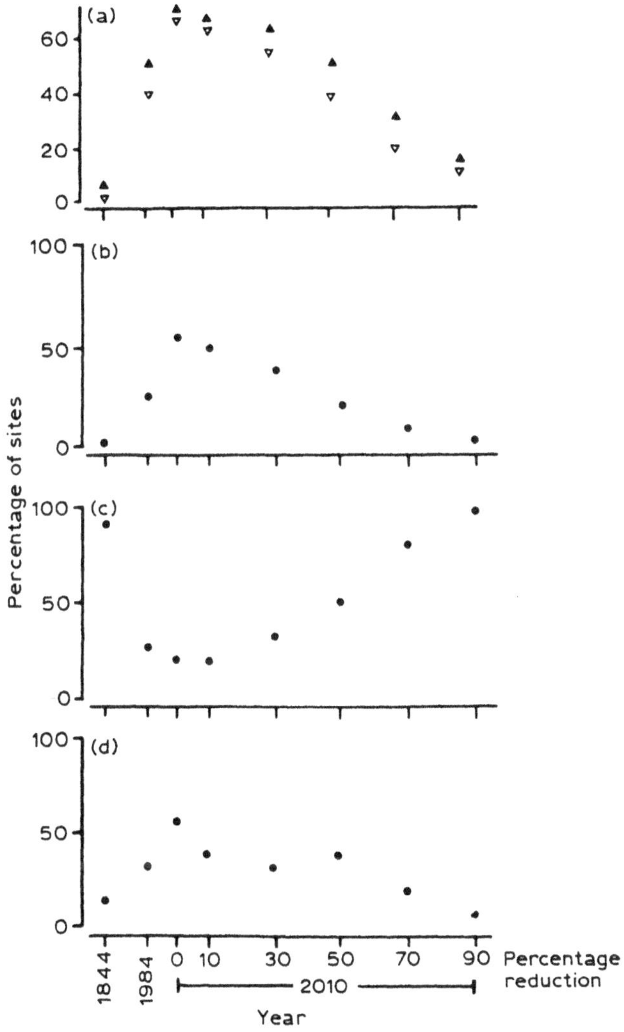

Figure 11.15. Changes in the chemistry and biology of streams in the acid-sensitive region of upland Wales according to the MAGIC model. Values are given for 1844, 1984 and 2010 following different percentage reductions in sulphate deposition. The panels show the percentage of sites with (a) (▽) pH < 5·5 and (▲) aluminium > 0·1 mg litre^{-1}; (b) with invertebrate communities typical of highly acid streams; (c) with invertebrate communities typical of streams which are soft but not markedly acidified; and (d) with invertebrate communities typical of circum-neutral streams. (Adapted from Ref. 80.)

Work in Wales provides an example of modelling the chemical and biological effects of acid deposition, with results illustrated in Figs 11.14 and 11.15. Patterns of chemical change for individual streams in the Welsh region depend on catchment sensitivity, but the typical response from a sensitive case at Llyn Brianne was a progressive decrease in pH and increase in Al concentration from the early 20th century onwards (Fig. 11.14). These changes were most pronounced under simulated forest, particularly for Al, where they persisted even with a 50% reduction in SO_4^{2-} deposition from 1984 onwards. Such elevated Al concentrations would clearly have biological consequences, and this was demonstrated by the fish survival, fish density and invertebrate models. These modelling simulations included a marked decline between 1844 and 1984 in trout density and survival, which was most severe under simulated forest. Deposition reduction of 50% from 1984 onwards permitted the maintenance of some fish under moorland conditions, but such a reduction did not restore the fish population in the forested catchment, indicating that the presence of conifer forest could cancel the effects of the reduced SO_4^{2-} deposition. Without deposition reductions (i.e. if no action were taken), continued fishery decline occurred even under the moorland simulation.

According to the modelling simulations, past chemical conditions in streams during the 19th century were atypical by present day standards in Wales, having low hardness (< 3 mg total hardness per litre as $CaCO_3$), but pH > 5·7 and low Al. As a result, invertebrate predictions for this period involve extrapolations beyond the range over which the invertebrate model was calibrated. From 1940 onwards, there was no apparent change in the modelled invertebrate fauna in streams at Llyn Brianne despite some acidification. However, forest presence advanced the appearance of the most impoverished invertebrate fauna, which did not recover in spite of deposition reductions of 50%.

On a regional scale, MAGIC indicated a pronounced increase between 1844 and 1984 in the percentage of sites with total Al concentrations > 0·1 mg litre^{-1} and pH < 5·5, conditions to which many organisms would be sensitive. The median pH fell from 6·2 to 5·8 over the same period. The consequences of such changes in chemistry for the biology of Welsh streams would, according to the models, be marked. Only 25% of sites had trout densities < 10 fish per 100 m^2 in 1844, whereas over 45% had densities below this value by 1984. Similarly, the survival time of trout fell markedly at many sites, and by 1984, over 20% had values less than 15 days. Streams around Llyn Brianne which have survival times this low are typically fishless.

In 1844, invertebrate communities typical of present-day acidic sites did not occur anywhere in the model region, but occupied 30% of sites by 1984. The percentage of sites with community types intermediate between acid and circum-neutral streams fell correspondingly.

Interestingly, the occurrence of the faunal community typical of circum-neutral streams in the present day increased between 1844 and 1984, probably due to increased calcium concentrations at some sites.

In the regional predictions, reductions in SO_4^{2-} deposition of at least 50% were required to arrest further decline in pH and increase in aluminium. Continued deposition at 1984 levels led to a further increase in the percentage of sites which were highly acidic, and changes such as fishery decline or reductions in the populations of dippers continued if no action was taken. A reduction in deposition of 50% was required to prevent further acidification with concomitant biological change. At greater deposition reductions than this, there was some return to former conditions, although only at around 10% of 1984 deposition values did they begin to approach those of the last century. At this deposition level, the model indicated some loss of the mayfly-rich community (type C), as calcium concentrations declined at some sites.

Like all models, which are imperfect representations of real world phenomena, there are many uncertainties involved in modelling the hydrochemistry and biology of surface water acidification. For example, potential uncertainties in the empirical approach to biological modelling include the following.

(1) An assumption that relationships between biology and chemistry represents a cause–effect link. This implicitly includes the assumption that relationships from site-to-site also equate with changes in time.

(2) A lack of complete understanding of the relative biological effects of chronic and episodic chemical conditions. Most biological modelling hitherto has related stream biology to annual mean chemical conditions. Hydrochemical models for long-term simulation often provide only annual mean chemistry, in turn the only input possible for biological prediction. If episodes were biologically damaging in ways which were not represented by, for example, mean pH, inaccuracy in modelling could result. This problem has recently been addressed.[28]

(3) A lack of understanding of the complexity of important features in the models, such as Al chemistry. Mean Al concentration is an important predictor in many biological models of acidification, but the catchment release and biological effects of different Al species are not fully understood.

Attempts to assess some of these uncertainties have involved both validation exercises, performed by manipulations of stream chemistry, by model application over wide geographical areas, and by an assessment of the empirical modelling approach over other geographical areas. Data on stream chemistry and biology has also been used pragmatically to examine some of these specific problems.[87]

Notwithstanding the above caveats, several features of all the modelling results are of interest. First, they indicate a pronounced acidification and Al mobilization in soft-water streams in upland Wales between the 19th century and present day. Such simulated change caused a considerable reduction in trout survival and density, a marked change in invertebrate and macrofloral communities, and a reduction in the number of streams suitable for dippers. Secondly, the most pronounced acidification occurred under the forest scenarios, with fish either totally eliminated or present at low density. Major changes in the model invertebrate fauna were advanced under forest conditions. By contrast, the absence of forest with reduced deposition could at least support some salmonids in the streams studied.

According to the model forecasts, the number of currently acidified streams might be reduced only if SO_4^{2-} deposition was reduced by over 50% of 1984 levels. Biological changes consistent with further acidification would, according to the models, occur in the absence of such a deposition reduction. Even greater reductions would be required to begin a return to chemical and biological conditions similar to those prior to acidification.

Air pollution models have recently shown that a significant proportion of the sulphate reaching mid-Wales originates in the UK, particularly in the British Midlands.[4] However, the same models show that an emission reduction of 30% in Britain might lead to a deposition reduction in mid-Wales of only 10%, and even emission reductions of 90% could reduce Welsh deposition by only 30–40% of full emission values. Pollutant sources in France, Belgium and Germany become increasingly important in the air pollution model as UK sources are reduced. Restoration of acidic Welsh streams by the control of air pollution alone would thus require, according to model forecasts, concerted action across Europe. Moreover, the presence of forest cover on sensitive Welsh catchments, due to their effect of scavenging airborne pollutants, could effectively offset a reduction in European emissions of 25% or UK emissions alone of 75%.[6] Such substantial action clearly places emphasis on other strategies of restoring or protecting other acidified surface waters (see Chapter 14). Such strategies will have their own mix of costs and benefits.

ACKNOWLEDGEMENTS

The author thanks the numerous colleagues from the Welsh region of the NRA, the Institutes of Hydrology and Terrestrial Ecology, and UWCC who helped to provide information for this review. The work was funded by the Department of the Environment (under contract PECD 7/10/73) and the Welsh Office, and formed part of the Llyn Brianne project.

REFERENCES

1. Drablos, D. & Tollan, A., *Acid precipitation: effects on forests and fish,* Proceedings International Conference on Acid Precipitation. SNSF Project, Oslo, Norway, 1980.
2. Altshuller, A. P. & Linthurst, R. A., *The Acidic Deposition Phenomenon and its Effects: Critical Assessment and Review Papers, Vol. 2: Effects Sciences.* North Carolina State University Acid Deposition Programme, EPA, Washington, DC, USA, 1984.
3. Perry, R., Harrison, R. M., Bell, J. N. B. & Lester, J. N., *Acid Rain: Scientific and Technical Advances.* Selper, London, UK, 1987.
4. Edwards, R. W., Gee, A. S. & Stoner, J. H., *Acid Waters in Wales.* Kluwer, Dordrecht, The Netherlands, 1990.
5. Scottish Development Department. *Acidification in Scotland.* Scottish Development Department, Edinburgh, UK, 1989.
6. Warren, S. C., Alexander, G. C., Bache, B. W., Battarbee, R. W., Crawshaw, D. H., Edmunds, W. M., Egglishaw, H. J., Gee, A. S., Hildrew, A. G., Hornung, M., Howells, G. D., Hunt, D. T. E., Pugh K. B., Wells, D. E., Whitehead, P. G., Cape, J. N., Harriman, R., Neal, C. & Ormerod, S. J., *Acidity in United Kingdom Fresh Waters,* United Kingdom Acid Waters Review Group Final Report. Department of the Environment, London, UK, 1988.
7. Haines, T. A., Acidic precipitation and its consequences for aquatic ecosystems: a review. *Trans. Amer. Fisheries Soc.*, **110** (1981) 669–707.
8. Howells, G. D., Acid waters—the effect of low pH and associated factors on fisheries. *Adv. Appl. Biol.*, **9** (1983) 143–255.
9. Morris, R., Taylor, E. W., Brown, D. J. A. & Brown, J. A., *Acid Toxicity and Aquatic Animals.* CUP, Cambridge, UK, 1989.
10. Witters, H. & Wanderborght, O., Ecophysiology of acid stress in aquatic organisms. *Ann. Soc. Royal Zool. Belgium,* **117** (1987) Suppl. 1.
11. Reuss, J. O., Cosby, B. J. & Wright, R. F., Chemical processes governing soil and water acidification. *Nature* (Lond.), 329 (1987) 27–32.
12. Rosseland, B. O., Eldhuset, T. D. & Staurnes, M., Environmental effects of aluminium. *Environ. Geochem. Health,* **12** (1990) 17–27.
13. Norton, S. A., Geologic factors controlling the sensitivity of aquatic ecosystems to acidic precipitation. In *Atmospheric Sulphur Deposition,* ed. D. S. Shriner, C. R. Richmond & S. E. Lindberg. Ann Arbor Science, MI, USA, 1980, pp.

521–31.
14. Kinniburgh, D. G. & Edmunds, W. M. *The Susceptibility of UK Groundwaters to Acid Deposition*, Hydrogeology Report No 86/3. British Geological Survey, Wallingford, UK, 1986.
15. Hornung, M., Le-Grice, S. & Norris, D., The role of geology and soils in controlling surface water acidity in Wales. In *Acid Waters in Wales*, ed. R. W. Edwards, A. S. Gee & J. H. Stoner. Kluwer, Dordrecht, The Netherlands, 1990, pp. 55–66.
16. Krug, E. C. & Frink, C. R., Acid rain on acid soil: a new perspective. *Science*, **221** (1983) 520–5.
17. Ford, M. S., A 10000-yr history of natural ecosystem acidification. *Ecol. Monogr.*, **60** (1990) 57–89.
18. Clymo, R. S., The growth of Sphagnum: some effects of environment. *J. Ecol.*, **61** (1973) 849–69.
19. Rosenqvist, T., Alternative sources for acidification of river water in Norway. *Sci. Total Environ.*, **10** (1978) 39–47.
20. Pearce, F., *Acid Rain*. Penguin, Harmondsworth, UK, 1987.
21. Wright, R. F., RAIN project: results after two years of treatment. In *Reversibility of Acidification*, ed. H. Barth. Elsevier, London, UK, 1987, pp. 14–29.
22. Langan, S. J., Episodic acidification of streams at Loch Dee, S. W. Scotland. *Trans. Royal Soc. Edinburgh: Earth Sciences*, **78** (1987) 393–7.
23. Lawrence, G. B., Fuller, R. D. & Driscoll, C. T., Release of aluminium following whole tree harvesting at the Hubbard Brook experimental forest, New Hampshire. *J. Environ. Quality*, **16** (1987) 383–90.
24. Henriksen, A., Lien, L., Traaen, T. S., Rosseland, B. O. & Sevaldrud, I. S., The 1000-lake survey in Norway. In *Proceedings of the Final Conference of the Surface Water Acidification Programme*. In press.
25. Skeffington, R. A., ed., Excess nitrogen deposition. *Environ. Pollut.* (Special Issue), **54** (1988).
26. Hornung, M., Reynolds, B., Stevens, P. A. & Hughes, S., Water quality changes from input to stream. In *Acid Waters in Wales*, ed. R. W. Edwards, A. S. Gee & J. H. Stoner. Kluwer, Dordrecht, The Netherlands, 1990, pp. 223–40.
27. Dickson, W., Some effects of the acidification of Swedish Lakes. *Internationale Vereinigung fuer Theoretische und Angewandte Limnologie Verhandlungen*, **20** (1978) 851–6.
28. Weatherley, N. S. & Ormerod, S. J., The importance of acid episodes in determining faunal distributions in Welsh streams. *Freshwater Biol.*, **25** (1991) 71–84.
29. Christophersen, N., Hauhs, M., Mulder, J., Seip, H. M. & Vogt, R. D., Hydrochemical processes in the Birkenes catchment. In *Proceedings of the Final Conference of the Surface Water Acidification Programme*. In press.
30. Bishop, K. H. & Grip, H., The significance of flow pathways in an episodically acid stream. In *Proceedings of the Final Conference of the Surface Water Acidification Programme*. In press.
31. Bird, S. C., Walsh, R. P. D. & Littlewood, I. G., Catchment characteristics and their effects on basic hydrology. In *Acid Waters in Wales*, ed. R. W. Edwards,

A. S. Gee & J. H. Stoner. Kluwer, Dordrecht, The Netherlands, 1990, pp. 203–21.

32. Reader, J. P. & Dempsey, C. H., Episodic changes in water quality and their effects on fish. In *Acid Toxicity and Aquatic Animals*, ed. R. Morris, E. W. Taylor, D. J. A. Brown & J. A. Brown. CUP, Cambridge, UK, 1989, pp. 67–84.
33. Ormerod, S. J., Donald, A. P. & Brown, S. J., The influence of plantation forestry on the pH and aluminium concentration of upland Welsh streams: a re-examination. *Environ. Pollut.*, **62** (1989) 47–62.
34. Ormerod, S. J. & Edwards, R. W., Stream acidity in some areas of Wales in relation to historical trends in afforestation and the usage of agricultural limestone. *J. Environ. Management*, **20** (1985) 189–97.
35. Harriman, R. & Morrison, B. R. S., Ecology of streams draining forested and non-forested catchments in an area of central Scotland subject to acid precipitation. *Hydrobiologia*, **88** (1982) 251–63.
36. Stoner, J. H., Wade, K. R. & Gee, A. S., The effects of acidification on the ecology of streams in the upper Tywi catchment in west Wales. *Environ. Pollut.* (Ser. A), **36** (1984) 125–57.
37. Nisbett, T., *Forests and Surface Water Acidification*, Forestry Commission Bulletin 86. HMSO, London, UK 1990.
38. Kreiser, A. M., Appleby, P. G., Natkanski, J., Rippey, B. & Battarbee, R. W., Afforestation and lake acidification: a comparison of four sites in Scotland. *Phil. Trans. Royal Soc.* (Lond.), **B327** (1990) 377–83.
39. Cosby, B. J., Wright, R. F., Hornberger, G. M. & Galloway, J. N., Modelling the effects of acid deposition: estimation of long-term water quality responses in a small forested catchment. *Water Resources Res.*, **21** (1985) 1591–601.
40. Battarbee, R. W. & Charles, D. F., Diatom based pH reconstruction studies in Europe and North America: a synthesis. *Water Air Soil Pollut.*, **30** (1986) 347–54.
41. Jenkins, A., Whitehead, P. G., Cosby, B. J. & Birks, H. J. B., Modelling long-term acidification: a comparison with diatom reconstructions and the implications for reversibility. *Phil. Trans. Royal Soc.* (Lond.), **B327** (1990) 435–40.
42. Charles, D. F. & Norton, S. A., Paleolimnological evidence for trends in the atmospheric deposition of acids and heavy metals. In *Atmospheric Deposition: Historic Trends and Spatial Patterns*, ed. S. A. Norton. National Academy Press, Washington, DC, USA, 1985, pp. 86–105.
43. Battarbee, R. W., Flower, R. J., Stevenson, A. C., Harriman, R. & Appleby, P. G., Diatom and chemical evidence for reversibility of acidification of Scottish Lochs. *Nature* (Lond.), **332** (1988) 530–2.
44. Dillon, P. J., Reid, R. A. & Grosbois, E., The rate of acidification in aquatic ecosystems in Ontario, Canada. *Nature* (Lond.), **329** (1987) 45–8.
45. Caines, L. A., Watt, A. W. & Wells, D. E., The uptake and release of some trace metals by aquatic bryophytes in acidified waters in Scotland. *Environ. Pollut.* (Ser. B), **10** (1985) 1–18.
46. Fisher, S. G. & Likens, G. E., Energy flow in Bear Brook, New Hampshire: an integrated approach to stream ecosystem metabolism. *Ecol. Monogr.*, **43** (1973) 421–39.

47. Barlocher, F. & Kendrick, B., Dynamics of the fungal population on leaves in a stream. *J. Ecol.*, **62** (1974) 761–91.
48. Chamier, A.-C., Cell-wall degrading enzymes of aquatic hyphomycetes: a review. *Bot. J. Linnaean Soc.*, **91** (1985) 67–81.
49. Chamier, A.-C., Effect of pH on microbial degradation of leaf litter in seven streams of the English Lake District. *Oecologia*, **71** (1987) 491–500.
50. Kaushik, N. K. & Hynes, H. B. N., The fate of the dead leaves that fall into streams. *Arch. Hydrobiol.*, **68** (1974) 465–515.
51. Lock, M. A., Wallace, R. R., Costerton, J. W., Ventullo, R. M. & Charlton, S. E., River epilithon: towards a structural–functional model. *Oikos*, **42** (1984) 10–22.
52. Burton, T. M. & Likens, G. E., Salamander populations and biomass in the Hubbard Brook experimental forest, New Hampshire. *Copeia*, (1975) 541–6.
53. Grahn, O., Hultberg, H. & Landner, L., Oligotrophication: a self accelerating process in lakes subject to excessive supply of acid substances. *Ambio*, **3** (1974) 93–4.
54. Stokes, P. M., Ecological effects of acidification on primary producers in aquatic systems. *Water Air Soil Pollut.*, **30** (1986) 421–38.
55. Hall, R. J., Likens, G. E., Fiance, S. B. & Hendrey, G. R., Experimental acidification of a stream in the Hubbard Brook experimental forest, New Hampshire. *Ecology*, **61** (1980) 976–89.
56. Mullholland, P. J., Elwood, J. W., Palumbo, A. V. & Stevenson, R. J., Effect of stream acidification on periphyton composition, chlorophyll and productivity. *Can. J. Fish. Aquatic Sci.*, **43** (1986) 1846–58.
57. Parent, L., Allard, M., Planas, D. & Moreau, G., The effects of short-term and continuous experimental acidification on biomass and productivity of running-water periphytic algae. In *Impact of Acid Rain and Deposition on Aquatic Biological Systems*, ed B. G. Isom, S. D. Dennis & J. M. Bates. American Society for Testing and Materials, Philadelphia, PA, USA, 1986, pp. 28–41.
58. Findlay, D. L. & Kaisan, S. E. M., Phytoplankton community responses to acidification of Lake 223 experimental lakes area, northwestern Ontario. *Water Air Soil Pollut.*, **30** (1986) 719–26.
59. Havas, M., Effects of aluminium on aquatic biota. In *Aluminium in the Canadian Environment*, ed. M. Havas & J. F. Jaworski. National Research Council of Canada, Ottawa, Ontario, Canada, 1986, pp. 79–127.
60. Stokes, P. M., Benthic algal communities in acidic lakes. In *Effects of Acidic Precipitation on Benthos*, ed. R. Singer. North American Benthological Society, Hamilton, NY, USA, 1986, pp. 119–38.
61. Ormerod, S. J., Wade, K. R. & Gee, A. S., Macro-floral assemblages in upland Welsh streams in relation to acidity and their importance to invertebrates. *Freshwater Biol.*, **18** (1987) 545–58.
62. Mackay, R. J. & Kersey, K. E., A preliminary study of the aquatic insect communities and leaf decomposition in acid streams near Dorset, Ontario. *Hydrobiologia*, **122** (1985) 3–11.
63. Traaen, T. S., Effects of acidity on decomposition of organic matter in aquatic environments. In *Ecological Impact of Acid Precipitation*, ed. D. Drablos & A. Tollan. SNSF, Oslo, Norway, 1980, pp. 340–1.

64. Schindler, D., Mills, K. H., Malley, D. F., Findlay, D. L., Shearer, J. A., Davies, I. J., Turner, M. A., Linsey, G. A. & Cruikshank, D. R., Long-term ecosystem stress: the effects of years of experimental acidification on a small lake. *Science*, **228** (1985) 1395–401.
65. Boylen, C. W., Schick, M. O., Roberts, D. A. & Singer, R., Microbiological survey of Adirondack lakes with various pH values. *Appl. Environ. Microbiol.*, **45** (1983) 1538–44.
66. Schindler, D., The significance of in-lake production of alkalinity. *Water Air Soil Pollut.*, **30** (1986) 931–44.
67. Winterbourn, M. J., Hildrew, A. G. & Box, A., Structure and grazing of stone-surface organic layers in some acid streams in southern England. *Freshwater Biol.*, **17** (1985) 363–74.
68. Goulder, R., Evaluation of the saturation approach to measurement of V_{max} for glucose mineralization by epilithic freshwater bacteria. *Lett. Appl. Microbiol.*, **4** (1987) 29–32.
69. Rimes, C. A. & Goulder, R., Suspended bacteria in calcareous and acid head-streams: abundance, heterotrophic activity and downstream change. *Freshwater Biol.*, **16** (1986) 633–51.
70. Sutcliffe, D. W. & Carrick, T. R., Studies on mountain streams in the English Lake District. 1. pH, calcium and the distribution of invertebrates in the River Duddon. *Freshwater Biol.*, **3** (1973) 437–62.
71. Townsend, C. R., Hildrew, A. G. & Francis, J., Community structure in some southern English streams: the influence of physiochemical factors. *Freshwater Biol.*, **13** (1983) 521–44.
72. Wood, C. M., The physiological problems of fish in acid waters. In *Acid Toxicity and Aquatic Animals*, ed. R. Morris, E. W. Taylor, D. J. A. Brown & J. A. Brown. CUP, Cambridge, UK, 1989, pp. 125–53.
73. Weatherley, N. S., Rogers, A. P., Goenaga, X. & Ormerod, S. J., The survival and early life stages of brown trout (*Salmo trutta* L.) in relation to aluminium speciation in upland Welsh streams. *Aquatic Toxicol.*, **17** (1990) 213–30.
74. Schofield, K., Townsend, C. R. & Hildrew, A. G., Predation and the community of a headwater stream. *Freshwater Biol.*, **20** (1988) 85–96.
75. Pierce, R., Acid tolerance in amphibians. *BioScience*, **35** (1986) 239–43.
76. Cummins, C. P., Effects of aluminium and low pH on growth and development in *Rana temporaria* tadpoles. *Oecologia*, **69** (1986) 248–52.
77. Ormerod, S. J. & Tyler, S. J., Dippers *Cinclus cinclus* and Grey Wagtails *Motacilla cinerea* as indicators of stream acidity in upland Wales. In *The Value of Birds*, ed. A. W. Diamond & F. L. Filion. International Council for Bird Preservation, Cambridge, UK (1987), pp. 191–208.
78. Ormerod, S. J., Halloran, J., Gribbin, S. D. & Tyler, S. J., The ecology of Dippers *Cinclus cinclus* in relation to stream acidity in upland Wales: breeding performance, calcium physiology and nestling growth. *J. Appl. Ecol.*, **28** (1991) 419–33.
79. Ormerod, S. J. & Tyler, S. J. The influence of acidification and riparian land use on the foraging ecology of breeding Grey Wagtails *Motacilla cinerea* in Wales. *Ibis*, **133** (1991) 53–61.
80. Kumari, J., Brakke, D. F., Jenkins, A., Norton, S. A. & Wright, R. F., *Regional*

Acidification Models: Geographic Extent and Time Development. Springer Verlag, Heidleberg, Germany, 1990.

81. Minns, C. K., Kelso, J. R. M. & Johnson, M. G., Large scale risk assessment of acid rain impacts on fisheries: models and lessons. *Can. J. Fish. Aquatic Sci.*, **43** (1986) 900–21.
82. Neal, C., Whitehead, P. G., Neale, R. & Cosby, B. J., Modelling the effects of acidic deposition and conifer afforestation on stream acidity in the British uplands. *J. Hydrol.*, **86** (1986) 15–26.
83. Ormerod, S. J. & Tyler, S. J., Long-term change in the suitability of Welsh streams for Dippers *Cinclus cinclus* as a result of acidification and recovery: a modelling study. *Environ. Pollut.*, **62** (1989) 171–82.
84. Ormerod, S. J., Weatherley, N. S., Varallo, P. V. & Whitehead, P. G., Preliminary empirical models of the historical and future impact of acidification on the ecology of Welsh streams. *Freshwater Biol.*, **20** (1988) 127–40.
85. Weatherley, N. S. & Ormerod, S. J., The impact of acidification on macroinvertebrate assemblages in Welsh streams: towards an empirical model. *Environ. Pollut.*, **46** (1987) 223–40.
86. Weatherley, N. S. & Ormerod, S. J., Modelling ecological impacts of the acidification of Welsh streams: temporal changes in the occurrence of macroflora and macroinvertebrates. *Hydrobiologia*, **185** (1989), 163–74.
87. Ormerod, S. J. & Edwards, R. W., Modelling the ecological impact of surface water acidification: problems and possibilities. *Internationale Vereinigung fuer Theoretische und Angewandte Limnologie Verhandlungen*, **24** (1991) 1738–41.

Chapter 12

Corrosion

G. O. LLOYD
Division of Materials Metrology, National Physical Laboratory, Teddington, Middlesex, TW11 0LW, UK

and

R. N. BUTLIN
Building Research Establishment, Garston, Watford, Hertfordshire, WD2 7JR, UK

12.1 INTRODUCTION

This chapter discusses present knowledge of the effect of industrial pollutants on the degradation of building materials. A general account of degradation processes and of accelerating factors is followed by a discussion of trends in pollutant concentrations and consequent changes in rates of degradation, a summary of present-day rates for the most important materials and information on protection and preservation. Finally, the question of predicting rates for particular sites, either from compilations of data or from statistical or mechanistic models is considered. Most of the information contained in this chapter was first assembled for the Building Effects Review Group (BERG) of the Department of the Environment.

Space does not permit a full treatment of the corrosion of metals or of stone, and only a brief outline of general principles will be presented here. For fuller treatment of these matters, reference should be made to general monographs and compilations of data.[1–4] In particular, the report of the BERG contains the result of extensive enquiries into current understanding of effects on building materials and techniques for protection and preservation.

12.2 DEGRADATION PROCESSES

12.2.1 Metals

12.2.1.1 General considerations

Much of the available knowledge on degradation of materials is summarised in the Gravedigger's remark to Hamlet 'your water's a sore decayer......'. Even such corruptible materials as textiles, paper and leather appear to survive almost indefinitely in the arid climate of the Egyptian desert, while deterioration is rapid in damp conditions, especially if one of a variety of exacerbating factors is also present. This is true for the biodeterioration of organic materials, for the corrosion of metals and for the decay of such building materials as stone, concrete and mortar, although the accelerating factors differ considerably. The importance of water lies in its effectiveness as a medium for transport processes, accelerating a vast range of chemical and electrochemical reactions, and also creating mechanical stresses through cycles of solution and crystallisation. These processes are in turn modified by dissolved substances, most obviously by increasing acidity, but also by increasing electrical conductivity and bringing about changes in volume. Thus, degradation may take place through solution or by mechanical fragmentation due to stress, or in porous materials, by freeze–thaw cycles.

The importance of wetness in degradation was recognised very early: metal clamps for masonry were protected with lead coverings, and many materials with layers of bitumen. The effects of industrial pollutants and of sea-spray were also known at an early stage: for an account of campaigns against corrosion by industrial pollutants in the 17th century see Ref. 5. Pioneering investigations into rates of corrosion of metals in the 1920s and 1930s [1,6,7] led to the systematisation of knowledge on such pollutants as sulphur dioxide (SO_2) and sodium chloride (NaCl), and to the recognition of a principle of critical relative humidity, based on the observation that unless an atmosphere was sufficiently moist to cause condensation of water in the prevailing conditions, even quite high concentrations of pollutants produced only minor effects on metals shielded from rain. The deciding factor was the hygroscopicity of the salt hydrates, and hence the relative humidity at which a layer of salt solution was formed.

Field trials in a wide variety of climates produced a great deal of qualitative evidence to support this general approach: corrosion was rapid on tropical surf beaches, but extremely slow in arid conditions, and in temperate climates it was considerably increased by industrial pollutants and by seawater aerosol. At this stage, however, quantitative prediction of rates from measurements of atmospheric conditions was not achieved, partly because atmosphere monitoring was very limited, but also because a full understanding of the role of 'wetness' had not yet emerged. It was recognised later that the 'time of wetness' (i.e. the time during which the relative humidity exceeded the critical value assigned in laboratory experiments) made it possible to make increasingly successful predictions of rates of corrosion from measured concentrations of pollutants. Further discussion on this point will be found in Section 12.6.

All metals except gold are thermodynamically unstable with respect to their compounds: metals exist in nature as oxides, chlorides, sulphides etc. and considerable inputs of energy are needed for smelting these compounds. The driving force for the reverse reaction converting the metal back to the compound can be expressed as the electrical potential of an electrode of the metal, immersed in a solution of the salt, measured relative to a suitable reference electrode. Large negative values of this potential indicate a large driving force for corrosion and corresponding difficulty in obtaining the metal from its compounds. This does not usually imply that the metal will be rapidly converted back to the compound in dry conditions, or even in wet ones: a thin surface layer of reaction product may provide a very effective barrier to continued reaction. In the presence of liquid water, however, the primary product of reaction may dissolve, leaving the site of the initial attack available for further dissolution, and even if an insoluble product is ultimately formed, it may not be deposited in a form that effectively protects the underlying material. Certain metals, including some that are in principle extremely reactive (for example, chromium, aluminium and titanium) do form coherent 'passive' layers of corrosion product which protect the surface in a wide range of conditions.

Since the primary act of dissolution of a metal is electrochemical (the formation of a cation from a neutral atom, with concomitant release of free electrons)

$$M \rightarrow M^{n+} + ne^- \qquad (12.1)$$

a balancing reaction is necessary to consume these electrons if the process is not to be stifled electrochemically. One such process is the discharge of hydrogen cations:

$$H^+ + e^- \rightarrow \tfrac{1}{2}H_2 \qquad (12.2)$$

which proceeds readily in many cases in acid solutions, but is slow in neutral or alkaline solutions. The reaction most commonly involved in these conditions is the reduction of dissolved oxygen and water molecules:

$$H_20 + {}^1/_2O_2 + 2e^- \rightarrow 2OH^- \qquad (12.3)$$

followed by precipitation of the metal cations as the hydroxide or hydrated oxide. The corrosion of many metals in neutral solution is limited by the rate of this step because diffusion of oxygen through the solution is relatively slow.[8]

Because of the instability of metals relative to their oxides, all metals are covered by a layer of oxide which often provides a protective or passivating effect. Dissolved substances can, however, bring about the dissolution of such protective oxide films, thereby initiating the corrosion process, and can maintain it at a high rate subsequently by increasing the electrolytic conductivity of the solution.

Because of the electrochemical complexity of the corrosion process, a considerable number of incidental factors can affect the rate of corrosion reactions in practical cases. These include contact with metals of different electrode potential, the presence of crevices and contact with acids and oxidising agents. Certain chemical agents can encourage the formation of insoluble products (anodic inhibitors) or can prevent this by forming soluble complexes. Reference will be made to these matters in considering particular metals.

12.2.1.2 Iron and steel

Iron and its alloys constitute the most important class of constructional metals since they are strong and versatile, available in a wide range of forms, and reasonably cheap. Although a passive layer can easily be formed on iron, it is readily broken down by a wide range of corrosive agents and almost all samples of water contain sufficient quantities of such materials to bring about appreciable corrosion. Dissolution is by

$$Fe \rightarrow Fe^{2+} + 2e^- \qquad (12.4)$$

and the balancing cathodic reaction is eqn (12.3) above. Precipitation of ferrous hydroxide is followed by oxidation to hydrated ferric oxides goethite and lepidocrocite (FeOOH). The layer of solid reaction product thus formed provides some protection from further reaction in the sense that rates tend to diminish over time, but in many atmospheres long-term rates are very substantial and corrosion product is lost by spalling and solution. The overall reaction is very complex and involves cycles of oxidation and reduction of the scale as it is wetted and dried.

The reaction is normally controlled by the rate of supply of the oxygen required for eqn (12.3) and this may be greatly accelerated in the thin liquid layers formed in atmospheric exposure. This provides scope for acceleration of the reaction by corrosive pollutants such as SO_2; rates in deep immersion are limited by the oxygen supply and are insensitive to the concentration of corrosive agent.[8]

It has been shown that SO_2 is a particularly effective stimulant for the corrosion of iron since it is readily absorbed and is stored in the scale as ferrous sulphate ($FeSO_4$) and recycled, so that one molecule of SO_2 can bring about the corrosion of 150 atoms of iron.[9] In most exposure conditions, NaCl is much less effective, although it still contributes significantly to rates of corrosion.

The corrosion of iron is therefore encouraged by exposure to wetness and by factors that prolong wetness, and also by such atmospheric contaminants as SO_2, seawater aerosol and acid fumes. Rates can be reduced by drying, by excluding oxygen, by surface coatings that exclude corrosive agents, by alloying additions that encourage the formation of a passive film and by corrosion inhibitors. It is also possible to prevent corrosion electrochemically by providing a potential sufficiently negative to halt eqn (12.4). These factors will be discussed further in section 12.4.

12.2.1.3 Zinc and its alloys

From the point of view of its electrode potential, zinc is inherently a more reactive metal than iron. In atmospheric exposure, however, it corrodes at only about one-tenth of the rate. This is due mainly to the formation of coherent layers of insoluble corrosion products, for example, basic chlorides, sulphates and carbonates. For this reason, SO_2 is not recycled as it is with iron, and one molecule corrodes only one atom of zinc.[9] The basic mechanism, by reactions analogous with eqns (12.3) and (12.4), results in the formation of a film of zinc hydroxide, which is converted to the basic carbonate by atmospheric carbon dioxide (CO_2), and atmospheric pollutants act mainly by bringing about partial dissolution of this film. A recent theoretical treatment of rates of corrosion is based on the solution chemistry of the various products[10] and leads to equations similar in form to successful empirical models which can be used to predict rates in terms of rainfall and pollutant concentrations.[11]

Zinc is used as sheet for roofing and flashings, as a protective coating for iron and steel and in die-casting alloys.

12.2.1.4 Cadmium

Cadmium is used almost exclusively as a protective coating for iron and steel, and because of toxic hazards its use for these purposes is increasingly avoided. Mechanisms and rates are essentially similar to those for zinc.

12.2.1.5 Lead and its alloys

Mechanisms are similar to those for zinc, with formation ultimately of carbonates, sulphates, sulphides or chlorides, or basic compounds. Rates of corrosion in polluted atmospheres are extremely low, although performance may be somewhat impaired by contact with chlorides or organic acids, formed, for example, by decomposing organic materials.

12.2.1.6 Copper and its alloys

Copper is a more noble metal than the others commonly used in construction. The primary product of corrosion in moist atmospheres is cuprous oxide, which is converted over time into basic salts, chloride, sulphate or carbonate according to the locality. Rates of corrosion are small, even though about 50% of the copper corroded goes into soluble products. Rates may be increased considerably by concentrations of corrosive agent due to sea spray or organic acids, but copper roofs have lasted for hundreds of years in reasonably clean atmospheres.

12.2.1.7 Aluminium and its alloys

Aluminium exposed to air produces a thin, coherent layer of oxide which confers great resistance to further reaction, and corrosion in wet conditions is strongly impeded by the very low electronic conductivity of the oxide. It therefore provides one of the classic examples of a highly reactive metal which is effectively protected by its corrosion product. Dissolved salts can cause breakdown of the oxide film, leading to formation of hydrated oxide by reactions analogous with eqns (12.3) and 12.4). These layers are themselves highly protective in solutions not too far from neutrality, except in the presence of soluble copper salts which increase the conductivity of the oxide, or mercury salts which undermine it. There is, however, a tendency to localised pitting in the presence of high concentrations of corrosive agents, particularly of chlorides.

Aluminium therefore shows high resistance to atmospheric corrosion even in polluted atmospheres, and this can be further improved by artificially thickened oxide layers formed by electrolytic action ('anodising'). These layers can also incorporate corrosion inhibitors which improve performance still more.

A large variety of alloys have been developed for increased mechanical strength. An important class of these rely on copper-containing precipitates, which increase rates of corrosion and susceptibility to intergranular attack.

12.2.1.8 Tin and its alloys

Tin is more noble than steel or zinc, and it is readily passivated in a wide variety of conditions. A film of reaction product consisting mainly of stannous oxide gives considerable protection even in polluted atmospheres except in highly acid conditions. Tin is used mainly as a coating for base metals such as iron and steel, and if the coating is breached it may stimulate corrosion of the underlying metal by electrochemical action, essentially by providing a large area of efficient cathode to discharge the electrons produced by the relatively small area of exposed substrate. Tin is also used in a variety of alloys for special purposes, particularly for statuary and architectural metalwork, and for protective coatings. These have excellent resistance to most atmospheres, but slowly develop layers of corrosion product, which are often considered attractive.

12.2.1.9 Magnesium and its alloys

As with aluminium, magnesium is an inherently reactive metal which forms a particularly insoluble oxide and hydroxide. Attack results in a loose deposit of the hydroxide with evolution of hydrogen. These reactions do not proceed readily at low relative humidity and are not very significant in unpolluted atmospheres, but they may be exacerbated by acidic pollutants. Alloys are mainly with zinc and aluminium.

12.2.1.10 Nickel and its alloys

Nickel is relatively easily passivated and hence has much better resistance to atmospheric corrosion than iron, although mechanisms of corrosion are broadly similar. It is capable, however, of catalysing the oxidation of SO_2 to sulphuric acid (H_2SO_4), and in contaminated atmospheres the surface is dulled by a nickel sulphate layer. It is used to some extent as a coating for iron and steel. Nickel is also a constituent of a large range of alloys, usually containing chromium, and these show high resistance to atmosphereic corrosion through the formation of passive layers. Oxygen is required to maintain passivity and rapid corrosion is possible in crevices where access is restricted.

12.2.2 Stone

Evidence for the corrosion or decay of building stone in the atmosphere is extensive (for example see Proceedings of International Conferences on the decay and preservation of stone).[12] The general characteristics of stone decay were described in 1932.[13] In the main, the decay of a particular building stone depends on its chemical and geological composition[14] and on the pollution and meteorological climate in which it is used. Damage to stone in buildings can also arise from natural decay (frost, dissolution in rain) or by the effects of acid deposition by either wet or dry mechanisms.

A series of publications by the Building Research Establishment in the UK has addressed the problem of the utilisation of indigenous stone types in the British Isles, namely limestones, sandstones, magnesian limestones and slates.[15–18] These publications indicate availability, geological origin, examples of buildings where such stones can be found and where appropriate, methods of testing and test results. A separate reference gives a summary of test methods for durability and measurement of porosity and other relevant properties.[19] In addition, for limestones, methods of classification associated with a particular test (salt crystallisation) are given to indicate appropriate uses in different parts of a building, related to exposure and pollution levels.

In general, much of the literature concentrates on limestones and marbles, especially in relation to pollution attack since granites, and to a lesser extent sandstones (except the dolomitic and calcareous varieties) are less susceptible. The decay of stone may result in the removal of large pieces from the surface (frost attack), general removal of details of features on detailed carvings (pollution, dissolution, contour scaling of sandstones) and the formation of black gypsum crusts on sheltered areas of buildings, often containing particulate matter from the combustion of finely divided fuel particles for power generation.[20]

More detail of evidence of damage to stone and other materials by acid deposition is given in the report of BERG.[4] Studies of the stone at St Paul's Cathedral give relative values of historical and current rates of stone decay.[21] The conclusions from the BERG review are that 'evidence is not unequivocal that present rates of weathering in historic buildings are significantly different from those in the past'. As the trend in pollution levels in the UK, especially for SO_2, is downwards, perhaps the rate of decay may be expected to follow the downward trend. The main mechanism of damage to calcareous stone is the reaction of calcium carbonate ($CaCO_3$) with SO_2 to form calcium sulphate ($CaSO_4$), which is more soluble in rainwater than the original carbonate. This can either wash off the surface, leaving a clean and possibly more porous surface, or can wash into the stone where it recrystallises and causes further disruption, possibly over a very long period of time. Since pollution levels in historic time are impossible to derive from measurements, attempts have been made to assess levels from records of fuel usage and other archival sources for specific areas (e.g. London and York).[22]

$CaSO_4$ is probably involved in other damage mechanisms to stone, for example the contour scaling of sandstone. In this case it may be derived from mortar used with dimension stone in buildings. Other known causes of decay include salt spray in marine environments, although the mechanism of such decay is not well established. It may be related to hygroscopic salts keeping the surface moist and susceptible to attack by other means.

The natural weathering of porous stone occurs mainly by frost or the effect of soluble salts. In both cases the mechanism can be seen as the effect of the expansion either of water on freezing or of crystals of salt on rehydration inside the surface of a stone. The expansion of water on freezing is known to be too simplistic a description for the effect on stone as a similar effect has been observed with liquids that do not expand on freezing.[23] Salt weathering occurs either naturally or as a result of pollution damage. The pore size distribution of stone, its chemical composition and the chemical properties affect the susceptibility to frost and salt damage. CO_2 in the atmosphere, which gives rise to a pH of unpolluted rainfall of approximately 5.6, will be involved in the dissolution of calcareous stone in rainwater. The presence of pollution mainly in the form of SO_2 will enhance the rates of decay of calcareous stone by either wet or dry deposition.

12.2.3 Brickwork, concrete, mortar and renderings

12.2.3.1 Brickwork and mortar

There is little evidence that brickwork is damaged by industrial pollutants apart from the rather special case of volatile fluorides liberated by kilns and furnaces firing clay ceramics. As with any porous material, damage may result from volume changes caused by frost, or by penetration of soluble salts. The most common form of attack on brickwork is on the mortar rather than on the bricks themselves, resulting from reaction between soluble sulphates and tricalcium aluminate, with formation of calcium sulphoaluminate (ettringite): the increase in volume may cause surface crumbling or may disrupt the brickwork.

Damage is, however, quite rare, since some mechanism such as wet–dry cycling is needed to transport the sulphate into the mortar. There appears to be no evidence that this form of attack is stimulated by the sulphate produced by industrial pollution: although acidified rain may cause some superficial attack on mortar, this does not affect structural stability.

Renderings may be damaged by acidity (and by sulphate) by the same mechanisms, and to much the same degree, as cement and mortar.

12.2.3.2 Concrete

Superficial attack by SO_2 or soluble sulphates may also occur on concrete, but is seldom serious. By far the most significant problem concerns the corrosion of reinforcements, which may lead to loss of strength by the reduction of section or by debonding the reinforcing rods from the matrix. The increase in volume may also break the concrete, opening up the structure to further attack.

New cement paste is sufficiently alkaline (pH about 13) to prevent corrosion of steel, but in the course of time atmospheric CO_2 reacts with calcium hydroxide ($Ca(OH)_2$) and sets up a layer of nearly neutral $CaCO_3$ which is not protective, and the carbonation front advances into the concrete. There is little evidence that this process is significantly accelerated by the much less abundant SO_2, though once the reinforcement has been exposed by cracking it may accelerate subsequent corrosion.

12.2.4 Organic materials

12.2.4.1 Paint

Paints consist of inorganic fillers and pigments dispersed in a film of organic material which may be a polymer deposited from solution, or formed by oxidative cross-linking or by reaction between polymeric intermediates. Paint films may be degraded by one or more of several mechanisms. The polymer film may be attacked by atmospheric pollutants such as ozone, or by photo-oxidation by ultraviolet (UV) radiation, or the pigments may be attacked, for instance, by acidified rain. The paint may debond once the substrate has become exposed, and begin to deteriorate, and the degradation products may also damage the film by mechanical or chemical action. At this stage, the partially detached film will accelerate corrosion of the substrate by prolonging wetting.

Degradation processes may be indirectly assisted by industrial pollutants which make it difficult to secure the correct surface conditions for applying the paint. This is particularly true of porous sulphate stored in deep pits under the rust film on steel: if these deposits are not completely removed, sufficient water and oxygen will penetrate the paint film to allow renewed corrosion.

Some pigments and extenders, particularly those based on $CaCO_3$, are attacked by SO_2, and in highly acid conditions the polymer film itself may be hydrolysed and the drying process may be impeded. There is, however, little evidence that the life of paint films is greatly affected by industrial pollutants. A number of published damage functions suggest that the effects are small, and a survey concluded that repainting was usually done for reasons of prestige or appearance rather than at the economically optimal intervals.

The factor limiting the life of paint films is usually considered to be oxidation due to ozone or UV radiation, leading to loss of thickness, chalking or embrittlement. It has been shown that the life of paint was reduced in heavily polluted atmospheres in Czechoslovakia, but no pollution data were quoted.[24]

12.2.4.2 Polymers

Polymers are not used for load-bearing elements in buildings but are used extensively for decorative components, protective coatings, trim, infill panels, etc. Degradation takes place mainly through photo-oxidation or ozonolysis, leading to general erosion or to the formation of a brittle, cross-linked surface layer which initiates cracks which then propagate into the sound bulk of the material. Similar effects can occur at the surface of reinforcing fibres and filler particles, leading to loss of strength as well as disfigurement. Little information is available on the effect of industrial pollutants on these processes, but it seems unlikely that they add greatly to the damage caused by oxidative processes.

12.3 TIME TRENDS IN MATERIALS DEGRADATION

The effects of coal combustion products on buildings were already being discussed in the 17th century,[5] and regulations were enforced to limit the use of sea-coal in London. No quantitative measurements were made of concentrations of atmospheric pollutants before the work of Smith in the third quarter of the 19th century, which points to extremely high concentrations of SO_2, ammonia (NH_3) and organic pollutants.[25] SO_2 concentrations averaged over the whole built-up area of London have been calculated from estimates of fuel deliveries from 1600 onwards,[22] and it is suggested that concentrations (estimated from coal consumption per unit area) remained roughly constant from 1800 to 1900, and then fell by 50% by 1950.

Comparative trials of competing metal products were carried out in the 1890s, but this work used sites classified into broad categories such as 'industrial', 'rural', 'marine', etc. with little atmospheric monitoring, and quantitative information suitable for estimating the relevant concentrations is extremely scanty.

The British Iron and Steel Research Association (BISRA) began an important series of investigations at a wide range of sites in the UK and abroad in the early 1930s. These continued until 1960, and some follow-up work by British Steel is still proceeding. These researches provided excellent verification of the broad picture of atmospheric corrosion established by the laboratory studies of W. H. J. Vernon and U. R. Evans and their collaborators and showed the importance of wetness as well as of SO_2 and such natural contaminants as seawater aerosol. Atmospheric monitoring was minimal by present-day standards, relying mainly on limited analyses of the rainwater collected in deposit gauges, although some measurements of SO_2 concentrations were made in the 1950s and 1960s.

As well as providing a sound basis for selection of materials and protective systems, preparation for painting and the protection of metals in transport and storage, this body of research gives one of the few sources of quantitative information on pollution at British sites before the recent expansion in pollution monitoring. It did not, however, provide a very comprehensive or systematic body of measurements, and in particular did not include a really clean rural site. The most remote site used, Llanwrtyd Wells in central Wales, was very exposed, and its rainfall contained a high content of chloride ion, although it was 25 miles from the sea, and rates of corrosion there were similar to those at such seaside sites as Calshot (Hampshire) or Brixham (Devon): the lowest rate measured at this period was in the small town of Godalming (Surrey).

Typical rates of corrosion of plain carbon steel or ingot iron in the period 1933–1960 were 100–170 $\mu m\ year^{-1}$ at the most highly polluted industrial sites such as Sheffield, Woolwich or Derby, about 60 $\mu m\ year^{-1}$ at seaside sites such as Calshot (or Llanwrtyd Wells) and 48 $\mu m\ year^{-1}$ at Godalming. Rates were only a few micrometres per year in the arid climate of Khartoum and 5 $\mu m\ year^{-1}$ at a very clean sub-arctic site in northern Sweden, where although there was copious precipitation, water remained frozen for much of the year. At tropical surf beaches, rates approached 1 μm, $year^{-1}$.

Little additional measurement of rates of atmospheric corrosion was carried out in the UK after World War II because of the technical success of BISRA's work. Little evidence has therefore been available until very recently on rates of corrosion in the period of rapidly changing pollutant concentrations since 1950. A few more recent measurements made by British Steel were, however, included in the Watt Committee report on acid rain in 1988.[26]

As is reported in Chapter 8, there was a large increase in emissions of SO_2, associated with increased consumption of coal and fuel oil, until about 1970, and there is some evidence that the rate of corrosion of steel increased at such sites as Derby and Sheffield in the 1950s. At that period, the pollution of town atmospheres was heavily influenced by local heating plant, including domestic fires, discharging combustion products through low-level flues with high short-range deposition of SO_2. After 1970, however, coal was increasingly replaced by natural gas, which contains very little sulphur, with a corresponding reduction in emissions, and electricity generating plant was concentrated in large units situated mainly to the east of the country with tall stacks which reduced ground-level concentrations of combustion products. These trends have also been reinforced by improvements in fuel economy.

It has been widely assumed that the concentration of power stations in the Vale of York and the Trent Valley must have had an adverse effect on atmospheric pollution and hence on the degradation of building materials.[27] Power stations certainly contribute to SO_2 concentrations,[4] but owing to the concomitant changes noted above, the trend in concentrations has been strongly downwards. Little evidence has been available until recently on rates of corrosion.

In an analysis of trends in Central London, the BERG report[4] showed that while annual average concentrations of SO_2 were in the region of 400 $\mu g\ m^{-3}$ between 1930 and the mid-1950s, they then fell sharply, reaching about 40 $\mu g\ m^{-3}$ by the mid-1980s, and current values are still lower. In the City of Lincoln, measured values in the early 1960s, influenced mainly by local low-level sources, were 120–130 $\mu g\ m^{-3}$, while for a group of rural sites surrounding the city the value was about 40 $\mu g\ m^{-3}$. Changes since that period reduced the concentration in the city centre to about that in the surrounding countryside by 1983. This value had changed little in spite of the construction of three large power stations on the River Trent about 15 miles to the west. The BERG report estimates that the contribution of power stations to the concentration of SO_2 in Lincoln was about 40–50% in 1983, confirming that the large decrease has been due to the reduction in short-range deposition from local sources.

The effect of pollutant concentrations and meteorological factors on the deterioration of a range of building materials has been assessed since 1987 in the National Materials Exposure Programme (NMEP), organised by the Department of the Envioronment's BERG. Samples of plain mild steel, painted steel, galvanised steel, copper and aluminium and two types of limestone (with a third at selected sites) have been exposed at 29 sites around the UK with extensive meteorological monitoring and measuring facilities for pollutant concentrations. Samples have been withdrawn after 1, 2 and 4 years, and the programme is intended to continue until at least 1994. The sites cover a wide range of pollution climates, including city centres and polluted industrial areas, historic monuments, small towns, marine sites and remote areas in the Scottish Highlands and Northern Ireland.

The results[11,28] show that rates of corrosion of steel are much lower, even in the most polluted city centres and industrial areas, than those reported by BISRA in 1933–1960.[6,7] SO_2 concentration is still the most significant factor determining rates, and the measured rates reflect the great reduction in pollutant concentrations since 1970. Reductions in rate have also been found for galvanised steel, copper and aluminium. The data have been fitted to a variety of statistical models which make it possible to predict rates of corrosion on the basis of annual rainfall and average concentrations of SO_2 and nitrogen dioxide (NO_2). At most sites, the predictions are accurate to better than 20%. Further discussion of the data is included in section 12.5.

One significant feature of the results is that rates of corrosion decline as the film of corrosion product thickens, even though a high proportion of the corrosion product is lost during exposure by spalling or solution in rainwater.

It has been suggested in recent years that although concentrations of SO_2 in town atmospheres have fallen sharply, rates of corrosion may not have responded proportionately if concentrations of NO_2 have risen at the same time.[29] There is some laboratory evidence to the effect that although NO_2 alone has little effect on metals, it can increase the effect of a given concentration of SO_2, either by oxidising the latter to H_2SO_4 or by prolonging wetness by producing a hygroscopic nitrate.[30] The concentrations of both pollutants used in this work were, however, well above those present even in the most aggressive British conditions. It has been suggested from an analysis of corrosion data for zinc from sites in the Midlands that rates had not declined so rapidly since 1960 as SO_2 concentrations.[29] No NO_2 data were, however, included in this work. The BERG report noted also that among published functional relations between rates of corrosion of zinc and SO_2 concentration, the gradient d(rate)/d(SO_2) was greater the later the date of exposure.[4]

The results from the NMEP show that, for all the metals investigated, the rates of corrosion are well represented by functions of the form

$$\text{rate} = a + b[SO_2] + c[\text{rain}] \qquad (12.5)$$

where a, b and c are constants, $[SO_2]$ is the annual average concentration of SO_2 and [rain] is the total annual rainfall.[11,28] Only in the case of copper is the fit improved by adding an extra term for NO_2 concentration, and even here the coefficient is negative and the improvement small. The coefficients were also negative for all the other metals except aluminium, and in this case the coefficient was very small and there was no significant improvement in fit.

It therefore appears that NO_2 does not have a significant influence in increasing the corrosive effect of SO_2.

Other investigations forming part of the NMEP have shown that there was no increase in the short-term rate of corrosion of steel when concentrations of both pollutants were abnormally high together, either during a wet period or in the preceding few days, and rates of corrosion were no higher at York Minster, where the two concentrations correlated closely, than at Lincoln Cathedral, where the relation was much weaker.[31,32] Also, the regression equations tended to overpredict rates rather than underpredict them at town centres where NO_2 concentrations were high.

The results from the NMEP provide a considerable body of evidence that rates of corrosion of metals have responded strongly to the reduction in the concentration of SO_2 in town atmospheres in the UK since the 1930s. The reduction has been particularly marked in Central London. It is more difficult to make a comparison for rural conditions because of the lack of past data for relevant sites. It is relevant to note, however, that the rate of corrosion of steel at very wet temperate sites in the Scottish Highlands and Western Ireland in 1987–1991 was not very much greater than that for a sub-arctic site in the 1930s.

The position is less clear for stone. The BERG report noted that, for rain-washed Portland stone at St Paul's Cathedral, the mean rate of erosion was 220 $\mu m\ year^{-1}$ in 1980–1981, while dimensional measurements led to an estimate of 139 $\mu m\ year^{-1}$.[4] The former value may be increased because it includes material dissolved from the pore structure. Estimates of the mean rate between 1727 and 1982, based on the protrusion of lead plugs, were about 78 $\mu m\ year^{-1}$. It is therefore by no means clear that the rate of erosion of stone exposed for long periods to highly polluted atmospheres has yet responded to reductions in pollutant concentrations.

12.4 MEANS OF PROTECTION AND PRESERVATION

This section is concerned with treatments for improving the durability of materials under atmospheric attack, either in new components ('protection') or for prolonging the life of equipment already damaged by weathering ('preservation').

Many materials will of course be damaged to some extent by weather even in the absence of industrial pollutants, because water itself, UV radiation, ozone, temperature-cycling and biological agents all produce deterioration by a variety of routes, and all rainwater contains sufficient dissolved electrolyte from natural sources to cause appreciable corrosion.

The range of materials in use is very large, and many processes are involved in degradation. Certain materials show such poor resistance to weather that they would never be considered for outdoor use in the UK if they could not be satisfactorily protected. Others can survive for long periods without applied protection and need only occasional maintenance or cleaning. In many cases, damage to a protective coating can permit damage to the underlying material which then accelerates the original damage to the coating, for example, by causing loss of adhesion. In other cases, the coating may intensify damage to the substrate exposed at gaps or damage spots (section 12.2.1.8). Conversely, the coating may be chosen to protect the substrate sacrificially (section 12.2.1.2). An attempt will be made to classify protective schemes in general terms before briefly considering individual materials.

12.4.1 Protective schemes

12.4.1.1 General

Since so many degradation processes are caused or promoted by the presence of liquid water, barrier layers of several types are employed as means of protection. These may be completely impervious, such as metallic coatings, vitreous enamels or certain polymer coatings, or may exclude some constituents of the atmosphere but not others, as do paint coatings. Coatings may be used to apply such protective agents as corrosion inhibitors, biocides or UV absorbers.

The barrier type of coating will not usually protect the underlying material if it is mechanically damaged, and the resulting degradation may produce bulky corrosion products which further disrupt the coating. Otherwise excellent coatings such as vitreous enamels may fail in this way if the application involves excessive demands on their mechanical strength. Some metallic coatings may fail because they are inherently porous and may accelerate the corrosion of the substrate by electrochemical means. Rates of corrosion are normally limited by the rate of the cathodic process (see section 12.2.1) and a porous coating may produce small areas of anodic attack in the substrate while allowing a large cathodic current to flow, determined by the large area of the cathode. Intense corrosion will then result at the anodic areas, as often occurs with chromium coatings. On the other hand, coatings of a metal such as zinc, which although it is inherently reactive corrodes only slowly because of the insolubility of the corrosion product, not only protect steel by acting as a barrier, but can also protect at gaps by reducing the electrical potential to a level where iron cannot react. Small gaps protected in this way will then be further protected by the corrosion product of the zinc.

Apart from the use of coatings of resistant materials, certain metals, such as chromium and aluminium, are protected by extremely thin and stable films of oxide, even though they are not inherently unreactive in the sense that they can form no stable compounds. Fortunately, many metals which are not themselves easily or reliably passivated can be made so by relatively small alloying additions of elements such as chromium, which oxidise selectively to produce a suitable film of oxide. In this way, alloys can be made combining to a considerable extent the low price, workability and favourable mechanical properties of such materials as mild steel with the resistance to corrosion of chromium or aluminium. Coatings of materials with superior resistance to corrosion can be applied to susceptible components even if the coating material is expensive or has mechanical properties that would be unacceptable for the basic design. In this way, zinc can be applied to steel or pure aluminium to high-strength alloys with poor resistance, and polymeric or ceramic coatings can be applied to a wide range of materials.

The next sections contain brief notes on protective schemes and their applications, coating techniques, thickness, etc.

12.4.1.2 Metallic coatings

Coatings may be of precious metals (decorative or for electrical contacts) chromium, aluminium, zinc, cadmium, nickel, tin or lead. Some are highly passivating but may allow intense corrosion at pinholes or cracks. Some confer cathodic protection at gaps (e.g. zinc and to a lesser degree aluminium). They may be applied by electrodeposition, hot-dipping, flame-spraying, thermal diffusion, chemical reduction, etc. and have a wide range of thickness, mechanical properties and appearance. Coatings can be used alone or as a foundation for painting.

12.4.1.3 Ceramic coatings

These include vitreous enamel, other oxides, carbides, and nitrides, applied by a range of thermal diffusion and vapour deposition techniques. Protective action depends on coating thickness and integrity.

12.4.1.4 Organic coatings

A wide range of polymers can be applied from solution, by powder deposition and lamination. Protective action depends on mechanical integrity.

12.4.1.5 Paints

Brief notes on paints are provided in section 12.2.4.1. Paints do not protect metals by excluding moisture or oxygen, but partly by excluding ionic compounds, partly by providing an electrically insulating barrier and partly by incorporating corrosion inhibitors. This last action gives some ability to protect at gaps, pinholes and damaged areas.

12.4.1.6 Conversion coatings

The development of a passive layer on metals can be simulated by chemical or electrolytic treatment, by reaction with such agents as phosphoric or chromic acids or by forming a thick layer of oxide which may incorporate corrosion inhibitors. Such layers can be used as protective schemes in their own right or as a foundation for painting.

12.4.2 Protection for materials of construction

The following sections contain notes on the corrosion resistance of the common materials of construction and on means of improving performance in environments where the performance of the unprotected material is inadequate. Some notes on the preservation of damaged materials are also included. The choice of a protective scheme must be carefully considered in the light of all the factors, including exposure to rainwater and to particulate contaminants, gaps and crevices, the possibility of bimetallic effects, temperature cycling, etc. Further information on rates is contained in section 12.5.

12.4.2.1 Iron and steel

Rates of corrosion of plain carbon steels can be as high as 1 μm $year^{-1}$, although in present-day British conditions rates more like 20–100 μm $year^{-1}$ are to be expected. These rates are too high to permit the use of plain steels in outdoor exposure without some protective measures, except where metal sections are large and no problems arise with discoloration, seizure of screw threads, etc. Steels containing small additions of copper and other elements are somewhat more resistant and are used unprotected for some architectural purposes. Stainless steels have a high degree of resistance and the better grades are used unprotected (see section 12.5).

Protection. Painting is the most commonly used means of protection and in severe environments may be applied over metallic coatings. Priming paints usually contain corrosion inhibitors which can protect damaged or porous areas. Thorough preparation is essential since rust contains corrosive salts which can remain active under paint films. This is particularly important when repainting. Similar organic coatings can be factory applied ('coil-coated' steel), often in conjunction with zinc coatings. Metallic coatings, particularly of zinc or aluminium, can be applied by hot-dipping, electrolysis or cementation, and may be used alone or as a foundation for paint. Barrier-type layers of highly resistant metals such as chromium are used for protective and decorative applications. Vitreous enamels are extensively used but are susceptible to mechanical damage.

Preservation. Corroded iron and steel must be thoroughly cleaned to remove all traces of rust before applying further protective treatments. Painted steel should therefore be repainted before serious corrosion sets in, since adequate re-preparation of rusted components may be difficult or impossible. Preparation should be by grit-blasting to remove all rust, followed by a metallic coating and/or rust-inhibiting primer and appropriate finishing coats. Galvanised or aluminised steel can be left unprotected until the coating approaches failure and then cleaned and painted.

12.4.2.2 Zinc and galvanising

Zinc is used in the form of sheet for roofing and flashings, as die-casting alloys and as coatings for steel. It can also be applied as a pigment in priming paints, when it confers some of the properties of galvanised coatings. It corrodes much less rapidly than iron and steel, and so gives barrier-type protection combined with some sacrificial protection at damaged areas.

Protection and preservation. Zinc can often be used without protection, but can be painted. To prevent corrosion due to condensation in sheltered conditions, it can be treated with chromate passivating baths. Many zinc-coated articles are painted only after long weathering. Some light-section items such as fencing are simply replaced when they begin to fail, rather than being restored. For severely rusted steelwork, grit-blasting followed by metal spraying and repainting provides the most effective means of restoration, but this will often be impracticable. Zinc-pigmented paints may then provide an acceptable alternative.

12.4.2.3 Copper

Copper and its alloys have excellent resistance to atmospheric corrosion, at least an order of magnitude better than that of steel. These materials are therefore seldom painted, particularly since the corrosion product is often considered attractive. Lacquers containing the corrosion inhibitor benzotriazole may be used when it is desired to maintain the original bright appearance.

Protection and preservation. The excellent corrosion resistance makes preservative or restorative treatment unnecessary in many cases. Occasionally, contact with organic acids (for instance, in packaging materials) has caused deep penetration of corrosion product and special electrochemical techniques have been devised for neutralising these deposits.

12.4.2.4 Lead

Rates of corrosion are usually low, because of the presence of a highly insoluble film of sulphate. Problems usually arise, if at all, from contact with organic acids or from condensed and/or trapped moisture. Painting is possible though not usually necessary.

12.4.2.5 Aluminium

Rates are usually low, although SO_2, sea salt and alkaline dust are deleterious. Salts of copper or mercury, and contact with metallic copper, can cause serious problems, as can rain contaminated by flowing over copper sheeting or wood treated with preservatives. A range of anodised finishes is available with increased resistance to atmospheric corrosion, and also polymer-coated components. Painting is satisfactory but not usually necessary.

12.4.2.6 Wood

There is little evidence that atmospheric pollutants add greatly to the damage caused by prolonged wetting, resulting in bacterial and fungal degradation with increased porosity and accelerated attack. Preservation of wood involves the use of fungicides and a wide variety of organic coatings and impregnants. There is little evidence that the life of paint films is greatly affected by SO_2 or NO_2. Most problems are due to embrittlement through photo-oxidation, cracking due to dimensional changes in wet–dry cycles and consequent waterlogging. A poorly adherent film that admits water and delays drying may be worse than no film at all. Affected wood can be treated with fungicides after removal of damaged material and filled with adhesive polymer-based materials before repainting.

12.4.2.7 Stone

There is no simple way of protecting porous stone by painting or by the surface treatments used for metals and wood. As most building stones are porous (with the exceptions of marbles and granites) and often in contact with permanent sources of moisture (soil, especially in old buildings) and rain, obstructing the evaporation of such moisture could lead to blistering and surface removal as in salt damage or frost attack. Prolonged contact with the atmospheric pollutants, both gaseous and particulate, can lead to decay even where rain washing does not occur. The removal of surface disfigurements in the form of black deposits and stains is dealt with elsewhere.[33]

To assist design for long life, the overall durability of a stone is assessable in a comparative way and test results for durability and mechanical strength should be taken into account in the design of modern buildings. Standards exist for the design of stone claddings for use in buildings, and similar standards are currently under discussion in Europe. The use of silicone-based waterproofing agents is not usually recommended for preservation treatment especially not those known to block the pores of the stone rather than line them. One formulation designed in the UK which lines the pores rather than blocks them has performed satisfactorily on friable stonework. The use of lime-water treatments of limestone surfaces after cleaning has been shown in scientific studies to have little value in consolidating or preserving stone as it only penetrates to a depth of about 2 mm.[34] A penetration depth of around 50 mm is needed for satisfactory treatment.

12.4.2.8 Brick, mortar, cement, tiles and slate

There is little evidence of damage by SO_2 or NO_2. Most deterioration is due to frost or crystallisation of salts. No special protective measures appear to be called for.

12.4.2.9 Concrete

Superficial deterioration due to acid rain or dry deposition of SO_2 does not appear to affect structural integrity. Carbonation of cement paste establishes a deepening layer where pH is too low to protect reinforcements, which can cause loss of strength and cracking of the concrete matrix. Protective measures include the use of anti-carbonation coatings and galvanised, stainless steel or epoxy-coated reinforcing rods and addition of corrosion inhibitors to the cement mix. BS8110:1985 specifies minimum concrete cover and grades of concrete for reinforcements in various environments. Preservative treatments include removal and replacement of cracked or salt-laden concrete and cathodic protection of the reinforcement by an impressed-current system.

12.4.2.10 Glass

There is no evidence that modern glass is affected by SO_2 apart from leaching of a shallow surface layer in polluted humid atmospheres. Medieval glass, containing less silica and higher levels of potassium and calcium, suffers in damp, polluted conditions by formation of hygroscopic sulphates, and this process weakens the structure of the glass by removing ions of potassium, calcium and sodium. Coatings and external glazing cemented to the glass need prolonged assessment before they can be recommended in cases where their use can be avoided, and the best preservative measure for important ancient windows appears to be to install external glazing with separate ventilation.

12.4.2.11 Polymers

Polymers used as glazing bars, sheeting, rainwater goods, sealing compounds, etc. are protected with antioxidants, UV absorbers and pigments, and there appear to be no reports that their performance is worse in industrial areas as a result of pollution.

12.5. RATES OF DEGRADATION

A very large body of data is available on rates of corrosion of the metals and alloys in widespread use. Most of these measurements, however, cannot be referred to adequately monitored atmospheric conditions. Many of the measurements in industrial areas were carried out in the period 1930–1960, when concentrations of SO_2 were very much higher than they are at present. Also, as was noted in section 12.3, no really clean rural site was included in the UK programme in that period, and although rural sites were used in other countries, notably in the US, climatic differences make it difficult to transfer the measured rates to other localities. Rates differ between indoor and outdoor exposure in response to changes in the supply of corrosive agents and wetting experience, and also between the skyward and groundward surfaces of inclined specimens and between horizontal and vertical surfaces. There may be important differences between specimens exposed at different seasons. The design of metal objects, particularly the presence of joints and crevices, and the possibility of retaining moisture in insulating materials or accumulations of debris or under partially detached paint films, must always be borne in mind. This section contains a brief indication of the range of published rates for a broad selection of metals used in construction and manufacturing industry, with notes on contemporary rates in the UK.

Extensive reviews have been published (see, particularly, Refs 2 and 35–40). These should be referred to for more detailed information, and more limited reviews appear in Refs 4 and 26.

12.5.1 Iron and steel

In the 1930s, rates of atmospheric corrosion of iron and steel were up to 170 μm $year^{-1}$ in the most heavily polluted industrial areas, about 50–100 μm $year^{-1}$ in town centres, and probably 30–40 μm $year^{-1}$ at rural and 40–80 μm $year^{-1}$ at marine sites. Rates in most indoor environments were in the range

2.5–10 μm year^{-1}, reflecting protection from rain, though not always from condensation, and lower concentrations of pollutants. Up to 500 μm year^{-1} was experienced in the worst factory conditions. World-wide extremes for outdoor exposure were about 1 mm year^{-1} at tropical surf beaches and a few μm year^{-1} at Khartoum.

Present-day UK rates are a good deal lower than in the 1930s, about 70 μm year^{-1} at the most polluted sites, about 30 μm year^{-1} in many town centres and as little as 20 μm year^{-1} at clean rural sites.

Low-alloy steels containing about 0.3% copper corrode typically at about one-third the rate of plain steels, although rates are somewhat variable. Since they produce an attractively coloured scale they are sometimes used unprotected for architectural effect.

Martensitic stainless steels corrode at less than one-tenth the rate of plain carbon steels, and ferritic steels with up to 22% chromium still more slowly. They can be used unprotected in many environments, although they show some rust-staining. The austentic grades containing about 18% chromium and 8% nickel corrode about an order of magnitude slower than the martensitics, and higher alloys containing molybdenum are almost completely unaffected by industrial atmospheres.

12.5.2 Zinc and galvanised steel

Typical rates in the 1930s were 3–12 μm year^{-1} in industrial atmospheres, 1.2–1.5 μm year^{-1} in marine and 1–2 μm year^{-1} in rural exposure. On tropical surf beaches, rates were 40–75 μm year^{-1}. Present-day UK rates are 4.25 μm year^{-1} at the most polluted sites, about 2 μm year^{-1} in many town atmospheres and also at rural and coastal sites.

12.5.3 Copper and its alloys

Rates are very similar for most copper alloys and published values lie in the range 0.9–4.0 μm year^{-1} for industrial sites, 0.13–2.6 μm year^{-1} for marine and 0.05–1.0 μm year^{-1} for rural areas. Present-day UK rates appear to lie in a narrow band about 1 μm year^{-1} with even the most polluted sites less than 2 μm year^{-1}. Rates are considerably affected by rainfall.

12.5.4 Aluminium and its alloys

Compositions of aluminium alloys cover a wide range, and the high-strength alloys containing copper are considerably more susceptible to atmospheric corrosion than pure aluminium. Average loss of material, determined by loss of mass during exposure, is usually below 10 μm year^{-1}, but there may be localised pitting. Published rates are 1 μm $year^{-1}$ or less at rural and marine sites. Present-day UK trials show rates well below 1 μm $year^{-1}$, by far the highest (0.4–0.85) being at a small group of sites with heavy particulate contamination. Resistance to SO_2-containing atmospheres is very good, and little pitting has been found.

12.5.5 Titanium and its alloys

Corrosion resistance is comparable to that of the better stainless steels.

12.5.6 Lead and its alloys

In an extensive trial, rates of penetration were no more than 2 μm $year^{-1}$ in industrial areas, and problems arise mainly from crevice corrosion and from contamination with organic materials.

12.5.7 Nickel and its alloys

Nickel is not much used in atmospheric exposure except as plating, and then mainly as a foundation for more protective metals. A wide range of alloys with chromium, copper, molybdenum and iron have been developed for high strength and resistance to corrosion in industrial process equipment, but have little advantage for atmospheric corrosion over cheaper stainless steels.

12.5.8 Stone

As has already been indicated (section 12.4), it is difficult to quote data for stone which will be representative for both in-situ material on old buildings and for new construction in present-day conditions. The BERG report[4] provides data for carboniferous limestones in relatively unpolluted conditions, ranging from 9 to 88 μm $year^{-1}$, and rates at St Paul's Cathedral in the range 78–220 μm $year^{-1}$ (section 12.4). Comparative data are also quoted for a range of sites in Western Europe.[4] Further data are given in Refs 13–18.

Recent results from the NMEP provide rates for 29 sites in the UK, including remote locations in the Scottish Highlands and highly polluted regions.[11] The highest rate of erosion in these trials was 33 μm $year^{-1}$ at the site with the highest concentration of SO_2 (artificially fumigated), and the lowest at Wells Cathedral (Somerset) (10 μm $year^{-1}$), with low rainfall and low SO_2. Most city centres gave rates in the region of 15 μm $year^{-1}$, while in remote areas with high rainfall, rates could be almost as high as at the most polluted sites.

12.6 PREDICTION OF RATES

As has already been pointed out, guidelines drawn from published rates of corrosion make it possible to predict rates only by analogy from results obtained in rather broadly defined categories of atmosphere. This is inevitable, since most published investigations pre-date the recent emphasis on extensive atmospheric monitoring. A number of more recent corrosion trials have, however, provided comprehensive data on rainfall, relative humidity and concentrations of atmospheric pollutants, and some have included analysis of run-off water and have even differentiated between wet-deposited and dry-deposited pollutant.[41] Thus, while earlier trials provided only limited rainwater analysis, with perhaps a monthly average SO_2 concentration, daily pollutant concentrations or even records from continuous monitors are now often available, with records of temperature, relative humidity, 'time of wetness', wind speed and direction, etc. Some sites now also provide output from electronic instruments giving a continuous measure of rates of corrosion.

The factors controlling rates of corrosion are complex. Metals do not actively corrode except in the presence of sensible moisture, but it is possible for the corrosion product to absorb corrosive agents in dry conditions which will accelerate corrosion on subsequent wetting.[9] Other processes occurring during dry spells may also affect rates of corrosion later, for example, mechanical contraction leading to cracking of the scale.

Also, since the electrochemical processes involved in corrosion depend on the concentration of corrosive agents dissolved in the moisture film, and also on the supply of oxygen, rates will vary with changes in the thickness of the liquid film during wet–dry cycles and will also be different in heavy rain and in slight wetting by rain or dew. Other factors include the effect of the time of year when the specimen is first exposed, early wetting experience, and the effect of partial screening of the specimen surface.

It is impossible to deal with these points in the space available here. For more extensive reviews, see Refs 42 and 43. No formal theory taking account of these effects is available at present, and if such a theory were elaborated, approximations would be needed before actual site results could be predicted, since the site data would almost never include direct measurements of the parameters appearing in the equations.

In these circumstances, predictive equations must at present be written in terms of available data, which at best include some measure of wetness and measurements (at greater or less time resolution) of some of the relevant pollutant concentrations. Such datasets are used in multivariate statistical analyses to determine the best-fit coefficients for empirical models for rates of corrosion. A considerable number of such predictive equations have been published.[44–46] The datasets tend to be less comprehensive for site trials than for experiments in laboratory chambers: the latter, however, present difficulties of their own in the sense that conditions are often not very realistic.

It has often been found possible to devise an empirical model which represents a given dataset reasonably closely, and one group of authors has asserted that rates can be predicted to within 20% for any site using available data.[47] Attempts to combine the results of different empirical models have, however, revealed large discrepancies and it is now generally accepted that the relatively simple equations at present available work as well as they do partly because variables not included in the model do not cover too wide a range and partly because of covariances between the climatic and pollution data within a particular dataset. These conditions will not necessarily be fulfilled in a wider universe of discourse.

An instructive example of this effect is provided by the results from the British NMEP which includes rates of corrosion and climatic and pollution data for 1-year and 2-year exposure periods.[11] At present, however, analysis of the site measurements is incomplete, and no rainwater analyses or measurements of particulate contamination or chloride ion deposition are available, and the only measure of wetness is total annual rainfall. Despite these limitations, equations of the type

$$\text{corrosion} = a+b[\text{rain}]+c[\text{SO}_2]+d[\text{NO}_2]+e[\text{acidity}] \qquad (12.6)$$

make it possible to predict rates of corrosion to within ± 10% at the majority of the sites and to within ± 20% at almost all of them, with only a small number of outliers. *a*, *b*, *c*, *d* and *e* are constants, [rain] is total annual rainfall, [SO_2] and [NO_2] are annual average concentrations in $\mu g\ m^{-3}$, and [acidity] is deposition-weighted average hydrogen ion content of the rainfall in $\mu eq\ litre^{-1}$. These equations could clearly be made to fit better if data were available for chloride ion deposition, particulate contamination, etc., and a more directly relevant wetness parameter would probably also improve the situation.

These equations (for mild steel, galvanised steel, copper and aluminium) do not provide a satisfactory fit for the results of international trials covering a wider range of site conditions, almost certainly because the British site data show strong co-linearities between pollutant concentrations and rainfall. Concentrations of SO_2 and NO_2 tend to be high in industrial and urban areas, and these are not situated in the wettest parts of the country. Also, in most British conditions, the contribution of sea salt to rates of corrosion, though appreciable, is not overwhelming,[48] but the same is not true, for instance, at tropical surf beaches, where copious salt spray is rapidly concentrated by evaporation. Further problems arise from the choice and definition of wetness parameters. Because of the historic development of atmospheric corrosion research, the most common measure of wetness has been the time spent at relative humidity greater than 80%. A large variety of alternative wetness parameters have, however, been used at one time or another, either because they worked better or because nothing else was available. These include times of actual wetness recorded by electrochemical devices, annual rainfall, annual average relative humidity, number of days on which rain fell and number of wettings.

In a review of predictive equations, it has been argued that, since there are reasonably good statistical correlations between these various quantities, any one of them could be used for deriving approximate empirical equations.[42] The possibilities for deriving useful equations along these lines do, however, depend on the availability of climatic data measured at the precise position of the specimens, rather than, for example, at the nearest airport, as is sometimes done. This is particularly important for corrosion trials on buildings, which are particularly liable to localised disturbances to both rainfall data and pollutant concentrations.

The most promising route to improved predictive equations appears to be to develop mechanistic models in terms of the complex physical, chemical and electrochemical processes involved in the deposition of corrosive agents and their subsequent reactions and then to introduce simplifications to allow the use in these equations of data available for sites of interest. Average pollutant concentrations can, for instance, be obtained from large-scale computer simulations, and could be used in economic calculations on the benefits of reductions in pollutant emissions.

A promising recent development has been the development of a theoretical model for zinc in terms of deposition velocities for pollutants and the solution chemistry of the corrosion products.[10]

In terms of present and future abatement strategies on an international level, it is important to know what the quantitative relationships (damage functions) are for a range of building stone types in a range of pollution and meteorological climates. The BERG report concluded that existing quantitative information was unreliable and that further studies were needed. Site studies in the UK in the NMEP and on an international scale (UNECE) have existed for 4 years. Results are beginning to confirm the major factors involved in stone decay, notably the SO_2 concentration (dry deposition), the hydrogen ion concentration in rainwater, and the volume of unpolluted rain falling on the stone.[11,49] Similar findings are reported in an American study although the range of pollution climates was limited.[50]

Additional studies have been made in the laboratory where combinations of pollutants (NO_x, SO_2) have been used at levels higher than atmospheric to determine damage functions under more controlled conditions than site exposure studies.[51, 52] Early results show little sign of synergism between NO_x and SO_2 but studies are continuing.

REFERENCES

1. Evans, U. R., *Corrosion and Oxidation of Metals.* Edward Arnold, London, UK, 1960, supplements, 1968 and 1976.
2. Shrier, L. L., ed., *Corrosion*, 2nd edn. Newnes-Butterworth, London, UK, 1976.
3. Rozenfeld, I. L., *Atmospheric Corrosion.* NACE, Philadelphia, PA, USA, 1972.
4. Building Effects Review Group, *The Effects of Acid Deposition on Building Materials in the UK*, Department of the Environment, HMSO, London, UK, 1989.
5. Brimblecombe, P., *The Big Smoke.* London, UK, 1987.
6. British Iron and Steel Research Association, *Fifth Report of the Corrosion Committee*, Special Report 21. Iron and Steel Institute, London, UK, 1938.
7. British Iron and Steel Research Association, *Sixth Report of the Corrosion Committee*, Special Report 66. Iron and Steel Institute, London, UK, 1959.
8. Brasher, D. M., Stability of the oxide film on metals in relation to inhibition of corrosion. I. Mild steel in the presence of aggressive ions. *Brit. Corrosion J.*, **2** (1967) 95.
9. Schikorr, G., Über den Mechanismus des atmosphärischen Rostens des Eisens. *Werkstoffe u. Korros*, **14** (1963) 69–80.
10. Spence, J. W. & Haynie, F. M., *Derivation of a Damage Function for Galvanised Steel Structures*, Atmospheric Research and Exposure Assessment Laboratory Report no. EPA/600D- 90/030. EPA, Research Triangle Park, NC, USA, 1990.
11. Butlin, R. N., Coote, A. T., Devenish, M., Hughes, I., Hutchens, C., Irwin, J. G., Lloyd, G. O., Massey, S. W., Webb, A. H. & Yates, T. J. S., Preliminary Results from the Analysis of Metal Samples from the National Materials Exposure Programme. *International Conference on Acidic Deposition, Glasgow, 16–21 September, 1990. Atmos. Environ.*, **26B** (1992) 199–206.
12. Butlin, R. N., Effects of air pollution on buildings and materials. *Proc. Royal Soc. Edinburgh*, **97B** (1991) 255–72.
13. Schaffer, R. J., *The Weathering of Natural Building Stones, DSIR Building Research Station Special Report* no. 18. HMSO, London, UK, 1932.
14. Whalley, W. B. & McGreery, J. P., Weathering. *Prog. Phys. Geog.*, (1985) 560–86.
15. Leary, E. A., *The Building Limestones of the British Isles.* HMSO, London, UK, 1983.

16. Leary, E. A., *The Building Sandstones of the British Isles*. BRE, Watford, UK, 1986.
17. Hart, D., *The Building Magnesian Limestones of the British Isles*. BRE, Watford, UK, 1968.
18. Hart, D., *The Building Slates of the British Isles*. BRE, Watford, UK, 1991.
19. Ross, K. D. & Butlin, R. N., *Durability Tests for Building Stone*. BRE, Watford, UK, 1989.
20. Del Monte, M., Sabbioni, C. & Vittori, O., Airborne carbon particles and marble deterioration. *Atmos. Environ.*, **15** (1981) 645–51.
21. Butlin, R. N., *et al.*, Research on limestone decay in the United Kingdom. In *Fifth International Congress on Deterioration and Conservation of Stone*, p. 537, et seq., 1985.
22. Brimblecombe, P., London air pollution, 1500–1900. *Atmos. Environ.*, **11** (1971) 1157–63.
23. Honeybourne, D. B. & Harris, B., Structure of porous building stone and its relation to weathering behaviour. In *Proceedings 10th symposium of the Colston Research Society*, p. 343.
24. Kulis, M., Technisch-ökonomische Auswertung der Korrosionsverluste an Stahlkonstruktionen in Gebiet Nordbohmen. *Werkstoffe u Korros.*, **28** (1979) 197–205.
25. Smith, R. A., *Air and Rain: the Beginnings of a Chemical Climatology*. MacMillan, London, UK, 1872.
26. The Watt Committee on Energy, *Air Pollution, Acid Rain and the Environment*, Report no. 18. Elsevier, London, 1989.
27. House of Commons Environment Committee, *Acid Rain*, Fourth Report Session 1983–84. HMSO, London, UK, 1984.
28. Bawden, R. J., Butlin, R. N., Coote, A. T. & Lloyd, G. O., Effects of Acid Deposition on Metals. *Colloque Europeén Corrosion-Dégradation-Bâtiments CSTB/CEFRACOR, CNIT, Paris, 13–16 November, 1990*, CSTB, Paris.
29. Bawden, R. J. & Ferguson, J. M., Trends in materials degradation rates in the UK. *Corrosion '87*, I Corr. S.T., Birmingham, pp. 382–402.
30. Johansson, L. G., A laboratory study of the influence of NO_2 and SO_2 on the corrosion of steel, copper, zinc and aluminium. *Electrochem. Soc. Extended Abstr.*, **8262** (1985) 221–2.
31. Lloyd, G. O., Lloyd, B. and Crouch, A. F., Development of two types of Continuous Monitor for Atmospheric Corrosion. *Corrosion '87*, I Corr. S.T., Birmingham, pp. 403–23.
32. Lloyd, G. O., The Use of Galvanic Monitors in Research on Atmospheric Corrosion. *Corrosion '90*, I Corr. S.T., Birmingham, pp. 11–44.
33. Ashurst, J. & Ashurst, N., *Stone Masonry in Practical Building Construction*, Vol. 1, English Heritage Technical Handbook. Gower Technical Press, 1988.
34. Price, C. A., *et al.*, A further appraisal of the lime technique for limestone conservation using a radioactive tracer. *Studies Conserv.*, **33** (1988) 1978–81.
35. *Metal Corrosion in the Atmosphere*, STP 435. ASTM, Philadelphia, PA, USA, 1968.
36. *Corrosion in Natural Environments*, STP 558. ASTM, Philadelphia, PA, USA, 1974.

37. *Atmospheric Corrosion of Metals*, STP 767. ASTM, Philadelphia, PA, USA, 1982.
38. *Degradation of Metals in the Atmosphere*, STP 965. ASTM, Philadelphia, PA, USA, 1988.
39. Ailor, W. H., ed., *Atmospheric Corrosion.* John Wiley, New York, USA, 1982.
40. *Dechema Corrosion Handbook*, Vol. 7. VCH Verlagsgesellschaft, Weinheim, Germany, 1990.
41. Edney, E. O., Stiles, D. C., Spence, J. W., Haynie, F. H. & Wilson, W. E., A laboratory investigation of the impact of dry deposition of SO_2 and wet deposition of acidic species on the atmospheric corrosion of galvanised steel. *Atmos. Environ.*, **20** (1986) 541.
42. Benarie, M. & Lipfert, F. W., A general corrosion function in terms of atmospheric pollutant concentration and rain pH. *Atmos. Environ.*, **20** (1986) 1947–58.
43. Mansfeld, F., New approaches to atmospheric corrosion research using electrochemical techniques. In *Corrosion Processes*, ed. R. N. Parkins. Applied Science, London, UK, 1982, p. 21.
44. United Nations Economic Commission for Europe, *Effects of Sulphur Compounds on Materials, including Historic and Cultural Monuments*, Report ENV/IEB/R16.
45. Umweltbundesamt (FRG), *Luftverschmutzung durch Schwefeldioxid.* Berlin Bericht.
46. Jorg, F., Schmitt, D. & Ziegahn, K. F., *Materialschaden durch Luftverunreinigungen*, Fraunhofer-Institut fur Treib und Explosivstoffe, Report UBA- FB- 84-106 08 010. 1985.
47. Mikhailovskii, Yu. N., Theoretical and engineering principles of atmospheric corrosion of metals. In *Atmospheric Corrosion*, ed. W. H. Ailor. John Wiley, New York, USA, 1982, pp. 85–105.
48. Ambler, H. R., Atmospheric salinity and metal corrosion at various places in Great Britain. *J. Appl. Chem.*, **10** (1960) 213–25.
49. Lipfert, F. W., Atmospheric damage to calcareous stones: comparison and reconciliation of recent experimental findings. *Atmos. Environ.*, **23** (1989) 415–29.
50. National Acid Deposition Assessment Programme, Acidic Deposition: state of Science and Technology Report 19. *Effects of Acid Deposition on Materials*, Washington, DC, 1990.
51. Johansson, L. G., A study of the corrosion of calcareous stones exposed to humid air containing SO_2 and NO_2. *Durability of Building Materials*, **5** (1968) 439–50.
52. Johnson, J. B., *et al*, Laboratory exposure systems to simulate degradation of building stone under dry and wet deposition conditions. *Atmos. Environ.*, **24A** (1990) 2585–92.

Chapter 13

Effects on Visibility, Weather and Climate

HELMUTH HORVATH
Institute of Experimental Physics, University of Vienna, Boltzmanngasse 5, Vienna, Austria

13.1 INTRODUCTION

All substances in the atmosphere, both gases and particles, interact with electromagnetic radiation, especially light. This interaction leads to easily observable optical effects such as appearence of haze, colored sunrise and sunset, blue or white sky, brown haze, visibility reduction amongst others. Interaction of gases and particles with the solar and terrestrial radiation can have an influence on the radiation budget and can lead to temperature changes. Also particles participate in the formation of clouds, therefore climate and weather modifications are possible if the particle concentration and properties change.

Atmospheric acidity directly or indirectly is involved in all the above-mentioned processes to a greater or lesser degree. In this chapter the contributions of atmospheric acidity will be estimated and discussed.

13.2 OPTICS OF THE ATMOSPHERE, SCATTERING AND ABSORPTION BY GASES AND PARTICLES

The interaction of light with any matter in the atmosphere, both particles and gases, can be described by two processes, which usually occur simultaneously.

(1) *Light scattering.* Under the influence of the electromagnetic field of an incident radiation, a gas molecule or particle becomes a source of secondary radiation, which is emitted in all directions with varying strength. For scattering, the total energy of the radiation is conserved, it is only distributed in different directions. All particles or gases scatter light.

(2) *Light absorption.* Particles or gases remove some of the energy of the electromagnetic radiation, which is finally converted into heat. Absorption only occurs if a system has different energy levels and if the energy difference between the levels agrees with the energy of the photon interacting with the molecule or particle. Depending on the number of possible states the light may be absorbed only at a few wavelengths, at broad bands, or almost continuously. For most light absorbing atmospheric particles absorption occurs at a broad wavelength range, since it is due to graphitic carbon. Water is transparent in the visible, but exhibits considerable light absorption in the infrared (IR) region. Light-absorbing substances have a complex refractive index (m).

These interactions of light with particles and gas molecules are determined by applying the Maxwellian equations to the propagation of electromagnetic waves in a medium containing gas molecules and particles. Exact solutions are only possible for particles of simple shapes, such as spheres, ellipsoids, shells;[1] for complicated irregular shapes of particles microwave analogy experiments permit the determination of the desired quantities.[2]

For considerations below, the following quantities, frequently employed in atmospheric optics will be used.

- *Volume scattering function* ($\beta(\psi)$). Consider a volume of dV of aerosol or gas illuminated by a beam of unpolarized light having a flux density E (Fig. 13.1). The scattered radiation is observed in a direction given by the scattering angle ψ. The flux $d\phi$ of light scattered in a solid angle $d\Omega$ in direction ψ is given by $d\phi = E\ \beta(\psi)\ dV\ d\Omega$ with $\beta(\psi)$ being the volume scattering function. The light may be partly polarized, but for the following this is not important.
- *Extinction coefficient* (σ_e), *scattering coefficient* (σ_s), *and absorption coefficient* (σ_a). The attenuation of a light flux passing through a layer of air of length x is obtained by the well-known exponential law (Beer–Lambert or Bouguer law) $\phi = \phi_o \times \exp(-\sigma_e \times x)$, with σ_e being the extinction or attenuation coefficient. It has the dimension of inverse length. The removal of light flux from the incident beam is both by scattering and absorption. The total amount of light scattered by the

particles or gas is obtained by integrating the volume scattering function over the full range of scattering angles. For transparent particles and gases the attenuation is only due to scattering. For absorbing particles and gases the attenuation due to absorption has to be added. It is usual to write the extinction coefficient (σ_e) as the sum of the scattering coefficient (σ_s) and the absorption coefficient (σ_a).

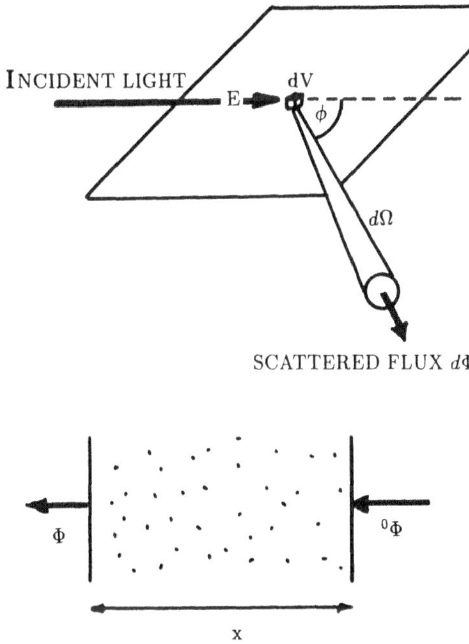

Figure 13.1. Definition of the volume scattering function (top) and the extinction or attenuation coefficient (bottom).

The volume scattering function and the extinction (scattering, absorption) coefficients of aerosols and gases can be calculated using standard procedures, which will not be described in detail here; the interested reader is referred to the relevant literature.[1, 3–5]

When interacting with electromagnetic radiation, the gas molecules can be considered as dipoles, emitting scattered light. For a gas with refractive index m and N molecules per unit volume the following formulae can be used (Rayleigh scattering):

$$\beta(\psi) = \frac{9\pi^2}{2N\lambda^4}\left(\frac{m^2-1}{m^2+2}\right)^2(1+\cos^2\psi)$$

and

$$\sigma_s = \frac{24\pi^3}{N\lambda^4}\left(\frac{m^2-1}{m^2+2}\right)^2 \times F_d \tag{13.1}$$

F_d is the depolarisation constant, which is close to unity.

For particles smaller than one-tenth of the wavelength similar formulae can be derived. If the particles are larger, a rigorous solution of the Maxwellian equations is possible for simple shapes such as spheres, ellipsoids or spherical shells. The result is obtained as an infinite series of functions, easily accessible to numerical calculations, and computer codes are available.[1,3] Usually the results of Mie calculations are the intensity function $i_1 = |S_1|^2$ and $i_2 = |S_2|^2$ and the efficiency factors Q_e, Q_s and Q_a. Their values depend on the refractive index of the material of the particles, the ratio of the radius of the particle to the wavelength, given by the size parameter $x = 2r\pi/\lambda$ and the scattering angle ψ. For N spherical particles per unit volume air with radius r, $\beta(\psi)$ and σ can be obtained from

$$\beta(\psi) = N\times\frac{\lambda^2}{4\pi^2}\times\left(\frac{i_1+i_2}{2}\right) \quad \text{and } \sigma_x = N\times r^2\pi\times Q_x \quad \text{with x} = \text{e, s, or a} \tag{13.2}$$

If the particles are polydisperse, the number size distribution function $n(r)$ gives the number of particles dN in the radius interval $[r, r+dr]$ by $dN = n(r)\times dr$. By replacing N by dN in eqn (13.2) and integrating over all particles size, the scattering function and extinction coefficient can be found:

$$\sigma_x = \int r^2\pi Q_x n(r)\,dr \quad \text{with x} = \text{e, s, or a} \tag{13.3}$$

13.3 OPTICAL CONSTANTS OF SUBSTANCES INVOLVED IN ATMOSPHERIC ACIDITY

In Table 13.1 the refractive index and the light scattering coefficient is shown for air and the precursor gases of atmospheric acidity. The scattering coefficient of the trace gases is negligible compared to the one of air. The only light-absorbing gas normally present in air is nitrogen dioxide (NO_2). It has a yellowish-reddish-brownish color, therefore high absorption in blue, and low absorption in red. In polluted air, normally its absorption is higher than the Rayleigh scattering of the air molecules.

Table 13.1
Optical properties of particle free air and oxides of sulfur and nitrogen

Gas	*Density*	*Wavelength (nm)*	*Scattering coefficient (Mm^{-1})*	*Absorption coefficient (Mm^{-1})*	*Reference*
Air	1.293 kg m^{-3}	450	26·44	–	5
		550	11·62	–	
		650	5·89	–	
SO_2	100 μg m^{-3} (= 38 ppb)	546	2.1×10^{-6}	–	6
NO	100 μg m^{-3} (= 78 ppb)	546	0.9×10^{-6}	–	6
NO_2	100 μg m^{-3} (= 52 ppb)	450	3.8×10^{-6}	71.2	6,7
		550	1.7×10^{-6}	17.2	
		650	0.9×10^{-6}	2.7	

The refractive indices of substances frequently found in atmospheric particles are given in Table 13.2; for crystalline substances the average refractive index is listed. The refractive index depends slightly on wavelength. Considering the other uncertainties in calculations of light extinction and scattering, e.g. replacing crystals by spherical particles, it is permissible to use an average refractive index for visible light. The refractive index of most inorganic materials is in the range between 1·45 and 1·6, and water has a lower refractive index. Many particles in the atmosphere are a mixture of different substances. For example, at high humidities salts become aqueous solutions, or particles that have undergone several transformation processes in clouds are inhomogeneous mixtures of several substances. In order to calculate the light scattering or extinction, an average refractive index is used. Several formulae to obtain the average refractive index exist;[12] none of them is theoretically well founded. Usually, volume additivity is assumed, i.e. if m_i and V_i/V are the refractive index and volume fraction of the ith substance, the average refractive index is obtained from

$$m = \sum m_i \times V_i/V \qquad (13.4)$$

Many of the substances forming particles in the atmosphere are crystalline, i.e. it is unlikely that all particles are spherical. Approximating particles as spherical may still give a reasonably good approximation, since in a large number of particles all orientations are possible.[13] For the considerations below, volumes containing at least 10^4 particles are used, therefore all orientations will scan but there is no general agreement on this subject.[2] At humidities above 60% most particles have some water on their surface, so that they can be considered as spheres or shells of spheres.

Table 13.2
Refractive indices of some substances forming atmospheric particles

Substance	*Density (kg m^{-3})*	*Refractive index* 450 nm	550 nm	650 nm	*Reference*
Substances involved in atmospheric acidity					
$(NH_4)_2SO_4$	1769	1·536	1·528	1·521	8
NH_4HSO_4	1780		1·482	(589 nm)	8
NH_4NO_3	1725	1·572	1·559	1·550	8
H_2SO_4	1841	1·435	1·430	1·425	8
Crustal materials					
$CaCO_3$	2930		1·586		
SiO_2	2200–2660		1·478		
Other substances					
Water	999	1·337	1·333	1·331	5
Organic matter	800–1100		1·44–1·60		
Urban aerosol[a]	1600		1·5–0·1		27
Soot[b]		1·75–0·45	1·75–0·44	1·75–0·43	9
	2000		1·96–0·66		10
			(1·20–1·35)–(0·1–0·22)		11
Soot aggregates	1000		1·56–0·47		10

[a] Calculated with the following assumptions: homogeneous mixture, using the volume mixing rule for the constituents listed for the Vienna aerosol in Table 13.4, and rounded to two significant figures.

[b] There is no general agreement on the optical data of soot. A selection of values is given in this table. For the calculations the data of soot aggregates have been used since soot particles usually have a fractal structure.

For a few model size distributions the spectral extinction and absorption coefficients of particles having a mass concentration of 100μg m^{-3} have been calculated. Results are shown in Fig. 13.2. Lognormal size distributions are a good representation of the size distributions of atmospheric particles. For

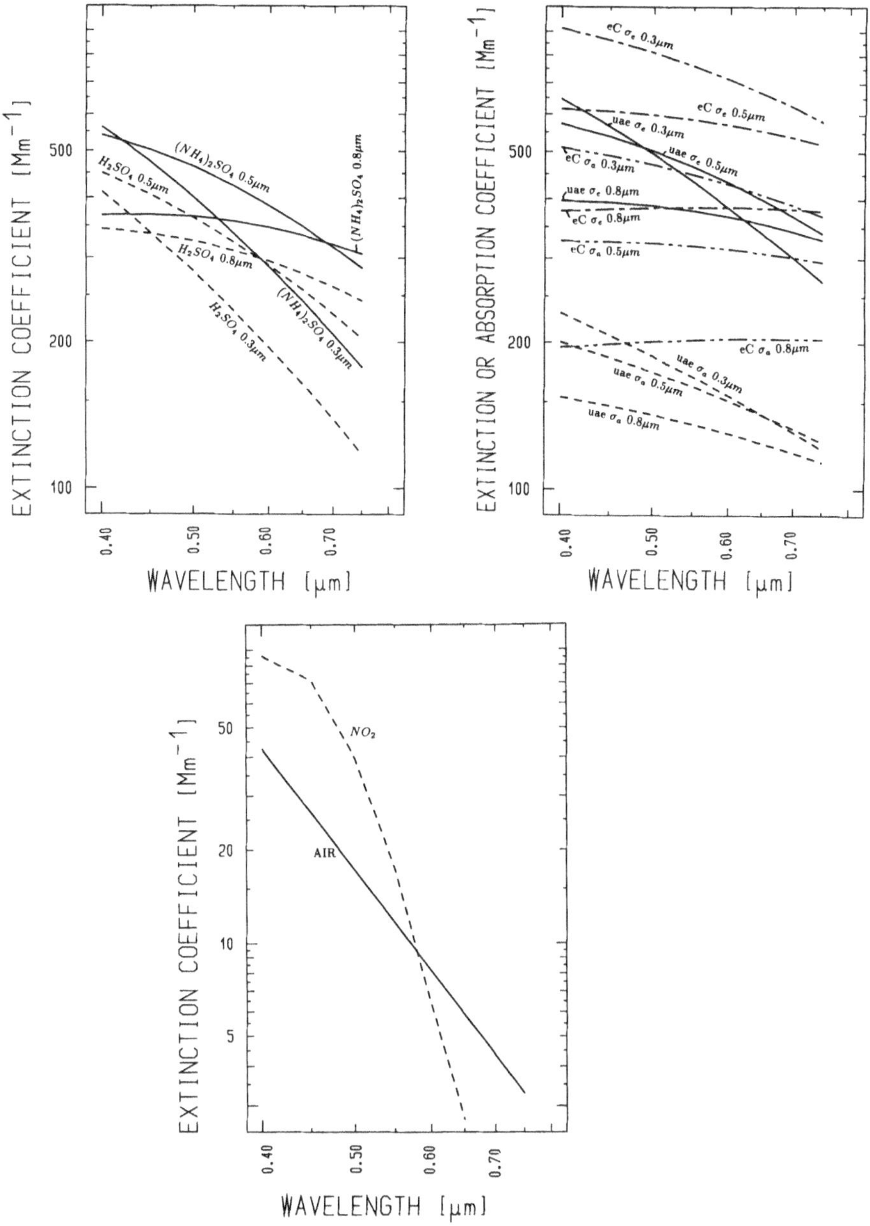

Figure 13.2. Spectral extinction coefficients of a suspension of particles (100 μg m^{-3}) having lognormal distributions with a geometric standard deviation of 1·7 and the geometric volume mean diameters given as parameters. (Upper left) $(NH_4)_2SO_4$ and H_2SO_4. NH_4NO_3 the latter is almost identical with $(NH_4)_2SO_4$, therefore not shown. (Upper right) Elemental carbon (eC– the main constituent of soot) and urban aerosol (uae), both the extinction coefficient (σ_e) and the absorption coefficient (σ_a) are given. (Bottom) Standard air and NO_2(100 μg m^{-3}). The refractiveindices for the substances are listed inTable 13.2.

this figure a geometric standard deviation of 1.7 has been used. The geometric volume mean radii have been chosen to be 0.3, 0.5 and 0.7 μm. The mean radii of atmospheric sulfate and nitrate particles are in this range;[14,15] the smaller ones are representative of 'fresh' particles, the larger ones of aged particles.

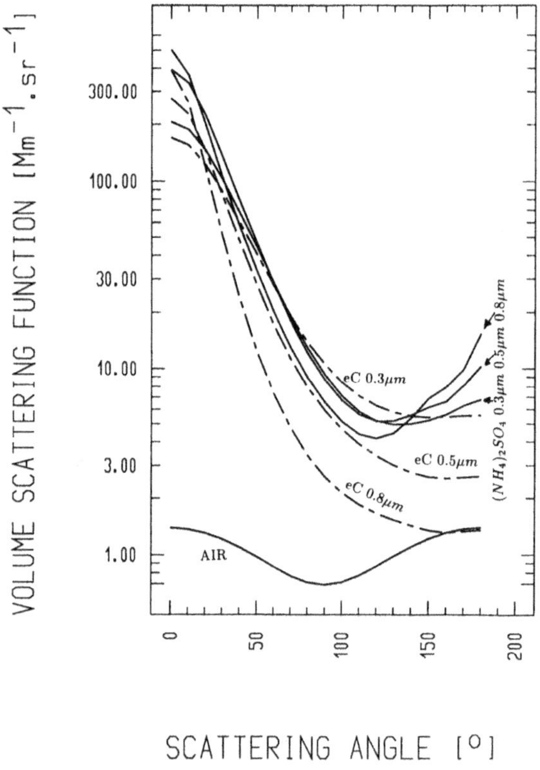

Figure 13.3. Volume scattering function of air and suspensions of particles having the same size distributions and refractive indices as in Fig. 13.2. All data are for a wavelength of 550 nm and for a mass concentration of 100 μg m^{-3}).

All spectral extinction coefficients show a decrease with increasing wavelength. This is more distinct for sulfates and nitrates which do not absorb light, especially for the size distributions with the mean diameter of 0·3 μm, where the Rayleigh scatter of the small particles is important. Ammonium sulfate ($(NH_4)_2SO_4$) and nitrate (NH_4NO_3) have very similar curves, since both refractive indices and densities are similar, sulfuric acid is less efficient in attenuating light, since it has both a higher density and smaller refractive index. The values for a 'dirty urban aerosol' are slightly higher but still show a similar trend. Soot has less wavelength dependence and due to the strong absorption, especially for particles with sizes below 0·1 μm, it also has higher extinction coefficients. The spectral extinction coefficient for air is considerably smaller than the extinction coefficients of the suspensions. It shows a strong dependence on wavelength (proportional to λ^{-4}). The extinction coefficient of NO_2 is also small compared to the one of the particles, but in the blue larger than for air. It is pure light absorption, the scattering of the gas is more than five orders of magnitudes less.

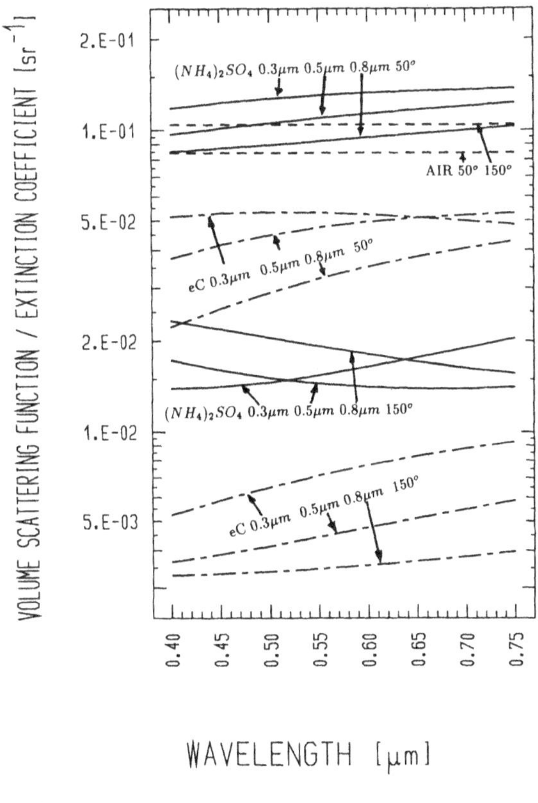

Figure 13.4. Ratio of the volume scattering function to the extinction coefficient as a function of wavelength. The same materials and size distributions as in the previous figures have been used.

For the same size distributions and materials the volume scattering function is shown in Fig. 13.3. All suspensions of particles have a considerable scatter in the forward direction and a minimum around 100°. For absorbing particles this is less distinct. Particle-free air has a completely different scattering function, with equal amounts of light scattered in the forward and backward direction and a decrease to half of this value at 90°.

The volume scattering function depends on wavelength much in the same way as the extinction coefficient. For later purposes, the ratio of the volume scattering function and the extinction coefficient $\beta(\psi)/\sigma_e$ will be needed. For the same size distributions as above, this is plotted as a function of wavelength in Fig. 13.4. The dependence on wavelength is small, and no definite trend can be seen.

13.4 VISIBILITY AND OTHER OPTICAL EFFECTS OF PARTICLES AND GASES

When looking through the atmosphere distant objects may appear less distinct, with little or no color, and sometimes even with a different color. This is caused by the interaction of gas molecules and particles with the light coming from the target and the daylight: light going from the target to the observer is attenuated and additional light from the sun, clouds, sky and ground is scattered into the eye of the observer. The further the target away, the less light comes from the target and the more light comes from the atmosphere.

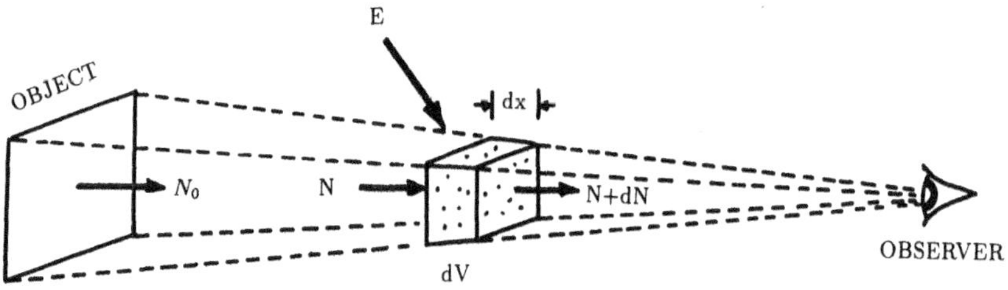

Figure 13.5. When looking at a target, a volume element (d*V*) attenuates light coming from the target and adds some light by scattering of light from the sun, the atmosphere and the ground.

Consider an observer looking at a distant object, having a radiance N_o (Fig. 13.5). The light flux emitted by this object towards the observer is partly scattered and absorbed, at a distance dx the amount of $d\phi = -\sigma_e \times \phi \times dx$ is lost. Since flux and radiance are proportional, the radiance is reduced by $dN_1 = -\sigma_e \times N \times dx$. On the other hand, the volume element dV of air is illuminated by the irradiance E, therefore a flux of $d\phi = E \times \beta(\psi) \times dV \times d\Omega$ is scattered by the particles. Since the volume element extending the distance dx in the horizontal has a cross-section of $dA = dV/dx$, the radiance dN_2 seen by the observer is $dN_2 = d\phi/(dA \times d\Omega) = E \times \beta(\psi) \times dx$. In total at the distance dx the radiance is changed by $dN = dN_1 + dN_2 = E \times \beta(\psi) \times dx - N \times \sigma_e \times dx$ with $N(0) = N_o$.

Since $\beta(\psi)$, E, and σ_e can vary with distance, an explicit solution of this equation cannot be given. Only the simplest case will be treated, i.e. that the atmosphere and illumination is homogeneous, i.e. $\beta(\psi)$, E, and σ_e are constant, therefore independent from the distance x from the observer. With this simplification one obtains

$$N(x) = E \times \frac{\beta(\psi)}{\sigma_e}[1 - \exp(-\sigma_e \times x)] + N_0 \times \exp(-\sigma_e \times x) \qquad (13.5)$$

If the observer looks towards the horizon, this can be considered as a target at infinity; the exponentials become zero, and the radiance of the horizon N_h is

$$N(\infty) = N_h = E \times \frac{\beta(\psi)}{\sigma_e} \qquad (13.6)$$

In order to see an object of radiance N, it has to have a difference δN to the radiance of its background N_b. For recognition only the relative radiance difference is of importance, i.e. $\delta N/N_b = C$, which is also called the contrast. The contrast can be both positive and negative, depending on whether the object is brighter or darker than the background. For absolute values of contrasts larger than the threshold ε the object is visible. For daylight vision the contrast threshold of the human eye is almost independent of the level of illumination. The traditional value first used in visibility theory is 0·02;[16] smaller values have been measured by later authors;[17,18] for many purposes a threshold of 0·05 is preferred.[19]

The contrast of a target of radiance N_o seen at distance x against the horizon as background is

$$C(x) = \frac{N(x) - N_h}{N_h} = \frac{N_o - N_h}{N_h} \times \exp(-\sigma_e \times x) = C_0 \times \exp(-\sigma_e \times x) \qquad (13.7)$$

The contrast at distance zero from the target (no interferences from the atmosphere) is called the intrinsic or inherent contrast, C_0. It depends both on the radiance of the target and the radiance of the horizon. During a day the object receives daylight from different directions, thus its radiance changes. Also the radiance of the horizon changes due to different scattering angles.

The distance V, where the object is just visible, is called the visibility, at this distance the contrast equals the contrast threshold. Therefore $|C(V)| = \varepsilon$, or

$$V = \frac{1}{\sigma_e} \times \ln \left| \frac{\varepsilon}{C_0} \right| \tag{13.8}$$

It depends both on the contrast threshold of the observer, the inherent contrast of the object and the extinction coefficient of the atmosphere. The first part of eqn(13.8) is solely due to atmospheric extinction, the second part contains properties of the object, the observer, the illumination and the way the light is scattered in the atmosphere. In order to be independent of these properties which are difficult to predict, a standard visibility V_{st} is introduced, being the visibility an observer with the standard contrast threshold of 0·02 will obtain with a perfect black object ($N_o = 0$.) In this case,

$$V_{st} = \frac{\ln 0.02}{\sigma_e} = \frac{3.91}{\sigma_e} \tag{13.9}$$

The visibility of eqn (13.8) is obtained from the standard visibility by

$$V = V_{st} \times \left[\frac{1}{3.91} \times \ln \frac{\varepsilon}{C_0} \right] \tag{13.10}$$

i.e.it has to be multiplied by a factor. Also for other visibility cases the observed visibility can be obtained by multiplying the standard visibility with a constant factor, which will depend on the specific viewing task.

In the following, only standard visibilities will be considered. Most of the other visibilities are obtained by multiplication of V_{st} with a factor. Therefore, if the standard visibility changes, the other visibility also changes by the same factor.

13.5 VISIBILITY AND ATMOSPHERIC ACIDITY

Since the standard visibility is proportional to the inverse extinction coefficient, the importance of any constitutent of the atmosphere can be estimated by calculating its extinction coefficient, or even better its ratio of extinction coefficient to the mass concentration. The mass extinction coefficient of particles depends strongly on their size distribution, generally particles in the radius range between 0·1 and 0·5 μm are most efficient, as can be seen in Fig. 13.6 for monodisperse particles. For particles of substances most commonly occuring in the atmosphere and having lognormal size distributions data are shown in Table 13.3. Both sulfates and nitrates are very efficient in attenuating light, since they usually have mean diameters between 0·3 and 0·6 μm. Freshly formed particles are slightly less efficient. Crustal materials usually occuring in the coarse mode are a factor of approximately 10 less efficient. The contribution of the precursor gases (except NO_2) can be neglected, see also Table 13.1.

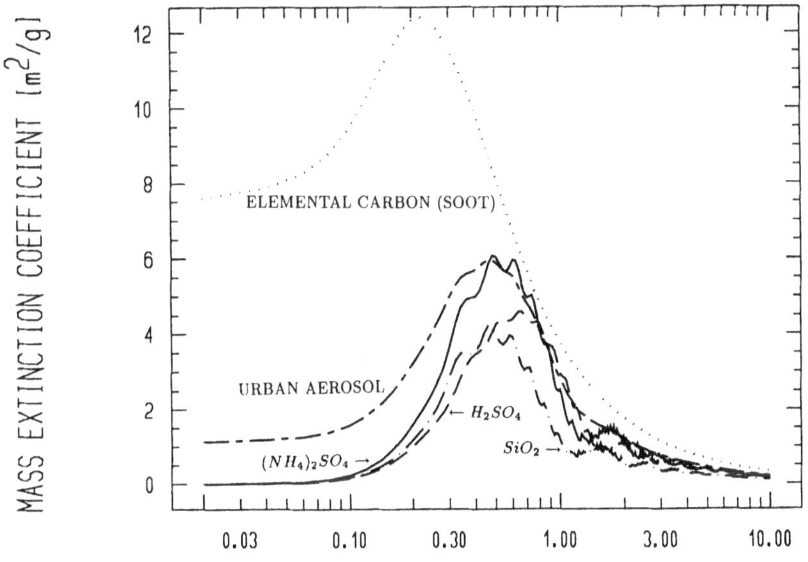

Figure 13.6. Mass extinction coefficient (ratio of extinction coefficient to mass concentration) for monodisperse particles of substances occurring in the atmosphere.

Using the values of Table 13.3 and published data on chemical composition of atmospheric aerosols, the contribution of each species can be calculated. A few examples are given in Table 13.4. Unfortunately, not all data on the chemical analysis of the airborne particulates include size distributions. Therefore reasonable values for the sizes had to be assumed, using

Table 13.3
Mass extinction coefficient of suspensions of particles and gases[a]

Substance	Diameter			
	0·3 μm	*0·5 μm*	*0·8 μm*	*5·0 μm*
$(NH_4)_2SO_4$	3·4	4·2	3·5	0·43
NH_4NO_3	3·8	4·6	3·7	0·44
H_2SO_4	2·3	3·2	3·1	0·41
$CaCO_3$	2·4	2·8	2·2	0·26
Silica	2·2	2·9	2·6	0·33
Soot	10·2	7·9	5·4	0·75
Organic material	5·8	7·3	6·3	0·76
Urban aerosol	4·3	4·7	3·8	0·51
Gases				
SO_2		2.13×10^{-8} (546 nm)		
NO_2	0.71 (450 nm)	0.17 (550 nm)	0.027 (700 nm)	

[a] The values are given in m^2g^{-1}, for the particles a lognormal size distribution with a geometric standard deviation of 1·7 and a mean volume diameter as given in the table was assumed.

the sum of the analyzed chemical species exists, and also some estimates had to be made on the composition of the fraction not analyzed.

The table covers both polluted and unpolluted areas, with visibilities ranging from 4 to 175 km. In all cases the contribution of sulfates and nitrates to the extinction coefficient is considerable and usually is at least 30%. Normally , the sulfates are more dominant but there are locations where the aerosol is nitrate dominated, see, for example, Phoenix, line 1. In remote regions the contribution of sulfates and nitrates is smaller, since the air molecules give a considerable contribution to σ_e; considering only the suspended particles the contribution of sulfates and nitrates is in the same range. The Arctic haze is dominated by sulfates, since the particles orginate mostly from industrial and urban regions in Europe, Northern America and Asia. Organic acids, even in the Amazonian basin seem to be of minor importance. On the one hand, the concentrations are low, and on the other hand they are usually in the gaseous state. (It has been reported that less than 2% of the formic and acetic acid of the Amazonian basin occur as particles[34]), where light extinction is negligible. Concentrations of organic acids measured in European conifer forest are between 12 and 21 μg m^{-3},[35] and there are indications, that at least part of it is in the particluate state, which could give a contribution comparable to (sulfates + nitrates).

Table 13.4
Examples from different locations in the world, giving the contribution of different species of particles to the light extinction coefficient of the atmosphere[a]

Location	*Visibility (km)*	*$(NH_4)_2SO_4$ (%)*	*NH_4NO_3 (%)*	*Elemental carbon (%)*	*Organic carbon (%)*	*Other substances (%)*	*Air (%)*	*NO_2 (%)*	*Footnote*
Upwind Vienna (rural)	32·2	33·3	3·0	26·5	19·0	8·6	9·6	ND	b
Vienna central	10·2	25·5	10·8	31·2	22·3	7·2	3·0	ND	c
Athens (Greece)	9·9	14·6	3·0	10·9	31·3	31·8	2·9	5.5	d
Arctic haze	41	73	ND	5·1	9·7	ND	12·2	ND	e
Japan summer (urban)	44·3	30·8	3·1	27·3	4·9	20·6	13·2	ND	f
Japan winter (urban)	11·4	8·6	9·2	31·8	23·2	23·8	3·4	ND	g
USA (remote)	175	13·1	ND	4·5	13·2	15·4	43·8	ND	h
Denver winter study	24·1	8·8	11·6	41·4	28·7	3·4	6·0	ND	i
Denver winter aerosol	ND	20·2	17·2	37·7	12·5	6·6	ND	5.7	j
Los Angeles clear day	41·8	12·8	4·4	10·8	42·1	3·2	12·4	14.1	k
Los Angeles heavy smog	9·4	12·1	1·6	15·5	45·5	14·4	2·8	8.1	l
Phoenix winter haze	9·9	4·9	14·7	34·8	29·9	10·2	10·2	ND	m
Santiago de Chile winter	4·15	23·1	6·7	21·7	ND	46·3	1·2	1.0	n
Amazonian Basin biomass burning	48·6	6·5	5·3 organic acids	14·0	60·0	ND	12·2	ND	o
Amazonian Basin boundary layer	53·5	3·4	3·3 organic acids	9·8	66·8	0.8	15·9	ND	p
'Average' urban	24·5	25·2	6·1	18·6	36·1	6·7	7·3	ND	q
'Average' non-urban	53·2	33·8	3·2	10·4	28·5	8·3	18·8	ND	q
'Remote'	133·8	20·2	3·1	0·4	10·0	26·5	39·8	ND	q

[a] The numbers presented in this table have been calculated using data available in the literature and with the following assumptions: dry particles, humidity growth of particles will be discussed below; fine particles have a mass mean diameter of 0·5 μm

(unless otherwise stated, or given in the reference); for elemental carbon a value of 0·3 μm has been used; any sulfate reported was assumed to be ammonium sulfate, nitrate was taken as ammonium nitrate; the difference between the substances given explicitly below and the total mass of the suspended particles is taken as other; if no further information was available, crustal particles were assumed to be coarse mode particles with 5 μm mass mean diameter; the visibility given is the standard visibility. ND means no data for the substance were available.

[b] Data taken from Ref. 20. Location Exelsberg, north-west of Vienna, no influence from local sources. The aerosol measured there is probably representative of rural Central Europe. Particles were sampled with cascade impactors for sizes between 0·1 and 6·5 μm aerodynamic equivalent diameter. The following masses were used: 9·6 μg m^{-3} of $(NH_4)_2SO_4$, 0·8 μg m^{-3} of NH_4NO_3, 3·2 μg m^{-3} of elemental carbon, 3·2 μg m^{-3} of organic carbon, 2·8 μg m^{-3} of other fine particles, 8·1 μg m^{-3} of crustal substances, no data for NO_2. Carbon was only analyzed totally, it was assumed that equal amounts of elemental and organic carbon were present.

[c] Data taken from Ref. 20. Location central Vienna. Same sampling procedure as *b*, the following concentrations were used: 23·1 μg m^{-3} of $(NH_4)_2SO_4$, 9 μg m^{-3} of NH_4NO_3, 11·7 μg m^{-3} of elemental carbon, 11·7 μg m^{-3} of organic carbon, 6·25 μg m^{-3} of other fine particles, 30·35 μg m^{-3} of crustal substances, no data for NO_2.

[d] Data from Ref. 21. Particles below 2·5 μm in diameter have been sampled, therefore there is no information on coarse mode particles. The following mass concentrations have been used to calculate values: 13·7 μg m^{-3} of $(NH_4)_2SO_4$, 2·6 μg m^{-3} of NH_4NO_3, 4·2 μg m^{-3} of elemental carbon, 16·9 μg m^{-3} of organic carbon, 43·3 μg m^{-3} of other fine particles, 126 μg m^{-3} of NO_2.

[e] Data from Ref. 22. Location is Ny-Ålesund, Spitzbergen. Size distributions were measured with impactors in the size range between 0·01 and 0·5 μm, the size distributions had a peak around 0·25 μm. The aerosol was measured during an arctic haze event, and was mainly from industrial areas. Only the following chemical data were available: 20·4 μg m^{-3} of $(NH_4)_2SO_4$, 0·474 μg m^{-3} of elemental carbon, 1·6 μg m^{-3} of other fine particles (assumed to be organic carbon).

[f] Data from Ref. 23. measured in the Chiba Prefecture, near Tokyo in July 1987. Chemical composition of the fine particles (below 2 μm) and coarse particles were reported. Since no data on carbon were available, measurements reported in Ref. 24 of the same area were used to obtain the necessary data, by multiplying the mass of the TSP with the fraction obtained from the other dataset. The following values were used: fine particles: 6·33 μg m^{-3} of $(NH_4)_2SO_4$, 0·66 μg m^{-3} of NH_4NO_3, 2·3 μg m^{-3} of elemental carbon, 0·41 μg m^{-3} of organic carbon, 2·86 μg m^{-3} of other fine particles. Coarse particles: 3·43 μg m^{-3} of $(NH_4)_2SO_4$, 1·76 μg m^{-3} of NH_4NO_3, 0·8 μg m^{-3} of elemental carbon, 1·76 μg m^{-3} of organic carbon, 29·9 μg m^{-3} of other coarse particles, no data for NO_2.

[g] Same location and sampling as *f*, data for Dec 1987: fine particles: 6·8 μg m^{-3} of $(NH_4)_2SO_4$, 6·7 μg m^{-3} of NH_4NO_3, 10·5 μg m^{-3} of elemental carbon, 10·5 μg m^{-3} of organic carbon, 24·9 μg m^{-3} of other fine particles. Coarse particles: 2·0 μg m^{-3} of $(NH_4)_2SO_4$, 1·56 μg m^{-3} of NH_4NO_3, 1·7 μg m^{-3} of elemental carbon, 3·4 μg m^{-3} of organic carbon, 28·4 μg m^{-3} of other coarse particles. No data for NO_2.

[h] Data taken from Ref. 25. Location Spirit Mountain, Nevada, elevation 1500 m, with no influences from nearby local sources. Representative for the arid south-west

of the US. Contributions calculated by means of the scattering efficiencies given by the authors. Concentrations are 1·61 μg m^{-3} of $(NH_4)_2SO_4$, 0·1 μg m^{-3} of elemental carbon (estimate by the present author), 0·84 μg m^{-3} of organic carbon, 0·15 μg m^{-3} of other fine particles, 1·4 μg m^{-3} of crustal substances, no data for NO_2.

[i] Winter study of 1978,[26] average of all data. Size distribution of the fine aerosol gave a mass mean diameter of 0·3 μm. Concentrations of the fine particles: 4·1 μg m^{-3} of $(NH_4)_2SO_4$, 4·9 μg m^{-3} of NH_4NO_3, 6·4 μg m^{-3} of elemental carbon, 7·5 μg m^{-3} of organic carbon, 2·5 μg m^{-3} of other fine particles (obtained by subtracting the sum of the above listed masses from the mass obtained by integration of the impactor data given in Fig. 13.3). Coarse particles: 0·8 μg m^{-3} of $(NH_4)_2SO_4$, 0·42 μg m^{-3} of NH_4NO_3, 2·1 μg m^{-3} of elemental carbon, 3·7 μg m^{-3} of organic carbon, no values for other coarse particles or crustal substances, no data for NO_2.

[j] Data from Ref. 27.

[k] Los Angeles 'clear day' from Ref. 28. Fine particles with mean diameter of 0·3 μm having the following concentrations: 3·54 μg m^{-3} of $(NH_4)_2SO_4$, 1·23 μg m^{-3} of NH_4NO_3, 0·99 μg m^{-3} of elemental carbon, 6·78 μg m^{-3} of organic carbon, 9·1 μg m^{-3} of crustal substances, 76·9 μg m^{-3} of NO_2.

[l] Los Angeles 'heavy smog' from Ref. 28. Fine particles with mean diameter of 0·3 μm having the following concentrations: 14·9 μg m^{-3} of $(NH_4)_2SO_4$, 1·79 μg m^{-3} of NH_4NO_3, 6·37 μg m^{-3} of elemental carbon, 32·74 μg m^{-3} of organic carbon, 12·8 μg m^{-3} of other fine particles, 65·48 μg m^{-3} of crustal substances, 200 μg m^{-3} of NO_2.

[m] Data from Ref. 29. Only fine particles were reported, mean diameter of 0·3 μm assumed: 5·7 μg m^{-3} of $(NH_4)_2SO_4$, 15·2 μg m^{-3} of NH_4NO_3, 13·4 μg m^{-3} of elemental carbon, 20·2 μg m^{-3} of organic carbon, 12·6 μg m^{-3} of other fine particles, 5·3 μg m^{-3} of crustal substances, no NO_2 reported.

[n] Data from Ref. 30. Elemental carbon concentrations were not determined; therefore data for elemental carbon were taken from Ref. 31, a different study performed in the same area but at a different time (spring instead of winter), therefore elemental carbon most likely is too low: 36·9 μg m^{-3} of $(NH_4)_2SO_4$, 13·8 μg m^{-3} of NH_4NO_3, 20 μg m^{-3} of elemental carbon, 87·3 μg m^{-3} of crustal substances, the remaining difference to the total suspended particulates of 122·5 μg m^{-3} was assumed to be half fine particles (organic carbon), and half other coarse mode particles, 57 μg m^{-3} of NO_2 .

[o] Data collected from an aged plume of biomass burning, flight 16 of Ref. 32. Particles assumed to have a mass mean diameter of 0·3 μm, with the following chemical species: 1·54 μg m^{-3} of $(NH_4)_2SO_4$, 1·0 μg m^{-3} of organic acids (MSA, formates, oxalates, acetates), 1.1 μg m^{-3} of elemental carbon, 8.3 μg m^{-3} of organic carbon, no other fine particles, no crustal substances, no NO_2.

[p] Data of the boundary layer of Ref. 32: 0·75 μg m^{-3} of $(NH_4)_2SO_4$, 0.4 μg m^{-3} of organic carbon, 0·25 μg m^{-3} of other fine particles, no crustal substances, no NO_2.

[q] Data from Ref. 33. Averages of many locations listed in the paper, only fine particles. Urban and non-urban locations mostly in the US, remote from 11 sampling locations in a marine environment (other particles dominated by sea salt). Mean diameters assumed to be 0·3 μm for urban and non-urban, 0·5 μm for remote. The concentrations for urban, non-urban and remote were: 11·8, 7·3 and 1·4 μg m^{-3} of $(NH_4)_2SO_4$, 2·5, 0·6, and 0·2 μg m^{-3} of NH_4NO_3, 2·9, 0·75 and 0·014 μg m^{-3} of elemental carbon, 9·9, 3·6 and 0·5 μg m^{-3} of organic carbon, 4·9, 2·75 and 2·7 μg m^{-3} of other fine particles, no data for NO_2.

Another important contributor to the light extinction is elemental carbon. It is a by-product of any combustion and thus is normally associated with sulfates and nitrates. Due to its light absorption soot is far more efficient in attenuating light than any other substance (see Table 13.3). Also important is organic carbon. In an urban environment it can originate from emissions of gasoline-powered cars (e.g. Athens[21]), for Los Angeles 30% of the mass of the fine particlulates is reported to be organic carbon, originating from charcoal broilers, road dust and non-catalyst vehicles, 50% of the organic carbon is secondary aerosol.[36] Since the density of organic materials is low, they are more efficient in scattering light than, for example , sulfates.

The precursor gas NO_2, although present in concentration comparable to the particulates, is of minor importance for the attenuation of light and the visibility.

It should be noted that all figures given in Table 13.4 are for the dry aerosol. With increasing humidity sulfates, nitrates and other products of atmospheric acidity grow to a larger size due to uptake of water in comparison to other particles of the aerosol, therefore the contribution of sulfates and nitrates to the attenuation of light can easily be more than 60 % (see also below).[37]

As already mentioned, the precursor gas SO_2 has no effect and NO_2 little effect on the attenuation of light in the atmosphere or the visibility. After being converted to sulfate the effects on visibility can be considerable. A hypothetical example shall demonstrate this: assume a steam generator, burning 1000 t of fuel per hour with 1% of sulfur. Without desulfurisation 10t of sulfur are emitted per hour. Assume all sulfur is converted to $((NH_4)_2SO_4)$, which gives 1000 t day^{-1}. This product shall be distributed evenly over an area of 500 × 500 km up to an altitude of 300 m, giving a mass concentration of 13·4 μg m^{-3}. For a mass mean diameter of 0·5 μm of the particles, they will have an extinction coefficient of 56·3 Mm^{-1}, which is a considerable contribution. If the area normally has visibilities of 100 km, the addition of this aerosol will make the visibility drop to 41 km. At high humidities the effect will be even higher (see below). The eastern US has a large number of sources of SO_2 and visibilities became lower over the last decades, most likely for this reason, since extinction coefficients and sulfur emissions correlate excellently, especially in the humid summer-time, since humidity growth increases the contribution of sulfates as well as nitrates which are also formed during combustion of sulfur-containing fuel.[38]

In many industrial regions the air has a dirty appearance, sometimes a brownish haze can be seen. This can in part be due to the presence of NO_2. When looking at a dense haze layer, the radiance is that of the horizon. The radiance of the horizon has been determined previously and is given by eqn (13.6). It is obtained by multiplying the irradiance of the sun with the ratio of the volume scattering function and the extinction coefficient. Values for β/σ_e are given in Fig. 13.4. For all the aerosols used for the calculations, β/σ_e shows little dependence on wavelength. Therefore, the spectral energy distribution of the light coming from the haze layer is similar to the one of the sun, i.e. the haze layer will appear in the same color as sunlight. The weak dependence of the radiance of a layer of haze (relative to the irradiance) on wavelength has a very simple explanation: for transparent particles the extinction is due to scattering and an increased scattering by the particles means also increased extinction, therefore β/σ_e remains constant. To some extent this is also the case for soot, since its refractive index does not depend strongly on wavelength.

Using β/σ_e for one scattering angle means that light reaching a person looking at the haze will only be scattered once. This is a useful first approximation, but multiple scattering cannot be neglected. In that case scattering processes with different angles have to be considered and less dependence on wavelength occurs. Therefore the haze layer will have almost the same color as sunlight.

The volume scattering function depends very much on the scattering angle, therefore a haze layer is bright when observed in the forward scattering, i.e. towards the sun, and dark or gray when observed away from the sun. This is both the case for non-aborbing aerosol particles such as $(NH_4)_2SO_4$, as well as for absorbing particles such as soot but for absorbing particles the amount of scattered light is lower in all directions.

In contrast to this an infinite layer of air shows little difference in radiance at different angles, since the volume scattering function of air only varies by a factor of two (see Fig. 13.3). Therefore, the clean air above a haze layer looks bright when looking away from the sun, making the haze appear dark, even if it contains only non-absorbing particles such as sulfates. For a non-horizontal sight path, pure air does not form an optically thick layer, therefore the scattering of the light dominates the extinction, and the sky appears blue. If the air contains additional particles, the layer has a higher optical density, and more white light is added to the blue sky, giving the well-known appearance of a whitish sky in hazy areas. Since sulfates and nitrates dominate the light extinction, they are also responsible for the discoloration of the sky to a large degree.

For an atmosphere containing NO_2 this is different, since the light absorption of this gas amounts to more than 99.999% of the extinction coefficient. Scattering can be neglected and for an aerosol with volume scattering function β, extinction coefficient σ_e and absorption coefficient σ_a of NO_2 the radiance of an optically thick layer of haze is :

$$N = E\times\frac{\beta(\psi)}{\sigma_e+\sigma_a} \tag{13.11}$$

The ratio of the spectral radiance of a haze layer containing NO_2 to a haze layer containing no NO_2 is shown in Fig. 13.7. There is a decrease in radiance in the blue part of the spectrum, the haze layer with NO_2 will appear in a different color. If darker than the surroundings, it will appear brown, if brighter it will be seen as orange or yellow. For comparison the relative radiance of a brown plume emitted by a power plant is shown too. The plume was barely visible, when leaving the stack, since almost no particulates were contained in the flue gases. After a few kilometers the plume appeared brown, probably due to the oxidation of nitric oxide (NO) to NO_2. The radiance is given relative to the radiance of the horizon without plume and was measured with a telephotometer from a distance of approximately 15 km. The dependence on wavelength is very similar to the calculated values given in the same figure.

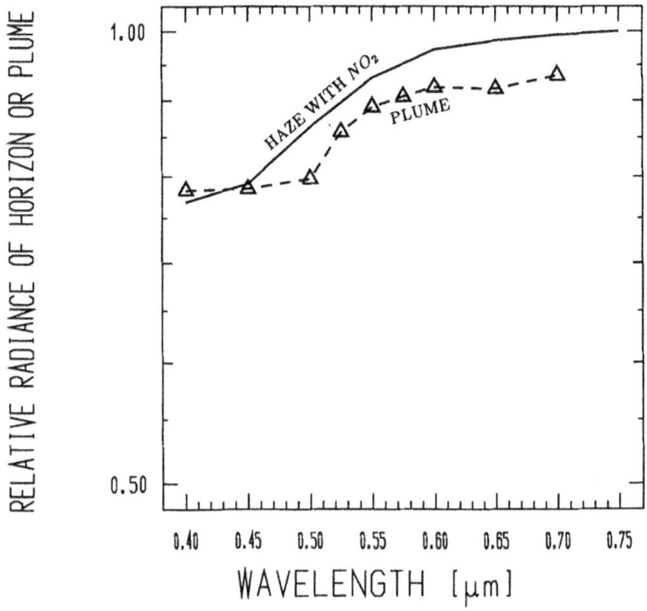

Figure 13.7. Radiance of a layer of haze containing 100 μg m^{-3} particles with a volume mean diameter of 0·5 μm and 200 μg m^{-3} NO_2 relative to the radiance of the same haze without NO_2. The radiance of a brownish power-plant plume measured with a telephotometer is shown relative to the radiance of the horizon. The similarities are striking (unpublished data measured by the author).

A rigorous treatment of the color of brown hazes needs a quantification of the 'brownness' of light seen. This can be done by determining the chromaticity co-ordinates (which can be understood as the intensities of three colored lights which mixed additively will give the same color). It is also important to consider multiple scattered light. The interested reader is referred to Refs 39–41. Detailed considerations show, that in all cases NO_2 causes a brownish appearance of haze. To some extent the aerosol, especially light-absorbing particles, can also give a positive or a negatve contribution.

13.6 THE EFFECT OF HUMIDITY ON AEROSOL PARTICLES

It is a well-known fact, that some salts, e.g. NaCl, when stored at high humidities will become moist, i.e. form a solution. Many particles in the atmosphere have the tendecy to become droplets when surpassing a certain relative humidity. Depending on the material, the humidity of the phase transition can be fairly low, e.g. 33% for magnesium chloride ($MgCl_2$), a minority compoment of sea salt, or high, e.g. 91% for sodium sulphate (Na_2SO_4).[42,43] When dealing with pure substances, the phase transition is abrupt and large hysteresis effects occur; for the atmospheric aerosol, containing a mixture of many particles and substances, the increase of a parameter such as the scattering coefficient is continuous with increasing humidity and is well documented by experiments (see, for example, Refs 44–46).

The atmospheric aerosol contains both particles consisting of pure substances, which show a well-defined phase transition[47] as well as particles where no exact point of the phase transition can be found, which has led to the introduction of the concept of mixed particles.[48]

Almost no change in the properties occurs at humidities below 60%. At high humidities the majority of the particles contain water, therefore it is reaonable to call it a wet aerosol. Size increases can be measured by collecting size-segregated samples of the wet and the dry aerosol (heated 5°C above ambient) with a cascade impactor[49] and growth factors up to five have been observed. Therefore, with increasing humidity the optical properties of the particles can change considerably. It should be noted that there are also materials, which will show no growth with increasing humidity, e.g. fresh soot, many organic materials, and insoluble mineral particles. [43,50]

A large piece of salt exposed to moist air will absorb water vapor and form a solution, until an eqilibrium state is reached and the saturation vapor pressure $e_{sat,s}$ over the solution equals the vapor pressure in the air. It is customary to use the water activity a_w defined as the ratio of the equilibrium vapor pressure over a plane surface of the solution to the equilibrium vapor pressure over a plane surface of pure water $e_{sat,w}$ therefore,

$$a_w = \frac{e_{sat,s}}{e_{sat,w}} \tag{13.12}$$

For ideal solutions, the water activity is the mole fraction of the water in the solution. For most substances of interest this is not the case, therefore one has to rely on empirical determinations. A large collection of data can be found.[51] Activities for a few substances are shown in Fig. 13.8 as a function of molality (moles of solute per kg of solvent).

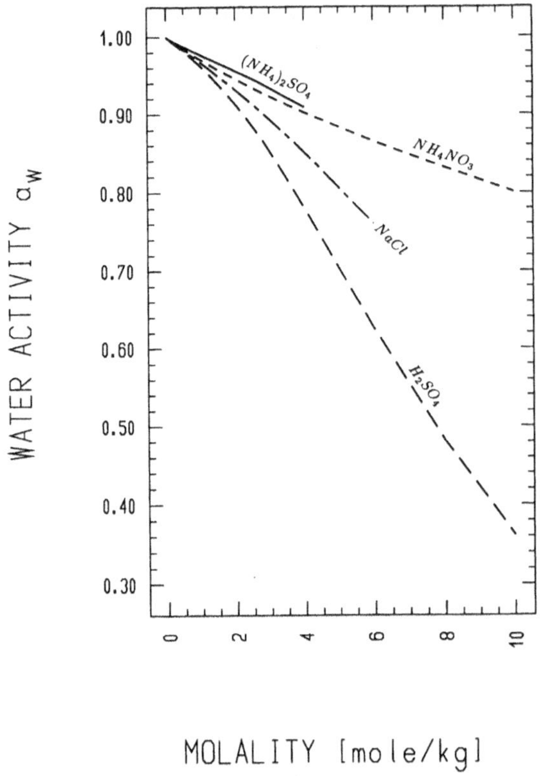

Figure 13.8. Water activity (a_w) for different salts as a function of molality (moles of solute per kg of solvent).

If the ambient air has a certain relative humidity an electrolyte in equilibrium will reach the concentration, where its water activity equals the relative humidity. For small particles the equilibrium vapor pressure is obtained by multiplying the value over a plane surface with the curvature correction term obtained from the Kelvin equation (see, for example, Ref. 52), yielding

$$\text{Relative humidity} = a_w \times \exp\frac{2M\sigma}{RT\rho r} \qquad (13.13)$$

with M the molecular weight of water, σ its surface tension, R the gas constant, I the absolute temperature and ρ the density of the liquid.

From this equation the equilibrium radius for a given humidity can be obtained by solving this equation with respect to r. But it must be borne in mind that surface tension, density and water activity vary with the radius of the solution droplet, since an increase in size means uptake of water, and thus different values for a_w, ρ and σ. Fortunately for particles larger than 0·1 μm the exponential function is 1·01 or less, so if, for example, critical supersaturations are not considered the exponential expression can be neglected.

With these simplifications the radius of a particle is obtained in the following way: for a given relative humidity the molality M of the solute is found by equating the water acitvity with the relative humidity, for example, from Fig. 13.8. The mass of the solute m_s is obtained by multiplying the molality with the molecular weight M_w, therefore $m_s = \text{M}\times M_w$. The mass of the water m_w in the solution is $m_w = 1000$ g by the definition of the molality. Therefore, the mass of the humid particle (m_h) is $m_h = m_s + m_w$. Assuming volume additivity, the volume v_h and the radius r_h of the humid particle are obtained from the volume v_o and r_o of the dry particles by

$$\frac{v_h}{v_0} = 1+\frac{\rho_m}{\rho_w}\times\frac{m_w}{m_m} \quad \text{and} \quad \frac{r_h}{r_0} = k = \left(1+\frac{\rho_m}{\rho_w}\times\frac{m_w}{m_m}\right)^{1/3} \qquad (13.14)$$

with ρ_w and ρ_m being the densities of water and the solute.

The radii of all particles increase by the same factor k given by Ref. 16, but only under the limiting conditions of particle radii larger than 0·1 μm and humidities below 95 %.[49] If the dry particles have a number size distribution $n(r_o)$, i.e. in a radius interval dr the number dN equals $dN = n_o(r_o) \times dr_o$, and all particles grow by the factor k, the number dN remains the same, but radius interval change by the factor k, therefore $dN = n(r_h/k) \times dr_h/k$, which means, that the number size distribution $n_h(r)$ of the humid aerosol is

$$n_h(r) = n\,(r/k)/k \qquad (13.15)$$

For a representation on logarithmic intervals d$\ln r$ does not change if all the particle radii are increased by a factor, therefore the number distribution curve is just moved parallel in the horizontal direction. Obviously, the volume and the mass of the particles increase. Examples for the dry and wet mass size distributions of selected substances are shown in Fig. 13.9. One can see that substances with a lower a_w grow to larger particles, and that NH_4NO_3 has a higher size increase.

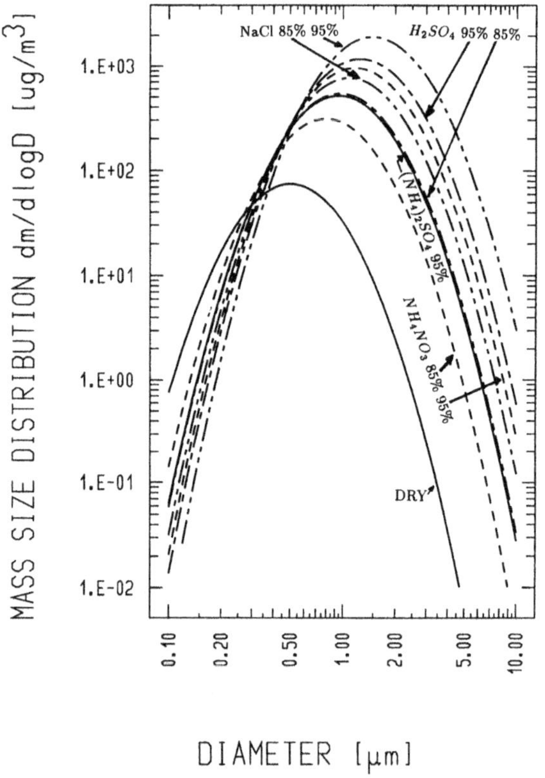

Figure 13.9. Change of mass size distribution of a dry aerosol containing 100 μg m^{-3} with humidity.

Of great importance for atmospheric optics is the increase in extinction coefficient with increasing humidity. This can easily be calculated by using the number size distribution for the increased particles, eqn 13.15, the refractive index from eqn 13.4 and inserting in eqn 13.3. Calculated increases of extinction coefficients for a few model size distributions with pure substances are shown in Table 13.5

Table 13.5
Extinction coefficient of aerosols formed of pure substances at different humidities[a]

Humidity (%)	$(NH_4)_2SO_4$	NH_4NO_3	H_2SO_4	$NaCl$
0	423 (100%)	458 (100%)	324 (100%)	360 (100%)
85	423 (100%)	1267 (276%)	1792 (553%)	2060 (572%)
95	1839 (434%)	3072 (671%)	3329 (1027%)	4024 (1117%)

[a] A lognormal size distribution of the dry particles with a geometric volume mean diameter of 0·5 μm and a geometric standard deviation of 1·7 and a mass concentration of 100 μg m^{-3} of the dry aerosol has been used. The extinction coefficient is given in Mm^{-1}.

One can see a considerable increase in extinction coefficient, as could also be expected from experimental data, see, for example, Refs 44–46 and 49. Particles that are easily soluble in water grow much faster than organic materials, minerals or soot, e.g. Arizona road dust shows no increase in particle size up to humidities of 95 %.[53] When grouping data of the Denver winter study in periods with relative humidities below 40 % and above 70 % relative humidity, an increase in diameter has been measured, which amounts to 1·45 and 1·68 for sulfates and nitrates and 1·23 and 1·16 for elemental carbon and hydrocarbons.[26] Assuming the extinction coefficient to be proportional to the square of the radius, the importance of sulfates and nitrates has roughly doubled with respect to carbon. In an aged aerosol particles containing elemental carbon show a size increase with increasing humidity, since soot particles are also incorporated in particles containing sulfate (internally mixed, see Ref. 54). Therefore, one could say that atmospheric acidity is indirectly also responsible for the humidity growth of light absorbing particles.

13.6 INFLUENCE ON WEATHER AND CLIMATE

To a large fraction the end-product of emitted SO_2, NO_x, or organic acids are particles which are eventually deposited on the surface of the earth. In between, the gases and particles can undergo various transformations and can participate in processes, which may have an influence on weather and climate. Acidic particles may actively participate in in-cloud processes possibly changing the droplet size or the stability of clouds and inducing rainfall. Light scattering and absorption by acidic particles and gases can alter the radiation balance of the earth and induce temperature changes and thus climatic changes.

Since all acidic particles are hygrosopic, they acquire water with increasing humidity. In order to form a cloud droplet, the particle has to surpass its critical radius, when surpassing the critical supersaturation in the cloud. After this the particle grows to infinity, or until the available humidity is used up. The critical supersaturation strongly depends on the size of the particle and its chemical composition. For example, a 0·1 μm particle of $(NH_4)_2SO_4$ has a critical radius of 1·32 μm at a supersaturation of 0·00053, a NaCl particle of the same size needs a supersaturation of 0·00037 to reach the critical radius of 1·92 μm. If the initial particle size is 1·0 μm, the corresponding critical supersaturations and sizes are 0·000016, 0·000012, 4·5 μm and 6·07 μm respectively.[46] In a cloud or during cloud formation the supersaturation is not constant, so particles may grow and shrink alternately. It may even be that one kind of particle will grow, and the other will shrink (in the above example this would be at a supersaturation of 0·005). Whether a particle surpasses the critical radius depends also on the time-span available for the growth. Due to this complex behavior it is difficult to make exact predictions on the influence of particles on the cloud droplet formation. Generally, it can be said, that particles requiring a lower supersaturation will have a greater chance to become activated and form cloud droplets. Non-activated particles will still grow with increasing humidity, and form the interstitial aerosol. Although not growing to cloud droplets they may compete for the available water vapor, increase the stability of the clouds and even decrease precipitation.

In order to estimate the influence of atmospheric acidity on cloud formation and maybe rainfall experimentally, one has to compare conditions with and without acidic particles. Unfortunately this is next to impossible, since there are both natural and man-made particles in the atmosphere, and due to the lifetime of about 10 days[55] particles produced at one location can be found even thousands of kilometers away. Another possibility for estimating the contributions of atmospheric acidity, is to analyze historic weather data, since without doubt, the concentration of acidic particles in the atmosphere has increased with the industrialization, e.g. demonstrated by ice core samples taken in Greenland,[56,57] showing a three-fold increase in sulfates in the ice between 1890 and now. When taking weather records of the northern hemisphere, one can see no significant change in precipitation over the last century, although sulfates and nitrates have increased drastically. On the other hand, the various experiments trying to artificially introduce rain have not been very successful. Downwind of paper mills an increase in rain has been observed, but it is not certain, whether this is due to the emitted giant particles or the moisture and heat release of the plant. One can therefore conclude, that it appears likely, that increased presence of acidic particles may have some effect on rain (both positive and negative) on a local scale, but has little or no effect on a global scale.

Radiation from the sun is scattered and absorbed by the particles and water droplets in the atmosphere. For particles larger than 0·1 μm most of the light is scattered in the forward direction (see Fig. 13.3), therefore a large part of the sunlight eventually will reach the ground. A certain amount is scattered in the backwards direction, and therefore is radiated into space. If the particles are light absorbing, part of the light energy will be used to heat the atmosphere and less light is scattered. Since particles originating from atmospheric acidity do not absorb light, their presence will decrease the amount of direct sunlight, increase the skylight, but decrease the net amount of solar radiation available, since part of the radiation is scattered into space. Therefore, a cooling could be expected with increasing particle numbers originating from atmospheric acidity.

The albedo of clouds, especially dense clouds is very high, due to multiple scattering of light on the droplets, as can easily be observed from an airplane flying above the clouds. Even a dense cumulus, which looks dark from underneath is bright white from above. Using many simplifications not discussed here the optical thickness of a cloud extending a vertical distance z and containing N particles with diameter d is given by $\tau = 0{\cdot}5 d^2 \pi N z$. Assume a fixed liquid water content of a cloud. If the number of droplets increases due to a larger amount of cloud condensation nuclei, the droplet size decreases. Since the diameter of the droplets is inversely proportional to the third root of the droplet number, the optical thickness of the cloud increases proportionally to the third root of the droplet number. An increase in the particle concentration thus increases the optical density of the cloud, and thus also its albedo. The same is also true if the particles are not activated and grow to interstitial aerosol particles of a few micrometers. Therefore, an increased number of deliquescent particles will increase the amount of solar radiation reflected from the cloud into space. This effect is larger for thin clouds, which occur more frequently. Experimental evidence for this exists too, since cloud reflectivity differences have been observed from space in the vicinity of shiptracks, which can at least in part be an effect of sulfur emitted with the exhaust gases.[58]

This fact has led to the postulation, that the biosphere may act as a self-regulating mechanism to keep a constant temperature on earth:[59] planktonic algae produce dimethylsulfide (DMS), which is oxidized in the atmosphere and eventually forms sulfate particles, which serve as cloud condensation nuclei. If the temperature on earth rises, e.g. due to greenhouse gases, the productivity of the phytoplankton increases, the number of condensation nuclei increase, decreasing the flux of sunlight, and eventually decreasing the temperature. According to this hypothesis, the earth's temperature is mainly regulated by the temperature-dependent DMS emissions of the microorganisms, which thus could be considered as the earth's thermostat.

Although it sounds very reasonable, that acidic particles of natural origin control the temperature, it is not yet possible to prove or disprove this hypothesis, since only a limited amount of data, such as the temperature dependence of DMS emissions, and the fate of the sulfate particles in air and cloud are available. Therefore, objections against this hypothesis have been published, see, for example, Ref. 60. The main argument against the thermostat hypothesis is the large contribution of anthropogenic sulfur in the Northern hemisphere: 22 TgS year^{-1} of marine biogenic sulfur, 32 TgS year^{-1} of terrestrial biogenic and 98 TgS year^{-1} of anthropogenic sulfur are emitted. For the Southern hemisphere the numbers are 26, 16 and 6 TgS year^{-1}. Presently, the Nothern hemisphere is dominated by anthropogenic sulfur and has a three-times higher emission rate, whereas the Southern hemisphere is dominated by biogenic sulfur. Therefore, one would expect a higher cloud albedo in the Northern hemisphere and lower temperatures. But annual zonal means of the cloud albedo have shown, that the albedo is lower in the Northern hemisphere, despite the higher sulfate concentrations. On the other hand, a comparison of the sea surface and marine air temperatures indicate a lower temperature of the Northern hemisphere by about 0·3°,[61] which would be in agreement with Ref. 59. The lower cloud albedo in the Northern hemisphere could also be explained by the presence of light-absorbing soot, a product of all combustion processes. When incorporated in a sulfate particle it can be even more absorbing than when mixed externally.[62] The non-observed cooling of the atmosphere due to the postulated increase in cloud albedo due to increased sulfur emissions could therefore be compensated by the heating effect of the greenhouse gases and soot.

This short discussion shows that a variety of possible mechanisms exist. At present it is not known which mechanism dominates, so one is limited to speculations and hypotheses. Further research will show which assumptions are realistic and which are not. But already, one can clearly see the important role of atmospheric acidic particles in weather and climate.

ACKNOWLEDGEMENT

Part of this work was supported by the Fonds zur Förderung der wissenschaftlichen Forschung in Österreich, grant numbers 4849 and 7616 and by the US EPA.

REFERENCES

1. Bohren, C. F. & Huffman, D. R., *Absorption and Scattering of Light by Small Particles*. John Wiley, New York, USA, 1983.
2. Zerull, R. H., Giese, R. H., & Weiss, K., Scattering measurements of irregular particles vs Mie Theory. SPIE 112, *Optical Polarimetry*, **112** (1977) 191–9.
3. Reist P. C., & Wilson, W., An interative model for predicting scattering functions of polydisperse aerosols. Paper presented at the International Conference on Aerosols and Background Pollution, University College, Galway, Republic of Ireland, 13–15 June 1989.
4. Kerker, M., *The Scattering of Light and Other Electromagnetic Radiation*. Academic Press, New York, USA, 1969.
5. McCartney, E. J., *Optics of the Atmosphere, Scattering by Molecules and Particles*. Wiley, New York, USA, 1976.
6. Landolt-Börnstein *Zahlenwerte und Funktionen*, Vol. 2, part 8. Springer, Berlin, Germany, 1962, ed. K.H. Hellwege and A.M. Hellwege pp. 873–4.
7. Hodkinson, R. J., Calculation of color and visibility in urban atmospheres polluted by gaseous NO2. *Int. J. Air and Water Poll.* **10**, 137–144, 1966.
8. Landolt-Börnstein *Zahlenwerte und Funktionen*, Vol. 2, part 8. Springer, Berlin, Germany, 1962, ed. K.H. Hellwege and A.M. Hellwege p. 261.
9. Jaenicke, R., Properties of atmospheric aerosols. In *Numerical Data and Functinal Relationships in Science and Technology*, New series, Springer, Berlin, Vol. 4b, ed. Börnstein, 1988, p. 417.
10. Ouimette, J. R. & Flagan, R. C., The extinction coefficient of multicomponent aerosols. *Atmos. Environ.*, **16** (1982) 2045–19.
11. Batten, C. E., Spectral optical constants of soot, from polarized angular reflection measurements. *Appl. Optics*, **24** (1985) 1193–9.
12. Chylek, P., Srivastava, V., Pinnick, R. G. & Wang, R. T., Scattering of electromagnetic waves of composite spherical particles: experiment and effective medium approximations. *Appl. Optics*, **27** (1988) 2396–404.
13. Chylek, P., Grams, G. W. & Pinnick, R. G., Light scattering by irregular randomly oriented particle. *Science*, **193** (1976) 480–2.
14. Meszaros, E., *Atmospheric Chemistry*, Elsevier, Amsterdam, The Netherlands, 1981.
15. Whiby, K. T. & Svendrup, G. M., *The Character and Origin of Smoke*, ed. G. M. Hidy & P. K. Mueller, Wiley, New York, USA, 1980, Table 12.
16. Koschmieder, H., Theorie der horizontalen Sichtweite. *Beitr. Phys. d. Atmos.*, **12** (1925) 33–53 and 171–181.

17. Blackwell, H. R., Contrast thresholds of the human eye. *J. Opt. Soc. Amer.*, **36**,(1946). 624–643,
18. Gorraiz, J., Habenreich, T. & Horvath, H., Der Einfuß kleiner Farbdifferenzen auf die Kontrastschwellen des menschlichen Auges bei verschiedenen Lichtarten. *Die Farbe*, **32/33** (1987) 355–79.
19. Johnson, R. W., Daytime visibility and nepholometer measurements related to its determination. *Atmos. Environ.*, **15** (1981) 1835–45.
20. Puxbaum, H., Ellinger, R., Baumann, H. & Wolpenka, B., Untersuchungen der Schadstoffvorbelastung für Wein. In *Die Luft in und über Wien*, eds H Puxbaum & A. Hack, Symposium 1983. Technical University of Vienna, Austria, 1983, pp. 42–54.
21. Scheff, P. A. & Valiozis, C., Characterisation and source identification of respirable particulate matter in Athens, Greece. *Atmos. Environ.*, **24A** (1990) 203–11.
22. Heintzenberg, J. & Covert, D. S., Chemically resolved submicrometric size distribution and external mixing of the Arctic haze aerosols. *Tellus,* **39B** (1987) 374–82.
23. S. Uchiyama. Seasonal variation in size for major ionic species in the atmospheric aerosol. *J. Japan Soc. Air Pollut.,* **25** (1990) 77–84.
24. Arai, H., Samata, M. & Shirasuna, Y., Research Report of Source Contribution of Suspended Particles, Yokohama City, Japan, 1989.
25. White, H. & Macias, S., Light scattering by haze and dust at Spirit Mountain, Nevada. *In Visibility and Fine Particles,* ed C.V. Mathai, Air and Waste Management Assoc. Pittsburgh, 1990, pp. 914–22
26. Countess, R.J., Cadle, S.H., Groblicki, P.J. & Wolff, G.T., Chemical analysis of size segregated samples of Denver's ambient particlulates. *J. Air Pollut. Control Assoc.*, **31** (1981) 247–52.
27. Groblicki, P.J., Wolff, G.T. & Countess, R.J., Visibility reducing species in the Denver 'Brown Cloud'. I. Relationships between extinction and chemical composition. *Atmos. Environ.*, **15** (1981) 2473–84.
28. Seinfeld, J.H., *Atmospheric Chemistry and Physics of Air Pollution.* Wiley, New York, USA, 1986, p. 297.
29. Solomon, P.A. & Moyers, J.L., A chemical characterization of wintertime haze in Phoenix, Arizona, *Atmos. Environ,*, **20** (1986) 207–13.
30. Vega, V. de la, Fuentes, H., Ortiz, J. & Prendez, Y.M., Balance de masa en aerosoles atmosphericos de Santiago de Chile. *Bol. Soc. Chil. Quim.*, **32** (1987) 187–98.
31. Horvath, H. Trier, A. & Catalan, L. The relationship between outdoor and indoor soot concentration in an urban environment. *J. Aerosol Sci.*, **20** (1989) 1273–6.
32. Andrae, M.O., *et al.* Biomass burning emissions and associated haze layers over Amazonia. *J. Geophys. Res.*, **93** (1988) 1509–27
33. Heintzenberg, J., Fine particles in the global troposphere. A review. *Tellus*, **41B** (1989) 149–60.
34. Andreae, M.O., Talbot, R.W., Andreae, T.W. & Hariss, R.C., Formic and acetic acid over the central Amazon region, Brasil. 1. Dry season. *J. Geophys. Res.*, **93** (1988) 1616–24.

35. Bingmer, H. Fischer, M., Ernst, L., Kesselmier, J., Andreae, M.O. & Helas, G., Volatile organic acids and ammonia measurements in a spruce forest. Paper BIATER 24, presented at the EUROTRAC Symposion, Germisch Patenkirchen, Germany, April 1990.
36. Hildemann, L. M., Mazurek, M. A. & Cass, G. R., Distinctive features of fine organic aerosols emitted from various combustion sources. Paper ID5, presented at the AAAR 1989 annual meeting at Reno, NV, USA, 9–13 October, 1989.
37. Appel, B. R., Tokiwa, Y., Hsu, J., Kothny, E. L. & Hahn, E., Visibility as related to atmospheric aerosol constituents. *Atmos. Environ.*, **19** (1985) 1525–34.
38. Husar, R. B. & Wilson, W. E., Trends of seasonal haziness and sulfur emissions over the eastern U.S. In *Visibility and Fine Particles,* Air and Waste Management Association, Pittsburgh, USA, 1990, pp 318–27.
39. Horvath, H., On the brown color of atmospheric haze. *Atmos. Environ,* **5** (1971) 333–44.
40. Husar, R.B. & White, W.H., On the color of the Los Angeles smog. *Atmos. Environ.* **19** (1976) 199–204.
41. Sloane, C.S. Contribution of soot and NO_2 to the discoloration of urban skies. *Atmos. Environ.* **22** (1988) 2021–31.
42. Pruppacher, H.R. & Klett, D., *Microphysics of Clouds and Precipitation.* D. Reidel, Dordrecht, The Netherlands, 1980, p. 84.
43. Twoomey, S., *Atmospheric Aerosols.* Elsevier, Amsterdam, The Netherlands, 1977, p. 81.
44. Covert, D.S., Charlson, R.J. & Ahlquist, N.C., A study of the relationship of chemical composition and humidity of the light scattering by aerosols. *J. Appl. Meteorol.* **11** (1972) 968–76
45. Fitzgerald, J.W., The relative contribution of fluctuations in relative humidity and particulate concentrations to the variability of the scattering coefficient over the North Atlantic. *Atmos. Environ.,* **14** (1980) 71–70.
46. Hänel, G., The properties of atmospheric aerosol particles as a function of the relative humidity at the thermodynamic equilibrium with the surrounding moist air. In *Advances in Geophysics,* Vol. 19, Academic Press, New York, USA, 1976, pp. 73–188.
47. Rood, M.J., On the metastability of atmospheric aerosols. PhD Dissertation, University of Washington, WA, USA, 1985
48. Winkler, P., The growth of atmospheric aerosol particles as a function of the relative humidity. II. An improved concept of mixed nuclei. *J. Aerosol Sci.*, **4** (1973) 372–87.
49. Berner, A., Haze and its relation to atmospheric accumulation aerosol. *Sci, Total Environ.,* **86** (1989) 251–63.
50. Heintzenberg, J. & Covert, D.J., On the distribution of physical and chemical particle properties in the atmospheric aerosol. *J. Atmos. Chem.* 383–97 (1990).
51. Laube, M. & Höller, H. Condensation and sublimation. In Landold Börnsein, *numerical data and functional relationships in Science and Technology,* New Series, Springer Berlin, Val. 4b, ed. G. Fischer. p. 46, 1988
52. Pruppacher, H.R. & Klett, D. *Microphysics of Clouds and Precipitation.* D.

Reidel, Dordtecht, The Netherlands, 1980, p. 139.

53. Lundgren, D.A. & Cooper, D.W., Effect of humidity on light scattering methods of measuring particle concentrations. *APCA Journal* **19** (1969) 243–7.
54. Ferek, R.J., Lazrus, A.L. & Winchester, J.W., Electron microscopy of acidic aerosols collected over the northeastern United States. *Atmos. Environ,* **17** (1983) 1545–61.
55. Jaenicke, R., Atmospheric aerosols and global climate. *J. Aerosol Sci.*, **11** (1980) 577–88.
56. Neftel, A., Beer, J., Oeschger, H., Züecher, F. & Finkel, R.Z., Sulfate and nitrate concentrations in snow from south Greenland 1895–1978. *Nature*, **314** (1985) 611–13.
57. Mayewski, P.A. *et al.* Sulfate and nitrate concentrations from a south Greenland ice core *Science*, **232** (1986) 975–7.
58. Twomey, S.A., Piepgrass, M. & Wolf, T.L., An assessment of the impact of pollution on global cloud albedos, *Tellus*, **36B** (1984) 356–66.
59. Charlson, R.J., Lovelock, J.E., Andreae, M.O. & Warren, S.G., Oceanic phytoplankton, atmosheric sulfur, cloud albedo and climate. *Nature*, **326** (1987) 655–61.
60. Schwartz, S.E., Are global cloud albedo and climate controlled by marine phytoplakton? *Science*, **336** (1988) 441–5.
61. Wigley, T.M.L., Possible climatic change due to SO_2 derived cloud condensation nuclei. *Nature*, **339** (1989) 365–7.
62. Heintzenberg, J. Light scattering parameters of internal and external mixtures of soot and non absorbing material in the atmosphere. In *Proceedings of the Conference on Carbonaceous Particles in the Atmosphere*, ed. T. Novakov, CONF 7803201. University of California, Berkeley CA. 1978 pp. 278–81.

Chapter 14

Technologies of Control

RICHARD M. DOWNER
Department of Chemistry, UMIST,
P.O. Box 88, Manchester, M60 1QD, UK

and

MIROSLAV RADOJEVIĆ*
Department of Fuel and Energy,
University of Leeds, Leeds, LS2 9JT, UK

14.1 INTRODUCTION

The major anthropogenic source of gaseous acidic deposition in the atmosphere is the combustion of fossil fuels for electrical power generation and transportation, and the implementation of any acid rain control strategy presupposes the availability of effective technologies of control. These may include the minimising of pollution at source by modifying the combustion process, by removing the pollutants from the fuel or the effluent gas, and by minimising the effects of acidic pollution in receptor areas (e.g. liming of lakes). Figure 14.1 illustrates the various stages at which controls can be applied. The acidity problem may also be abated by switching to alternative (non-fossil fuel) sources of energy (e.g. nuclear, solar, wind, etc.).

The control of acidic precursors from stationary and mobile sources involves primarily sulphur dioxide (SO_2) and the oxides of nitrogen (NO_x). Current control strategies aim to reduce the emissions of these gases at source. However, many other pollutants released by the same, or other sources, may contribute to the acid rain problem through their involvement

* Present address: Department of Chemistry, University of Brunei Darussalam, Bandar Seri Begawan 3186, Brunei Darussalam.

in the general chemistry of the troposphere (Chapter 3). For example, many compounds may oxidise SO_2 or give rise to species capable of reacting with SO_2. A discussion of control technologies for these pollutants is beyond the scope of the present review.

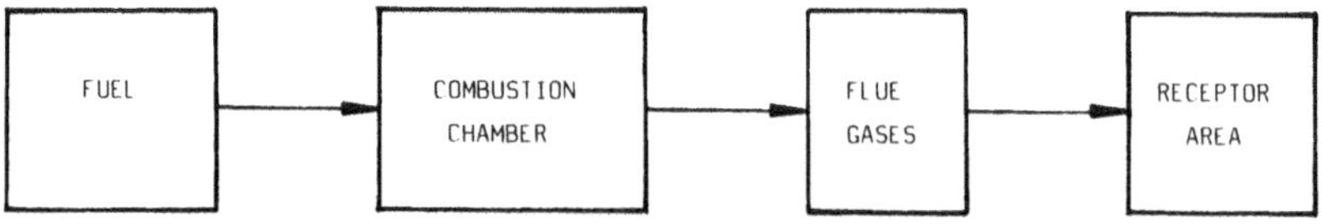

Figure 14.1. Various stages at which controls can be applied to abate the acid rain problem.

14.2 STATIONARY SOURCES

The absence of pollution controls from stationary sources in the first half of the century was reflected in the poor air quality of many cities with the occasional incidence of specific disasters associated with acidic 'smog', most notably in London in 1952. This led to the implementation of the 1956 Clean Air Act which primarily aimed to eliminate smoke pollution by designating 'smoke-free' zones. Furthemore, as part of the Clean Air Act, a memorandum on chimney heights was issued by the Department of the Environment.

The chimney stack may be viewed as a pollution-control device enabling the dispersion and mixing of the effluent gases with ambient air. Consideration must be given to a variety of factors (e.g. meteorology, topography, effluent properties, etc.) when designing chimney stacks, and the effective chimney height is usually related to some of these factors by means of a formula. Increasing the height of the chimney stack (H) results in a decrease in the ground-level concentration (C) of pollution ($C \propto 1/H^2$) and the use of taller chimney stacks, together with the implementation of 'smoke-free' zones, resulted in a significant improvement in urban air quality and the elimination of severe acid 'smog' incidents. The higher stacks enabled pollution to be dispersed away from the source areas; however, the pollution was now transported by air currents to distant, unpolluted regions giving rise to the current concerns of acid rain. This has led to stringent legislation on emissions of acidic gases in many industrialised countries and the deployment of pollution-control technologies by industry.

The technology for controlling particulate air pollution is fairly well developed and it is not discussed in detail here because only a minor fraction of acidifying species may be associated with particles in the flue gas.

Furthermore, in view of the stringent legislation on smoke, which has been in force for three decades, many stationary sources are equipped with effective particle-control devices.

The reduction and control of pollution from stationary sources may be made by several methods.[1–3] There are economic, technical and environmental advantages to furnishing a chemical plant or power station from the outset. Retrofitting is another matter and is a major control in deciding the adopted method of cutting polluting acid emissions. In emphasis there are four main routes that may be considered, consisting of the use of low-pollutant fuels, the prevention of pollutant formation during the combustion process, the screening of the pollutants from exhaust and flue gases, and energy conservation. This section concentrates on the techniques for removing gaseous pollutants from exhaust and flue streams from industrial plants.[4]

14.2.1 Flue gas desulphurisation (FGD)

Flue gases can be scrubbed to remove gaseous pollutants. FGD is a straightforward and widely used method for reducing emissions of SO_2 and is currently used in over a thousand plants world-wide. The technique involves the spraying or scrubbing of the flue gases with a chemical absorbent that removes much of the SO_2 produced during the combustion process. There are three types of FGD: wet, dry, and a combination of the two. Table 14.1 summarises the main commercially viable processes currently in use. Of these, wet processes predominate.

14.2.1.1 Wet scrubbing FGD methods

Wet scrubbing processes were developed and first introduced in Japan and the US in the early 1960s. Wet processes can further be divided into those that involve the regeneration of the absorbent that is then recirculated, and those involving no recovery of the absorbent. Non-recovery methods dominate the FGD processes and use either lime or limestone, sodium alkali, or dilute sulphuric acid (H_2SO_4) as an absorbent and can remove 90–95% of the SO_2 formed during combustion.

The main advantage of the wet non-recovery processes are that they have comparatively low installation costs and can be easily retrofitted dependent upon local conditions.

The disadvantage with wet FGD is the waste by-products, usually in the form of a thick sludge, which is difficult to dehydrate, store, handle and dispose of. A typical 1000 MW power plant, burning coal with a 3·5% sulphur content, fitted with wet limestone FGD produces about 225 000 t of sludge annually. Disposal of the sludge poses a great problem. The sludge may also be contaminated with heavy metals from the fly ash, and may pose a threat to the local hydrological network.

Table 14.1
Summary of the main commercially viable desulphurisation processes

Wet processes	
Process by / Supplier	*Absorbent*
ASARCO	Dimethylaniline
Babcock–Hitachi	Limestone
Babcock–Kawasaki	Hydrated lime
Badische Mascinenfabrik	Hydrated lime
Bechtel	Seawater
Chemico	Magnesium oxide
Chisso Engineering	Ammonium hydroxide
Fläkt-Hydro	Seawater
Fläkt	Limestone
Hodogaya Chemical	Sodium lye
Ishikawajima Harima	Limestone
Kellogg	Lime
Lurgi Sulphadine	Xylidine
Mitsubishi	Ammonia
Mitsubi SS	Various
Norsk	Seawater
Osaka Sode	Sodium lye
Thyssen Eng. GmbH–Mitsubishi	Limestone
Wet–dry processes	
Process by / Supplier	*Absorbent / Adsorbent*
Davy McKee (Wellman Lord)	Sodium sulphite
Exxon	Ammonia / reduction
Fläkt Drypack	Hydrated lime
Hitachi	Ammonia / reduction
Uhde-Kobe Steel S.C.R.	Lime / coal
Leisgang	Lime
Lurgi Kureha	Sodium acetate / reduction
Niro Atomiser	Hydrated lime
Saarberg Hoelter	Hydrated lime
Salzgitter Lummus	Hydrated lime
Steinmüller Mitsui	Lime / magnesium oxide
Whellabrator Fyre	Lime / ammonium salts
Dry processes	
Process by / Supplier	*Adsorbent*
Bergbau-Forschung / Uhde	Activated carbon
Bischoff	Limestone
Knauf	Limestone / hydrated lime
Linde	Organic adsorbents
Lurgi Sulphacid	Activated carbon
Mining Research	Activated carbon
Steuler	Hydrated lime
Walther & Co.A.G.	Lime / ammonia

The use of naturally occurring water with slight alkalinity has also been considered. Battersea Power Station was a major example, whereby Thames river water fortified with the addition of chalk was used as the scrubbing liquor.[5] The water was then passed back to the Thames hence, arguably removing the problem of waste disposal.

Scrubbing of the flue gases with seawater (or water taken from estuaries) has also been used to significantly reduce SO_2 emissions.[6–9] The spent water is returned to the sea (or estuary) where arguably it is thought that the large diluting effect of the sea negates any environmental consequences of waste disposal.

Most wet FGD processes use lime or limestone as the scrubbing reagent. Limestone systems are more economic than lime due to there being no energy costs for calcination and it is also possible to obtain gypsum ($CaSO_4 . 2H_2O$), a useful building material, as a by-product of the reaction:

$$SO_2 + CaCO_3 \rightarrow CaSO_3 + CO_2 \quad (14.1)$$

$$CaSO_3 + 2H_2O + {}^1/_2O_2 \rightarrow CaSO_4 . 2H_2O \quad (14.2)$$

However, the performance of limestone systems is limited by reagent reactivity and low soluble alkalinity. The use of additives to enhance the performance has been studied.[10] Adipic acid and dibasic acid (a by-product from adipic acid manufacture) have been used as a buffering agent to reduce the dissolved alkalinity limitation. The enhanced limestone FGD system results in improved SO_2 removal (50–84%), but overall limestone utilisation is increased. The main advantage is that limestone rather than the more expensive and energy-intensive lime can be used to obtain a similar scrubbing efficiency.

In the Sodium–Alkali method,[9,11] SO_2 is scrubbed with a solution of sodium carbonate (Na_2CO_3), sodium bicarbonate ($NaHCO_3$) and sodium hydroxide ($NaOH$) producing large quantities of liquid waste which has to be treated before disposal. New chemical methods have allowed the water content of the sludge to be reduced and its compressive strength improved.

Recovery methods are less popular, initially being more expensive to install and more energy intensive in operation. Aromatic amine–water mixtures have been used in the scrubbing of combustion gases from smelters with high concentrations of SO_2.[4] In the ASARCO (American Smelting & Refining Company) process SO_2 is absorbed in dimethylaniline to form dimethylaniline sulphate which can then be regenerated by steam and the addition of sodium sulphide. The resultant SO_2 can be dried by washing with 98% H_2SO_4 and then compressed for storage. The Lurgi–Sulfidine process is similar but uses xylidine as the chemical absorbent.

Pressure-hydrated dolomitic limestone with hydrated lime is also used as a wet scrubbing agent, and has been found to be a more effective FGD reagent than lime or limestone alone.[12] Magnesium sulphite ($MgSO_3$) is used as the main scrubbing agent which reacts with the SO_2 and calcium oxide (CaO) (in the fly ash) to form magnesium sulphate ($MgSO_4$) and calcium sulphate ($CaSO_4$). The gypsum is removed to a settling pool. The magnesium sulphite is regenerated by reaction of the $MgSO_4$ (and the remaining $CaSO_4$) with the calcium hydroxide ($Ca(OH)_2$) and magnesium hydroxide ($Mg(OH)_2$) obtained from the hydrated lime and dolomite. Pressure-hydrated dolomite is a low-cost commodity and is therefore an economical source of $Mg(OH)_2$. Due to the ability to recover the scrubbing reagent the method has a moderate liquid to gas (L/G) ratio. A typical L/G ratio for the process using coal with a 4% sulphur content is 200 litres of slurry per 100 m^3 of flue gas resulting in a > 90% SO_2 removal rate. This ratio is half that used in limestone scrubbers and is lower than that for lime scrubbing.

Several processes utilise ammoniacal solutions to scrub flue gases.[13–15] Ammonium sulphate ($(NH_4)_2SO_4$) is a useful by-product and the unreacted ammonia (NH_3) is cycled back into the scrubber.

Recovery methods therefore allow for the extraction of useful resources like H_2SO_4 or sulphur from the waste sludge which can then be marketed. The waste sludge can be converted to gypsum by air oxidation and is a useful resource in countries with little natural gypsum such as Japan.

14.2.1.2 Wet–dry scrubbing FGD methods

Two different processes dominate wet–dry scrubbing FGD; the spray dry absorption (SDA) process[16–19] and the Walther process. The SDA process has been developed by several companies including Niro Atomiser, Fläkt and Lodge–Cottrell. Lime slurry or other absorbent is sprayed into the flue gas absorbing the SO_2 and being dried at the same time. Water is simultaneously evaporated by the heat of the flue gas as the SO_2 is absorbed. A dry powder consisting of gypsum and fly ash is removed from the gas stream using a filter or precipitator. This product has found applications as a filler in the concrete and brick-making industries as well as for normal land reclamation purposes. The method is most suitable for coals containing 0·3–2% sulphur where 90–95% SO_2 removal is routinely achieved. The Fläkt and Niro Atomiser technologies account for 80% of the world-wide share of such wet–dry FGD plants. A development is the E-SO_x process which combines existing FGD with advanced electrostatic precipitation technology.[20,21] Usually the sorbent is sprayed into the inlet of the electrostatic precipitator where SO_2 absorbtion and drop evaporation occur simultaneously. The particulate is then precipitated. A 50–60% SO_2 removal efficiency is typically obtained using lime slurry as the absorbent. Other absorbents are being considered.

The Walther process involves an NH_3 absorbent which is sprayed into the flue gas, reacts with the SO_2 to form $(NH_4)_2SO_4$ which on drying forms a pelletised end-product.

A major advantage of wet–dry scrubbers is that they produce a solid product that is easier to handle and use than a slurry.

14.2.1.3 Dry FGD methods

Dry scrubbers can remove 70–90% of the SO_2 from the flue gases and have lower capital, maintenance and operating costs. Several totally dry processes exist which usually entail passing the flue gas over a fixed bed of reactive scrubber. Commercial plants are currently operating world-wide which often generate useful by-products.

The Lurgi–Sulfacid process involves the passage of the cooled (60–70°C) waste gases over a catalyst bed of activated carbon.[22] The acid formed in the catalyst is washed out and concentrated. It is a commercially significant by-product. Other cheaper forms of activated carbons have been suggested.

Flue gases may also be passed over a vanadium pentoxide (V_2O_5) catalyst resulting in oxidation of SO_2 to SO_3. The latter is then reacted with gaseous NH_3 to give $(NH_4)_2SO_4$, which can be used as a fertiliser.

A sodium aluminium silicate waste product from the extraction of alumina from bauxite has also been found to be an active reagent in absorbing SO_2. Treatment of the resultant material can also lead to commercial by-products. NaOH and sodium sulphate (Na_2SO_4) can be removed and alumina and silica can be separated as raw industrial feedstock. In a similar fashion, activated manganese oxide (MnO) has been considered as an absorbent forming manganese sulphate ($MnSO_4$). On the addition of NaOH, MnO is regenerated with the formation of sodium sulphate (Na_2SO_4). The Na_2SO_4 is electrolysed to regenerate the NaOH and H_2SO_4 can be recovered. The method is only suitable where cheap electrical power is available.

A novel scrubbing process uses a molten mixture of alkali metal carbonates (Li, Na, K) at around 425°C as the scrubbing medium.[23] Up to 99% is readily absorbed from gas mixtures of 3% SO_2. The regeneration system, operating at 600°C, produces hydrogen sulphide (H_2S) which can be used to produce elemental sulphur.

14.2.1.4 Dry additive processes

Several processes have been developed whereby an absorbent is added to the fuel prior to combustion. Lime injection in multistage burners (LIMB) is a technique rapidly being developed to control both SO_2 and NO_x emissions.[20,24,25] It combines a low-NO_x burning method with a sulphur-removal method. Powdered coal is used which is combined with an adsorbent (usually lime or pulverised limestone) before injection into the burner. Combustion is carried out in stages as it is not very effective when limestone is injected directly into the flame, as the temperature is too high for the lime to react with the SO_2. The limestone reacts with the SO_2 to form gypsum which can readily be removed. Promoters (e.g. sodium compounds) may also be used to increase the activity of conventional sorbents such as limestone.

Similar processes have been developed in Austria and Germany whereby calcium carbonate ($CaCO_3$), CaO, or $Ca(OH)_2$ are added before combustion to the lignite or brown coals that are abundant in these areas; 40–60% of the sulphur can be removed by these means, the higher figure being reached by the use of high surface area sorbents and/or promoters.

Fluidised-bed boilers, because of their relatively low operating temperatures and good mixing of gas and solids are particularly suited to SO_2 reduction by the addition of limestone or other sorbents to the fuel.[26] A 70% reduction is usually obtained. A discussion of fluidised-bed combustion (FBC) techniques is given later.

14.2.2 Control of NO_x

NO_x is formed by the effect of combustion of nitrogen in the air and in the fuel. Conventional pulverised fuel burners produce about 1000 mg m^{-3} (~ 500 ppm) of NO_x in the waste gas. This is made up of nitric oxide (NO) (about 95%) and nitrogen dioxide (NO_2) (about 5%).[27] Of this, some 20–40% of the NO_x originates from the thermal oxidation of nitrogen in the air (i.e. 'thermal NO_x') with the remainder produced by the oxidation of nitrogen present in coal and fuel oil at levels of 1–2% (i.e. 'fuel NO_x'), (see Chapter 2). This reaction is dependent on the temperature, oxygen concentration, pressure and residence time of the gases in the combustion chamber. Two main approaches for NO_x control exist.[2,28] By modifying the combustion regime, such as lowering the temperature and reducing the time air stays in the combustion chamber, NO_x levels can be reduced. The other approach involves selectively reducing the NO_x in the flue gases. Both approaches are considered.

14.2.2.1 Combustion modification

NO_x reduction can best be achieved by modifying the burner so that the fuel and air mix slowly, hence reducing the intensity and temperature of combustion. The combustion conditions can be modified in a number of ways such as improved burner design, burner location and modification, flue gas recirculation, staged air combustion, steam or water injection, reburning and FBC.[29–35]

Efficiences and costs of some combustion modification technologies suitable for retrofitting onto existing power plants are given in Table 14.2.

Table 14.2
Efficiencies and costs of methods used to control NO_x emissions from power plants

Technology	*NO_x reduced (%)*	*Capital cost (US\$ kW^{-1})*
Combustion modifications		
Overfire air	< 30	~ 5
Low NO_x burner with overfire air	< 50	10–30
Reburning	40–50	25–45
Selective reduction		
Selective catalytic reduction	60–80	60–180
Urea injection	30–50	5–15
Ammonia injection	60–90	20–30

Data from Ref. 29.

Low-NO_x burners are specially designed to control fuel and air mixing to promote low peak flame temperatures and oxygen concentrations, thereby reducing NO_x formation. Several different approaches are possible, including low excess air, staged air and staged fuel burners. Low excess air burners operate at oxygen concentrations of 0·5–1·5% compared to the 3–6% used in conventional burners.[30] In staged-air burners the combustion air is split into primary and secondary streams. Primary combustion is achieved in fuel-rich conditions which lower the peak flame temperature. Additional air is added in the secondary combustion region where combustion is completed.

In staged-fuel burners it is the fuel which is divided into primary and secondary flows. In the primary combustion zone the fuel is burned under lean conditions (i.e. high excess-air conditions) and this reduces the peak flame temperature and hence NO_x formation. In the secondary combustion region the remainder of the fuel is burned.

Staged-air combustion is similar in principle to the staged fuel burners except that the method is applied over the whole of the combustion chamber.

With this technique some of the combustion air is diverted from the burners and injected into the region beyond them, giving rise to fuel-rich primary combustion and fuel-lean secondary combustion therefore reducing NO_x formation. The method can be applied in several ways. With 'overfire air', 10–20% of the combustion air is injected through ports located above the uppermost burner. With 'burners out-of-service', the fuel supply is eliminated from the top burners and replaced with air only whilst the lower burners are operated under fuel-rich conditions. This method is suitable for retrofitting to multiburners and the technique is used in LIMB. Air used in the combustion is divided into a number of streams and introduced into the combustion chamber in stages. NO formation is minimised if combustion begins in a region where the air supply is restricted. More air can then be progressively mixed with the burning fuel until combustion is complete. Using this method NO_x levels in the flue gas can be halved.

Another technique for reducing peak flame temperatures and oxygen concentrations is flue gas recirculation in which up to 20% of the flue gas can be returned to the burners. Reburning is similar to staged-fuel combustion. It involves injecting a portion of the fuel, which can be either coal, oil or natural gas, above the main burners creating a fuel-rich secondary combustion region. Reburning can reduce NO_x levels significantly, by 50–70% in pilot-scale studies and probably by 40–50% in field applications.[29]

Combustion modifications can reduce NO_x levels by up to 50%, however, they generally have little effect on 'fuel-NO_x'. To achieve higher efficiencies it is necessary to apply post-combustion controls or rely upon non-retrofit techniques such as FBC.

14.2.2.2 FBC

In FBC pulverised fuel is burnt in a perforated bed containing limestone and/or dolomite and residual ash from previously burned coal. Air is blown up into the combustion chamber from underneath, causing the particles to float freely and behave as if fluid. Combustion occurs at a much lower temperature, 850–900°C compared to 1050°C of conventional coal-fired systems. Formation of NO_x is therefore substantially reduced due to the lower combustion temperature. As mentioned earlier, the presence of the limestone and/or dolomite in the bed material removes SO_2 (up to 90%) producing gypsum.

14.2.2.3 Selective reduction

Selective reduction can be either of the catalytic or non-catalytic variety.

Selective catalytic reduction (SCR). The most widely used and the most expensive selective reduction process is SCR. It is a flue gas denitrification process whereby NH_3 is added to the flue gas which is then passed over a fixed–bed catalyst. Around 80–90% of NO_x in the flue gas can be converted to elementary nitrogen in this way by means of heterogenous, surface-catalysed reactions such as

$$4NO+4NH_3 + O_2 \rightarrow 4N_2+6H_2O \tag{14.3}$$

$$6NO_2+8NH_3 \rightarrow 7N_2+12H_2O \tag{14.4}$$

Some of the NH_3 may survive the reaction giving rise to NH_3 concentrations in the flue gas of up to 5 ppm.[29] Catalysts are generally based on titanium or vanadium and they are manufactured either in honeycomb or plate forms. The most common catalyst is V_2O_5 on a finely divided titanium oxide substrate.[36] Selective catalytic conversion of the NO_x is optimised at temperatures between 300 and 400°C. There are two variations of SCR dependent upon where the catalytic reactor is situated. High-dust SCR systems are installed between the combustion chamber and gas scrubbing systems, therefore, they suffer from the high-dust concentration but require no additional heating of the flue gas. Tail-gas SCR systems are installed after fly ash collection and desulphurisation equipment.[16] At this position the flue gas temperature has to be increased to the required level for reaction, but the low SO_2 and dust levels increase the lifetime of the catalyst. The method is useful for fuels with low sulphur contents (< 1% S). With a higher sulphur content $(NH_4)_2SO_4$ aerosols may be formed by the reaction of the NH_3 with SO_2. Although SCR is very efficient in reducing NO_x emissions it is the most expensive technology available as shown in Table 14.2. It is used in some oil–powered plants in Japan and coal-fired plants in Germany.

Selective non- catalytic reduction (SNCR). Some research has been done on the reduction of NO_x by NH_3 and/or urea injection without the need for catalysts. The optimum temperature for reduction is between 900 and 1100°C.[37] At lower temperatures the reaction is negligible while at higher temperatures NH_3 may be oxidised to NO. The detailed mechanism of the reaction is complex and it involves a chain reaction propagated by free–radicals.[38] Formation of $(NH_4)_2SO_4$ and ammonium chlorides (NH_4Cl) by reaction of NH_3 with SO_2 and hydrochloric acid (HCl) in the flue gas is of potential concern. However, this may be remedied by removing these aerosols using dust control equipment before discharging the flue gas to the atmosphere. NH_3 concentrations in the flue gas are less than 10 ppm.[29,30] The cost of NH_3 /urea injection is less than that of SCR as there is no need for a catalyst bed, however further development is required before the technique is readily adopted.

14.3 MOBILE SOURCES

Emissions from mobile sources are summarised in Table 14.3, and this section is concerned mainly with automative pollution. The contribution of automotive sources to air pollution was recognised in the early 1940s when photochemical smog was first identified, and since then extensive measures have been introduced in the US to reduce emissions from these sources. Recently, concern has also been expressed about the contribution of these sources to the problem of acid rain, primarily as a result of NO_x emissions.

Control of automotive emissions may be achieved by

- engine modification,
- exhaust treatment, or
- fuel modification.

Table 14.3
Emissions from mobile sources

Engine	*Fuel*	*Vehicle*	*Emissions*
Spark ignition (SI)	Gasoline (petrol)	Automobile bus, truck, aircraft, marine	CO, NO_x, HC
Compression ignition (CI)	Diesel	Bus, truck, tractor, marine, locomotive	NO_x, SO_2, particulate
Steam	Coal / oil	Marine	NO_x, SO_2, particulate
Gas turbine (jet)	Turbine (e.g. kerosine)	Aircraft, marine, locomotive	NO_x, particulate

The contribution of automotive sources to air pollution may also be regulated by limiting traffic and providing better and cheaper public transport.

The development of control technologies has been stimulated by evermore stringent US legislation on automotive emissions (see Chapter 15). The evo-

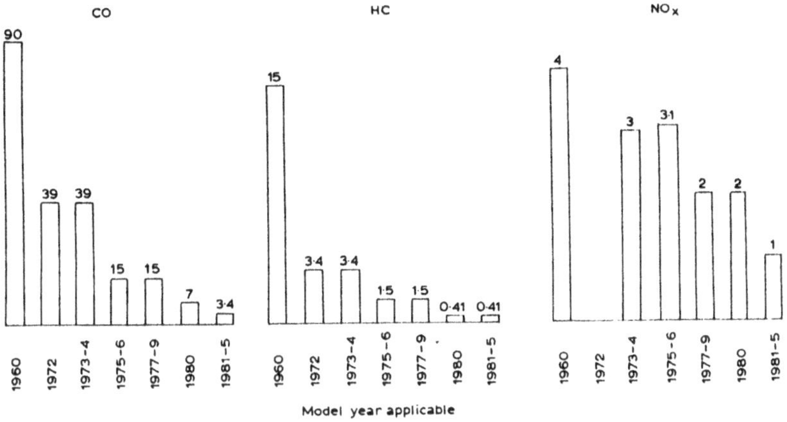

Figure 14.2. United States emission standards for new light-duty vehicles in g mile^{-1}. Data for 1960 are for pre-control exhaust emission levels.

lution of US emission limits is illustrated in Fig. 14.2 and pre-control emission levels are shown for comparison.

14.3.1 Petrol engines

The main pollutants in automobile exhaust are carbon monoxide (CO) and carbon dioxide (CO_2), oxides of nitrogen (NO_x), hydrocarbons (HC) and particulate matter. Only traces of SO_2 are present in the exhaust and these result from the combustion of sulphur compounds naturally occurring in petrol. Most of the sulphur compounds are removed during the refining process and the average sulphur content of petrol is approximately 0·03% by weight.[39] Automobiles therefore contribute less than 1% to the total anthropogenic emission of SO_2.

Current automotive emission control strategies aim to minimise CO, NO_x and hydrocarbons in the car exhaust. Only the NO_x emissions contribute directly to atmospheric acidity by being converted in the atmosphere to nitric acid (HNO_3) and nitrates. However, current control strategies have evolved out of a concern for urban air quality and the contribution of automotive NO_x emissions to acid rain has been recognised only in the last 10 years.

Hydrocarbons, CO and NO_x emitted by automobiles can react in a sunlit atmosphere, under appropriate meteorological and topographic conditions, to produce what has generally become known as 'photochemical smog'.

Photochemical smog episodes are characterised by high concentrations of ozone (O_3), other oxidants and aerosol particles, and may cause damage to human health, plants and materials as well as result in visibility degradation. Photochemical smog is generally associated with large urban centres with heavy traffic and it was first detected in the early 1940s in Los Angeles. Concern over the harmful effects of photochemical smog has prompted early legislation in the US to control automotive emissions; and similar measures are currently being adopted by other industrialised nations including member states of the EEC. Present-day pollution control strategies recognise the contribution of automotive pollutants to both photochemical smog and acid rain. Furthermore, many of the components of photochemical smog (e.g. O_3 and OH radicals) may also contribute to the acid rain problem by oxidising SO_2 and NO_x in the atmosphere.

Automotive emissions may be minimised using pre-combustion and post-combustion systems; or a combination of both. Pre-combustion systems consist of improved combustion chambers whereas post-combustion systems remove the pollutants in the exhaust. Oxidising and reducing catalytic converters are typical examples of the latter systems.

The earliest method of automotive emission control was the positive crankcase ventilation (PCV) introduced in 1963 to control crankcase emissions. This consists of taking the gases from the crankcase, via a flow control valve, into the intake manifold. The gases burn in the combustion chamber with the fuel–air mixture, eliminating the crankcase as a source of pollution.

Evaporative emissions of hydrocarbons from the fuel tank and carburettor have been controlled by adsorption on activated carbon; fresh air may be purged through the canister containing the carbon in order to carry the vapours to the combustion chamber for burning, thus regenerating the activated carbon.

A variety of methods can be used to control automotive emissions from petrol engines and these are summarised in Table 14.4. In the present discussion, controls for NO_x emissions will be emphasised, although controls for hydrocarbons and CO will also be mentioned, because within the general context of automotive pollution all three pollutants have to be considered.

The major NO_x species in exhaust gas is NO which may be oxidised to NO_2. Other NO_x species (N_2O, N_2O_3, etc.) can be ignored for the purpose of this discussion as they are not important constituents of vehicle exhausts. NO is formed from nitrogen in the air during combustion via the Zeldovich chain reaction mechanism. Oxidation of nitrogen containing compounds in the fuel is only of minor significance for petrol-driven vehicles.

Earlier techniques for controlling emissions from petrol engines were capable of removing either NO_x and/or CO and/or hydrocarbons, but not all three pollutants simultaneously, and in fact some control techniques could actually increase the emissions of one pollutant whilst lowering the emissions of another. Nevertheless, many of these techniques were capable of reducing enormously emission of CO and hydrocarbons in comparison to uncontrolled vehicles. The early literature on automotive emission controls has been reviewed.[40]

Table 14.4
Control of emissions from automotive petrol engines

Control method	*Pollutant controlled*			*Comments*
	HC	*CO*	*NO_x*	
Combustion modification				
Reduced compression ratio	Yes	Yes	No	Loss of power
Lean burn	Yes	Yes	No	May increase NO_x
Ceramics	Yes	Yes	Yes	Being developed
Catalytic conversion				
Oxidising	Yes	Yes	No	Pt or Pt/Pd catalyst
Reducing	No	No	Yes	Rh catalyst
Dual Redox	Yes	Yes	Yes	Redox catalysts
Three-way catalyst	Yes	Yes	Yes	Pt/Rh catalyst
Other methods				
Positive crankcase	Yes	No	No	Eliminates 30% of HC
Activated charcoal	Yes	No	No	No evaporation losses
Exhaust gas recirculation	No	No	Yes	Loss of power
Air injection	Yes	Yes	No	May increase NO_x
Afterburner	Yes	Yes	No	Problem with size
Turbocharging	No	No	Yes	May increase NO_x
Water injection	No	No	Yes	May increase CO/HC

Effective reduction of NO_x in the exhaust gas has been achieved only fairly recently and it is now possible to remove all three offending pollutants using a single device; the three-way catalytic (TWC) converter.[41]

14.3.1.1 Engine modification

Over the years there have been considerable advances in engine design, and these have resulted in better performance, higher efficiency, and reduced emission of pollutants. Engine modifications which may reduce emissions include redesigning of the combustion chamber and valves, improvements in the fuel–air ratio, distribution, spark ignition, timing and temperature control. For example, it has been found that the shape of the combustion chamber and spark location may have a substantial effect on NO and hydrocarbon exhaust emissions.[42] NO_x species are formed at peak temperatures during combustion and their emission increases with increasing availability of oxygen. Emissions of NO_x may therefore be reduced by lowering the peak temperature and by limiting the supply of oxygen at the peak temperature.

Water injection. Peak combustion temperature may be reduced by retarding the spark or by diluting the fuel–air mixture. Dilution may be achieved by water injection. A water–fuel ratio of 1 (by weight) may reduce NO_x emission by 60%;[40] however, this method is impractical as it requires a large water tank which would require frequent refilling. Freezing and rust are additional problems with this method.

Exhaust gas recirculation. Another method of diluting the incoming fuel–air mixture is by means of exhaust gas recirculation (EGR). In this method a portion of the exhaust gas (about 20%) is recirculated via the intake manifold enabling up to a 90% reduction in NO_x emissions. Loss of power is one of the drawbacks with EGR, however, with improved electronic control it is possible to limit recirculation to periods of high NO_x formation when some power loss may be tolerated. This method is the most effective means of NO_x emission control by engine modification and it was successful in meeting early US emission standards. However, the more stringent limits in force at present in the US require the use of exhaust gas treatment devices.

Lean burn engines. Recent advances in combustion science have led to the development of lean burn technology which involves engine operations at high air–fuel ratios, typically in excess of 20:1. This form of advanced combustion is still in its infancy and it cannot yet meet the stringent US or proposed EEC standards. Tests have shown that present-day lean burn engines can meet these emission standards only when equipped with catalytic converters.[43]

Other modifications. Other promising techniques include variable time injection, the use of ceramics in engine construction and catalytic combustion as an alternative to standard combustion although these are still in their infancy.

14.3.1.2 Exhaust treatment

Air injection. An early method of exhaust treatment adopted in the late 1960s was to introduce fresh air into the exhaust stream by means of an air pump. This provided oxygen for the oxidation of hydrocarbons and CO to give CO_2 and water. While this was a cheap and effective technique it could not meet the more stringent US standards of the late 1970s. Furthermore, this method may increase NO_x emissions.

Afterburners. To achieve a high degree of hydrocarbon and CO oxidation in the exhaust a high temperature is required, as well as a sufficient supply of oxygen, and a long residence time. This heat may be provided by means of a flame inside a thermal reactor mounted in the exhaust system and generally known as an afterburner. Afterburners perform most effectively on rich mixtures with air injection as rich combustion conditions produce low NO_x. Hydrocarbons and CO are then oxidised in the afterburner. Thermal reactors are generally larger than catalytic converters because of the required residence time for the oxidation reactions. They are not popular with manufacturers.

Catalytic converters. Catalytic converters are an exhaust treatment device based on the principle of heterogeneous catalysis. Catalysts are substances which speed up the rate of a chemical reaction without being consumed in the process by lowering the activation energy of the reaction. Heterogeneous catalysts are solid materials with large surface areas on which reactions can take place. The reactant gases are chemisorbed onto the active surface sites where they may react with each other. Chemisorption is an adsorption process in which weak chemical bonds are formed between the gas molecules and the solid surface, as opposed to physical adsorption in which surface retention of gaseous molecules is achieved by means of purely physical forces. The products of the surface-catalysed reaction are then desorbed freeing the active sites for further reaction. Catalytic converters may broadly be classified as oxidising or reducing catalysts. Oxidising catalysts promote the reactions of hydrocarbons and CO with oxygen to form CO_2 and water. The polluting gases and oxygen are chemisorbed onto the surface resulting in a weakening of molecular bonds and facilitating oxidation. In reducing converters NO is reduced to nitrogen. A variety of base-metal catalysts have been evaluated in the past, however, these cannot be employed to achieve the stringent emission standards. Platinum, palladium, rhodium and their mixtures are the most popular catalysts in present-day use.[44] Oxidation catalysts generally consist of platinum or platinum–palladium mixtures and these are usually combined with air injection to provide sufficient oxygen for the reaction. This method cannot, however, meet the stringent US emission limits for NO_x and it has to be combined with an effective NO_x removal technique such as EGR. Catalytic reduction of NO_x is more complicated. While both platinum and palladium can reduce NO_x in a reducing atmosphere, they may also reduce a significant amount of NO_x to NH_3. It is possible to combine a reducing catalyst with an oxidising catalyst in series in what is known as a dual converter.[44] The exhaust gas is passed through the first converter to reduce NO_x and then through an oxidising converter, after addition of air, to oxidise the hydrocarbons and CO. In some of these systems NH_3 produced in the first converter, the reducing converter, is re-oxidised back to NO_x in the second oxidising converter. Platinum–rhodium catalysts can reduce NO_x in a rich or stoichiometric environment without significant NH_3 production and these can be effectively combined with an oxidising catalyst in a dual converter, enabling removal of all three offending pollutants. The most effective method of simultaneously reducing emissions of NO_x, CO and hydrocarbons is by means of a TWC[41,45–47] When operated under stoichiometric conditions this catalyst enables reaction between the three pollutant gases to produce harmless products (water, nitrogen and CO_2). The characteristics of a TWC are shown in Fig. 14.3, illustrating that all three pollutants can be reduced simultaneously only within a narrow 'window' of air–fuel ratios close to stoichiometric. With lean mixtures hydrocarbons and CO are reduced efficiently, however, control of NO_x is poor. Conversely, with rich mixtures control of NO_x is good, but the control of hydrocarbons and CO is not. TWC therefore require very good control to maintain stoichiometric conditions (air/fuel = 14·7) and this is achieved by means of electronic multipoint fuel injection and oxygen sensors. Oxygen probes in the exhaust can measure the amount of excess oxygen and then this is used to control the air/fuel ratio.[46–50] At present three-way catalysis employs platinum and rhodium mixtures although the use of palladium catalysts is also being investigated. TWC enable the removal of all three pollutants inside a single converter and they are the most popular method currently in use for meeting stringent emission standards. Early problems with catalytic converters included poisoning of the catalyst by lead and other additives in petrol.[51–53] These additives may occupy the active sites making them unavailable for the catalysed reactions.

In the case of lead, about half of the lead added to the fuel is deposited on the catalyst in the form of a surface layer 15 μm thick, and consisting of lead sulphate, oxysulphate and chlorobromide, hence, deactivating the catalyst.

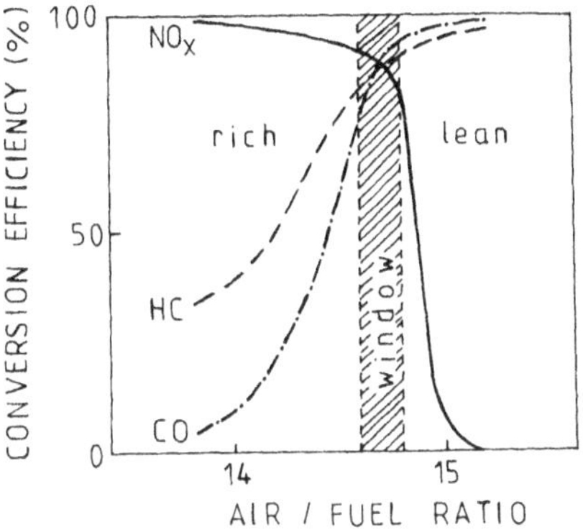

Figure 14.3. Characteristics of three-way catalysts.

The presence of lead in fuel results in a rapid decrease in hydrocarbon conversion efficiency but it does not, apparently, affect CO conversion. Nevertheless, lead-free petrol is required for control systems employing catalytic converters. In the US unleaded petrol was introduced on the market simultaneously with catalytic converters. The phasing-out of lead in European petrol is also being accompanied with the introduction of catalytic converters. Lead in combination with other additives can also adversely affect thermal reactors (afterburners) by enhancing corrosion rates. Lead may also cause problems in EGR systems by forming deposits which may clog the injection orifices. The hardware for catalytic converters consists of catalyst supported on a substrate inside a converter canister which can be fitted to the exhaust system of the automobile. Ceramic substrate has proved to be the most effective and cordeorite is the most popular support. Cordeorite is a poor support by itself and it is coated with alumina in order to provide a large surface area for adsorption. Substrates are manufactured in the form of pellets or monoliths and the active catalyst metals are deposited on the alumina-coated surface. Monoliths are honeycombed structures washed with a slurry of aluminium hydrate and baked dry to produce an activated gamma-alumina (γ-Al_2O_3) surface approximately 20 μm thick. Pellets are cylindrical or spherical γ-Al_2O_3 particles approximately 3 mm in diameter. Monoliths and pellets are both rinsed in a solution of the salt of the catalyst metals to produce a monomolecular layer of the active catalyst on the surface. Other substrates such as metallic meshes have also been tried but ceramics are preferred by the manufacturers. Coating the inside of the exhaust pipe with the catalyst and the insertion of catalytically coated spiral metallic inserts in the exhaust pipe have also been proposed.[41] Although catalytic converters are fairly simple devices, the catalysts employed are expensive, and precious metals are found in only a few strategic deposits. Approximately 0·5 t of rock has to be mined in order to produce the 1·5 g of metal required for each vehicle equipped with a catalytic converter. Another problem with catalytic converters is the emission of metals as the support breaks up. The catalyst is expected to last the lifetime of the vehicle, and there is the possibility of recovering and recycling the precious metals. The benefits of minimising emissions from individual vehicles are offset by the increasing number of vehicles on the road. Perhaps a better approach to urban air quality would be the introduction of car-free zones in urban centres along the lines of the smoke-free zones in the 1950s. This would require better town planning, efficient cheap public transport systems and an increasing reliance on non-polluting forms of transport. These measures would necessitate changing social attitudes. Nevertheless, several European cities are having some success with such schemes.

14.3.2 Diesel engines

Diesel or compression ignition (CI) engines are employed in heavy duty vehicles (e.g. trucks, buses, tractors) and their contribution to air pollution and atmospheric acidity is negligible in comparison to petrol-fuelled spark ignition (SI) engines.[40] Concentrations of hydrocarbons, CO and NO_x in diesel fuel are normally lower than in petrol exhaust and it does not contain lead additives.[54] Diesel exhaust, however, contains high concentrations of particulate, sulphur and malodorous compounds. The black smoke, odours, and noise pollution may cause public annoyance in restricted areas such as bus terminals and concern has been expressed about the potentially carcinogenic aromatic compounds present in particulate matter emitted by diesel engines.

Diesel exhaust contains high concentrations of sulphur compounds as a result of the high sulphur content of the fuel. On average, diesel fuel contains 0·2% S, this being emitted as SO_2 or particulate sulphate present predominantly as H_2SO_4.[55] Between 5 and 10% of the total particulate matter emitted in the exhaust may consist of sulphate aerosol. Nevertheless, the contribution of diesel engines to total anthropogenic SO_2 and sulphate pollution is of minor importance in comparison to emissions from stationary sources. Sulphur in diesel fuel may, however, cause engine wear due to its corrosive nature and it may adversely effect the performance of exhaust treatment devices.

Although the environmental impact of diesel exhaust is considerably lower than that of petrol exhaust, techniques have been developed to reduce diesel emissions and more stringent emission standards for diesel engines are being introduced. The approach to emission controls for diesel engines is similar to that for petrol engines and includes engine modification, diesel fuel modification and exhaust treatment devices. As with petrol devices, the reduction of one pollutant may result in an increase of another. Particulate matter and NO_x exhibit this inverse relationship in the case of diesel engines.

Emissions of NO_x may be reduced by retarding injection timing, the injection of water and the use of EGR. Particulate emissions may be controlled with exhaust treatment devices which include particle traps, flow-through oxidation catalysts and particle agglomerators.[56–63]

Ceramic monolith traps and stainless-steel meshes have been used as filter media. A problem with traps is that the substrate is easily damaged (e.g. cracks, melts), but this has been ameliorated with the use of traps coated with precious-metal catalysts and/or the use of fuel additives.[57,64–69] Particulate filtration efficiencies as high as 95% have been achieved with monolithic ceramic traps.[70] Traps may however oxidise the SO_2 in the exhaust to H_2SO_4 and sulphates. The use of low-sulphur diesel fuel could increase the lifetime of the traps.

Electrostatic particle agglomerators allow fine soot particles (< 10 μm in diameter) to conglomerate into larger particles which more readily sediment out of the atmosphere. Agglomerators based on a combination electrostatic/cyclone collector have also been developed.

Engine modifications for emission control from diesel vehicles include improved combustion chamber design, electronic control of the combustion process and the use of ceramic technologies. Future emission control strategies will include a combination of several technologies; engine, fuel and lubricant modification as well as exhaust treatment to achieve the anticipated stringent emission standards. Emissions of SO_2 and sulphates are best limited by lowering the sulphur content of the fuel.

Diesel-powered passenger vehicles have been produced by several manufacturers, mainly for use as taxis. Although diesel engines have the advantage over petrol engines of emitting lower levels of hydrocarbons, CO and NO_x, more research and development is required in order to make them acceptable to the general public. Amongst others, problems of smoke, noise and odour pollution will need to be addressed.

14.4 OTHER METHODS OF ACID RAIN CONTROL

14.4.1 Liming

Liming of lakes and their catchment areas has been a popular method of countering the effects of acidification in receptor areas. Treatment involves addition of powdered limestone or dolomite into the affected lake or surrounding catchment area in order to neutralise acidity. With this method it is difficult to judge when, where and how much reagent to apply. As the water becomes acidic with time the treatment has to be repeated. Furthermore, there are potential environmental consequences of liming to consider. Lime essentially treats the symptoms of acidification and although some success has been achieved with the method it does not represent a long-term solution to the problem. This would require elimination of the cause, i.e. the reduction of acidifying pollution at source.

A novel approach has been suggested whereby the 'liming' agent is added to the flue gas before release from the chimney.[71,72] Finely divided (3 μm diameter) $Mg(OH)_2$ powder is injected into the flue gas at the outlet from the electrostatic precipitators fitted on power stations. Due to the small particle size of the additive, little will be lost as dry deposition, but a high percentage will precipitate with the rain over the areas which suffer from acidification. The $Mg(OH)_2$ will not react with the acids in the plume but will neutralise acids in the lakes and their catchments. Full-scale trials need to be carried out.

14.4.2 Fuel switching and synthetic fuels

One approach to pollution control could be the switching of fuels for both mobile and stationary sources. In the case of internal combustion engines it is possible to substitute alcohols, alcohol–petrol mixtures, liquified natural gas (LNG), liquified petroleum gas (LPG) and compressed natural gas (CNG) for petrol to give less polluting exhaust gases. So far these substitutes have found limited use in motor vehicles. There is also the possibility of using biomass fuels, such as vegetable oils, to power diesel engines. Electric motor vehicles powered by a variety of batteries have been in existence for some time and despite considerable advances in this area of technology they have not come into widespread use due to many disadvantages including size, weight, and the recharging requirements of the batteries.

So called 'synthetic' fuels are alternatives to conventional fossil fuels and they can be manufactured from coal or oil shales.[73] Although their use for power generation would produce cleaner flue gases, coal-based synthetic fuel manufacturing plants would present problems of SO_2, NO_x and waste-water pollution that will require controlling.[74,75] Furthermore, there could be problems with accidental spillage of these fuels during transportation as synthetic fuels are more toxic than conventional fuel oils.[76] Natural gas is a cleaner fuel than either coal or oil and an increasing reliance on its use for electricity generation would reduce emissions of acidic gases from power stations.

14.4.3 Alternative energy sources and policy

Research into the economic costs of environmental damage and control is discussed in Chapter 16.

Alternatives to clean-up technologies discussed so far are energy policies which favour energy conservation, increasing efficiency and increased use of alternative energy sources. Reducing energy losses in industry and implementing advanced building technologies could result in considerable energy savings. Alternative energy sources such as solar, wind, wave, hydroelectric, geothermal and even oceanic heat pumps could be used in many parts of the world but currently they satisfy only a minor fraction of the global energy requirement. It is estimated that by the year 2025, only 10% of these requirements will be met by alternative energies. Generally, these resources are only relevant on the local or regional scale due to the high cost of electricity transmission or transport costs.

Nuclear energy produced by fission is the only technology capable of competing with fossil fuels for power generation. Nevertheless, the nuclear industry is currently experiencing serious setbacks in the wake of recent accidents, most notably Three-Mile Island and Chernobyl. Although nuclear energy does not contribute to the phenomena of acid rain or global warming, radioactive releases to the environment have consequences as drastic, if not more so, than the use of coal or oil, dubbed 'dirty' fuels by the nuclear industry. In addition to accidental releases, the nuclear industry has not solved the problems of radioactive waste disposal or decommissioning of redundant plants. Another form of nuclear power, fusion, has still to be harnessed for industrial purposes.

It should be noted that alternatives to fossil-fuel combustion are not necessarily 'environmentally friendly', and in fact environmental problems have been noted with many of these. Fossil-fuel combustion is the most widespread technology and it has been in use for a long time–its environmental consequences are therefore relatively well established. Environmental problems associated with such large scale use of other energy sources could also be quite serious. The energy industry has evolved historically in a haphazard way determined by economic and not environmental considerations. It would be gratifying to see this change in the future with detailed and careful assessment studies being required before adopting new technologies on a vast scale.

REFERENCES

1. Vernon, J., Technologies for control of sulphur dioxide emissions. In *Acid Deposition: Sources, Effects, and Controls*, ed. J. W. S. Longhurst. British Library, London, 1989, pp. 287–99.
2. Reiling, W., Technologies for controlling acid deposition precursor emissions from large power plants. In *Acid Deposition: Sources, Effects and Controls*, ed. J. W. S. Longhurst. British Library, London, 1989, pp. 301–8.
3. Sawyer, R. F., Central power plants. In *Combustion-Generated Air Pollution*, ed. E. S. Starkman. Plenum Press, New York, USA, 1971, pp. 275–90.
4. Strauss, W., *Industrial Gas Cleaning*. Pergamon Press, New York, USA, 1975, pp. 80–171.
5. Rees, R. L., The removal of oxides of sulphur from flue gases. *J. Inst. Fuel.*, **25** (1953) 350–7.
6. Bromley, L. A., Use of sea-water to scrub sulphur dioxide from stack gases. *Int. J. Sulfur Chem.*, Part B, **7** (1972) 77–84.
7. Kelly, F. H. C., Design and operation of a counter current gas scrubbing system. *Proc. Aust. Inst. Min. Metall.*, **152** (1949) 17–39.
8. Radojević, M., The use of seawater for flue gas desulphurisation. *Environ. Technol. Lett.*, **10** (1989) 71–6.
9. Yamamoto, H. & Uchino, T., Control of atmospheric emissions from glass manufacture. In *International Congress Glass, 10th Meeting 1974, Ceram. Soc. Japan*, Vol. 4, Ceramic Society of Japan, Tokyo, 1974, pp. 42–51.
10. Mobley, J. D., Control of SO_2 emissions by the adipic acid enhanced limestone flue gas desulfurisation process. Paper presented at the 75th Annual Meeting of the Air Pollution Control Association, New Orleans, LA, USA, 20–25 June 1982.
11. Berst, A. H., Sulphur dioxide control with double alkali scrubbing systems. Paper presented at the 75th Annual Meeting of the Air Pollution Control Association, New Orleans, LA, USA, 20–25 June 1982.
12. Abrams, J. Z., Monahan, W. J. & King, A. M., Application of dolomitic lime process for flue gas cleaning at Colstrip 3 & 4. In *Proceedings of the American Power Conference*, Vol. 46, 1984, pp. 132–40.
13. Kennaway, T., The Fulham-Simon Carves process for the recovery of sulphur from flue gases. *J. Air Pollut. Control Assoc.*, **7** (1958) 266–83.
14. Pakrasi, A., Davis, W. T., Reed, G. D. & Keener, T. C., A combined $Ca(OH)_2/NH_3$ flue gas desulphurisation process for high sulphur coal: results of a pilot plant study. *J. Air Waste Management*, **40** (1990) 987–92.
15. Wallis, E., The recovery of sulphur in a marketable form from flue gases. *Brit. Chem. Eng.*, **7** (1962) 833–6.

16. Bentgsson, S., Fläkt technologies for reduction of sulphur dioxide and nitrogen oxides emissions. In *Acid Rain: Scientific and Technical Advances*, ed. R. Perry, R. M. Harrison, J. N. B. Bell & J. N. Lester. Selper, London, UK, 1987, pp. 808–17.
17. Anon., Control of acid rain by dry purification. *Chim. Oggi.* (1985) 41–4.
18. Felsvang, K., Morsing, G. & Veltman, P., Acid rain prevention thru new sulfur oxide (SO_x)/nitrogen oxide (NO_x) dry scrubbing process. In *Proceedings of the 8th Symposium on Flue Gas Desulfurisation*, EPRI CS-3706, Vol. 2, Electrical Power Research Institute, EPRI, Palo Alto, California, 1983.
19. Karlsson, H. T., Klingspor, J., Linne, M. & Bjerle, I., Activated wet–dry scrubbing of sulphur dioxide. *J. Air Pollut. Control Assoc.*, **33** (1983) 23–8.
20. Martin, G. B., Abbott, J. H. & Sparks, L. E., Advances in control technology for acid deposition. In *Aerosols*, ed. S. D. Lee, T. Schneider, L. D. Grant & P. J. Verkerk. Lewis, Chelsea, MI, USA, 1986, pp. 403–15.
21. Plaks, N., New technology for the control of aerosols from stationary sources. In *Aerosols*, ed. S. D. Lee, T. Schneider, L. D. Grant & P. J. Verkerk. Lewis, Chelsea, MI, USA, 1986.
22. Hartman, M., Polek, J. R. & Coughlin, R. W., Removal of sulfur dioxide from flue gas sorption and catalytic reaction on carbon. *A. I. Chem. Eng. Symp. Series*, **67** (1971) 7–22.
23. Olderkamp, R. D. & McKenzie, D. E. 'The molten carbonate process for control of sulphur dioxide emissions'. APCA Meeting, St. Paul June 1968.
24. Martin, G. B. & Abbot, J. H., EPA's LIMB R&D program–evolution, status and plans. In *Proceedings of the First Joint Symposium on Dry SO_2 and Simultaneous SO_2/NO_x Control Technologies*, Vol 1, EPA-600/9-85-020a and -020b (NTIS No. PB85-232353 and -232361). 1985.
25. Green, A. E. & Pamidimukkala, K. M., Oil backout and acid rain. In *Acid Deposition Causes Effects*, State Assess. Model, Proc. Workshop, ed. A. E. S. Green & W. Smith. Government Institute, Rockville, MD, USA, 1983, pp. 130–52.
26. Dennis, S., Alternative sorbents for flue gas desulphurisation, especially in fluidised-bed combustors. *J. Inst. Energy*, (1989) 202–7.
27. Leibfritz, R., Reiling, W. & Rentz, O., Application, status and developments in the control of acid deposition from major power generation sources in Europe. In *Acid Rain: Scientific and Technical Advances*, ed. R. Perry, R. M. Harrison, J. N. B. Bell & J. N. Lester. Selper, London, UK, 1987, pp. 791–800.
28. Sarofin, A. F. & Flagan, R. C., NO_x control for stationary combustion sources. *Prog. Energy Combust. Sci.*, **2** (1976) 1–25.
29. Eskinazi, D. & Cichanowicz, J. E., Stationary combustion NO_x control. *J. Air Pollut. Control Assoc.*, **39** (1989) 1131–9.
30. McInnes, R. & Wormer, M. B. van, Cleaning up NO_x emissions. *Chem. Engng*, **97** (1990) 130–5.
31. Allen, J. W. & Hufton, P. F., The development of low NO_x burners. In *The Problems of Acid Emissions: an Opportunity for British Industry*. The Institute of Chemical Engineers, Rugby, UK, 1988, pp. 197–215.
32. Lim, C., Szymanski, T., Gerdes, W. H. & Mills, K. J., Recent developments in Norton's practical flue gas treatment systems. In *The Problems of Acid*

Emissions: an Opportunity for British Industry. The Institute of Chemical Engineers, Rugby, UK, 1988, pp. 171–96.
33. Kyte, W. S., Possible emission control technologies for coal fired power stations in the UK. In *The Problems of Acid Emissions: an Opportunity for British Industry.* The Institute of Chemical Engineers, Rugby, UK, 1988, pp. 11–25.
34. Cato, G. A., Hall, R. E. & Muzio, L. J., *Reduction of Pollutant Emissions from Industrial Boilers by Combustion Modification*, KVB Report, 76-WA/FU-5, 1976.
35. Hunter, S. C., Hall, R. E. & Carter, W. A., *Evaluation of Combustion Modification for Emission Reduction on Industrial Processes*, KVB Report 78–WA/APC-8, 1978.
36. Herrlander, B., Comparison of four catalysts used in selective catalytic reduction of NO_x. Paper presented at the 1987 Joint Symposium on Stationary Combustion NO_x Control, 1987, 23–27 March, New Orleans, Louisiana.
37. Lyon, R. K., Thermal $DeNO_x$. *Environ. Sci. Technol.*, **21** (1987) 231–6.
38. Lyon, R. K. & Bonn, D., Kinetics of the NO-NH_3-O_2 reaction. In *Proceedings of the 17th International symposium on Combustion*, Combustion Institute, Pittsburgh, PA, USA, 1979, pp. 601–9.
39. Somer, J. H., Garbe, R. J., Lawrence, R. D. & Baines, T. M., Automotive sulphate emissions, SAE paper 770166. *Ann. Index/Abstr. SAE Tech. Papers*, **86** (1977) pp. 38–9.
40. Patterson, D. J. & Henein, N. A. *Emissions from Combustion Engines and their Control.* Ann Arbor Science, Ann Arbor, MI, USA, 1972.
41. Öser, P., Catalyst systems with an cmphasis on three-way conversion and novel concepts, SAE paper 790306. *SAE Trans.*, **88** (1979) 1057–70.
42. Wentworth, J. T., Effects of combustion-chamber shape and spark location on exhaust nitric oxide and hydrocarbon emissions, SAE Paper 740529. *SAE Trans.*, **83** (1974) 1966–75.
43. Evans, W. D. J. & Wilkins, A. J. J., Catalytic emission control strategies for Europe. *Sci. Total Environ.*, **59** (1987) 97–112.
44. Demmler, A. W., Automotive catalysis. *Automotive Engng.*, **85** (1977) 26–32.
45. Searles, R. A., Motor vehicle emission control by catalysis. In *Pollution in the Urban Environment, Polmet '85*, ed. M. W. H. Chan, R. M. M. Hoare, P. R. Holmes, R. J. S. Law & S. B Reed. Elsevier, London, UK, pp. 265–70.
46. Seiter, R. E. & Clark, R. J., Ford three-way catalyst and feedback fuel control system, SAE Paper 780203. *SAE Trans.*, **87** (1978) 828–42.
47. Canale, R. P., Winegarden, S. R., Carlson, C. R. & Miles, D. L., General Motors phase II catalyst system, SAE Paper 780205. *SAE Trans.*, **87** (1978) 834–52.
48. Toyoda, T., Yamakara, Y., Inove, T., Oishi, K. & Hattori, K., Development of closed loop secondary air control three-way catalyst systems, SAE Paper 800399. *SAE Trans.*, **89** (1980) 1564–72.
49. Walber, W. G., A worldwide overview of automotive engine control sensor technology, SAE Paper +78027. *SAE Trans.*, **87** (1978) 853–70.
50. Howarth, D. S. & Wilkhelm Jr, R. V., A zirconia based lean air-fuel ratio sensor, SAE Paper 780212. *SAE Trans.*, **87** (1978) 871–6.
51. White, J. E. & Zakrajsek, C. E., Implications of precious metal catalysts with

leaded fuels, SAE Paper 810086. *SAE Trans.*, **90** (1981) 353–68.
52. Williamson, W. B., Gandhi, H. S., Heyde, M. E. & Zawacki, G. A., Deactivation of three-way catalysts by fuel contaminants–lead, phosphorus and sulphur, SAE Paper 790942. *SAE Trans.*, **88** (1979) 3196–204.
53. Duncan, J., Combustion study of the deactivation of a three-way catalyst by lead and manganese, SAE Paper 841408. *Ann. Index/Abstr. SAE Tech. Papers*, **93** (1984) 200.
54. Walder, C. J., Broom, D. & Ritchie, P. S. J., The reciprocating diesel engine. In *Vehicle Engines–Fuel Consumption and Air Pollution*, ed. M. S. Janota. Peter Peregrinus, Stevenage, UK, 1974, pp. 34–55.
55. Truex, T., Pierson, W. R. & McKee, D. E., Sulphate in diesel exhaust. *Environ. Sci. Technol.*. **14** (1980) 1118–21.
56. Urban, C. M., Landman, L. C. & Wagner, R. D., Diesel car particulate control methods, SAE Paper 830084. *SAE Trans.*, **92** (1983) 1278–308.
57. Wong, V. M., Yu, M. L., Mogaka, Z. N. & Shaked, S. M., Effects of catalytic wire-mesh traps on the level and measurement of heavy duty diesel particulate emissions, SAE Paper 840172. *Ann. Index/Abstr. SAE Techn. Papers*, **93** (1984) 53.
58. Ullman, T. L., Hare, C. T. & Baines, T. M., Primary particulate trap tests on a 2-stroke diesel bus engine, SAE Paper 840079. *SAE Trans.*, **93** (1984) 1470–84.
59. Rao, D., White, J. E., Wade, W. R., Aimone, M. G. & Crkanek, H. A., Advanced techniques for thermal and catalytic diesel particulate trap regeneration, SAE Paper 850014. *SAE Trans.*, **94** (1985) 1111–27.
60. Abthoff, J., Schuster, H. D., Langer, H. J. & Loose, G., The regenerable trap oxidiser–an emission control technique for diesel engines, SAE Paper 850015. *SAE Trans.*, **94** (1985) 1128–38.
61. Sachdev, R., Wong, V. W. & Shaked, S. M., Analysis of regeneration data for cellular ceramic particulate trap, SAE Paper 840076. *SAE Trans.*, **93** (1984) 1432–7.
62. Buchman, M. F. & Enga, B. E., Regeneration behaviour of light-duty catalytic trap oxidiser systems, SAE Paper 840080. *SAE Trans.*, **93** (1984) 1485–92.
63. Budd, E. R., Catalytic control for off-highway diesels, SAE Paper 840170. *Ann. Index/Abstr. SAE Tech. Papers*, **93** (1984) 53.
64. Montierth, M. R., Fuel additive effect upon diesel particulate filters, SAE Paper 840072. *SAE Trans.*, **93** (1984) 1408–21.
65. Ludecke, D. A. & Bly, K. B., Diesel exhaust particulate control by monolith trap and fuel additive regeneration, SAE Paper 840077. *SAE Trans.*, **93** (1984) 1444–52.
66. Wiedman, B., Doerges, U., Engeler, W. & Poettner, M., Application of particulate traps and fuel additives for reduction of exhaust emissions, SAE Paper 840078. *SAE Trans.*, **93** (1984) 1453–69.
67. Buchman, M. F. & Enga, B. E., Catalytic diesel particulate control system design and operation, SAE Paper 830080. *SAE Trans.*, **92** (1983) 1224–37.
68. Wade, W. R., White, J. E., Florek, J. J. & Cikanek, H. A., Thermal and catalytic regeneration of diesel particulate traps, SAE Paper 830083. *SAE Trans.*, **92** (1983) 1253–77.

69. Öser, P. & Thoms, U., Particulate control systems for diesel engines using catalytically coated and uncoated traps with consideration to regeneration techniques, SAE Paper 830087. *SAE Trans.*, **92** (1983) 1329–42.
70. Lepperhoff, G. & Kroon, G., Impact of particulate traps on the hydrocarbon fraction of diesel particles, SAE Paper 850013. *SAE Trans.*, **94** (1985) 1098–100.
71. Lees, B., Desulphurisation of flue gases: is there another way?, *Energy World*, **158** (1988) 1–2.
72. Lees, B., The chemistry of atmospheric pollution. *Chem. Brit.*, Jan. (1989) 38.
73. Stenby, E. W., Hoskins, W. W. & Donaldson, A. M., Environmental control impacts of selected alternate fuels on existing power plants. In *Environmental Problems and Solutions*, ed. T. N. Vezoglu. Hemisphere, New York, 1990, pp. 207–31.
74. Josephson, J., Synth fuels scale-up. *Environ. Sci. Technol.*, **14** (1980) 652–6.
75. Felder, R. M., Kelly, R. M., Ferrell, J. K. & Rousseau, R. W., How clean gas is made from coal? *Environ. Sci. Technol.*, **14** (1980) 658–66.
76. Giddings, J. M. & Washington, J. N., Coal-liquefaction products, shale oil and petroleum. Acute toxicity to freshwater algae. *Environ. Sci. Technol.*, **15** (1981) 106–13.

Chapter 15

Legislative Measures of Control

D. J. HUGHES
Faculty of Law, University of Leicester, Leicester, LE1 7RH, UK

15.1 INTRODUCTION

Eradicating environmentally deleterious activity by law requires the following:

- agreed perception of perceived hazards;
- technology or other appropriate modes of action to deal with the problem;
- convergence of interest amongst those creating the hazard and those suffering from it so eradication becomes acceptable, either economically, politically or socially, or as a result of a combination of these factors;
- perception of the issue 'in the round' so there can be co-ordinated and integrated approaches to the problem harmonising and directing all means of regulation towards eradication; and
- strong and effective agencies of enforcement.

Without these conditions being satisfied the most to be hoped for is palliation of a problem—control rather than eradication. Laws designed to cope with acidification frequently do not meet the above criteria, and so are frequently mere palliatives.

15.2 THE PROBLEM

As noted by the Royal Commission on Environmental Pollution, acidification, commonly known as 'acid rain' began causing concern in the early 1970s, particularly in Scandinavia and the north-eastern US.[1] The

Organisation for Economic Cooperation and Development (OECD) set up a research programme into the phenomenon which reported in 1977 while the Cooperative Programme for the Monitoring and Evaluation of the Long-Range Transmission of Air Pollutants in Europe (EMEP) commenced its research into sulphur dioxide (SO_2) fluxes in that year.[2] There has been much subsequent research (see, for example, Refs 3–5).

The prime acidifying emissions are SO_2 and the various nitrogen oxides (NO_x) particularly nitrogen dioxide (NO_2). These come mainly from urban areas, industrial emitters and motor vehicles. Some 50% of NO_x emissions come from vehicle engines. The gases may be absorbed by plants and other matter—'dry deposition'—or oxidised to acids, particularly sulphuric acid (H_2SO_4) and nitric acid (HNO_3). These dissolve in atmospheric moisture to form various ions of hydrogen, sulphate and nitrate which contaminate precipitation, though even uncontaminated rainfall is slightly acidic. Acid deposition is claimed to affect land, structures and surface water deleteriously. It has been claimed that fisheries have been affected, forests have been reduced by acidification of soils, and buildings have suffered increased weathering in consequence of acidic emissions into the atmosphere.[6–8] Though there remain uncertainties about how, and to what extent, acid deposition causes damage, there is general agreement that sulphur and NO_x emission caused by human action are resulting in damage to terrestrial and aquatic ecosystems. Furthermore there is real, if not totally unequivocal evidence, that weathering and damage to buildings accelerates as a result of acid deposition. It is also generally accepted that acid deposition causes particular international problems because the implicated gases disperse over a long distance from points of emission: those who emit most are frequently not those who suffer the most from the consequences of emissions. Scandinavia suffers from emissions originating in other European regions; Canada argues emissions from the industrial northern US affect her.[9] However, it can hardly be entirely convincingly argued this acceptance is entirely coincident with the first requirement for eradication of a perceived hazard stated above. It cannot be argued there is an agreed perception of a perceived hazard.

15.3 SOURCE OF AND LEGAL RESPONSES TO THE PROBLEM

The Royal Commission on Environmental Pollution identified in 1984 emissions from fossil fuel burning power stations as the chief sources of acidifying gases in the UK.[10] In 1984 some $2{\cdot}4 \times 10^6$ t of SO_2 were emitted from Central Electricity Generating Board (CEGB) power stations, accounting for some 63% of UK SO_2 emissions; 46% of UK NO_x emissions came from such stations then. However, power stations are not the only sources of SO_2 and NO_x. In 1978 UK total emissions of SO_2 were $5{\cdot}02 \times 10^6$ t ($4{\cdot}67 \times 10^6$ t in 1980).[11] Motor vehicles in the UK produce some 40% of the nation's NO_x emissions.[12] Clearly, acidifying gases derive from both mobile and stationary combustion sources.

A legal response to this situation which desires to, at least, reduce emissions, must therefore seek to promote a variety of controls over the emission sources. It is difficult to promote a single, unified regulatory approach, for a variety of technical, political, social and economic reasons: but it is desirable that controls should be harmonised, co-ordinated and integrated wherever possible so that deficiencies in one form of control do not undermine good results achieved by another. Such harmonisation, co-ordination and integration are often absent from legal systems: this is sufficient to militate against laws being effective.

It is furthermore clear that legal responses to the problem cannot exist at solely a national level. The atmosphere knows no frontiers and acidification in one nation can easily result from another's emissions. To illustrate this there are figures for the period October 1978 to September 1980 indicating that Switzerland suffered 78% of its acid deposition from foreign sources, Austria 76%, Luxembourg, 73%, the Netherlands 71%, Norway 63%, Sweden 58% and Finland 55%. By way of contrast the UK suffered 79% of its deposition from indigenous sources, Italy 70%, the former GDR 65%, Spain 63%, the former USSR 53%, France 52% and Yugoslavia 51%.[13] Too crude a list of 'exporters' and 'importers' of acidification should not be extrapolated from just one set of figures, but they establish the problem is international, and that one nation's benefits result in clear disbenefits to others. More recent figures, derived from research, and international information exhanges, have confirmed SO_2 as the prime cause of acidification, aided and abetted by NO_x—see, for example, Ref. 14. Furthermore, improved mathematical models, evaluated in practice by actual frontier flux measurements, have enabled 'sulphur budgets' to be drawn up. EMEP has drawn up a table for Europe for 1978 to 1982 using such a model. This takes into account the size and population of states and estimates emissions per capita per year and emissions and depositions per square metre per year. The 1984 EMEP Report, as abstracted in Ref. 5, shows a 'balance sheet' of SO_2

emissions for Europe. It is becoming more feasible to produce evidence that a given nation is receiving specified proportions of acidifying agents from another given state. Consequently, legal controls should not only be harmonised, co-ordinated and integrated nationally, a similar pattern is needed internationally. This again will currently be found in many respects to be lacking.

15.4 INTERNATIONAL LAW

International Law has been defined as the 'indispensible body of rules regulating for the most part the relation between States without which it would be virtually impossible for them to have steady and frequent intercourse'.[15] It tends to be a weak system of law lacking, in many cases, an effective system of sanctions.

International Law is derived, *inter alia*, from custom, i.e. rules internationally recognised resulting from long historical practice; treaties, i.e. 'international conventions, whether general or particular, establishing rules expressly recognised by . . . States',[16] and decisions of judicial or arbitral tribunals.

Customary International Law is ill-equipped to deal with acidification, for it recognises the notion of 'territorial sovereignty', the right to carry on state activities—and to use state resources for the benefit of the state—a notion wide enough to encompass burning fossil fuels with their consequent emissions. Against this there is the belief, held by some, that states must not allow transboundary pollution from their activities to damage another state's territory.[17] This is reinforced by the statement in the United Nations (UN) Charter of Economic Rights and Duties of States that states should only exercise sovereign rights: 'without causing damage to the legitimate interests of others'.[8,18] It is further reinforced by the argument of the International Law Association that: 'States are under an obligation to abate transfrontier pollution [so] . . . that no substantial damage is caused in the territory of another state'.[19,20] The Experts' Group on Environmental Law of The World Commission on Environment and Development has recommended that states should establish criminal sanctions to punish those of their nationals, who damage the environment and in particular those who damage ecosystems of international significance, especially where their actions are in areas outside the limits of national jurisdiction.[21] The group has also recommended states should establish environmental standards (i.e. environmental quality standards, emission, technological and product standards) to prevent or abate interference with the environment,[22] and that they should prevent or abate transboundary environmental interference, or any significant *risk* thereof, which causes harm that is *more than* minor or insignificant.[23]

Even so, notions of *state* responsibility for transnational environmental damage are still relatively little developed, despite the decision in the *Trail Smelter Arbitration* (1938–1941).[24] A smelter at Trail, British Columbia, emitted pollutants which damaged parts of the State of Washington. An international arbitration between Canada and the US declared: 'under the principles of international law . . . no state has the right to use or permit the use of its territory in such a manner as to cause injury by fumes in or to the territory of another . . . when the case is of serious consequence and the injury is established by clear and convincing evidence'.[25] Damages were paid following the decision and an emission abatement programme established. However, the decision establishes no absolute rule of liability. There is, for instance, no obligation to prevent or reduce *every* harm; indeed the duty only applies where substantial or significant harm arises. Furthermore, 'clear and convincing' evidence of a breach of duty is required before liability can be established. In the *Trail Smelter* case the pollution travelled only 10 miles, and Canada admitted liability. Not all transboundary issues are so clear. For this reason the Experts Group on Environmental Law argued for establishing a principle that states should prevent and abate transboundary environmental interferences, or *significant risks thereof*, on the basis that states should be liable when they have caused a polluting event, either intentionally or negligently, or have failed, intentionally or negligently, to prevent persons within their national jurisdictions from causing pollution or failing to abate it. This principle, it is argued, should entail a further obligation to prevent activity entailing significant risk of environmental interference. The significance of risk would be established by reference to the probability of a risk materialising and the magnitude of likely harm should damage eventuate.

The improved modelling techniques for 'sulphur budgets' described above *may* make it possible to achieve enhanced state liability for transboundary pollution by development of the *Trail Smelter* principle. A major weakness of International Law has been the difficulty, indeed the impossibility, of allocating responsibility for particular acidification within a given state to another state(s). Pollution from a given source in one state causing specific pollution in another is unlawful currently. This unlawfulness could extend to situations where a variety of sources in various states give rise to emissions producing damage in another, *provided* it is accepted that the new modelling techniques enable the attribution of acidification damage to particular states. There could, however, be qualifications on a state's liability in that it might be necessary for it to be aware of the likelihood of damage from its emissions, and an obligation to take remedial action could be limited to abating transboundary polluting fluxes. Additionally, the polluted state can claim no greater degree of protection against such fluxes than it lays down by national law in respect of its own international pollution. As nearly all 'acidified states' are partly to blame for their condition, the standards they set to control themselves would have a determining influence on the content of the duty others owe to them. In addition, the expedition with which an acidified state pursued internal emission controls would influence the performance of transboundary flux controls it could require of others.[5,26–28]

In consequence of the historic weakness of customary International Law, it is necessary to consider what provision has been made by treaty. Treaties are increasingly important as a source of International Law and have been classified as those which lay down generally applicable rules ('law-making' treaties) and those dealing with individual issues, concerning only two, or a very few, states ('treaty-contracts').[29]

Though treaties are increasingly commonly made on environmental issues, state practice with regard to implementing those treaties varies considerably. In the UK parliamentary approval, normally in the form of legislation, is required in respect of treaties affecting the private rights of British subjects, or which modify or change the law of the land, or which require the grant of new powers to the Crown, or impose new financial obligations on the Crown, or cede British territory to another state. Additionally, some treaties are made expressly subject to parliamentary approval. In the US, however, treaties which, either expressly or by necessary implication, require legislation to make them operative within the nation (i.e. 'non-self executing' treaties) are not regarded as binding American courts until the necessary legislation is passed. Treaties dealing with matters which fall within any area over which Congress has exclusive legislative power are considered, *prima facie*, non-self-executing. Treaties that are self-executing are taken to bind American courts and are the law of the land by virtue of Article VI Para 2 of the US Constitution. Other states follow a variety of practices with regard to this issue: some regard self-executing treaties as part of their law, others require legislative approval for virtually all treaties.[30] Thus, even where international accord on an environmental issue is reached, implementation will largely depend on legislative action in individual states.

The starting point for modern treaty-based International Law on atmospheric pollution is The Stockholm Conference of June 1972 on the Human Environment, called following a UN Resolution of 1968. This was attended by 114 nations and 37 inter-governmental organisations. Prior to the conference the Secretary General of the United Nations argued that international accord was appropriate in relation to, *inter alia*, problems of atmospheric pollution, particularly because it is often not possible to localise the effect of a pollutant, and the need for monitoring atmospheric environmentally risky industrial activities.

The conference adopted a Declaration on the Human Environment, a non-legally-binding code of environmental ethics.[31] This contained principles providing: 'states have . . . the sovereign right to exploit their own resources and the responsibility to ensure that activities within their jurisdiction or control do not cause damage to the environment of other states . . .' (Principle 21). Principle 22 continued: 'states shall cooperate to develop the international law regarding liability and compensation for the victims of pollution . . . caused by activities within the jurisdiction or control of such states to areas beyond their jurisdiction'. Principle 17 provided: 'appropriate national institutions must be entrusted with the task of planning, managing or controlling environmental resources . . .'. These principles are sufficiently imprecise to leave a great deal of discretion to states seeking to advance their own self interest, but they have served as a basis for subsequent international developments.

As alluded to above, the OECD, whose membership includes the Western European, North American and Australasian nations and Japan, undertook a study of transboundary air pollution by sulphur. This commenced in 1972, following the Stockholm conference, reporting in 1977, and succeeded in that year by the EMEP under the aegis of the UN Economic Commission for Europe (ECE), and the UN Environment Programme (UNEP), an organisation set up following the Stockholm conference. In due course ECE initiatives, coupled with pressure from the Northern European nations and support from the Eastern European bloc, led to the Convention on Long-Range Transboundary Air Pollution, which was done at Geneva on 13 November 1979, and came into force on 16 March 1983. Between 1980 and 1985 30 nations ratified the convention, as did the EEC. Most are European nations, but Turkey, Canada and the US are also signatories. In 1984, following the 1982 Stockholm

Conference on Acidification of the Environment, a protocol was added to the convention to provide long-term financial support for the programmes of air pollution monitoring, and, in 1985, the Helsinki protocol was added, coming into force in October 1985, under which the so-called '30% club' exists, to which subsequent references will be made.

The Geneva Convention is a general law making framework committing signatories to developing policies and strategies to reduce atmospheric pollution, including transboundary pollution, and its elimination where possible. Signatories commit themselves to use the best available technology that is economically feasible to achieve these ends.[32] The Helsinki protocol requires signatories to achieve a 30% reduction from 1980 levels in their national emissions, or cross-boundary flow, of sulphur by 1993. Twenty-one states signed the protocol; many have promised reductions substantially greater than 30% by dates other than 1993. France, for example, has promised a 50% reduction by 1990, and Sweden a 65% reduction by 1995.[33] Unfortunately, and this is a further weakness of International Environmental Law, there is no legal mechanism to compel states who are signatories of the Geneva Convention also to ratify the Helsinki protocol, though international pressure may achieve the object. Thus, Greece, Iceland, the Republic of Ireland, Italy, Poland, Portugal, Spain, Turkey and, most importantly, the UK and the US did *not* join the 30% club. The UK argued its emissions of SO_2 had reduced by 42% on 1970 levels by the mid-1980s, and that it would achieve a further 30% reduction on 1980 levels of emission by the late 1990s.

British reluctance to reduce emissions illustrates one problem inherent in *any* legal attempt to set pollution eradication standards, namely how to choose a point at which emissions are, as it were, 'pegged' so that a reduction from that level can be achieved. The UK's view was that the choice of 1980 as the 'pegging' date was not in its interests as its emissions of SO_2 were already reducing at that date—though this was not a direct result only of anti-pollution measures but also represented the consequences of the decline of traditional 'smoke stack' industries, general industrial decline and a change from coal and oil to natural gas as a major industrial and domestic fuel.[34]

The Geneva Convention is a further step forward with regard to international control of polluting emissions; unfortunately its provisions are weakened by vague phraseology. Statements such as 'taking due account of the facts and problems involved', 'endeavour to limit . . . as far as possible', 'without undue delay', 'compatible with balanced development', 'the best available technology which is economically feasible', tend to compromise the effectiveness of the Convention.[35] Nevertheless, there are some stricter requirements to carry out research in, exchange information on, and monitor and evaluate on a co-operative basis the emission problem.[36] Article 10 of the Convention established machinery to review implementation and development, and Article 13 provides for settlement of disputes under the Convention to be primarily by way of negotiation between signatory states, rather than by means of punitive or compensatory sanctions.

In addition to the Convention the signatory states adopted a Resolution on Long-Range Transboundary Air Pollution under which they are to harmonise national atmospheric pollution-control policies. These are to have the aims of limiting, and, as far as possible, reducing over time and preventing, pollution of the air, including transboundary pollution.[37]

In 1982, at its 60th Annual Conference in Montreal, The International Law Association gave further guidance on how transboundary pollution could be subjected to legal control. The Association defined the phenomenon as the introduction by human action of material or energy into the environment so as to deleteriously affect health, living resources, ecosystems, material property, amenity and the legitimate use of the environment where the pollution originates in whole or part in one state and then proceeds to affect another's territory. It was further declared that states should abate and control such pollution to ensure no substantial injury is caused to another's territory, and should limit new or increased pollution to the lowest level achievable by reasonably practicable measures. Furthermore, warning of activities likely to cause significant transboundary pollution should be given and consultation with those likely to be affected should take place. Additionally, in 1982 the General Assembly of the United Nations set up the Intergovernmental Committee on the Development and Utilization of New and Renewable Sources of Energy, and UNEP has complemented the work of this body by supporting projects for energy production from renewable, and non-acidifying-gas-producing sources.

International Environmental Law, however, continues to share the weakness of many national legal systems in that it concentrates on preventing or remedying harm to a state as a consequence of, for example, emissions: it does not recognise the environment as an entity in itself, something that has its own rights. The law is concerned to act only to protect the interests of states, institutions, corporations or humans. Furthermore, it appears that International Law is concerned only with 'wrongful acts', e.g. those in breach of international obligations which arise as a result of a state's fault, i.e. its acts done wilfully, maliciously or with culpable negligence. It is hard to prove such fault with regard to acidification. This is a cumulative process: where it arises as a result of transboundary 'invasions' of airborne acidifying material, it is difficult, to say the least, in many cases to identify the source pollutants.[38] Likewise, state liability is dependent upon a wrongful act being attributable to a state: many emissions of acidifying material inside a state will, of course, result from the activities of private individuals. It should finally be noted that there is no cohesive machinery for enforcing the principles of international legal doctrines. A person injured by a transboundary polluting act must either pursue his case as best he may in the courts, or, in an appropriate case he may be able to persuade his own state government to take action under International Law. Each course of action has its own drawbacks. Action in a foreign court raises linguistic difficulties; action in one's own courts raises the problem of enforcing any judgement obtained against a person outside the courts' jurisdiction; action *by* a state against another, though possible as the *Trail Smelter* case shows, is generally appropriate only in serious cases where the determinative evidence is clear and convincing.

A gathering of 105 states at Nairobi in 1982 reaffirmed the 'Stockholm principles' but accepted that current International Environmental Law lacks stringency. In 1983 the World Commission on Environment and Development (WCED) was created by the General Assembly of the United Nations (the work of the Experts Group on Environmental Law has already been referred to). Under the auspices of the WCED a new legal basis for environmental protection has been proposed in the form of a universal convention under the auspices of the United Nations which could consolidate existing, and establish new, legal environmental protection principles, and set out concomitant rights and responsibilities of states. In addition, a monitoring and investigative UN Commission for Environmental Protection and Sustainable Development, and a High Commissioner or 'Environmental Ombudsman' to act as a 'trustee' for the environment have been recommended. Further recommendations include the following: increased accession of states to global and/or regional environmental conventions; greater adherence by states to the letter and spirit of existing conventions; updating of existing conventions to take account of the most recent scientific and technical information; drawing up of new conventions to enhance co-operation and co-ordination with regard to environmental issues; greater legal input into environmental treaty negotiations, and adoption at national level of laws reflecting international principles with regard to environmental interferences.[39]

Currently, International Environmental Law is inadequate to control acidification partly because it recognises that states are sovereign and entrusts to them the task of controlling pollution within their own territories and partly because it does not meet the criteria set out at the head of this chapter.[40] But before considering individual national legal systems' contribution to controls, it is necessary to consider a major regional system of law.

15.5 EUROPEAN COMMUNITY LAW

European Community Law has as its primary source the various treaties made by the member states, especially The Treaty of Rome 1957, and the Single European Act 1986. Article 189 of the Treaty of Rome empowers the Council and Commission of the Community to issue binding directives to member states, and these have been the main Community measures in relation to the environment. These bind as to the end to be achieved, not as to the means to achieve the end: those are left to the choice of member states. Directives are issued *not* to bring about *total* uniformity of law and practice, but to harmonise the laws of member states. The development of the Community's environmental directives has been guided by non-binding regulations issued by the UNECE.[41]

As will also become apparent when consideration is given to US law and practice, Community Law suffers from a number of problems of an institutional and inherent structural character. These are inevitable given the close connection between the ability of the Community to develop policies and issues of European integration. It has been pointed out that shortcomings in air pollution laws in both the Community and the US can be traced to 'defective . . . intergovernmental relations between central and State governmental entities, though emanating from just the opposite directions, of course. The US suffers from central government top-heaviness while the Community struggles with a power and authority deficit stemming from Member State overbearance'.[42] Furthermore, the roots of Community environmental policy can be traced back to the years 1964–1973 when it was solely incidental to measures designed to promote trade between member states. Even when a more easily recognisable environmental programme emerged in 1972 its unsure legal base, for the Treaty of Rome did not expressly empower environmental protection by the institutions of the Community, enabled member states to resist environmental measures. The Community nevertheless persevered with the development of environmental policies, and achieved legislative competence which overlapped with that of the member states by reliance on what is known as the 'soft law' approach, i.e.'a process of incrementalism representing a new type of Community competency development through political consensus of member states'.[43,44]

The Single European Act (SEA) provides, however, a firm base for environmental action. Article 100A of the amended Treaty of Rome provides for environmental matters to be a component part of the harmonisation of legislation relating to 'the internal market'. Provision is further made for decisions of the Community to be made by a 'qualified majority' (i.e. 54 votes out of 76 in the Council of the Community) which prevents one or two member states from blocking measures agreed by the others. So far, however, majority voting has not been used in relation to setting emission standards for industry, though it has in respect of product standards, for example, emission standards for motor cars in 1989.[45] More particularly, Articles 130R, 130S and 130T provide that the Community has competence to 'preserve, protect and improve the quality of the environment, to contribute towards protecting human health, to ensure a prudent and rational utilization of natural resources'. Though it is further provided that the Community shall take action on the environment only 'to the extent to which the objectives referred to . . . can be attained better at Community level than at the level of individual member states' (the 'subsidiarity principle'), it is thought that there is no longer a need to prove a link between environmental protection action and the achievement of the internal market.[46] The Community is furthermore to act 'on the principles that preventive action should be taken, that environmental damage should as a priority be rectified at source, and that the polluter should pay. Environmental protection shall be a component of the Community's other policies'. The Community is further enjoined to 'take account of available scientific and technical data, environmental conditions in the various regions of the Community, the potential benefits and costs of [environmental] action or of lack of action, [and] the economic and social development of the Community as a whole and the balanced development of its regions'. As has been noted already, the Council has not adopted majority voting in respect of industrial emission standards for Article 130S provides that unanimity of voting is to be the norm in relation to proposed environmental action; however, power is given to 'define those matters on which decisions are to be taken by a qualified majority'.[47]

Community Law somewhat shares a deficiency with International Law in that, though the Treaty of Rome requires the Commission of the Community to ensure implementation of Community Law by member states and to take proceedings against defaulting states before the Court of Justice of the Communities, the means of enforcing obligations are weak. Failure to implement enviromental directives has been an increasingly worrying feature of the Community for some years, with rates of failure accelerating of late, and with the UK being a particularly bad offender.[48]

The Commission requires notification from member states that they have translated community obligations into national law, and must also inquire whether those national measures comply, formally and practically, with relevant Community obligations. Secondly, the Commission investigates whether directive requirements are being practically implemented in member states. Wide investigative powers are given by Article 155 of the Treaty of Rome. Use is also made of periodic national reports on directive implementation. Article 169 of the Treaty of Rome enables the Commission to take proceedings against a member state for an infraction of a directive (after giving the state in question an opportunity to comment on its position) by drawing up a 'reasoned opinion' on the issue. If conformation to this does not take place the offender can be proceeded against before the Court of Justice of the Communities. Article 171 of the Treaty of Rome further requires the state in question to take 'the necessary measures to comply with the judgement of the Court of Justice'. What 'the necessary measures' are is left to the state to decide, and though the infraction should conclude there is no clear obligation to undo past misdeeds.[49] In the past states have delayed incorporating directive requirements into their laws, or have only partially incorporated them, or have treated them as recommendations, not obligations. At times judgements of the Court of Justice under Article 171 have not been implemented by the relevant state for some years; defiance of the court by states is not unknown. In such cases compliance is likely to come about only as a result of political pressure applied by other member states. Formal proceedings before the court are rarely, however, pursued to the judgement stage for the majority of disputes have been settled. For the future, however, under the Community's 4th Action Programme on the Environment, greater effort will be put into raising public awareness of Community Environmental Law and into its application.[50]

Community Law enforcement remains weak, despite more rigorous enforcement policies and a rise in the number of proceedings taken against member states by the Commission of the Community, so that in 1986 60% of Commission v. Member States cases concerned failures to implement directives, usually through oversight or inefficiency—mere internal difficulties of implementation constitute no defence in such proceedings. The Commission lacks, however, the means to monitor the implementation of Community Law, and may have to rely on interviews and exchanges of letters to obtain relevant information. It has been suggested, nevertheless, that over-rigid attempts by the organs of the Community to ensure compliance with community obligations could lead to even more examples of non-compliance. In this connection it has been argued that the Community should consider implementation problems at the policy making stage, while the Community's judiciary should also consider compliance issues when coming to their decisions. A greater degree of public knowledge of Community legal obligations, more co-operation with relevant national institutions and bodies by the Community, enhanced studies of patterns of implementation, and an independent Community implementation enforcement agency have also been suggested as ways of ensuring enhanced compliance with Community obligations.[51]

Community Law on atmospheric pollution has been a 'late developer', growing rapidly, however, from the mid-1980s as member states, particularly Germany, began to suffer from acidification.[52] The law utilises a number of means of regulation of atmospheric pollution—air quality standards, vehicle emission standards, fuel composition standards and combusion plant standards.

'Air quality standards . . . are legal upper limits imposed on levels of pollutants in ambient air during a given period of time'.[53] 'A distinction needs to be made between values having the force of law, where any infringement leads as a matter of course to the application of measures to ensure limits are respected, and those serving for guidance . . .'[54] Standards may be based on absolute concentration limits integrated over a chosen period of time, or on average values or percentiles, the object being to fix an acceptable level of pollutants in the atmosphere consistent with the 'good health' (a notoriously indeterminate concept) of those exposed to them. The process of creating such a standard involves a mixed scientific and non-scientific process. The physical and chemical properties of pollutants are determined, including toxicity and absorbability. This enables estimates to be made of likely lethal levels of any given pollutant following experimentation on cell cultures and animals, and epidemiological studies. However, these will involve extrapolations from data and a number of observed uncertainties. The experimental figures have then to be divided further by a safety factor bringing the level at which no adverse effects are observed to a lower 'limit value' for exposure. The safety factor is determined evaluatively, considering factors such as varieties of sensitivity to exposure, ranges of human pathological conditions, possible consequences of pollutant recombinations, the quality of the data relied on. Limit values are thus empirically derived with ethical, economic, political, and philosophical factors playing a part alongside mathematical and biological considerations.

Reliance on air quality standards by the Community has caused some problems with member states for they have normally utilized, for stationary sources of polluting emissions, uniform emission standards as means of control. The Community standards also utilise a 'standstill' principle to ensure there is no significant deterioration in air quality from the standard set. In this, Community Law mirrors US practice—see further below.[55] Community policy on air pollution has also been influenced with regard to acidification issues by the 1979 Geneva Convention, see above. But no single standard has a monopoly in this area, and in addition to Air Quality Limit Values (AQLVs), controls also exist over particular products, over individual sources of emissions and by virtue of integrated cross media controls. AQLVs exist, *inter alia*, with regard to smoke and SO_2 and nitrogen dioxide (NO_2).

With regard to SO_2 and airborne particulates, uniform emission limits for their presence at ground level were set by 80/779/EEC, with limits being met, where possible, by 1 April 1983, and a mandatory date for compliance of 1 April 1993, together with a duty generally to implement more stringent limits contained in Annex II of the directive. Furthermore, air quality may not deteriorate by a significant degree, even where it is currently *better* than the standard of the directive.

A complex system of values sets limits of 80 $\mu g\ m^{-3}$ for smoke, over a year (taken from the median of daily values) with a winter limit (1 October to 31 March) of 130 $\mu g\ m^{-3}$ and a year (peak) limit of 250 $\mu g\ m^{-3}$. SO_2 limits are allowed to vary according to the measure of smoke, so where smoke is low more SO_2 may be present. This has been subject to severe criticism.[56]

NO_2 standards for the ambient atmosphere were set by directive 85/203/EEC. The date for compliance with the limit was set at 1 July

1987. The limit is 200 μg m^{-3} (calculated on the basis of the 98 percentile of mean values per hour recorded throughout the year) with more stringent requirements in specially protected zones. Member states remain free to set stricter limits on NO_2.

It is unclear how far the smoke and SO_2 limits will affect transboundary pollution. The directive only applies to ground-level concentrations, though action under the directive will, to some extent, reduce total emissions and thus transboundary fluxes. On the 15 July 1980 the Council of the Community declared that member states would endeavour in accordance with the directive to limit and reduce transboundary smoke and SO_2 pollution.[57] Revision of the SO_2 limits is due to take place by the end of 1992.

The implementation of AQLVs has met with resistance by some member states; both France and Germany have been threatened with action under Article 169. Problems of monitoring implementation have also been experienced as have practical problems on the part of member states.[58,59]

Community vehicle emission standards were introduced in 1970.[60] The basic notion was that in tests (the manner of performing which was specified) the mass of carbon monoxide (CO), the mass of hydrocarbons and the mass of NO_x had to be less for vehicles of a given reference weight than certain specified figures. The directive and its successors provide also for 'type approval' of vehicles conforming with specified design and construction standards. This allows for block approval by competent authorities of types of vehicle on the basis that they meet, for example, emission standards. Where such approval is given manufacturers may proceed to produce a series of vehicles conforming to the type specification approved. Provided the standards of the directives are met it is not competent for a member state of the Community to refuse national type approval to a vehicle, i.e. vehicles fulfilling directive requirements must be approved as satisfying member states' emission requirements. Type approval, however, allows individual production line vehicles to produce more emissions than an individual vehicle tested for compliance.

The emission requirements of directives have gradually been made more stringent.[61] The Community's approach to setting standards historically relied on adapting the law to suit existing technology. Stricter limits emerged only as technologies improved rather than by following the path of requiring exhaust emissions to be as low as technically achievable (ALATA) (the ALATA principle which tends to force the pace of technological improvement). Furthermore, the technical parameters within which emission standards are set are based in the Community on a maximum speed of 50·0 km h^{-1}, with an average speed of 18·7 km h^{-1} starting with a cold engine. Such test values are good for determining emissions of CO and hydrocarbons, but are less efficient with regard to NO_x whose emissions increase at higher speeds. By way of contrast, in the US the technical parameters are based on much higher average speeds.[54]

Since 1985 the Community has moved towards the introduction of vehicles with stricter emission limits based, not on the weight of the vehicle, but its engine's cubic capacity. Proposals were submitted by the Commission in 1984 on emissions. These were in response to Community realisation that its emission standards were less strict than those of the US and Japan. The Commission accordingly sought to establish the objective of reducing emissions of CO, NO_x and hydrocarbons, but did not specify the technical measures necessary to achieve this.[62] The Community was able to agree on standards for large motor vehicles weighing up to 3·8 t with engine cubic capacities of more than 2000 cc. These would meet US emission standards and be fitted with three way catalytic converters. For small cars, i.e. those of less than 1400 cc capacity, a two-stage reduction of emission limits would be introduced from 1990–1991 and from 1992–1993. Problems arose over medium cars between 1400 and 2000 cc: there was a dispute over the appropriate technology to achieve a reduction of emissions—catalytic converters or 'lean burn' engines. A compromise solution was reached in Luxembourg in 1985; subsequently Danish opposition prevented its implementation. Following the change in the Community's voting procedures from 1986 under the Single European Act to allow majority voting Danish opposition was overcome. The new directive on emission limits was adopted formally in December 1987. Under the requirements of directives 88/76/EEC for cars of 2000 cc or more capacity as from 1988–1989 CO emissions in g $test^{-1}$ would be 25, hydrocarbons and NO_x would be 6·5, and NO_x would be 3·5. For medium-sized cars the limits would be, as from 1991–1993, 30 g $test^{-1}$ for CO and 8·0 g $test^{-1}$ for hydrocarbons and NO_x with no specific limit for the latter. For small cars as from 1990–1991 the limits would be 45 g $test^{-1}$ for CO, 15 for hydrocarbons and NO_x and 6·0 for NO_x. A further reduction in small car limits from 1991–1993 would produce figures equivalent to those for medium-sized cars. However, following pressure from the European Parliament in June 1989 it was agreed that the emission limits for small cars should from 1992 be 19 g $test^{-1}$ for CO and 5 for hydrocarbons and NO_x.[63] There is also a Community commitment to implementing emission controls as strict as those of the US, to exclude non-conforming vehicles from member states and to allow states to introduce tax incentives for those who purchase the new 'smaller, cleaner' cars ahead of the time when compliance with stricter emission requirements is mandatory.[64]

The Community's objective in introducing stricter limits is, *inter alia*, to reduce car NO_x emissions from $3{\cdot}17 \times 10^6$ t $year^{-1}$ to $1{\cdot}67 \times 10^6$ t, with $1{\cdot}33 \times 10^6$ t of this coming from petrol-engined vehicles. However, such a reduction depends on factors additional to vehicle construction standards—how fast will the turnover be to the new 'clean' cars, what will increases in vehicle numbers and mileages be? Polluting emissions from mobile sources account for some 25–30% of the Community's atmospheric pollution; if this is to be reduced in real terms even stricter emission controls will be required in the future.

Fuel composition standards designed to reduce SO_2 pollution date from 1975 with regard to 'gas oil', i.e. fuels used primarily for domestic heating.[65] Reductions in the permitted sulphur content of gas oil were proposed in 1985 and in 1987 directive 87/219/EEC, setting limits to be met as from 1 January 1989, was adopted. The sulphur content of gas oil in all member states is now not to exceed 0·3%. Member states may enforce a 0·2% sulphur content limit in areas where this is necessary to prevent damage to the environment or national heritage. A further reduction in limit values may be initiated.

Currently, there are no Community limits with regard to 'fuel oil', i.e. oil used for industrial heating and in power stations. Work on drawing up such limits was commenced in 1975 but halted in 1981. However, in 1988 the Commission recommenced its work and extended it to cover the sulphur content of coal with a view to introducing measures to reduce acidifying emissions.[66]

The general directive of 1984 (84/360/EEC) on air pollution from industrial plants was an important initiative with regard to acidification, marking a move from directives fixing individual quality standards for particular pollutants to a framework under which a community-wide system of limit values for a whole range of emissions can be fixed. The directive requires that, as from 30 June 1987, member states must ensure certain listed industrial plants (including coke ovens, oil power stations and other combustion plants whose heat output is more than 50 MW) cannot be constructed without prior authorisation only to be issued in certain circumstances. These include requirements that all appropriate preventive pollution measures must be taken, including use of the best available technology not entailing the incurring of excessive costs; that emission limit values are not exceeded, taking into account AQLVs also, and that significant air pollution must not be caused by emissions. In particular, pollution caused by, *inter alia*, SO_2 and NO_x must not be caused.

Any substantial alteration of an existing plant is considered as a new plant needing authorisation, and the directive further requires that, in any case, existing plant should, over time, be adapted by use of the best available technology to reduce air pollution. Member states are further required to keep industrial plants up to date with developments in what constitutes 'the best available technology'. There are also provisions on fixing limit values, measurement and assessment techniques, information exchanges, and to ensure the public are informed of applications for plant authorisation.

The weakness of the directive is that any emission limits fixed under it must be unanimously agreed and cannot be fixed by a qualified majority vote. Furthermore, the requirement that the best available technology should be used is diminished by the rider that this should not entail 'excessive costs'. Nevertheless, any limits set must take account of the natures, quantity and harmful potential of emissions.

A further problem is that relevant plants have to meet emission standards that have been established for their type of installation, and this raises a procedural issue as to whether the views of the Commission of the Community 'should be expressed in informal guidance documents or specific directives. If directives are chosen, then the question arises as to whether those directives should be adapted to technical progress by the EC Commision itself . . . or only with EC Council approval'.[67]

Before the adoption of 84/360/EEC there was a further proposal to limit polluting emissions from large combustion plants. This related to all combustion plants rated with a heat output of over 50 MW, thermal power stations (normally those with an output of 300 MW or more) and other large industrial heat generators. Limit values for, *inter alia*, SO_2 and NO_x for new plants were proposed and limit values for existing plants to be applied on a reducing basis were further proposed. The object was, by 1995, to reduce SO_2 emissions by 60% compared with 1980, and NO_x by 40%. After debate by member states a further directive was issued by the Council of the Community on 24 November 1988 in respect of large combustion plants (LCP).[68]

The basic objects of this directive are

- to reduce from 1980 levels of emissions of SO_2 from existing plant in three stages, 1993 (25%), 1998 (43%), 2003 (60%);
- to reduce emissions of NO_x from such plant in two stages, 1993 (20%), 1998 (36%); and
- to set Community emission limits for, *inter alia*, SO_2 and NO_x from new plants (i.e. those commissioned after July 1987) so that, compared with emissions from unmodified plants, SO_2 emissions are to be reduced by 80% and NO_x emissions by some 50%.

Certain exemptions are granted to some member states, particularly Spain, whose indigenous fuel supplies are so 'rich' in sulphur that desulphurisation would be disproportionately expensive, or whose emissions annually are lower than those of other member states (e.g. the Republic of Ireland and Greece). Furthermore, the state of emission reduction will vary from state to state, allowing for economics, modes of energy production and environmental differences. Belgium, France, Germany and the Netherlands are to exceed overall requirements by achieving 70% SO_2 reductions by 2003, for example. Implementing the directive should reduce Community power and refinery plant emissions of SO_2 by 6×10^6 t year^{-1} by 1989, and halving them, compared to 1980, by 2003.

The 1988 directive has been hailed as signalling 'the solution to the Community's acid rain problem'.[69,70] That may be too sanguine: in particular reduced acidifying emissions from one source, e.g. thermal power stations, may be cancelled out, in part at least, by emissions from other sources, e.g. if growth in personal ownership and use of motor vehicles is allowed to continue unchecked. Failure to see acidification as a problem 'in the round' with a number of interlocking causes is still apparent in much of the Community's dealings with the issue. It cannot be said that currently community measures are as effectively co-ordinated and integrated as they could be.[71]

Similar criticisms may be levelled at national legal systems' attempts to deal with acidification.

15.6 THE UK

The UK is a member of the European Community. As such its laws are increasingly shaped by Community requirements in an attempt to ensure compliance with directives, though, as stated above, non-compliance as a charge has often been levelled at the UK.

General air quality standards have never been, historically, a feature of

British environmental regulation. There is no one, comprehensive British statute which relates to the issue, instead piecemeal legislative provision exists under The Clean Air Acts of 1956 and 1968, The Alkali etc. Works Regulations Act 1906, The Health and Safety at Work Act 1974, the Control of Pollution Act 1974, The Road Traffic Act 1988, and The Environmental Protection Act 1990.

The 1956 and 1968 Clean Air Acts are largely concerned with domestic smoke control. Smoke control, and changing heating habits, led to a reduction in domestic smoke emissions (in $\times 10^6$ t) from 1·50 to about 0·50 between 1953 and 1974. However, the Acts are primarily concerned with control of visible smoke: this includes soot, ash, grit and grimy particles, and visible vapours such as water vapour, but does not seem to extend to invisible material such as CO or free SO_2.[72,73] The other legislation mentioned above is more particularly relevant to the control of emissions from vehicles and power stations and the sulphur content of oil and so will be dealt with in greater detail subsequently. However, this highlights the, historically, fragmented approach of the UK with regard to air quality, despite the recommendation of the Royal Commission on Environmental Pollution, as long ago as 1976, that there should be a comprehensive Act regulating both combustion and non-combustion processes, with local authorities appointable as agents for central regulatory inspectorates.[74]

British practice in this area has been primarily based on the concept of reducing pollution by requiring individual polluters to use 'the best practicable means' to reduce emissions, rather than on setting mandatory quality standards for air generally. The UK has lacked co-ordinating provisions to bring together relevant statutory authorities so that all appropriate means of control are harmonised to work to one end. In 1982, however, a comprehensive review of air pollution legislation was initiated. A consultation paper issued in 1986 proposed that European Community standards for air quality, SO_2 and NO_2, should be translated into UK law, with ministers being empowered to set air quality standards (nationally and regionally) and to limit specified emissions or emissions from specific processes.[75] However, despite moves by the Department of the Environment (DoE) to move towards systems of integrated pollution control,[76] major legislation to implement the above proposals was needed. In the interim the UK's response to Community air quality directives was to argue that action taken under existing legislation could achieve the required objectives, and guidance in the form of DoE Circular 11/81 (27 March 1981) *encouraged* local authorities, the principal agents for implementing the Clean Air Acts, to utilise their powers to implement Community requirements. Of course, a circular is not 'law' and the normal rule is that no penalty can be imposed directly on those who do not comply with it.

All in all it is hard to see how air quality controls in Britain have played any real part in combatting acidification.[77] However, this will change; see further below.

Controls over the sulphur content of gas oil in the UK may also be briefly disposed of. Under Sections 75–77 of the Control of Pollution Act 1974 ministers are empowered to control, by way of regulations, the content of liquid fuels. Regulations have been introduced to meet the requirements of directives 75/716/EEC and 87/219 EEC on the sulphur content of gas oil.[78–80] Control does *not* extend to the sulphur content of fuel oil used in power stations and industrial heating plants.

Vehicle emissions have been increasingly controlled in line with the progressive tightening of emission limits set by the European Community. The emission requirements for vehicles generally depend upon their year of introduction to the roads and the limits in force in that year. The basic scheme of control was initially drawn up before the UK acceded to the Treaty of Rome: the form of regulation is that vehicles must comply with British standards and, in some cases, with those fixed by successive Community directives.

The Road Vehicles (Construction and Use) Regulations 1986,[81] as amended, are continued in force under Section 41 of The Road Traffic Act 1988. The statute enables the Secretary of State to make regulations detailing and specifying the construction of motor vehicles, i.e. mechanically propelled vehicles, other than motor cycles or invalid carriages, constructed to carry loads or passengers.[82] Regulation 61 of the Construction and Use Regulation currently lays down obligations that vehicles shall not emit avoidable smoke or avoidable visible vapours, and, in the case of spark-ignition-engined vehicles first used on or after 1 October 1982 they must comply at the time of first use with Community directives 78/665/EEC or 83/351/EEC as to emission standards. However, in the case of such vehicles *first used before 1 April 1991* they may alternatively comply at that first use time with directives 78/665/EEC, 83/351/EEC or 88/76/EEC. In any other case a vehicle may comply with either 83/351/EEC or 88/76/EEC. It is an offence under Section 42 of the 1988 Act to contravene or fail to comply with the Construction and Use Regulations.

It will be noted that British emission requirements do not yet translate the most recent, and strictest, Community emission directives into domestic law. The minimum emission requirements with which most vehicles in the UK must comply are those of 78/665/EEC. However, those first used before 1 April 1991 are allowed to comply with the stricter requirements of 83/351/EEC or 88/76/EEC. After 1 April 1991 vehicles coming into first use have to comply with the requirements of only either 83/351/EEC or 88/76/EEC. Action to implement directive 89/458/EEC is awaited.

The UK operates type approval schemes as provided for by Community directives and these in general mirror the emission requirements of the various directives.[83,84]

Control of industrial sources with acidifying emissions in the UK has principally been entrusted to central government inspectorates operating under the Health and Safety at Work Act 1974.[85,86] Control extended to premises falling within the Health and Safety (Emissions into the Atmosphere) Regulations,[87] in particular thermal electricity generation, gas, coke, petrochemical and petroleum works are covered. Under Section 5 of the Health and Safety At Work Act 1974 those controlling such works have had to use the 'best practicable means' for preventing emission into the atmosphere from their premises of noxious or offensive substances and for rendering harmless and inoffensive such substances as may be emitted. 'Noxious and offensive substances' are defined to include NO_x, SO_2 and sulphur trioxide (SO_3).[88]

'The best practicable means' referred to above was never exhaustively defined: it certainly included the manner of using plant provided for emission control purposes and also the supervision of any operation involving emissions of regulated substances.[89] The expression would seem to comprehend a duty to take account of the prevailing state of scientific knowledge[90] and to comply with proper standards of construction and reasonable use.[91] However, it also enabled taking into account the means of those subject to the requirement and the economic consequences to the community of control. The technological limitations of old equipment could also be taken into account and the traditional stance of the enforcing inspectorate was not to place a burden of taking precautions disproportionately heavy to the risks involved on operators of regulated premises.[92,93]

In order to comply with the requirements of Community directives 84/360/EEC and 88/609/EEC (see above) the UK needed to act firmly, and though legislation was promised to set specific air quality and emission standards, it also seemed that continued use would be made of the 'best practicable means' system, with emission limits forming part of guidance issued by Her Majesty's Inspectorate of Pollution (HMIP). The CEGB (now privatised as 'National Power' and 'Powergen') undertook to retrofit three of its largest existing thermal stations with flue gas desulphurisation equipment, but it was argued that this is not enough to reduce UK SO_2 emissions by 60% by 1993. It appeared, therefore, that other plants, e.g. oil refineries, would have to yet further reduce their SO_2 emissions if the UK's obligations overall were to be met. The reduction programme was to be submitted to the EC by the end of 1990. The scheme of the new legislation was to enable the Secretary of State to plan for reduction of specified pollutants which HMIP would then apply, plant-by-plant, in specific authorisations.[94,95]

That is the system now to be found in the Environmental Protection Act 1990. However, even before its enactment The Air Quality Standards Regulations SI 1989/317 were made under Section 2(2) of the European Communities Act 1972 in order to implement, *inter alia*, directives 80/779 and 85/203 on SO_2 and smoke and NO_x respectively, to set AQLV, to require monitoring of air quality, and to reduce specified pollutants in air below the specified limit. Certain industrial districts in the West Midlands, Yorkshire, the North East and, in Glasgow, were, however, exempted from regulation until 1 April 1993. The 1990 Act goes much further than this. The scheme of the Act is that, *inter alia*, the air is defined by Section 1 as an environmental medium, and as such receives protection from polluting releases from any process or substances capable of causing harm to humankind or other living organisms. 'Harm' includes damage to the health of organisms or interference with the ecological systems of which they are a part. Section 2 of the Act then empowers the Secretary of State to make regulations to prescribe processes needing authorisation—some (the most complex) being subject to oversight by HMIP who will apply integrated pollution control (IPC). These processes—the 'A' list—are thus subject to a system in which one body (HMIP) regulates *all* polluting emissions to *all* the environmental media–land and water as well as air. The 'B' list processes on the other hand, being less complex, are subject to control by local authorities who regulate them in respect of atmospheric pollution only. For the purposes of this present study it is 'A' list processes which are of prime importance, and these are prescribed by Schedule 1 of the Environmental Protection (Prescribed Processes and Substances) Regulations SI 1991/472. Chapter 1 of this includes production of fuel and power and associated processes in the 'A' list, in particular boilers and furnaces of 50 MW capacity or more.

No person is able to carry on a relevant prescribed process without an authorisation from HMIP, operated in accordance with any conditions imposed, see Section 6 of the 1990 Act. HMIP may refuse applications for authorisation, or may grant them subject to conditions, and periodic reviews of authorisation conditions are statutorily required. Section 20 of the Act further requires HMIP to maintain a public register of information relating to authorisations. However, the heart of IPC regulation is the BATNEEC system of Section 7 of the 1990 Act. 'The Best Available *Techniques* Not Entailing Excessive Costs' must be used in authorised processes to prevent the emission of prescribed substances, into the environmental media, or otherwise to reduce them to a minimum and render them harmless. The distinction between the wording of the UK Act and relevant EC directives which refer to 'technology' as opposed to 'techniques' should be noted; the change of wording is considered to make BATNEEC more flexible and adaptive to new situations. Guidance issued so far on BATNEEC indicates that it will be an *objective* standard set individually in relation to individual processes and that it will not depend upon the authorisee's ability to meet costs, though certain economic issues will be taken into account: for example, a marginal decrease in emissions will not be required if this entails a massively disproportionate expenditure. The use of the word 'techniques' indicates that it is not just equipment that must be used to prevent emissions; *how* staff are trained, and *how* equipment is used are contemplated as 'techniques', while an individual BATNEEC standard will be expressed in performance terms rather than as a requirement to use a particular technology, thus enabling adaptation to new devices, etc., as they appear. The BATNEEC system is further supplemented by Section 3 of the 1990 Act which enables the Secretary of State to make regulations in relation to releases from processes setting standard limits for the amount, or amounts at a given time, of prescribed substances, their concentrations and characteristics. The Secretary of State is furthermore empowered to create environmental quality objectives for the environmental media, and, further, to plan for establishing limits for total releases of substances into the environment, and to bring about a phased reduction of emissions.

It is under these powers, officially claimed to be the strictest European industrial pollution regime, that the UK will seek to meet its European obligations outlined above.[96] It is considered by the DoE that the target reductions of 40% (SO_2) and 30% (NO_x) from 1980 levels by 1998 in accordance with the LCP directive can be achieved by fitting flue gas desulphurisation (FGD) equipment at 8000 MW of power plant capacity, coupled with the burning of low-sulphur coal and the phasing-out of some old coal-fired plant and replacing it by gas-powered plant. It will, however, be more costly to reduce SO_2 emissions by the 60% figure required by the directive for 2003, and considerably more capital expenditure will be needed.[97]

15.7 THE US

Legislative air pollution controls in the US effectively date from the Clean Air Act of 1963 followed by The Motor Vehicles Air Pollution Control Act 1965 laying down federal emission limits for vehicles. This Act was extended to regulate stationary sources of pollution in 1967 and a radically new system of control, with responsibilities divided between the state and the federal government, acting via the Environmental Protection Agency, (EPA), was introduced by The Clean Air Act Amendments 1970, which was further amended in 1977.

The law in the US has been in many ways proactive, setting emission limits, particularly for vehicles, based on the most advanced technical possibilities and, indeed, on possible future developments. Consequently, vehicle makers, in particular, have had to adopt complex and expensive measures to meet legal requirements.[98,99]

National Ambient Air Quality Standards (NAAQS) have been set under the above legislation for, *inter alia*, SO_2 and NO_x, with a view to ensuring the health of the population and 'public welfare' which includes protection of soil. NAAQS themselves are not directly enforceable; they are guides used in determining degrees of emission control to be enforced with regard to existing polluting sources, and to restrict the location of new sources. Implementation of NAAQS is a state responsibility. They must draw up state implementation plans (SIP). These actually control, *inter alia*, pollutant emission levels for existing and new emission sources, with requirements that standards be attained 'as expeditiously as possible'. States are able to require use of 'reasonably available control technology' (RACT) on existing emission sources. Not only, however, should SIP achieve NAAQS, they ought also to maintain them. Steps are required for prevention of significant deterioration (PSD) of air quality. Each SIP must provide for controls over new polluting sources and reviews of existing ones, where they are modified, to ensure PSD. Thus, new or modified sources require permission in order to obtain which it must be shown there will be no violation of NAAQS, that

PSD requirements with regard to, *inter alia*, SO_2, are met and that the best available control technology, (BACT) is being used.

In addition to NAAQS US law provides for individual emission standards for new and modified polluting sources, the new source performance standard (NSPS). Under the 1977 amendments to clean air legislation, EPA is required to set NSPS for sources burning fossil fuels. FGD plant may have to be installed to meet the standards. NSPS now exists with regard to SO_2 and NO_x emitted from solid-fuel- and oil- or gas-fired boilers in electricity power stations.

With regard to motor vehicles, increasingly strict emission standards have been applied. The 1970 Act prescribed emission reductions of 90% in respect of NO_x emissions (as measured coming from 1971 vehicles) by 1976, though this requirement was subsequently suspended and modified. In the US, petrol-engined vehicles of up to 6000 lb (2700 kg) gross weight may only emit up to 0·41 g mile^{-1} of hydrocarbons, 3·4 g mile^{-1} of CO, 1·0 g mile^{-1} of NO_x and 0·6 g mile^{-1} of particulate matter. For light-duty vehicles of up to 8500 lb (3825 kg) gross weight the limits are 0·8 g mile^{-1} of hydrocarbons, 10·0 g mile^{-1} of CO, 2·3 g mile^{-1} of NO_x and 0·26 g mile^{-1} of particulate matter, and for heavy-duty vehicles (over 8500 lb (3825 kg)) the limits are 1·3 g bhp^{-1} h^{-1} of hydrocarbons, 15·5 g bhp^{-1} h^{-1} of CO, 10·7 g bhp^{-1} h^{-1} of NO_x and 0·25 g bhp^{-1} h^{-1} of particulates. California, though its emission limits are *overall* as strict as the rest of the nation, is allowed to vary its requirements because of its generating higher than average vehicle speeds—for example its NO_x requirements are stricter.[100]

New vehicles may not be introduced into service unless certified by EPA as conforming to emission standards, and prototype vehicles are tested by EPA for certification purposes. EPA also has powers to test assembly lines of production line vehicles following certification of the prototype. If, following checks on a sufficiently large number of vehicles chosen at random on the production line, it appears the type is not meeting standards the certificate of conformity may be revoked.

Vehicle manufacturers are statutorily required to warrant to purchasers that what they buy conforms, in terms of design, specification and construction, with relevant emission standards, and, further, that the vehicle contains no defects of material or workmanship which would result in non-compliance with standards. Additionally, manufacturers must warrant that compliance with standards will subsist throughout a vehicle's useful life, *provided* maintenance and operation is carried out in accordance with the manufacturers' instructions, and also that the failure to comply resulted in the owner of the vehicle being penalised subject to sanctions. The efficacy of these provisions, which are, of course, not absolute requirements, depends upon the existence of vehicle testing programmes, a state rather than a federal responsibility. EPA has taken action against states which have failed to implement satisfactory inspection and maintenance (I/M) vehicle testing programmes, a requirement for such a programme arising after a state has asked for more time to meet NAAQS in respect of CO and NO_x.[101] Since 1977 an increasing number of states have introduced I/M programmes, and by 1986 30 states had such programmes.

EPA has further powers to order vehicle makers to recall for repairs and/or modification groups of vehicles failing to meet emission requirements, and these powers have been used successfully in respect of some thousands of vehicles.[102] Indeed it seems where vehicles do not conform with standards *the* remedy is recall and repair.[103]

Despite the existence of all the foregoing powers, the law in the US has not been as effective as many could have wished. One major issue has been the lack of enforcement. The Air Quality Act 1967 was ineffectively enforced because it depended on action by individual states who had neither the will nor the resources to tackle the issue. As a reaction to this the clean air legislation of 1970 relied on federal action with provision for enforcement sanctions. However, that legislation has been criticised as being too ambitious, and conceived in a euphoria of public anti-pollution consciousness which resulted in 'fixed, unrealistic deadlines, frequent amending, and corresponding loss of faith in the Act by the public' the result of setting 'public health as an absolute goal disregarding cost' and 'precluding efficiency maximising trade-offs between health targets, economics and technological feasibility'.[104] The legislation has also been criticised for relying too heavily on technological solutions to pollution problems while paying scant heed to the development of alternative strategies of control based on a 'flexible multi-faceted control approach'.[104] Furthermore, the overall strategy of US clean air legislation has been criticized: 'There are four basic US federal implementation strategies for air pollution legislation: first, federal standards that are federally implemented; secondly, federal standards implemented and enforced by the states; thirdly, a free-market approach; and fourthly, federal incentives for state adoption and implementation of environmental measures. The Clean Air Act has utilised a mixture of all four . . . It was Congress's will that state and local governments be the main executors . . . because of limited federal EPA budgets and resources, difficulties of centralised information collection, greater familiarity of state and local officials with regulated facilities and the existence of varying local conditions affecting implementation and enforcement, and the desire of States to maintain a degree of flexibility and independence.'[105] This has been coupled with an under-resourced EPA whose success in limiting polluting emissions has been limited and which has been criticised as being too accommodating towards the states' own inadequate atmospheric pollution control strategies.

As a counter-argument to what is seen as the over sanction-based approach of the US legislation, there have been calls for a market-based strategy having 'one simple aim: to find the most efficient way to spend each anti-pollution dollar' and, while retaining ecological goals, allowing 'market forces to determine the best way to achieve these . . . rather than forcing solutions on polluters by law'.[106] Such a system would be based on a notion of issuing 'permits to polluters specifying an allowable amount of discharge. Some companies might sell surplus permits to other firms; others might buy extra permits . . . permit credits could be sold or leased to other companies with problem smokestacks . . . Companies which could afford to reduce their emissions fairly cheaply would sell their surplus reduction credits to companies crippled by emissions reductions mandated for them . . . permit amounts could then be gradually reduced until the air pollution limits set by the EPA and authorised by Congress are met'.[107] However, such a market-based philosophy is not reflected in the latest US laws as will become apparent later.

The US legislation is moreover not particularly directed towards acidification as a problem. EPA has considered it lacks powers to tackle the issue head on and, under President Reagan (1980–1988), was not vigorous in enforcing the law. Despite the existence of NAAQS, responsibility for implementing clean air policies has remained based on state-by-state strategies, and the regulation of individual polluting sources. What is required is a national strategy to reduce emissions, particularly of SO_2, from power stations in the eastern and mid-western US. Despite pressure, and hopes, for legislative change throughout the 1980s little was achieved.[108–111] In May 1989 it was reported that there would be proposals from the White House to amend the Clean Air Act, and President Bush's 1989 budget message to Congress stated that acidifying emissions would be reduced. This could involve a reduction of 10×10^6 t year^{-1} of SO_2 from coal-burning power stations, switch-overs to engines running on methanol and ethanol, and restrictions on car ownership and use. Such a controversial package was expected to be fiercely opposed.[112,113]

The US Clean Air Act was further revised by The Clean Air Act Amendments of 1990, Section 1630, which received the presidential signature on 15 November 1990. The new laws are highly prescriptive in form and are phrased in a highly detailed manner. US air pollution laws have been made more stringent and EPA's powers have been extended to allow it to create maximum achievable control technology standards for some 250 emission sources of 189 hazardous air pollutants. In addition to preconstruction permits for polluting sources already required under the previous law, emission sources will further require operating permits which will set compliance standards and which will need periodic recertification, monitoring and reporting of information. Furthermore, each source will have an emission limit. EPA's powers of enforcement have also been strengthened by the new law, as have those of the state agencies, with fines of US$1 million per offence being allowed and imprisonment of 1–15 years also being allowable for guilty individuals—a phrase which can apply to employees guilty of allowing polluting emissions, their supervisors, senior managers and even corporate officials. However, there are doubts as to how well this new package of laws can work. The cost of compliance has been estimated at between US $25 and 50 billion ($10^9$) year^{-1}, while EPA's ability to meet its new tasks has also been called into question.[114] It still remains to be seen whether the new US laws will be more effective than their predecessors in combatting acidification.

15.8 CONCLUSION

Acidification is a world-wide problem which can only be solved by world-wide emission control strategies implemented vigorously on a regional and national basis. Such implementation will not be without cost or pain. New technologies can reduce acidifying emissions, but their application will involve expense: furthermore, real reductions in emissions may involve restrictions on vehicle use and ownership leading to a diminution of ease of mobility, arguably one of the 20th century's principal freedoms. Nevertheless, there is the potential for law to enshrine increasingly effective atmospheric protection standards: they are already embryonically present. The real question is whether there is the political and social will to press forward with introducing and implementing those standards.

It must further be questioned whether the current legal approach in both the US and the EC, which is based on an increasingly tough sanctions-based model, is the most appropriate way of dealing with the issue. Some thought needs to be given to evaluating the alternative market-based permit approach advocated by some lawyers in the US, and increasingly by economists world-wide. This has the merit of attempting to couple economic and legal forces. Certainly, a purely based sanctions approach will prove ineffective unless it is adequately enforced for 'over-regulation and under-enforcement'; it is undoubtedly an evil which undermines public respect for the law. Such an eventuality would be most undesirable at a time when the public ethical climate clearly favours moves to be made towards enhanced environmental protection.[115]

REFERENCES AND NOTES

1. Royal Commission on Environmental Pollution, Tenth Report, Cmnd 9149. HMSO, London, UK, 1984.
2. *The Long-Range Transport of Air Pollutants: Measurements and Findings*. OECD, Paris, France, 1977.
3. International Electric Research Exchange Working Group II, *Effects of SO_2 and its Derivatives on Heath and Ecology*, Vol. 2. Central Electricity Generating Board Research Laboratories, Leatherhead, UK, 1981.
4. Elsom, D., *Atmospheric Pollution*. Blackwell, Oxford, UK, 1987, pp. 250–73.
5. *Interim Report of the Committee on Legal Problems of Continuous and Instantaneous Long Distance Air Pollution*, International Law Association's 62nd Conference. ILA, London, UK, 1987, p. 198.
6. Royal Commission on Environmental Pollution, *op. cit.*, paras 5.82–5.89.
7. Fremlin, J. H., *Power Production: What Are the Risks?* OUP, Oxford, UK, 1987, pp. 247–51.
8. Building Effects Review Group, *The Effects of Acid Deposition on Buildings and Building Materials*. HMSO, London, 1989.
9. Wetstone, G. S. & Rosencranz, A., *Acid Rain in Europe and North America: National Responses to an International Problem*. Environmental Law Institute, Washington, DC, USA, 1983.
10. Royal Commission on Environmental Pollution, *op. cit.*, paras 5.91–5.96.
11. Elsom, D., *op. cit.*, p. 260, Table 11.2.
12. *The Times*, 22 Mar. 1989.
13. Elsom, D., *op. cit.*, p. 259, Table 11.1.
14. Report of the International Conference on Acidification and its Policy Implications, Amsterdam, 9 May 1986, International Law Assoc., London, 1986.
15. Stark, J. G., *Introduction to International Law*, 10th edn. Butterworth, London, UK, 1989, p. 15.

16. *Ibid.*, p. 33.
17. Bothe, M., Developments and problems of co-operation with regard to transboundary air pollution in Western Europe. In *Transboundary Air Pollution: International Legal Aspects of the Co-operation of States*, ed. C. Flinterman, B. Kuriathowska & J. G. Lammers. Martinus Nijhoff, Dordrecht, The Netherlands, 1986, p. 120.
18. United Nations Charter of Economic Rights and Duties of States, 12 Dec. 1974.
19. Committee on Legal Aspects of the Conservation of the Environment, *Draft Rules of International Law*, Belgrade Conference 1980. ILA, OECD, Paris, France, art. 3, para. 1.
20. *ILA Montreal Rules of International Law Applicable to Transfrontier Pollution*, Report of the 60th Congress. ILA, OECD, Paris, France, art. 3.1.
21. Lammers, J. G., ed., *Environmental Protection and Sustainable Development*, Graham & Trotman/Martinus Nijhoff, London, UK, 1987, p. 9.
22. *Ibid.*, p. 54.
23. *Ibid.*, p. 75.
24. 1941 American Journal of International Law 684.
25. 3 Reports of International Arbitration Awards. 1911 and 1965
26. Lammers, J. G., *op. cit.*, pp. 78 –9.
27. *Yearbook of the International Law Commission*, Vol. 2.,1979, pp. 90, 176.
28. *Problems in Transfrontier Pollution*. OECD, Paris, France, 1979.
29. Stark, J. G., *op. cit.*, p. 42.
30. *Ibid.*, pp. 80–6.
31. *Ibid.*, pp. 403–10.
32. Johnson, S. P. & Corcelle, G., *The Environmental Policy of the European Communities*. Graham & Trotman, London, UK, 1989, pp. 304–5.
33. Elsom, D., *op. cit.*, p. 261.
34. *Ibid.*, pp. 263–4.
35. Geneva Convention, Articles 2 and 3.
36. *Ibid.*, Articles 7–9.
37. Gunding, Multilateral Co-operation of states under the ECE Convention on Long-Range Transboundary Air Pollution. In *Transboundary Air Pollution: International Legal Aspects of the Co-operation of States*, ed. C. Flinterman, B. Kuriathowska & J. G. Lammers. Martinus Nijhoff, Dordrecht, The Netherlands, 1986.
38. Levin, *Protecting the Human Environment*. UNITAR, New York, USA, 1977, p. 44.
39. Lammers, J. G., *op. cit.*, pp. 11–12, 15–17.
40. The author acknowledges his indebtedness to an unpublished degree dissertation (Acid rain—can the law effectively protect our environment, by Catherine Oliver, University of Leicester, UK, 1988), the research for which was carried out partly under his supervision.
41. Haigh, N., *EEC Environmental Policy and Britain*, 2nd rev. edn. Longman, London, UK, 1989, p. 221.
42. McGrory, Air pollution legislation in the United States and The Community. *Euro. Law Rev.*, **15** (1990) 316.
43. *Ibid.*, p. 305.

44. Wellens, & Borchardt, Soft law in European Community law. *Euro. Law Rev.*, **14** (1989) 267.
45. Haigh, N., EEC policy and implementation. In *European Environmental Yearbook*, 2nd edn. Doc Ter International London, UK, 1991.
46. Usher, The source and structure of EEC environmental law. In *Environment and Planning Law*. Butterworth, London, UK, 1991.
47. *Ibid.*, pp. 3–15.
48. Johnson, S. P. & Corcelle, G., *op. cit.*, p. 339, Table 16.1.
49. Hartley, T. C., *The Foundations of European Community Law*, 2nd edn. Clarendon Press, Oxford, UK, 1980, pp. 310–15.
50. Johnson, S. P. & Corcelle, G., *op. cit.*, pp. 340–1.
51. McGrory, *op. cit.*, pp. 308–9.
52. Johnson, S. P. & Corcelle, G., *op. cit.*, pp. 109–11.
53. Stern, A. C., ed., *Air Pollution*, Vol. VIII, 3rd edn. Academic Press, Orlando, FL, USA, 1986, p. 96.
54. Georgiades, Chiron, & Joumard, Establishment of atmospheric pollution standards for motor vehicles. *Sci. Total Environ.*, **77** (1988) 221–4.
55. McGrory, *op. cit.*, p. 306.
56. Haigh, N., *op. cit.*, pp. 186–8.
57. *Official Journal*, C222, 30 Aug. 1980.
58. Gentry, Emission controls. In *Environment and Planning Law*. Butterworth, London, UK, 1991.
59. For detailed information on the air pollution issue in the Member States, see *European Environmental Yearbook*, 2nd edn. Doc Ter International, London, UK, 1991, pp. 399–431.
60. Directive 70/220/EEC. *Official Journal*, L76, 6 Apr. 1970.
61. See Directives 78/665/EEC and 83/351/EEC.
62. For the history of the development of these proposals, see Johnson, S. P. & Corcelle, G., *op. cit.*, pp. 127–34.
63. Directive 89/458/EEC.
64. Gentry, *op. cit.*, pp. 84–5.
65. Directive 75/716/EEC.
66. Johnson, S. P. & Corcelle, G., *op. cit.*, p. 123.
67. Gentry, *op. cit.*, p. 87.
68. Directive 88/609/EEC.
69. Johnson, S. P. & Corcelle, G., *op. cit.*, pp. 142, 136–47.
70. Haigh, N., *op. cit.*, pp. 224–7.
71. See, also, limit values for, *inter alia*, SO_2 emissions from new municipal waste incinerators under Directive 89/369/EEC.
72. Garner, J. F. & Crow, R. K., *Clean Air Law and Practice*, 4th edn. Shaw & Sons, London, UK, 1976, p. 21.
73. Hughes, D., *Environmental Law*. Butterworth, London, UK, 1986, Ch. 9.
74. Royal Commission on Environmental Pollution, *Air Pollution Control: An Integrated Approach*, Cmnd 6371. HMSO, London, UK, 1976.
75. DoE, *Air Pollution Control in Great Britain: Review and Proposals*. HMSO, London, UK, 1986.
76. Environmental Data Services ENDS Report 167, Dec. 1988, pp. 3–4, 10–11.

77. Haigh, N., *op. cit.*, pp. 182–97.
78. SI 1976/1986, The Oil Fuel (Sulphur Content of Gas Oil) Regulations.
79. SI 1976/1989, The Motor Fuel (Sulphur Content of Gas Oil) Regulations.
80. SI 1990/1096.
81. SI 1986/1078.
82. The Road Traffic Act 1988, Section 185(1).
83. SI 1984/981, as amended.
84. To check the updating of such regulations, see *The Encyclopaedia of Road Traffic Law and Practice*, Sweet & Maxwell, London, UK, no date.
85. The Secretary of State is the responsible minister.
86. SI 1987/180, Control of Industrial Air Pollution (Transfer of Powers and Enforcement) Regulations.
87. SI 1983/943.
88. *Ibid.*, Reg. 4, Sched. 2.
89. Health and Safety At Work Act 1974, Section 5(2).
90. See *Adsett* v. *K and L Steelfounders and Engineers Ltd* [1953] 2 All ER 320 on 'all practicable measures'.
91. See *LCC* v. *GER* [1906] 2 KB 312 on 'so far as practicable'.
92. *West Bromwich B. S.* v. *Townsend* [1983] ICR 257.
93. *Manchester Corp.* v. *Farnworth* [1930] AC 171.
94. Haigh, N., *op. cit.*, pp. 226–7.
95. Environmental Data Services ENDS Report 176, Sept. 1989, pp. 25–6.
96. Hughes, Integrated pollution control and the water environment. *Water Law*, **2** (1991) 20.
97. Scottish Power forced to advance FGD plans. *Environ. Business*, 10 Apr. 1991, p. 1.
98. Georgiades, Chiron, & Joumard, *op. cit.*, pp. 224–5.
99. See *Union Electric Company* v. *EPA* 427 US 246 (1976), where the Supreme Court held that EPA is not required to review the economic and technological feasibility of a State Implementation Plan with regard to achieving National Ambient Air Quality Standards.
100. Georgiades, Chiron, & Joumard, *op. cit.*
101. *United States* v. *Ohio Department of Highway Safety* 635 F2d 1195 (6th Cir. 1980).
102. *Chrysler Corporation* v. *EPA* 631 F2d 865 (D.C. Cir. 1980) and 449 US 1021 (1980).
103. *Center for Autosafety* v. *Ruckelshaus* 21 E.R.C. 1985 (D.C. Cir. 1984).
104. McGrory, *op. cit.*, p. 301.
105. *Ibid.*, p. 311.
106. *Ibid.*, p. 302.
107. *Ibid.* pp. 302–3.
108. Frick, G. W. & Wegman, L. N., Air pollution control. In *Environmental Law Handbook*, 8th edn, ed. J. G. Arbuckle *et al.* Government Institutes, Rockville, MD, USA, 1985, Ch. 6.
109. Marchesani, V. J., Air quality management in the United States. In Stern, A. C., *op. cit.*, Ch. 1.
110. Campbell, W. A., United States Clean Air Act litigation. In Stern, A. C., *op.*

cit., Ch. 2.
111. EPA figures (Mar. 1989) that show that in 1987 in the US some $1{\cdot}1\times10^6$ t of dangerous chemicals were released into the atmosphere.
112. *The Times*, 16 May 1989.
113. *Ibid.*, 12 June 1989.
114. Emgelgau, Written on the wind. *Resources*, **13**(1) (1991) 7.
115. 'Over-regulation and under-enforcement' should be known as 'The Hawkins Principle' after Richard Hawkins, a founder member of the UK Environmental Law Association and a tireless campaigner against this particular evil.
116. Environmental Data Services ENDS Report 195, 1991, p. 28.

ADDENDUM

IPR 1/1 issued by HMIP in March 1991 gives guidance on LCP and associated emissions, principally to the atmosphere, with regard to IPC. The emission limits at the core of the new guidance are, in general, those set by the EC's 1988 directive on LCP (see above). For *new* plant the 1988 EC requirements are applied, and for existing plants a time-scale is laid down for compliance with the requirements, though a period of toleration is to be allowed until 2001 by when all relevant processes should meet the requirements that currently apply to new plant. The guidance has, however, been criticised as vague and not sufficiently specific in its requirements, leaving some room for manoeuvre for inspectors. It may be that a desire to allow flexibility with regard to the implementation of BATNEEC has had to be further tempered by the need to ensure that the UK's SO_2, etc. emissions are reduced in line with national commitments under EC law.[116]

Chapter 16

Strategies for Controlling Acid Rain: Economic Considerations

BRUCE A. FORSTER[a] and THOMAS D. CROCKER[b]
[a]*Department of Economics and Finance*, [b]*College of Business, University of Wyoming, Laramie, WY 82071, USA*

16.1 INTRODUCTION

Opponents of acid rain controls in the US emphasize that the benefits of such a program are highly uncertain while the control costs are known and large. The following two quotations are typical of this stance (note: authors' italics):[1, 2]

> With all of the uncertainty surrounding acid rain, *the one certain consequence is a staggering cost*–perhaps exceeding US$100 billion over 20 years.
>
> . . . the economics of acid rain involve *a known high level of reduction costs* coupled with a questionable level of benefits.

The physical effects of acid deposition on health, materials, ecosystems, etc. have been discussed in Chapters 9–13 of this book. The only published estimate of the aggregate economic benefits of reducing acid deposition in the US is that of US$5 billion annually (in 1978 constant dollars) (throughout this chapter, billion is 10^9).[3] The estimates given for separate categories are as follows: buildings and structures, US$2·0 billion; forest ecosystems, US$1·75 billion; agricultural crops, US$1·0 billion; aquatic ecosystems, US$0·25 billion; all other, US$0·10 billion.[3] Other economic estimates for individual receptor categories in the US and in Canada show the potential variability in control benefits referred to in the above quotations.[4]

However, the control-opponents' claims that the controls costs are *known* is too strong since these costs are not independent of the control strategy that is to be implemented. Even for a given strategy the total cost (including the social cost of an 'inappropriate' reduction–too little or too much) of a controls program is also not independent of the true benefits of those controls.

It has been argued that an informal coalition of electric power utilities in the Upper Ohio Valley, along with their high-sulfur coal suppliers, the high-sulfur coal miners and the electricity consumers in the region have an incentive to exaggerate the costs and disruption of such a control program.[4,5] The purpose of this chapter is to show that, quite apart from the incentive to exaggerate costs, the cost of controls depends upon the chosen control strategy and cannot be known without knowing the strategy contrary to the views espoused above.

The emphasis in this chapter is on the alternative economic strategies. The alternative available or emerging technological options are discussed in the first chapter in this section of the book and in Refs 6 and 7.

There are two competing approaches to reducing pollution such as the acid rain precursor pollutants of sulfur dioxide (SO_2) and nitrogen oxides (NO_x). In the command-and-control approach, pollution control legislation may dictate the technological method by which specified pollution reductions are to be achieved. A key feature of command-and-control regulations is their inflexibility. The 1977 Amendments to the US Clean Air Act are examples of this approach. The 1977 revision of the new source performance standards (NSPS) requires electric utilities to achieve a uniform percentage reduction in SO_2 emitted independent of the original sulfur content of the coal used. The prevention of significant deterioration (PSD) regulations require that new electric power installations in 'clean' regions of the US use 'best available control technologies' (BACT) to limit SO_2 emissions. The result of both the NSPS and PSD regulations is that scrubber technology is essentially mandated. The International Nitrogen Oxides Protocol recently signed by the US and several European countries similarly dictates a BACT approach to reducing NO_x.

The other approach relies on market mechanisms or incentives to induce firms to reduce pollution voluntarily. These include various *ex ante* mechanisms such as taxes, subsidies, and marketable permits to pollute and *ex post* mechanisms such as liability rules. Economists generally prefer market mechanisms because they permit firms flexibility in selecting abatement methods to minimize costs of abatement.

Sections 16.2 and 16.3 of this chapter deal with qualitative issues in determining and achieving an 'optimal' pollution level using various taxes, subsidies or quantitative restrictions. Section 16.4 considers alternative permit schemes for achieving regional pollution control. These three sections demonstrate that there are a number of conceptual issues that need to be addressed in choosing control strategies and several of these are still open to question. Section 16.5 deals with statistical information comparing the command-and-control approaches with the economic incentives approach. There are substantial cost differences found between approaches in most cases. Some researchers show a large cost saving from using market incentives while others lean toward the command-and-control approach. Section 16.6 examines 'institutional' features that may lead to more costly acid rain controls schemes being selected.

16.2 TAXES AND SUBSIDIES

Pigou, the first advocate of market mechanisms for controlling pollution, noted that perfectly competitive industries would overproduce relative to the social optimum and hence generate excessive amounts of pollution.[8] This overproduction and excessive pollution is caused because the firms neglect the costs imposed on society when pollution generated by production activities is discharged to the environment. The firm easily disposes of a useless waste product and saves control costs; the receptor suffers damage or is forced to make defensive adjustments to avoid or offset the damaging effects. To rectify the situation, Pigou advocated charging the firm a tax per unit of its output. The tax rate should equal the marginal social damage at the socially ideal level of output and pollution. This tax causes the firms to 'internalize' the social costs of their polluting activity since the pollution damage is a cost of production. This tax is commonly referred to as a Pigovian tax.

During the 1970s and 1980s economists have subjected the Pigovian tax to considerable analysis to determine the conditions under which it produces the desired outcome. Output taxes have also been compared to alternative market-based approaches such as subsidies to reduce pollution, input taxes, emission taxes, or marketable permits.

The situation for which Pigovian taxes work is very special. The use of such taxes assumes complete information by the agency. The Pigovian tax will produce the socially ideal resource allocation if the firms are in a perfectly competitive industry, if the pollution damage function is a linear function of the pollution level, and if the pollution level is linearly related to the firm's output level. Thus, the social damage of pollution is a linear function of the firm's output level; hence, the social pollution damage per unit of output is constant for all levels of output. In this case the long-run output level of the individual firms subjected to the tax is unchanged. The firm's average cost of production rises by the tax rate but the minimum average cost production level is unchanged. Since minimum average cost is higher, the market price of output is increased in long-run equilibrium and the price equals marginal social cost defined as marginal production cost plus marginal social damage. With a downward-sloping demand curve, industry output contracts and hence so does the aggregate pollution level. The adjustment to the long-run social optimum takes place through a reduction in the number of firms operating in the industry rather than a reduction in firm-level production.

If the damage function is not linearly related to output then the Pigovian tax will not achieve the optimal industry output. If marginal damages increase as output increases then the Pigovian tax levies an excessive tax burden on the polluting firms compared to actual damages they impose and the individual firm's average cost curve is pushed too high; hence, market price is driven too high and the industry contracts too much. On the other hand, if marginal damages decrease as output rises then the tax burden is too low and the market price does not rise sufficiently. The contraction in industry output in this case will be insufficient for the socially ideal outcome.[9]

The fact that Pigovian taxes cause some firms to shut-down led to an interest in using a subsidy scheme to induce firms to reduce pollution. Firms are offered a subsidy per unit of output that production is reduced below the non-intervention production level. The value of the unit subsidy is the marginal social damage at the optimal output level. The subsidy is treated by the firm as an opportunity cost of increasing output and the firm is induced to reduce production so as to minimize unreimbursed losses.[10] However, in this case the average cost of the firm is lowered rather than increased. The market price will fall as entry into the industry takes place to take advantage of profits generated by the subsidy. Thus the subsidy scheme produces a perverse result.[10,12] The subsidy may not work even in the short-run if the social ideal requires some firms to shut-down since the subsidy will keep all firms operating and hence aggregate pollution will be excessive.[12] However, if the subsidy is limited to existing firms and is paid as compensation for the loss of a property right rather than a reduction in output then the distortions of the entry/exit decision will not occur. However, this subsidy is different from that discussed above. In this scheme there is essentially a pollution tax with a lump-sum redistribution.[11] Nevertheless, subsidies would still limit existing firm's motivations to seek out environmentally benign technologies.

When the industry is not perfectly competitive then one cannot immediately conclude that the industry's output and pollution levels are excessive.[13] If the industry is a pure monopoly, i.e. a single producer, it is possible that the firm is still underproducing relative to the social optimum. In order to generate monopoly profits, the firm restricts output below that which would be produced by a competitive industry facing the same cost and demand parameters. This monopoly output may be lower than desirable even when the firm's pollution impacts are considered if the social costs created by the monopoly distortion (market price exceeding marginal production cost) exceeds the social costs of the pollution impacts. In this case, an expansion in output (and hence pollution) is required rather than a contraction. Even when a contraction in output is called for, the Pigovian tax will be too severe.[14,15] However, if there are rent-seeking costs incurred by the firm to protect its monopoly position (or to acquire a monopoly position), a tax higher than the Pigovian may be required.[16]

If the flow of pollution does not depend upon the level of output but rather the mix of factors of production, then a tax on output is inappropriate. It has been noted that an output tax could even induce an increase in pollution if the pollution is generated by an 'inferior' factor of production.[17] A tax on the offensive input is more appropriate than an output tax.

A major criticism of both the output-based and input-based tax is that such taxes do not provide an incentive for the firm(s) to develop alternative technological methods to reduce emissions. This suggests that the more appropriate tax is one on emissions which provides the firm with the incentive to reduce emissions in the most efficient manner in order to minimize the tax burden. The use of emissions fees for a number of different market structures and behavioral theories of the firm have been considered.[15] In the managerial models of maximizing firms, the emissions tax induces the firm to undertake abatement activities to minimize total pollution control expenditures unless the firm perceives 'good will' from the abatement. Similarly, public or government agencies interested in maximizing an objective function dependent upon agency output and its level of perquisites will undertake abatement activities so as to minimize total pollution abatement costs. This result may not be achieved, however, if the agency is protected by 'managerial slack'. It was noted that such managerial slack also occurs in organizational models of firm behavior that emphasize information costs, internal coalitions, bounded rationality and the internal dynamics and structure of the firm.[15] Such slack may lead firms to respond slowly to various price changes including the imposition of an emissions fee.

For regulated industries, it was concluded that in general the firms will not pursue a cost-minimizing pollution abatement strategy since the pollution abatement equipment adds to the firm's rate base.[15] Hence, the firm has an incentive to over-invest in pollution equipment. It was noted that, for a regulated firm, an increase in the emissions fee may not even lead to a reduction in emissions.

For the pure monopoly case it was noted that in general allocative efficiency will require the use of a production subsidy in addition to the emission fee in order to avoid excessive contraction in output.[15] A similar result was obtained earlier.[18]

16.3 UNCERTAINTY AND ASYMMETRIC INFORMATION

Several researchers have considered the choice of emission taxes versus quantitative restrictions (such as marketable permits or legislative standards) when the control agency faces uncertainty concerning the nature of the pollution damage function and/or the pollution control cost function.[19–23] The uncertainty may be caused by stochastic elements in the damage or control-cost functions or may reflect a lack of information by the control agency.

With uncertainty, neither command-and-control nor economic incentives are able to achieve the *ex post* optimum. The relevant *ex ante* choice is the policy that minimizes the welfare loss once the outcome is realized or which maximizes expected welfare.

The research shows that if the uncertainty appears in the damage function then the tax and quantitative restrictions impose exactly the same social welfare loss. Thus in this case the existence of uncertainty does not affect the choice between alternative control instruments. However, if the uncertainty appears only in the control-cost function then the alternatives do perform differently. In this case the choice between instruments depends upon the relative elasticities of the marginal damage and marginal control-cost functions. If these are equal in absolute value, then both approaches yield the same result. If the marginal damage function is steeper than the marginal control-cost function then the quantitative regime is preferred to the tax regime while the tax instrument is preferred if the marginal cost curve is steeper than the marginal damage curve.

The uncertainty concerning the control-cost function facing the agency may not be due to natural randomness but rather a lack of information concerning the structure of the relevant functions. If the firms have complete information concerning the structure of control costs while the agency does not, then this asymmetric information may lead to strategic behavior on the part of the firms. It has been shown that if the agency uses a tax system then the firm has an incentive to understate the control cost in order to obtain a lower tax.[24] On the other hand, if the agency is planning to use a marketable-license approach, the firm has an incentive to overstate the control costs in order to induce the government to issue a larger number of licenses. The larger the number of licenses the lower the market price of licenses that the firms will have to pay.

In order to counter this strategic behavior, a mixed emission charge and license system was proposed that balances the firm's incentive to overstate costs with a licensing approach and understate costs with a tax approach.[24] In this plan a number of licenses are issued and firms are paid a subsidy per license held in excess of their emissions level. This scheme induces firms to truthfully reveal their cost structure to the agency. Such a mixed system has also been advocated elsewhere;[25] however, that scheme also includes a unit tax on emissions in excess of the firm's license holdings.

The above schemes treat the emissions abatement issue independent of other operations of the firm. The emissions abatement has been considered as part of the firm's overall decision-making for the case in which the firms are subject to 'economic' regulation (i.e. price regulation) as well as environmental regulation.[26,27] Given the asymmetric information concerning the cost and effectiveness of abatement and since the agency knows that the firms have an incentive to misrepresent the information it can take this into account and design 'incentive compatible' policies. It has been shown that a constant emissions tax should not be used for abatement since the tax increases the information costs of the regulator.[26] It was concluded that standard-setting and output-price regulation are sufficient regulatory instruments for pollution abatement. The output-price regulation allows the regulator to tax the firm implicitly and achieve emissions reduction while avoiding the information costs.

The problem that the division of regulatory power between 'price' regulators and environmental agencies can create has been recognized.[27] The environmental agency desires a balance between the benefit and costs of pollution control while the price regulator is concerned with appropriate pricing policies given the firm's production and abatement costs. Concern for the welfare of the consumers of the firm's output leads the price regulator to have an incentive to avoid abatement which will increase costs and hence raise the firm's output price. If the two regulators co-operate then the full-information optimum can be achieved through the use of standard-setting and price regulation.[26] However, these regulatory agencies may not co-operate since their interests conflict. The environmental agency may then behave as a 'Stackelberg leader' serving its own interests. The non-cooperative equilibrium features the maximum emissions fee enforceable by the courts, a higher standard than would be imposed in the full-information optimum, and a lower output level for the firm.

A similar result has been found for firms operating in competitive product markets when those firms employ Bayes–Nash strategies in communicating with the environmental agency.[28] That is, aggregate output and emissions are lower with regulation than would be achieved at the full-information equilibrium.

A scheme that produces correct incentives for firm entry and exit when the agency faces unknown damages and unknown control costs has been proposed.[29] The simultaneous use of a modified Clarke[30] procedure and a Vickrey[31] auction was suggested. The Clarke procedure induces individuals to truthfully reveal their willingness-to-pay to avoid various pollution levels. This information is then used as an input to the Vickrey auction to determine the control costs of the polluting firms.

In the Clarke procedure the victim is charged any increased abatement cost (net of any additional benefits to other individuals) caused by that individual affecting the efficient pollution outcome. Truth-telling is the optimal response for the victim under the Clarke procedure. For this approach to work the victims must be well-informed of the various effects of the pollution in order to determine their willingness-to-pay to avoid the effects. It has been counseled that this is a time-consuming activity that may discourage participants.[29]

In the Vickrey auction the firm is to be charged the marginal social damage minus the sum of other firms demands for emission rights. In the absence of collusion an individual firm does not influence this charge and is induced to correctly reveal its demand for emission rights. Collusive behavior will still produce an efficient outcome in the Vickrey auction provided the emission permits are tradeable *and* the pollution damage function is convex. When non-convexities in the damage function occur then collusion between firms can defeat the truth-telling incentives. Since substantial non-convexities and irreversibilities have been noted in acid rain damages, this may reduce the usefulness of the proposed Vickrey auction.[32]

16.4 SPATIAL CONSIDERATIONS

Considerable effort has been devoted to examining the use of marketable pollution permits to control pollution in the face of spatial dispersion of pollutants.[33] In the seminal analytical contribution, Montgomery proposed two alternative schemes:[34] one is the emissions-based permit system (EPS) which defines permits in terms of allowable emissions at the source; the second is the ambient-based permit system (APS) which defines the allowable emissions in terms of pollution at the receptor rather than at the source.

Montgomery was concerned that trading under the EPS would be complicated since, in general, one-for-one permit trades could not be allowed. Such trades would alter environmental quality at various receptors if the trading firms have different dispersion coefficients. Montgomery specifies the rights to emit in terms of the locations of the firms. The existence of a market equilibrium that minimizes total abatement costs is dependent upon a restrictive (and perhaps unattainable) condition concerning the inital assignment of permits to satisfy environmental quality constraints.

For the APS, Montgomery demonstrated the existence of a cost-minimizing market equilibrium that is independent of the initial distribution of permits. Montgomery demonstrated that the equilibrium exists, not that it will necessarily be achieved.[35] Indeed the firm's problem is not straightforward. Under the APS the firm must determine how much of a given amount of emissions is received by all locations and obtain the requisite portfolio of permits. This requires that the firms determine their own pollution dispersion coefficients.

It has been concluded that Montgomery's APS is likely to impose higher transaction costs on polluting firms than the EPS.[36] A modification of the EPS that does possess a least-cost trading equilibrium that is independent of the initial assignment of permits was proposed. Again, it has been said that this demonstrates the existence not the attainment of such an equilibrium.[35] Permits are based upon emissions but are not traded on a one-for-one basis. Instead, the trades must not result in a violation of the environmental standard. For any planned increase in emissions, a firm must purchase a sufficient number of emission permits from other sources to 'offset' its own increase, hence this scheme is referred to as the pollution offset system (POS).

This POS scheme sounds no less complicated than the EPS; in fact, it seems to require more information. The firm planning an emissions increase must not only know its own diffusion coefficients, but must be able to identify other firms that impact the same receptors. It is hard to see why this scheme would afford lower transaction costs.

It is not surprising that it has been possible to demonstrate the mathematical equivalence of the POS and APS with perfect competition.[37] It has been argued that, under imperfect competition, the two systems perform differently with some polluters being able to be 'free riders' under the POS.[35]

The modeling performed in this research assumes that pollution concentrations at each receptor are linearly related to emissions from various sources. That is they assume

$$q_{ij} = d_{ij}e_i$$

where q_{ij} is the pollution concentration at receptor j from source i, e_i is the emission rate at source i and d_{ij} is the fraction of 1 unit of emissions in i that arrives at j assumed constant.

For acid rain the linearity hypothesis is still an open question.[38] For example, based upon empirical studies, it has been concluded that the relationship between smelter emissions in the western US and precipitation sulfate is in fact linear,[39] while elsewhere it was argued that such a conclusion is premature.[40]

Even if the relationship is linear the transfer coefficients are likely to be random (subject to meterological conditions for example). Such randomness of the transfer coefficients substantially increases the complexity of trading under either the APS or the POS. For the firms the problem of acquiring the requisite permits is analogous to firms investing in supply capacity with peak-load demand conditions. This issue has not been addressed in the context of the APS or POS.

Liability rules have received little or no attention in acid rain control discussions. These rules are *ex post* incentive devices because they are not activated until damages are realized. However, if a polluter knows that he will be held liable for a damage probability or magnitude, he will have an incentive to alter his *ex ante* behavior so as to reduce this *ex post* liability. It has been shown that, when only the polluter can act to reduce damages, strict liability rules are efficient.[41] Under strict liability the polluter is liable whenever his behavior can be shown to cause damages. Its result are the *ex post* equivalent of *ex ante* taxes and subsidies.

Another potential *ex post* device is a negligence rule. Negligence implies that the polluter's behavior has violated some 'due standard of care.' It is thus the *ex post* equivalent of *ex ante* quantitative restrictions such as emissions or input standards. The negligence rule has been shown to be inefficient if polluters enter or exit the industry.[42]

Probably because their relevance is thought to be sharply limited, neither liability nor negligence rules have attracted any attention in discussions of acid rain control strategies. This inattention might be justified by difficulties in identifying responsible parties, uncertainty about causation, inability of polluters to compensate for massive damages, and the costs to receptors of successfully bringing suit under liability and negligence rules.

16.5 STATISTICAL STUDIES

The economic effects of various modifications of a proposal by US Congressman Aspin to levy a sulfur-in-fuel tax to control acid rain have been considered.[43] The basic Aspin proposal levied a tax of US$0·50 per pound of sulfur contained in the coal with a rebate of US$0·50 per pound granted for each pound of sulfur removed prior to emission. A subsidy covered 90% of the cost of emissions reduction technology. It has been estimated that this plan would generate a nationwide SO_2 emission reduction of $6{\cdot}3\times10^6$ t annually.[43] The cost to the utilities is US$2·7 billion per year (1985 dollars). Most of the reduction in SO_2 comes from fuel switching and tax payments exceed rebates and subsidy payments by US$1·2 billion annually. The overall social cost of the Aspin proposal has been estimated at US$1·4 billion per year.

If the tax/rebate fee is increased to US$1·00 per pound of sulfur (while retaining a 90% subsidy) the emissions drop by $9{\cdot}7\times10^6$ t of SO_2 annually. The tax receipts and rebate/subsidy payments roughly balance and the social costs are the utility costs of US$3·5 billion annually.

If the tax/rebate fee is kept at US$0·50 but the subsidy is eliminated then emissions fall by only 5.3×10^6 t annually. Tax receipts exceed rebates by US$3·2 billion while utility costs are US$4·2 billion producing a social cost of US$1·0 billion annually. The reduction in emissions in this case is achieved by fuel-switching.

The Edison Electric Institute estimated that the sulfur and nitrogen emissions tax proposed by US Congressmen Gregg and Downey would cost the electric utilities US$16 billion dollars.[44] Their analysis assumed that there would be no reduction in emissions. This assumption is untenable.

It has been determined that the cost-saving from using marketable permits (based upon emissions) in Illinois would amount to 40–60% of the cost of requiring a uniform reduction in SO_2 emissions.[45] The restrictions placed upon the permit market by Diemer and Eheart lead them to believe that their permit costs estimates are biased downwards. They showed that their results depended upon the availability and costs of nuclear power as a means of reducing SO_2 emissions.

The cost-effectiveness of three alternative regulatory approaches to achieving an 85% overall reduction in hydrocarbon emissions from the DuPont company have been considered.[46] The first approach (source standards) required a uniform reduction of 85% across all sources. The second approach (plant standards) allowed the abatement to be varied across different sources within a given plant to obtain an overall plant reduction of 85%.

The final approach (regionally marketable permits) allows the abatement to be varied across plants to achieve an overall reduction of 85% for the company as a whole.

The study, using in-house DuPont data, showed that the plant standards approach achieved a cost-saving of 35% compared to the rigid source standards.[46] The cost-saving from using the regionally marketable permit scheme was 76% compared to the source standard approach.

A linear programming model was used to examine the use of scrubbing of emissions trading for achieving annual SO_2 emission reductions of 8, 10, and 12×10^6 t in a 31 state region of the eastern US.[47] Interstate trading would reduce overall abatement costs by only 2% since the reduction in scrubbing costs was offset by increased coal prices. It was felt that the cost-saving was too small, given the administrative complexities, to warrant using an emissions trading policy.

This contrasts with a study of the British electric power industry,[48] in which transferrable permits achieved an 88·9% reduction of SO_2 emissions at a cost-saving of 27·1% compared to a uniform reduction scheme.[48] The firms in the study either used scrubbing to achieve all of the abatement, or used no active abatement at all. These latter firms opted to purchase permits to cover their existing emissions.

This all-or-none choice of scrubbing is attributed to the nature of the scrubbing cost function. A cost function for scrubbing in American electric utilities has been determined.[49] The average cost (US$ per pound of SO_2 removed) was found to decrease with the level of SO_2 removal and with the sulfur content of the fuel. This feature would lead firms to exhibit all-or-nothing strategies. Firms exhibiting 'all-or-nothing' scrubber choices when faced with an emissions tax have also been found elsewhere.[50] With a tax of US$0·10 per pound of SO_2, all firms switched to lower sulfur coal. With a tax of US$0·30 per pound old plants would retrofit with a scrubber. New plants opt for scrubbers when the tax reaches US$0·20 per pound.

A uniform percentage reduction plan has been compared to two marketable permit schemes for controlling acid deposition in the Cleveland region.[51] One scheme based permits upon emissions while the other based permits on ambient concentrations. Surprisingly, the emission-based permit scheme proved more costly (US $158 million) than the uniform reduction scheme (US $123·5 million) while the ambient-based scheme was the least costly (US $83 million) of the three approaches. The cost savings from the uniform reduction and ambient permit schemes come at the expense of increased long-range transport of acid deposition since these schemes focus on ground-level concentrations. Firms may satisfy the local pollution conditions by building tall stacks to disperse SO_2. Once constraints were imposed upon long-range transport, the cost advantage of the ambient permit scheme is reduced.

Policies to achieve given sulfate deposition rates in eastern North America have been considered.[52] If these are obtained by requiring a 50% reduction in SO_2 emissions at all sources the annual control cost to Canadian sources is US$0·4 billion while for US sources the control cost is US$6·9 billion. If source regions are allowed to achieve a 50% reduction in the least-cost fashion then the Canadian costs drop to US$0·3 billion while the American costs drop to US$2 billion!

Still lower costs are obtained if the deposition target levels are achieved by identifying appropriate sources (through the use of transfer coefficients) and pursuing a least-cost approach across sources. In this case the Canadian costs drop marginally to US$0·2 billion but the American costs plummet to US$0·5 billion! The actual cost estimates were found to be sensitive to the choice of transfer coefficients from various long-range transport models.

It has been argued that additional social gains could be obtained by substituting the revenue from emissions taxes for general tax revenue raised from labor income taxes or corporation income taxes.[50] The emissions taxes improve economic efficiency while the income taxes reduce economic efficiency. The annual efficiency gains from replacing US labor income tax revenue with revenue from emissions taxes on particulates and sulfur oxides range from US$0·6 billion to US$3 billion while the gains from replacing corporate income tax revenue range from US$1 billion to US$4·9 billion (in 1982 dollars).

16.6 INSTITUTIONAL FACTORS BIASING COSTS UPWARDS

There are several economic and institutional factors that may cause the acid rain controls to be more costly than necessary. The electric utility industry is subject to 'rate-of-return' regulation by public utility commissions (PUCs). The utilities' profits are restricted to a specified fraction of their invested capital. This leads to a potential bias towards capital-intensive pollution control which increases the social cost of reducing emissions.[15,53] Pollution-control devices such as scrubbers expand the utilities' capital base and allow them to earn higher profits. The utilities may receive credit for construction costs during installation of pollution control devices.[54] The State of Ohio allows utilities to recover costs for coal research and development every 6 months.[55]

Pre-1986 tax policy also provides a bias toward scrubber technology rather than fuel-switching. The electric power industry has been claimed to be the major group benefiting from various pollution control tax breaks.[54] Since the various tax preferences (investment tax credits, industrial development bonds, and accelerated cost recovery) are granted only for identifiable pieces of equipment, these preferences create a bias in favor of scrubbers over modifications in fuel, for example. The effective tax rate for the electric utilities, given these tax breaks for pollution control, was only 20% of the average tax rate for the corporate sector between 1964 and the 1970s while the rates were roughly equal during the 1950s.[54] The annual cost saving to the industry exceeded US$4·3 billion by the late 1970s.[54] This is paid for by the tax-paying public and represents a social cost of pollution control.

The regulatory process has been seen as a bargaining game between the regulatory agency, the firms, and the citizens.[56] In that approach, the regulatory agency has an objective function that allows for trade-offs between environmental quality, local jobs, and firm profitability. Given the agency's limited resources for bargaining and enforcement it may go after those firms that have little political power such as small firms and plants rather than the larger, more politically powerful ones. Small plants would have higher abatement costs given the nature of the scrubbing cost function discussed above. A coalition of 1750 small power plants (the American Public Power Association) in the summer of 1983 advocated the use of an emissions tax on US SO_2 emitters to control acid rain rather than mandated scrubbing.[57] Targeting small plants with scrubbing would increase the cost of abatement, since they would prefer to switch fuels. Compliance with US environmental regulations has been found to place a greater burden on the small rather than large plants and reduced the number of plants in the affected industries.[58] With economies of scale in pollution control, smaller firms have difficulty competing with larger firms. It was also noted that the environmental regulation had the effect of increasing the use of capital relative to labor.[58] It also increased capital's share of output, although it was not clear if this was due to increased capital intensity or an increase in the rate of return earned by capital in the surviving plants.

Concern for local employment levels would lead to less strict enforcement where regulation would have more impact or to the design of policies that minimize such impacts. Mandated scrubbing protects the jobs of the coal miners in the high-sulfur coal regions in the mid-western and eastern US. The Indiana state legislature has considered a bill prohibiting the import of low-sulfur out-of-state coal.[59] In discussing acid rain control strategies, fuel-switching has been dismissed as being a politically unfeasible option because of the shifts in income distribution between high- and low-sulfur coal regions that could arise.[2] It has been argued that regulators may prefer design standards that specify technology over performance standards because the output level achieved for any given pollution level is lower under the performance standard and hence employment will be lower.[60]

The firms in the electric power industry that are in compliance with the Clean Air Act Amendments and hence have scrubbers will also prefer that the 'newly' regulated firms be required to install scrubbers rather than being permitted to pursue a more flexible, lower-cost strategy. Those plants operating under the 1970 regulations (but not the 1977 Amendments) are allowed to use low-sulfur coal to meet the standards. These firms will find their fuel costs driven up if these 'new' firms are allowed to switch to low-sulfur coal because the demand pressure will increase the price of low-sulfur coal. 'Very low'-sulfur coal prices could rise by US\$11 t^{-1} while 'low'-sulfur coal prices could rise between US\$1 and US\$8 t^{-1}.[61]

16.7 CONCLUSIONS

Neither the analytical nor the empirical literature relevant to evaluating the economic consequences of acid deposition control strategies implies that a single, universally preferable control strategy exists. Given the ambiguities in analytical and empirical understandings, a variety of efficient control strategies can exist simultaneously. At present, we appear to be able to say little more than that the benefits and the costs of control will differ with the type of strategy employed. This recognition, though by no means earthshaking, at least provides a basis to view skeptically discussions of the benefits and costs of acid deposition control independently of the control strategies likely to be adopted.

REFERENCES

1. Mallick, E., Report on the 15th APCA government affairs seminar: quest for clean air–challenges to the 100th congress. *J. Air Pollut. Control. Assoc.*, **37** (1987) 698–9.
2. Regens, J. L. & Rycroft, R. W., *The Acid Rain Controversy*. University of Pittsburgh Press, Pittsburgh, PA, USA, 1988.
3. Crocker, T. D., Economic impact of acid rain: Hearings before the Select Comm. on small business and the Community on Environment and Public Works, US Senate 96th Congress, 2nd Session 103–11. US Government

Printing Office, Washington, DC.
4. Forster, B. A., *The Acid Rain Debate: Science and Special Interests in Policy Formation.* Iowa State University Press, Iowa, CA, USA, 1993.
5. Forster, B. A., Acid rain games: incentives to exaggerate control costs and economic disruption. *J. Environ. Management*, **28** (1989) 349–60.
6. Moore, T., How advanced options stack up. *EPRI J.*, July/Aug. **12** (1987) 4–13.
7. Shepard, M., Coal technologies for a new age. *EPRI J.*, Jan./Feb. **13** (1988) 4–17.
8. Pigou, A. C., *The Economics of Welfare*. Macmillan, London, UK, 1920.
9. Carleton, D. W. & Loury, G. C., The limitations of Pigovian taxes as a long-run remedy for externalities. *Quart. J. Econ.* **95** (1980) 559–66.
10. Kneese, A. V. & Bower, B. T., *Managing Water Quality: Economics, Technology, Institutions*. Johns Hopkins Press, Baltimore, MD, USA, 1968.
11. Dewees, D. N. & Sims, W. A., The symmetry of effluent charges and subsidies for pollution control. *Can. J. Econ.*, **9** (1976) 323–30.
12. Polinsky, A. M., Notes on the symmetry of taxes and subsidies in pollution control. *Can. J. Econ.*, **12** (1979) 75–83.
13. Buchanan, J. M., External diseconomies, corrective taxes and market structure. *Amer. Econ. Rev.*, **59** (1969) 174–7.
14. Misiolek, W. S., Effluent taxation in monopoly markets. *J. Environ. Econ. Management*, **7** (1980) 103–7.
15. Oates, W. E. & Strassman, D. L., Effluent fees and market structure, *J. Public Econ.*, **24** (1984) 29–46.
16. Misiolek, W. S., Pollution control through price incentives: the role of rent seeking costs in monopoly markets. *J. Environ. Econ. Management*, **15** (1988) 1–8.
17. Plott, C. R., Externalities and corrective taxes. *Economica*, **33** (1966) 84–7.
18. Smith, V. K., A note on effluent charges and market structure. *J. Environ. Econ. Management*, **2** (1976) 309–11.
19. Weitzman, M., Prices vs. Quantities. *Rev. Econ. Studies*, **41** (1974) 477–91.
20. Yohe, G. W., Substitution and control of pollution. A comparison of effluent charges and quantity standards under uncertainty. *J. Environ. Econ. Management*, **3** (1976) 312–24.
21. Adar, Z. & Griffin, J. M., Uncertainty and the choice of pollution control instruments. *J. Environ. Econ. Management*, **3** (1976) 178–88.
22. Fishelson, G., Emission control policies under uncertainty. *J. Environ. Econ. Management*, **3** (1976) 189–97.
23. Crocker, T. D., ed., *Scientific Truths and Policy Truths in Acid Deposition Research, Economic Perspectives on Acid Deposition Control.* Butterworth, Boston, MA, USA, 1984.
24. Kwerel, W., To tell the truth: imperfect information and optimal pollution control. *Rev. Econ. Studies*, **44** (1977) 595–601.
25. Roberts, M. & Spence, M., Effluent charges and licenses under uncertainty. *J. Public Econ.*, **5** (1976) 193–208.
26. Baron, D. F., Regulation of prices and pollution under incomplete information. *J. Public Econ.*, **28** (1985) 211–31.
27. Baron, D. F., Noncooperative regulation of a nonlocalized externality. *Rand J. Econ.*, **16** (1985) 553–68.

28. Spulber, D. F., Optimal environmental regulation under asymmetric information. *J. Public. Econ.*, **35** (1988) 163–81.
29. Collinge, R. A. & Bailey, M. J., Optimal quasi-market choice in the presence of pollution externalities. *J. Environ. Econ. Management*, **10** (1983) 221–32.
30. Clarke, E., Multi-part pricing of public goods. *Public Choice*, **XI** (1971) 17–33.
31. Vickrey, W., Counterspeculation, auctions and competitive sealed tenders. *J. Finance*, **16** (1961) 8–37.
32. Crocker, T. D. & Forster, B. A., Decision problem in the control of acid precipitation: nonconvexities and irreversibilities. *J. Air Pollut. Control Assoc.*, **31** (1981) 31–8.
33. Crocker, T. D., Structuring of atmospheric pollution control systems. *The Economics of Air Pollution*, ed. H. Wolozin. Norton, New York, USA, 1966.
34. Montgomery, W. D., Markets in licenses and efficient pollution control programs. *J. Econ. Theory*, **5** (1972) 395–418.
35. McGartland, A., A comparison of the marketable discharge permit systems. *J. Environ. Econ. Management*, **15** (1988) 35–44.
36. Krupnick, A. J., Oates, W. E. & Van DeVerg, E., On marketable air-pollution permits: the case for a system of pollution offsets. *J. Environ. Econ. Management*, **10** (1983) 233–47.
37. McGartland, A. & Oates, W. E., Marketable permits for the prevention of environmental deterioration. *J. Environ. Econ. Management*, **12** (1985) 207–28.
38. Altshuller, A. P. & Linthurst, R. A., ed., *The Acidic Deposition Phenomenon and Its Effects: Critical Assessment Review Papers*, Vol. I, *Atmospheric Sciences*. US EPA, USA, 1984.
39. Oppenheimer, M., Epstein, C. B., Yunke, R. E., Acid deposition, smelter emissions and the linearity issue in the western US. *Science*, **229** (1985) 859–62.
40. Hidy, G. M., Acid deposition in the western United States. *Science*, **233** (1986) 10.
41. Shavell, S., Strict liability vs. negligence. *J. Legal Studies*, **9** (1980) 1–25.
42. Polinsky, A. M., Strict liability vs. negligence in a market setting. *Amer. Econ. Rev.*, Papers and Proceedings, **70** (1980) 363-7.
43. Streets, D. G. & Veselka, T. D., Economic incentives for the reduction of sulfur dioxide emissions. *Energy Systems and Policy*, **11** (1987) 39–59.
44. *Coal Outlook*, 14 Sept. 1987.
45. Diemer, J. S. & Eheart, J. W., Transferable discharge permits for control of SO_2 emissions from Illinois power plants. *J. Air Pollut. Control Assoc.*, **38** (1988) 997–1005.
46. Maloney, M. T. & Yandle, B., Estimation of the cost of air pollution control regulation. *J. Environ. Econ. Management*, **11** (1984) 244–63.
47. Morrison, M. B. & Rubin, E. S., A linear programming model for acid rain analysis. *J. Air Pollut. Control Assoc.*, (1985) 1137–48.
48. Welsch, H., A cost comparison of alternative policies for sulfur dioxide control: the case of the British power plant sector. *Energy Econ.*, **10** (1988) 287–97.
49. Reinert, K. A. & Ratick, S. J., A note on estimating a long-run average cost curve for flue gas de-sulfurization. *J. Environ. Econ. Management*, **15** (1988) 30–4.
50. Terkla, D., The efficiency value of effluent tax revenues. *J. Env. Econ. Mgt.*, **11**

(1984) 107–23.
51. Atkinson, S. E., Marketable pollution permits and acid rain externalities. *Can. J. Econ.*, **XVI** (1983) 704–22.
52. Castel, A. & Griffith, C., Cost effective management of wet sulfate deposition. *Water Air Soil Pollut*, **31** (1986) 1035–45.
53. Tschirhart, J. T., Transferable discharge permits and profit-maximizing behavior. *Economics Perspectives on Acid Deposition Control*, ed. T. D. Crocker. Butterworth, Boston, MA, USA, 1984, pp. 157–171.
54. Jones, D. N. & Tybout, R. A., Environmental regulation and electric utility regulation: compatibility and conflict. *Boston College Environ. Affairs Law Rev.*, **14** (1986) 31–59.
55. *Coal Age*, March 1987.
56. Farber, S. C. & Martin, R. E., Market structure and pollution control under imperfect surveillance. *J. Indust. Econ.*, **XXXV** (1986) 147–60.
57. *Coal Week*, Aug. 1983.
58. Pashigian, P. B., The effect of environmental regulation on optimal plant size and factor shares. *J. Law Econ.*, **XXVII** (1984) 1–28.
59. *Coal Outlook*, 6 Mar. 1989.
60. Besanko, D., Performance versus design standards in the regulation of pollution. *J. Public Econ.*, **34** (1987) 19–44.
61. *Coal Outlook*, 28 Sept. 1987.

Chapter 17

Acid Pollution and Public Policy: the Changing Climate of Environmental Decision-Making

ALAN IRWIN

Department of Human Sciences, Brunel University
Uxbridge, UB8 3PH, UK

17.1 INTRODUCTION

This chapter is written at a time when the environment has attained a prominent position on the world-wide social agenda. It seems that environmental issues will assume greater public significance during the 1990s than was the case even during the surge of interest experienced (firstly in the US but subsequently world-wide[1]) during the 1960s and early-1970s.[2] The emergence of a 'green consumer' movement is being portrayed as a major social trend[3]—and indeed this is already discernible in countries such as Germany[4] and Britain.[5] Across Europe, the green political movement has steadily grown in importance so that the June 1989 European Parliament elections saw strong support in countries such as Britain (14·5%), the former FRG (8·4%) and France (11%).[6] Further indications of increased environmental awareness can be found in political speeches (most notably by the former British Prime Minister[7]) and also in newspaper attention to these issues. All the evidence is, therefore, that as a public and political issue the protection of the environment will certainly not disappear—but instead retain at least its current salience. The strongly *international* flavour of current debates should further reinforce the durability and high visibility of environmental concern.

The purpose of this chapter is to consider the acid pollution problem against this background of changing social and political attitudes towards the environment. More particularly, now that concern for the environment is being expressed at every political and public level, serious attention needs to be focused on the practical obstacles to effective and timely policy responses.

It is argued here that one important obstacle to such progress relates to the various *uncertainties* which need to be tackled when making far-reaching, potentially expensive and sometimes irreversible decisions about environmental protection. Within public debate, issues are often presented in clear-cut and definitive terms—usually by those who are attempting to persuade others of the strength of their viewpoint.[8] However, it will be suggested that greater acknowledgement is needed of the complexities and difficult judgements involved in environmental management. Going further, it will be argued that—unless a more honest and open understanding of the processes of environmental decision-making can be reached—the chances of keeping pace with the next generation of hazard concerns will be greatly reduced.[9] The question which underlies discussion in this chapter is therefore *how are we to approach these urgent but uncertain issues in a rational and reasonable manner?* As is hopefully demonstrated here, such an apparently innocuous question conceals a variety of treacherous political, technical, moral and philosophical issues. Taken together, they suggest something of the *social* challenge represented by environmental threats.

The acid pollution issue is an excellent place to begin this discussion of effective environmental management since it serves as an embodiment of the difficulties raised by the new generation of global environmental threats (including not only acid pollution but also such problems as the greenhouse effect and damage to the ozone layer). As the Canadian Minister for Environment put it in 1981, acid damage is[10]

> . . . a striking paradigm for so many of our environmental problems. Its scope is international. Its causes are scattered in a variety of countries. Its consequences extend far beyond international boundaries. And its successful resolution will require co-operative international action.

Whilst acknowledging the significance of this international dimension, it is argued here that the lessons to be learnt from the acid rain case extend far beyond that dimension alone. Thus, for example, as David Cope has argued,[11]

> The entire saga does . . . provide a good illustration of the problems of trying to reach an acceptable environmental solution when both the environmental issue being addressed and the sector of the economy involved . . . are in a state of rapid change.

In order to tackle these issues, the analysis begins with the long-running technical debate (and, often, controversy) over the nature and effects of acid rain. As will be seen, one constant theme of the acid rain debate has been the claim by various parties that *they* have the best scientific understanding of the underlying questions.

17.2 ACID RAIN AS A TECHNICAL CONTROVERSY

In many ways the central issues at stake in the acid pollution controversy have been *technical* in nature. Of course, it seems obvious that before any national or international action can take place, such questions as the causes and consequences of acid deposition need to be fully resolved. Equally, it has been clear at various points in the debate that different parties have claimed to offer a 'scientific' version of the situation. Thus, references to the 'rhetoric and reality' or the 'fact and fiction' of the problem have become common-place. In this section, the nature of the technical dispute over acid deposition will be discussed and, in particular, the role of scientific evaluations in assisting the establishment of public policy will be considered. In so doing, only the most general aspects of the debate will be considered so that broad characteristics can be discerned. Rather than following the fine detail, the strengths and limitations of technical analysis in resolving environmental disputes will be discussed.

Throughout its history, the acid pollution issue has been characterised by a series of technical disagreements over its causation and consequences. Park, for example, noted in 1987 the following significant differences of opinion which were then still unresolved:[12]

- the extent to which man-made emissions of sulphur dioxide (SO_2) and nitrogen oxides (NO_x) were the cause of the problem;
- the extent to which observed damage could be definitely attributed to acid deposition;
- the significance of the damage;
- the extent to which emission controls might have a real effect;
- the cost-effectiveness of controls.

This list could possibly be extended (and certainly could be disputed by various parties). It is also clear that the precise focus of the debate has changed over time as particular areas are seen to be 'understood' and others become more problematic. However, what is equally interesting for this discussion has been the potential for such technical—and otherwise arcane—questions to take on major practical significance. This point is illustrated especially clearly by the recent history of the British debate over acid pollution. In particular, there has been a long-running disagreement between the former Central Electricity Generating Board (CEGB) (now privatised as National Power and PowerGen) and environmental groups over these issues. Thus, the general position of Sir Walter (now Lord) Marshall in July 1984 was stated as follows in CEGB literature.[13]

> 'If it is shown that acid rain is causing significant damage to the environment and if reducing sulphur emissions from power stations offers a real improvement and if it is the most cost-effective solution, we will take whatever action is necessary. However, I want to be confident that we know enough about the problem and, if sulphur is the problem, the most cost-effective way of dealing with it, before putting up the costs of producing electricity from our coal-fired power stations by hundreds of millions of pounds a year.'

This view can be sharply contrasted with the following statement from the House of Commons Environment Committee report on acid rain in the same year.[14]

> We have taken fully into account reservations which have been expressed to us, but we do not feel these outweigh the weight of evidence which we have received. We believe that our work demonstrates that sufficient evidence exists to show that the decision to reduce emissions of SO_2, NO_x and hydrocarbons should be taken now.

Although the House of Commons Committee was sharply critical of the CEGB attitude to the problems, international environmentalist groups and their supporters have been even more hostile. Such groups have stressed in particular that the 'more research is needed' argument has been an excuse for delay and CEGB intransigence.[15]

> In the past decade, chemists all over the continent have revelled in compiling ever longer lists of chemical reactions taking place in plumes of power-station smoke, along cold fronts and warm fronts, in clouds, over land and sea and even over pig farms. The chairmen of organisations which have been shown to be responsible for pollution, such as Lord

> Marshall . . . have paid our scientists handsomely to increase our knowledge of the complexities. The polluters hope that if they dig deep enough they will discover that their own pollution is doing somebody, somewhere, some good.

Such differences of opinion and the associated antagonisms between parties suggest something of the heated nature of disputes over environmental questions. However, what they also reveal is the extent to which scientific assessments—and particularly scientific uncertainties—have become 'centre-stage' in the controversy. Thus, the CEGB resolutely defended its position on *scientific* grounds. The same approach has been adopted by oppositional groups who have repeatedly claimed to offer a 'better' scientific account than that of the 'establishment'.

Such employment of technical expertise as a 'resource' for different stances within environmental disputes has become common-place over the last few decades. Certainly, parallels can be drawn with the nuclear debate and also with a whole series of controversies as analysed to date by social scientists (e.g. debates over chemical toxicity, major accident hazards, mandatory seat-belt wear[16]). Whilst, of course, it appears eminently rational to draw upon scientific analyses in this fashion, the question must arise of the *limitations* to science in resolving such disputes. Is it reasonable for us to expect *definite* technical solutions to pressing environmental problems such as acid pollution?

If one examines the technical debate over acid deposition in a little more detail, a number of areas of persistent doubt and controversy become evident. Thus, the whole question of linking cause (acidity levels attributable to various pollution sources) and effect (visible physical consequences) has been highly problematic (especially when contrasted with the common assumption that science can provide some definite 'proof' in such cases). In particular, uncertainties at this level surround the following.

- The *chemical complexity* of the processes under consideration. Understanding the minutiae of the photochemical reactions involved has always presented substantial difficulty.
- The sheer *range of the effects* being analysed (forests, plants and crops, lakes, rivers, fish, human health, corrosion). Whilst there are common strands to the causation of all of these effects, each also presents its own individual problems for scientific analysis.
- *Time-scale* problems present special difficulties for linking cause and effect. Thus, the episodic nature of acid pollution and the lag between peaks in, for example, SO_2 pollution and acidity levels add to the prob-

lem of identifying trends in the levels of international acid deposition.

- The *spatial diffusion* of effects undoubtedly represents a major area of complexity. Thus, the transboundary migration of the pollution problem has made it more difficult to relate a pollution effect (e.g. Norwegian fish deaths) to a specific cause (e.g. British power stations).
- The *variability of cause–effect response* needs also to be considered. Thus, no two locations seem identical in terms of the buffering capacity of the soil and various geological characteristics, the precise pH level, prevailing metereological conditions, habitat, wild-life distribution, etc.
- *Intervening variables* need also to be considered. Thus, for example, the effect of other pollutants, environmental agents or 'natural' biological processes (perhaps working synergistically) needs also to be considered.

When taken together, such factors suggest some of the inevitable complexities and uncertainties involved in demonstrating the linkage between the causes and consequences of acid deposition. Consequently, heated discussion has taken place around the 'linearity' question of whether a reduction in SO_2 and NO_x would lead to a proportional reduction in acidity levels. Equally, the precise processes of acidity damage have been open to debate—with attention increasingly being directed to the role of ozone, sunlight and the leaching of heavy metals. In all these respects, therefore, it becomes extremely difficult to link together pollutant sources, levels of acidity and observed environmental effects. Of course, it has been precisely these kinds of uncertainty which have led to charges from the CEGB, the British government and certain others that the acid deposition case is 'not proven'. In 1984, Mellanby made the case for caution as follows.[17]

> Acid rain is thus a real problem, but its importance has often been exaggerated. It does much less harm in the populous parts of Britain than did the air pollution we formerly suffered. It has, however, been taken up as a *cause celebre* by the so-called 'ecologists' who were disappointed because their grim prophesies of doom from pollution and starvation have not been fulfilled.

Before discussing the case of acid rain further, however, it is important to step back from the detail of this case in order to assess how much it has in common with other contemporary environmental debates. Cramer, for example, in a study of the role of ecologists within environmental decision-making, has noted that a number of forms of uncertainty will affect technical experts when asked to offer policy advice (whether or not they acknowledge or even recognise this).[18] She characterises these into three main types.

First of all, there is *pragmatic uncertainty*; thus, scientists may be asked to offer practical advice at very short notice and without sufficient equipment or staffing resources. Whilst the field of acid deposition research has advanced substantially over the last few years, there is no doubt that many international scientists outside the main institutional laboratories have experienced this problem.

Secondly, there is *theoretical uncertainty* as occurs when there is no strong theoretical framework (or 'paradigm' to use the Kuhnian term[19]) which unites a scientific field but rather a disparate pattern of disciplinary approaches and academic perspectives. This seems particularly relevant in the present context since acid deposition research draws upon a large variety of backgrounds (including chemistry, metereology, the biological sciences, physics and geology). It is to be expected that these different disciplines will each highlight different aspects of the acid deposition problem and will carry a separate 'tool-kit' of analytical techniques. Such a multidisciplinary pattern is a regular feature of research fields which have developed in a 'problem-orientated' fashion rather than emerging from a single established discipline.

Thirdly, there is *the uncertainty related to the complexity of the ecological predictions to be made*. Thus, ecologists are called upon to model highly complex kinds of 'real world' behaviour (i.e. 'open systems') rather than operating within the tightly controlled 'closed systems' of the laboratory. Acid deposition seems an excellent example of an 'open system' with numerous intervening variables in flux at any time, with consequently high levels of response variability and with modelling and trend-identification proving extremely difficult.

In a review of the relationship between scientific expertise and environmental issues, Yearley further notes that the fact that environmental scientists are so often working on the *margins of observability* adds considerably to these uncertainties.[20] Thus, environmental problems often involve subtle and barely tangible forms of damage, they may be difficult to monitor and their precise trends may be imperfectly understood. As Park has noted in this context,[21]

> There are very few really dramatic effects of acidification, and many of the inferred impacts are extremely subtle if not largely invisible (even to the trained eye), at least in the early stages.

It is because of uncertainties such as these that many commentators have seen technical advice as, if anything, exacerbating rather than eliminating policy disagreement.[22] In a careful study of the role of scientific evidence in controversies over chemical toxicity, Graham *et al.* conclude that,[23]

> On the one hand, when very simple questions are asked (by regulatory authorities), the conscientious scientist finds it difficult to know how to answer because of the ambiguity created by simplification. On the other hand, scientific knowledge is not adequate to answer the more elaborate questions with any measure of confidence. Our point is that scientists often seem to disagree with one another because they are caught between ambiguity and ignorance, between questions that are too hard and questions that are too simple.

Whilst within normal scientific debate such a condition of 'ambiguity and ignorance' might be entirely beneficial (e.g. in stimulating scientific discussion and further research), within a policy context where highly consequential decisions need to be speedily taken grave difficulties can occur. Accordingly, in practical policy terms, the particular *burden of proof* to be applied in such cases becomes crucial. Do we await absolute proof of environmental harm due to a specific cause (which may, of course, be effectively impossible to achieve given the uncertainties listed above) or do we operate on the basis of sufficient (but possibly inconclusive) evidence?[24] As the Brundtland Report put it in a discussion of the greenhouse effect,[25]

> How much certainty should governments require before agreeing to take action? If they wait until significant climate change is demonstrated, it may be too late for any countermeasures to be effective against the inertia by then stored in this massive global system.

At this stage the highly *judgemental* nature of technical decision-making becomes evident. Given that the scientific evidence cannot present us with definitive understanding of the situation, how best are we to evaluate uncertain evidence?

The question of 'how much certainty' has echoed throughout the acid rain debate in Britain. For example, the House of Commons Environment Committee put a similar question to those giving technical evidence in 1984. In Sir Hugh Rossi's words,[26]

> As scientists you are looking for 100% certainty before reaching a con-clusion. In so far as political decisions have to be taken, to what extent should politicians await complete certainty before they take action?

Dr Martin Holdgate (Chief Scientist at the Department of the Environment) replied to this by accepting the judgemental nature of these questions.[27]

> All one can do is say that one has to balance the probabilities of a problem, the probabilities of damage, against the implications it has for the community . . . and probably in any one decision this balancing act goes through successive stages.

Holdgate also stressed the kinds of technical uncertainty outlined above. Here, for example, in a discussion of fish deaths:[28]

> What I am saying is that I do not think that the causal chain, or the simple correlation with single factors, is anywhere near as clear in this area. This is, of course, our difficulty.

Marshall replied to a similar line of questioning in these terms:[29]

> I have a statutory responsibility, and the Board has a statutory duty, to produce electricity at the cheapest possible price consistent with protecting the environment. I am not saying that I want to maintain our present practice indefinitely. I am simply saying that at the moment I see enough arguments that discount that and I want to be sure . . . what the scientific facts are.

The argument of this chapter is that judgements as to 'how much certainty?' must be made by *all* parties to environmental disputes—whether they choose to acknowledge this or not. Thus, if we examine the opposing side to this dispute we can see such parties as Greenpeace making their own, rather different, evaluations of the situation. For example, in 1989, in the context of 'Waldsterben' (forest death):[30]

> Although the scientific evidence absolutely proving a link between air pollution and tree damage is not yet complete, the present risks presented by air pollution to our trees justifies and indeed cries out for swift and effective action to clean up power station emissions by the mid-1990s.

As the earlier quotation from the Brundtland report reminds us, this need for a judgemental aproach to the scientific evidence is not unique to the acid deposition issue. It is also important to emphasise that these levels of uncertainty are not a reflection of 'bad science' (or of inadequate scientists) but are instead an almost unavoidable characteristic of environmental issues such as acid deposition. Instead of wishing to criticise scientists, the argument is for a greater acknowledgement that science cannot always provide immediate and clear-cut answers to complex environmental problems. In that sense it is unreasonable (and unduly cynical) to expect scientists simply to 'come up with the goods'. Equally, we cannot afford to *reject* scientific evidence just because it cannot offer certainties. As Ravetz has put the case,[31]

> . . . now we must cope with the imperfections of science, with radical uncertainty, and even with ignorance, in forming policy decisions for the biosphere. Do we merely turn away from such problems as beneath the dignity of scientists, or do we learn somehow to make even our own ignorance usable in these new conditions?

It follows from the analysis so far in this chapter that practical policies need to take account of these inevitable technical uncertainties (and of the controversies and disagreements that may follow) whilst at the same time making full use of the understanding that scientific analysis *can* provide. These points will now be taken further by moving explicitly towards a discussion of political action in this area. Given the technical uncertainties discussed here, what special policy-making problems are raised by environmental issues such as acid deposition?

17.3 ACID RAIN AS A POLITICAL CONTROVERSY

It has already been argued in this chapter that the acid deposition debate has been characterised by a high degree of technical uncertainty and, going further, that various social groups have drawn rather different policy conclusions from this uncertainty. If the political debates over this issue within Britain are considered, a high level of doubt and controversy can be identified.[32] In particular, since the early 1970s there have been repeated challenges to the official British position on this issue. Park describes the ensuing controversy as follows:[33]

> . . . a political debate of national and international significance, as contentious as it has been prolonged. The debate has seen scientist set against scientist, scientist set against politician, and politician set against politician. Political moguls have faced a barrage of opposition from a curious and highly unusual alliance between some government scientists and the environmental lobby in Britain, and from leaders of most European countries, all of whom have grown increasingly frustrated by the Thatcher government's overt reluctance to grasp the nettle of reducing SO_2 emissions.

From 1971 onwards, Scandinavian nations began to accuse Britain of responsibility for acidity damage caused by the 'export' of SO_2 and NO_x.

Special allegations were directed towards the CEGB for its part in acid deposition. The British response to such charges was to highlight previous success in controlling air pollution and to maintain a cautious (and at times defensive) stance on the issues. Of course, it must be acknowledged that the position on acid deposition adopted by different national and international bodies (e.g. regarding the technical uncertainties discussed above) is likely to reflect their own perceived interests. Even as recently as 1985, McCormick noted that[34]

> The level of government interest and action is usually in direct proportion to the degree of damage identified at home. The Scandinavians and Canadians have been the most vocal; the West Germans changed course only when faced with proof of substantial damage within their own borders, and Britain—where relatively little damage is officially acknowledged and where acid pollution is barely a public issue yet—has so far steadfastly refused to act.

Whilst it was only since 1985 that a major change of policy within the CEGB and British government occurred, a number of other steps were taken from the early 1980s onwards. For example, in a highly publicised move in September 1983, the Royal Society announced a £5 million study of 'the causes of acidification of surface waters in affected areas of Norway and Sweden'—with funding provided by the CEGB and the National Coal Board (NCB). Other research around that time also highlighted the possibility that parts of *Britain* might be affected by acid pollution (e.g. freshwater lakes and public buildings).

International pressures on Britain to make a major public commitment to acidity reduction were steadily growing during the 1980s. In particular, from the late 1970s onwards a series of international conferences began to call for a co-ordinated programme of pollution control. The 1982 Stockholm Conference on Acidification saw Britain (and the US) falling into isolation—at least among Western nations. In 1984, this pressure increased further when the '30% club' was established. By the end of June 1984, 18 nations had agreed to reduce their SO_2 emissions by 30% by 1993. Once again, Britain found itself almost alone in the international debate. The defence offered—that the link between SO_2 and acidity was still unproven—was widely seen as mere self-interest given that Britain was a net exporter of acid rain and would appear to have less to gain than other countries such as Germany, Sweden or Norway. From 1983, the European Community began to exert its influence in international debate. In particular, a framework directive came under discussion which proposed to regulate vehicle exhausts and to reduce emissions of pollutants from 'large combustion plants'. Both these proposals were to assume great significance in shaping British responses to the acid deposition problem.

In Britain, criticism of the 'official' position on acid deposition reached a new pitch in 1984. In that year, the Royal Commission on Environmental Pollution in their 10th report highlighted acid deposition as 'one of the most important pollution issues of the present time'. In addition to recommending that a high priority be given to further research, the Royal Commission also advocated that SO_2 abatement measures should be introduced by the CEGB on a pilot basis.[35]

> Otherwise, international pressures may eventually dictate a significant reduction in emissions of SO_2, with the adoption of expensive 'crash programmes' against a background of technical experience and of a need for importation of equipment from countries that have already embarked on a programme of abatement.

Later in the same year, the House of Lords Select Committee on the European Communities produced a report which argued in favour of Britain's entry to the '30% club'. However, the greatest British political controversy over acid rain in 1984 surrounded the September publication of the House of Commons Environment Committee's report on *Acid Rain*. In a strong attack on government policy, this all-party committee stated that[36]

> Enough is now known to justify the development and application of technology for removing the causes of effects now abundantly apparent.

The report made very clear the committee's attitude to the 'burden of proof' question discussed above.[37]

> Manufacturing and energy-producing industry have used the reticence or uncertainty of scientists to push to the full their case for not embarking on large expenditure on emissions control.

The report recommended that Britain 'join the 30% club immediately' and also meet a draft EEC directive then proposing to reduce SO_2 emissions by 60% and NO_x by 40% by 1995.

The Government's reply was to reject these specific recommendations—using as justification once again the technical uncertainties surrounding acid deposition. In Parliament, the Minister—William Waldegrave—defended his position as follows.[38]

> When one is spending other people's money, in this case electricity con-

sumers' money, it is always easy for Ministers to push for crash programmes, but we are still ignorant about an essential part of the problem in terms of a genuine cost-benefit analysis not so much in terms of money, but in terms of the environmental gain to be achieved from reductions in deposition.

The Government was thus adopting, in January 1985, a cautious and gradual approach to the acid deposition problem—pointing to the progress since 1970 but refusing to engage in a large-scale programme of fitting pollution control equipment to existing power stations. Their stated goal as announced in December 1984 was to achieve a 30% reduction in 1980 levels of both SO_2 and NO_x by the end of the 1990s (which fell short of the '30% club' requirements).[39] Clearly, also, the question of technical uncertainty was a major factor in debate. As, for example, Allan Roberts MP argued in response to Waldegrave,[40]

They want conclusive proof provided scientifically. They want a linear correlation between a reduction in sulphur in one place and a reduction in damage in another. If they want that before they act, they will never act, because they will not get that evidence. It is impossible to produce such evidence.

Despite the Government's stance during this debate, many commentators have discerned a change in the official British attitude to the problem from 1985 onwards. Thus, in early 1986 Waldegrave admitted some British responsibility for acid rain in Norway. During a visit to London by the Norwegian Minister for the Environment, Waldegrave promised that Britain would be reviewing its policy on sulphur emissions from power stations. More significantly, in September of that year Margaret Thatcher (while on a visit to Norway) announced a new initiative to cut acid pollution. The British government thereby agreed to spend £600 million over the following 10 years on pollution control equipment (flue gas desulphurisation—FGD). This programme would take effect at three of its major 2000 MW coal-fired power stations and would begin with the newly completed Drax station. It was also announced that all new coal-fired stations would be fitted with the technology to control acid emissions. However, the political response to this move was less than enthusiastic. *New Scientist* commented that the programme 'may do little more than preserve the status quo'. Other groups also—such as the Labour Party and Friends of the Earth—saw it as falling short of the more radical changes that were needed. As one Friends of the Earth campaigner put it,[41]

> The best we can say about Mrs Thatcher's decision . . . is that it amounts to an admission of guilt.

Indeed, Greenpeace, went so far as to make a complaint to the Advertising Standards Authority that CEGB press advertising (which claimed that the Board was 'mounting an extensive programme of countermeasures') was inaccurate and misleading.[42]

Pressure on the government and the CEGB to change their stance on acid deposition was (somewhat ironically) increased also by the CEGB's release late in 1985 of a video, *Acid Rain*, which whilst claiming to be 'educational' in purpose succeeded in angering both environmental groups and many Scandinavian scientists. One group from the Norwegian Agriculture University considered the video to be both 'misleading and insulting'. A Swedish biologist, Hans Hultberg, complained bitterly that he had been misrepresented in the film. In addition to such criticisms of the video, two other points began to become clear. First of all, that there may be substantial disagreement amongst even CEGB scientists about the 'official' position. Secondly—and perhaps even more damagingly—that the public appearance of unity between the CEGB and the Department of the Environment may be collapsing. Thus, the environmental and scientific press reported that a separation of the two groups was being openly acknowledged.[43]

> Although the government had previously given wholehearted support to arguments of the kind used in the film, it now chose to disavow the CEGB. Last December [1985] the Minister for the Environment let it be known that he regretted that such a film should have been made. He admitted that it provided only a selective account of the scientific arguments, and that it was in fact a political disaster.

Although the CEGB resolutely defended the accuracy and integrity of their video ('in a short film made for a general audience it was not possible to cover every diverging viewpoint . . .'[44]), it was obvious that public relations damage had been done. Of course, the changes in government policy described above followed these incidents. It should also be noted that the 1988 video released by the CEGB on this subject—*Acid from the Sky*—represented a substantial move away from the 1985 'hard-line' approach.

Thus, by the end of 1986 it was clear that policy changes were taking place in Britain—albeit at a pace which continued to frustrate environmentalist groups and also foreign governments such as Norway. In May 1987, further incremental progress was made when the Government announced a 10-year action programme to control NO_x emissions through the installation of low-NO_x burners at the CEGB's 12 largest power stations. Combined with the 1986 decision regarding FGD at three existing stations, this represented a £1 billion retrofitting programme. Whilst this was clearly a major step towards reconciling the British stance with that of its critics, the Government was still not prepared to sign the '30% club' protocol—its reasons for this (as presented in a September 1987 Department of the Environment briefing paper) are worth quoting in full.[45]

> The UK Government has recognised the significance of the Club as a symbol of international concern, but regards the Club's objective as arbitrary as it does not take into account emission reductions achieved before 1980, when the UK record was better than that of other countries. The UK may be able to achieve the target (it has already managed 22% reduction since 1980), but cannot be sure of this, because rising energy demand could temporarily halt the downward trend. The retrofit programme . . . will not begin to affect the output figures until about 1994 or 1995.

The environmentalist response to this was to argue that SO_2 and NO_x emissions in Britain were actually *increasing* from 1985 onwards. Thus, Greenpeace argued in 1988 that sulphur emissions were likely to rise by 10% by 1993 then fall back to current levels by 1998 following pollution control at the three stations.[46] As *Acid News* described the situation in February 1988,[47]

> The UK will thus continue to spread acidifying sulphur and nitrogen pollutants over other countries during the nineties as well—thus further damaging its reputation internationally. The epithet, 'the dirty old man of Europe', will regrettably still be applicable, it seems, for quite a number of years to come.

Prominent on the international agenda by this stage was a draft European Community (EC) directive on the control of emissions from 'large combustion plants' which had been under discussion in Brussels since December 1983. As adopted by the Council of Ministers in March 1984, this proposal included power stations, metal manufacturing and processing, chemical works, the food industry and waste-disposal installations. For power stations, the reductions required at this stage were 60% for SO_2, 40% for NO_x and 40% for dust (based on 1980 figures). These were to be achieved by December 1995. Reactions to this proposal were predictably mixed; equally predictably, negotiations within Europe were protracted. By December 1987, the UK position was that the extensive retrofitting programme (involving FGD at another nine stations and 'at a possible cost of well over £2000 million' all to be carried out in a very short period of time) 'cannot be justified'.[48] However, negotiations were still continuing in the hope that a more 'reasonable' agreement could be reached. At the same time, whilst the UK was certainly not the only nation to be unwilling to accept the proposal, it was regularly singled out for criticism by other environment ministers and by environmentalist groups—thus, according to *Acid News*, prompting the British Minister, Nicholas Ridley, to react sharply.[49]

> We are tenth on the list of offenders, out of say almost fifteen at the top, so why should we be put in the dock and be pilloried like this. The figures are fixed. We have had our major reductions–30% before 1980 and since then a further 20%. We did it before them, so if you choose a baseline which suits the continentals they can make us look as if we are laggers, but it is they who are laggers because we started first.

Unsurprisingly, given the acrimony of the debate, *Acid News* demonstrated little sympathy for the British situation.[50]

> This emotional statement shows that the European-wide criticism is hitting the British government, but that they still prefer to react only with polemics and misleading statements.

What this exchange also underlines is the policy significance of such otherwise technical questions as the direction of pollution trends. Whilst the official position was to see the downward trend as a genuine reflection of emission control, acid rain campaigners saw it as an artefact of a weak economy and which would soon be overwhelmed by any subsequent industrial growth.

As of 1988 a number of compromise agreements were being attempted on the 'large combustion' draft directive–these in particular sought to circumvent the major obstacles to agreement (namely, concerning the size of new plant which would be covered, the timetable for implementation, and the precise percentages for reductions in SO_2 and NO_x). The existence of such negotiations as this (albeit at an earlier stage in the debate) led Park to comment as follows.[51]

> Acid rain provides a fascinating illustration of international politics at work. All the usual ingredients are there—hidden agendas, vested interests, frustration and elation, inspired initiatives, and raging rhetoric. Money and morality, as ever, are key forces in the political debate.

It was from exactly this political cauldron that the new EEC controls finally emerged. In June 1988, it was agreed that Britain would accept the directive—this would involve reductions in emissions of SO_2 and NO_x by 2003. Thus, SO_2 must be cut by 20% (on 1980 levels) by 1993, by 40% by 1998, and by 60% in 2003. NO_x emissions would be cut by 15 and 30% in two stages (1993 and 1998).

As *New Scientist* commented, much internal wrangling surrounded Lord Caithness' agreement of the directive:[52]

> . . . both Britain's energy secretary, Cecil Parkinson, and the CEGB chairman, Lord Marshall, lobbied hard to persuade the Department of the Environment not to make major concessions.
>
> Both men argued that stricter pollution controls would make the CEGB less inviting to investors. But the department apparently shaken by threats from other member states to take Britain to the European Court of Justice, decided to capitulate . . .

Discussing the situation in 1989, the *Financial Times* commented on the EEC agreement as follows.[53]

> The difficulty in obtaining such agreements reflects the different interests of individual states.
>
> For example, Britain, with its 80% reliance on coal-fired electricity, fears that such a programme would significantly increase its electricity costs.
>
> It is nevertheless trying to shake off its reputation of being the 'dirtiest' country in Europe.

Nevertheless, agreement to the EEC directive—whilst not appeasing the environmental critics[54]—represented a substantial compromise with European opinion.[55] The controversy over acid pollution was, however, far from over. Numerous press reports in early 1990 suggested that Britain was wavering in its commitment to the directive. *The Guardian*, for example, reported that the Department of the Environment was reneging on its agreement to spend approximately £2 billion on FGD.[56] This move was widely linked to the planned privatisation of electricity generation. In response to such criticisms, Margaret Thatcher re-confirmed the government intention to meet the requirement,[57] but stated that this would not rely[58]

> on a single method alone but [by] combining desulphurisation equipment, new gas-fired plant and other means such as the use of low-sulphur coal.

This new turn in the debate antagonised environmental groups once again—many of whom argued that Britain had been successful in negotiating a relatively lenient emission reduction precisely because of its need to burn domestic coal (with its high sulphur content). However, what the developments of 1990 illustrated once again was the close linkage between environmental decision-making and the changing economic and political climate in which it must take place. In the next section of this chapter, the wider issues for environmental policy-making raised by this whole history of events will be considered.

17.4 ACID RAIN AND ENVIRONMENTAL POLICY-MAKING

> Cultural constraints and natural constraints are both at work and have to be explored together if we are to do the best we can for the biosphere . . . To ask 'Can we manage the biosphere?' is to ask 'Can we manage our institutions?'[59]

The brief account given here of the acid deposition debate in Britain suggests something of the difficulties raised by environmental policy-making. Throughout this history, it has been seen that a number of areas of complexity and uncertainty have served to vex the decision-making process. Also, this process has been protracted and, at times, highly controversial—with the British government being the focus of substantial national and international criticism. It therefore seems important in view of this chapter's central concern with the effectiveness of environmental control to review the impediments and special difficulties (in comparison with what can loosely be termed more 'mainstream'[60] political issues) which have been in evidence. In particular, it will be argued that there are certain important characteristics of environmental disputes which make them especially intractable to policy resolution. Having established these policy characteristics, we will then be well-placed to address questions of future environmental strategies.

The first of these special difficulties—and which has already been extensively discussed here—concerns the *technical complexities and uncertainties* which have surrounded this issue. Whilst it is unnecessary at this stage to repeat the conclusions already reached above, the point must be emphasised that—whilst 'scientific proof' is often presented as the final arbitrator to such debates—technical argumentation has itself become a major (perhaps *the* major) focus of controversy and disagreement. Thus, the whole question of linking cause and effect has been open to a number of contrasting (and even contradictory) evaluations and perspectives. Such matters as the chemical complexity of the underlying processes, the range of acidification effects, the time delay between pollution control and actual reduction in acidity levels and the existence of intervening variables have served to prolong (and sometimes polarise) discussion. Even the identification of emission trends has proven contentious. Whether such technical uncertainty is an *inevitable* cause of acrimony and delay is a question which will be returned to below. At this stage, however, it can be noted that this characteristic of technical uncertainty has distinguished the acid deposition question from many of the more 'routine' issues on the British political agenda.

Alongside this identification of technical uncertainty as a policy difficulty, the *interlinkage of technical and social factors* within the policy process must also be emphasised. Questions of 'how much certainty?' or of the burden of proof to be applied in assessing evidence draw at the same time on both technical and non-technical factors. Thus, one's position on the acid deposition issue will be a matter of technical appreciation but also of perceived self-interest and of one's wider perspective on pollution issues. This point is quite evident when contrasting the stance on acid deposition adopted by different groups such as the CEGB, Greenpeace or the Norwegian government; the political perspective of these groups and their technical evaluation of the situation form part of the *same* web of concerns and commitments. No matter how problematic (and messy) this policy imbroglio may seem to scientists drawn in to environmental disputes (and who may prefer the relatively sheltered world of the laboratory), policy decisions not only cut across scientific boundaries but also draw upon such dimensions as economics, politics and social attitudes. In such a context, appeals to the 'real facts' are likely to be a rhetorical gesture rather than a helpful intervention to debate.

A third characteristic of this debate—and which again it would seem to share with many of the 'new generation' of environmental problems—is the *transboundary nature* of the pollution consequences. In the earlier stages of debate (i.e. before acidity damage within Britain became widely acknowledged) the fact that acid rain was seen as a 'foreign problem' seems to have served as a disincentive to political action and public pressure. In recognition of this, for example, Scandinavian environment officials have made special efforts to invite representatives from British organisations to witness acidification damage for themselves.[61] However, such efforts only underline the inherent difficulty of engaging public and political concern over a form of harm which may appear physically distant and which represents no part of everyday experience. It would appear that only a major national initiative (such as that over global warming subsequently launched by Margaret Thatcher) or a general increase in environmental awareness (such as has begun from 1988 onwards) can counteract this sense of remoteness and intangibility. Even then, it may be difficult to mobilise public pressure around issues which seem to have no potential for causing direct harm or detriment to that public.

Closely related to this transboundary dimension is the dependence of effective responses on *international collaboration* (since a purely domestic approach to such global issues is likely to prove ineffective, irrelevant or even counterproductive). It seems clear that questions of the environment will increasingly need to be addressed at this level (and indeed that the process of 'internationalisation' of environment protection is already well-established[62]). Of course, this requirement for international action can present difficulties but also opportunities. In particular, the need for concerted environmental action will place a new significance on larger questions of international politics and relations so that co-operation on issues of this kind may be dependent on agreements and organisations established primarily for other purposes.[63] The EC represents just one example of a forum which, despite its original (essentially economic), *raison d'être* has become central to European-wide environmental action.

Less positively, however, this 'internationalisation' also means that questions of the environment may become inextricably linked with wider international bargaining over monetary relations or diplomatic disputes. Thus, as Majone notes,[64] the present international political and economic order may impede the implementation of environmentally sound development strategies. The global context also raises difficult questions of the relationship between developed and developing nations—how are we to avoid a situation where environmental care is seen as a luxury for prosperous countries or (more cynically) as an attempt by the developed world to impose economic obstacles on newly industrialising nations? Quite clearly, Western nations have an international responsibility here to offer more than pious advice to the Third World.

Such discussion of international action leads us back to another major problem area within the acid deposition debate—the inherent difficulties of balancing *environmental benefit against economic cost*. Even if this question is approached simply at the national level, we can find constant reference during debate to its intractability. Of course, the full implications of electricity privatisation are still being debated in Britain—with many parties seeing such steps as a threat to environmental protection (especially where large-scale expenditure is required).

If we do approach such questions in purely financial terms then the clear problem arises of 'costing' environmental benefit—just how are we to balance social benefits against the large capital outlay involved in reducing SO_2

or NO_x emissions? During the 1970s, heated debate took place among economists concerning the possibilities of cost-benefit analysis as a technique for attaching a monetary value to such intangibles as pollution, environmental quality or even human life. That debate perhaps predictably foundered on the inherent difficulties of giving a price to matters (with human life as perhaps the most emotive example) which seemed somehow 'priceless'. Of course, the proponents of cost-benefit analysis were quick to argue that society already places implicit prices on such values. However, the general distaste for the attribution of a price-tag to human suffering has limited the explicit use of cost-benefit analysis in Britain. More recently, policy discussion of the costing of environmental harm has emerged once again—with the Department of the Environment reconsidering environmental economics and the possibilities for pricing measures which would encourage a more sustainable development route.[65] Whilst a more realistic costing of environmental degradation might be a progressive step, problems in this context seem unavoidable concerning, for example, the precise costs of acid deposition. Thus, whilst such pricing might prevent market forces from simply disregarding environmental harm, the allocation of monetary value would run into the substantial obstacle of those uncertainties and policy impediments outlined here.

Linked closely to these important economic issues there is also the problem of *technological inertia* to be considered (what Collingridge has termed the 'dilemma of control'[66]). Thus, the sheer scale of the financial investment and national significance of the technologies under discussion (whether power stations, motor vehicles or production plants) imposes severe restrictions on policy flexibility—as the Brundtland report also noted with regard to the greenhouse effect. Therefore, whilst incremental changes to these technologies (e.g. of the 'end of pipe' variety as perhaps for FGD) are possible—if expensive—more radical departures (e.g. towards 'clean technology') are effectively impossible without long lead times. In this context, Cope has argued that[67]

> FGD plants, like the dinosaurs, may once have ruled the emission control world but smaller, lither, more adaptable species have now evolved which look set to consign them to a passing phase of history.

What must also be taken into account in any evaluation of the difficulties of environmental policy-making is the existence of *competing interests and perspectives*. This point applies not only to overtly political actors such as environmentalist groups and industry organisations but also to governmental bodies and, as has been seen, to scientific evaluations of environmental threat. Whilst questions of the environment may not be entirely unique in this respect, it does appear that this is an area where contrasting social values, political preferences and scientific perspectives become especially clear—with pollution concerns issues often serving as a focus for otherwise disparate values concerning the future direction of our societies.[68] A special difficulty in such situations is that debate will be restricted to those parties with strongly polarised views rather than full account being taken of *public opinions on pollution.* In the case of acid deposition, it has often been difficult to gauge public opinions outside the small coterie of highly involved groups. The general expression of concern over this issue has been somewhat limited—probably due to the lack of any specific local constituencies and the difficulties for a layperson of separating acidification damage from other forms of pollution or industrial decay. If the public are not to be excluded from discussion, serious consideration will need to be given to improved environmental education and awareness building.

Before concluding this review of the special difficulties of environmental policy-making, mention must also be made of the particularly thorny *moral/philosophical questions* raised by such disputes. As Sagoff has argued,[69]

> The role of the policymaker . . . may be to balance what we believe in and stand for as a community with what we want and need to achieve as a functioning economy . . . It is not just a matter of balancing interests with interests; it is a matter of balancing interests with morality and balancing one morality with another morality.

The heat and emotion which environmental issues can engender strongly suggest that there is more to such disputes than technical analysis and financial calculation alone. As Sagoff observes, such issues also touch upon deep moral and ethical choices about the kind of society in which we wish to live and which we wish to bestow to future generations. Whilst again the stirring of popular emotions around such issues as the 'fate of the earth' may be distasteful to certain individuals and institutions,[70] there is also a pervasive sense of unreality (and folly) surrounding any efforts at stripping away all questions of moral choice and emotion from such matters (e.g. by reducing the situation wholly to cost-benefit calculations or decision theory). The acknowledgement of a strong moral and emotional underpinning to our views on the environment raises many practical difficulties; however, the alternative of attempting to suppress these in the hope of an unambiguous scientific rationality seems a recipe only for popular backlash and loss of political credibility.

17.5 THE SOCIAL MANAGEMENT OF UNCERTAINTY

The points argued so far have substantiated our original assertion that environ-mental management must somehow tackle a 'variety of treacherous political, technical, moral and philosophical issues'. Consequently (and unavoidably), the management of the environment is also the *management of uncertainty*. As the quotation at the beginning of the previous section suggests, this situation represents a substantial social challenge to our institutional structures and decision-making processes. As is also very clear, acid deposition represents just one example of the new pressures under which our environmental protection policies will be placed.

Rather than attempting to present a detailed blueprint for environmental policy, the remainder of this chapter briefly suggests a number of basic principles as a guide to future action. Hopefully, these will serve to stimulate future debate. Put simply, the argument here is for policies which are *adaptive*, which are *open* in their acknowledgement of technical and other forms of uncertainty, which *facilitate social dialogue* between competing outlooks, opinions and rationalities, and which are *global* in perspective.

The first of these general policy principles calls for a highly flexible and responsive approach to environmental management and decision-making—a point which applies equally, of course, not only to government bodies but also to industry and various pressure groups. Thus, there is a need for the constant re-evaluation of priorities and strategies in order to keep pace with the evolution of technical, economic and public assessments. Policy would, thereby, be envisaged not as a 'one-off' (which can be established and then safely stored away until it is forced back onto the agenda) but as a continuous loop of assessment and adjustment. A changing environment—and a changing understanding of our impact on that environment—requires continual adaptation and constant attention. Monitoring has an especially important part to play here. Flexible *technologies* will also be important—perhaps through a greater commitment to 'clean' technology—in order to match changing environmental demands.

The necessity for an adaptive environmental strategy also requires a less rigid attitude to the 'burden of proof' question than has often been visible during the acid deposition debate. Put bluntly, such a strategy cannot be maintained if key organisations refuse to act until conclusive evidence has been accumulated. Because we will inevitably be dealing with uncertainties, a 'balance of probabilities' attitude to new information and understanding will prove essential. Such an attitude will also be assisted by a greater openness in the decision-making process—as will now be considered.

This second policy principle can be linked especially closely with the discussion throughout this chapter of the strengths but also limitations of scientific testimony in the resolution of policy conflicts. Whilst technical experts in such arenas still enjoy a high degree of prestige, there is also the possibility that the apparent indecisiveness of science in providing clear answers to pressing social problems will engender a mood of disillusionment and hostility. One way of avoiding such a negative public reaction to technical uncertainty (and also of providing a more accurate account of the nature of the policy process) is through scientists ensuring that they do not make offers which they then cannot keep; for example, by suggesting that just one more research programme will tell us all we need to know or by denying the areas of doubt in their technical analyses. Graham *et al.* have pointed out the dangers of making overly ambitious claims about the role of science in such areas.[71]

> When science is oversold, non-experts, particularly interested citizens and politicians, tend to develop unrealistic expectations about what science can deliver. When it turns out that science cannot deliver on these expectations, public trust and confidence in science will suffer.

A similar point about 'overselling' might also be made about the processes of balancing costs and benefits within government institutions. Just as with the contribution of technical evidence, it would appear that excessive claims about the rationality and reasonableness of government policy can lead to a loss of perceived legitimacy and public credibility. Indeed, British policy on acid deposition seemed at times to be perilously close to this point—especially when this country was being progressively isolated from world opinion on this issue. One consequence of this 'overselling' is that subsequent policy initiatives will be greeted with an almost reflexive cynicism; for example, towards Margaret Thatcher's efforts at presenting her government as 'not merely friends of the Earth—we are its guardians and trustees for generations to come'.

In both the areas of technical advice and balancing-out cost and benefit, open acknowledgement is needed of the fine judgements and profound uncertainties involved. Whilst parliamentary bodies such as the Commons Environment Committee have been invaluable here in opening-up the policy options and encouraging intelligent discussion, governments and scientific organisations need also to consider how to take further initiatives in this direction.

Again linking closely to this point, there is the question of how best to establish positive social dialogues over environmental protection in order that the full spectrum of analyses and opinions can inform the policy process (and also that governments and industrial organisations do not become isolated from wider society). In order for such public discussion to occur in any meaningful fashion, the previous two policy principles must both be in operation but also be widely *seen* as in operation. The starting point for such a dialogue must also be the acceptance that environmental protection is an issue which legitimately draws upon just the range of social as well as technical dimensions which have been presented here. Difficult though it may be for certain decision-makers to accept, there is more than one rational approach to environmental policy and these 'competing rationalities' deserve (and will increasingly insist upon) a proper audience.[72]

The final point concerns the global/transboundary nature of environmental questions. It has been argued here that such a world-wide perspective is absolutely essential to any workable environmental management programme. In order for pollution consequences to be controlled and (hopefully) minimised, an effective international strategy is of paramount importance. This would include the establishment of international treaties but also the sharing of national expertise and experiences. However, as previously noted, such a strategy should also include a commitment from developed countries not only to provide an exemplar of good practice but also to offer practical (including economic) assistance to developing nations so that the requirements of industrial growth do not always collide with good environmental practice. One indicator of Western governments' genuine commitment to the natural environment will be their future willingness to listen not just to the voices of their own populations but also to those elsewhere who may be even less well-placed to enact real change.

ACKNOWLEDGEMENTS

The author acknowledges the help and support of the Science and Technology Policy group at the University of Manchester; gratitude is also expressed to the assistance and encouragement offered by Dave Gibbs, Jim Longhurst, Kim Webber and Steve Yearley.

REFERENCES AND NOTES

1. See, for example, Mitchell, R. C., Since *Silent Spring*: science, technology and the environmental movement in the United States. In *Scientific Expertise and the Public – Conference Proceedings*, ed. H. Skoie, Studies in Research and

Higher Education, The Norwegian Research Council for Science and the Humanities, Vol. 5, Oslo, 1979, pp. 171–207.
2. Lowe, P., Gathering greens. *Marxism Today*, Sept. (1989) 14–17.
3. Rawstorne, P., Emergence of 'the green consumer'. *Financial Times* 21 Apr. (1989) viii.
4. For example, one report includes that 78% of West German consumers have now switched to phosphate-free detergents. Button, J., Ecover, and over and over. . . . *Environment Now*, May (1989) 10.
5. See, for example, Elkington, J. & Hailes, J., *The Green Consumer Guide*. Victor Gollancz, London, UK, 1988.
6. MacKenzie, D. & Milne, R., A fresh Green tinge to Europe's research. *New Sci.*, 19 Aug. (1989) 23.
7. Two major speeches on this subject were given by Margaret Thatcher on 27 Sept.. 1988 (to the Royal Society) and on 14 Oct. 1988 (to the Conservative Party Conference).
8. For a discussion of this, see Yearley, S., Greens about science: the ambiguous attraction of scientific authority to the environmental movement. Mimeo, Department of Social Studies, Queen's University, Belfast, UK, July 1989.
9. For further discussion of this point, see Brown, J., ed., *Environmental Threats: perception, analysis and management*. Behaven Press, London, UK, 1989.
10. John Roberts, quoted in McCormick, J., *Acid Earth: the global threat of acid pollution*. Earthscan/International Institute for Environment and Development, London, UK, 1985, p. 6.
11. Cope, D., Fewer Giant Dinosaurs. *UK CEED Bull.*, July–Aug. (1990) 8.
12. Park, C. C., *Acid Rain: rhetoric and reality*. Methuen, London, UK, 1987, p. 8.
13. Quoted in Central Electricity Generating Board, *Acid Rain*. CEGB, London, UK, August 1984, p. 16.
14. Fourth Report from the Environment Committee, House of Commons, Session 1983–84, p. lxx.
15. Pearce, F., *Acid Rain*. Penguin, Harmondsworth, UK, 1987, p. 21.
16. Further references to this sociological literature are available in Irwin, A., Technical expertise and risk conflict: an institutional study of the British compulsory seat belt debate. *Policy Sci.*, **20** (1987) 339–64.
17. Mellanby, K., Acid comments on a cause celebre. *New Sci.*, 8 Nov. (1984) 37.
18. Cramer, J., *Mission-Orientation in Ecology: the case of Dutch fresh-water ecology*. Rodopi, Amsterdam, The Netherlands, 1987, pp. 49–52.
19. Kuhn, T. S., *The Structure of Scientific Revolutions*. University of Chicago Press, Chicago, IL, USA, 1970.
20. Yearley, S., *The Green Case*, Harper Collins, London, 1991.
21. Park, C. C., *op. cit.*, p. 63.
22. Collingridge, D. & Reeve, C., *Science speaks to Power: the role of experts in policymaking*. Frances Pinter, London, UK, 1986.
23. Graham, J. D., Green, L. C. & Roberts, M. J., *In Search of Safety: chemicals and cancer risk*. Harvard University Press, Cambridge, MA, USA, 1988, pp. 185–6.
24. This point is also discussed in Irwin, A., Deciding about risk: expert testimony and the regulation of hazard. In Brown, J., *op. cit.*, pp. 19–32.
25. Brundtland Report, Report of World Commission on Environment and

Development, *Our Common Future*, OUP, Oxford, UK, 1987. Quoted in Hutchison, R., The Greenhouse Debate. *Policy Studies*, **9**(4) (1989) 51.
26. Fourth Report from the Environment Committee, *Acid Rain*, Vol. II, Minutes of Evidence, House of Commons, 1983–84 Session, p. 91.
27. *Ibid.*, p. 92.
28. *Ibid.*
29. *Ibid.*, pp. 20–1.
30. *Greenpeace News*, Autumn (1989) 14.
31. Ravetz, J., Usable knowledge, usable ignorance: incomplete science with policy implications. In *Sustainable Development of the Biosphere*, ed. W. C. Clark & R. E. Munn. CUP/IIASA, Cambridge, UK, 1986, p. 417.
32. The following discussion focuses on the debate over power station emissions rather than that over other pollution sources (e.g. motor vehicles, petrochemical plants). This is partly for reasons of space but also due to the especially controversial nature of the CEGB case.
33. Park, C. C., *op. cit.*, p. 221.
34. McCormick, J., *op. cit.*, p. 56.
35. Royal Commission on Environmental Pollution, *Tenth Report: tackling pollution—experience and prospects*. HMSO, London, UK, Feb. 1984, p. 147.
36. Fourth Report from the Environment Committee, *op. cit.*, p. xi.
37. *Ibid.*, p. lxxi.
38. Waldegrave, W., House of Commons. Parliamentary Debates. Vol. 70, January 11, 1985, c. 1022.
39. Barnaby, F., Acid Rain: UK policies. *Ambio*, **17**(2) (1988) 162.
40. Roberts, A., House of Commons, Parliamentary Debates, Vol. 70, 11 January 1985, c. 1051.
41. Adan Markham, quoted in Agren, C., United Kingdom: At least making a start. *Acid News*, **3–4** (1986) 15. (*Acid News* is a newsletter published by the Swedish and Norwegian non-governmental secretariats on acid rain.)
42. *Acid News*, **4** (1987) 16.
43. Landin, B., Film exposed. *Acid Magazine*, Autumn (1986) 32.
44. *Ibid.*, p. 35.
45. DoE, Acid deposition and its effects. Sept. 1987, para. 51.
46. Steve Elsworth, quoted in *Acid News*, **2** (1988) 9.
47. Agren, C., United Kingdom—declining trend of emissions reversed. *Acid News*, **1** (1988) 11.
48. Quotations from DoE, *op. cit.*
49. Quoted in Pape, R., Large Combustors—directive gone astray. *Acid News*, **2** (1988) 15.
50. *Ibid.*
51. Park, C. C., *op. cit.*, p. 187.
52. Milne, R. & MacKenzie, D., Europe's worst polluter is tamed. *New Sci.*, 23 June (1988) 29.
53. *Financial Times*, 21 Apr. (1989) vii.
54. Friends of the Earth argued in 1989 that as a ten-year target at least an 80% SO_2 reduction is needed across Europe accompanied by a 75% cut in NO_x. (Secrett, C. & Porritt, J., *The Environment: the government's record*. Friends of the

Earth, London, UK, 1989.)

55. Thus, the press reported in 1989 that the electricity supply industry was planning to spend £1.9 billion on emission control equipment for its power stations. The CEGB was also planning to spend £180 million on clean coal technology. According to the Department of the Environment, this would achieve the reductions required by the EEC directive. (*The Independent*, 2 Sept. (1989) 2.)
56. *The Guardian*, 24 Feb. (1990), p. 7.
57. One commentator noted that Britain had by now little choice if it wished to avoid a re-run of the 9th century Viking invasion!
58. Speech to the Royal Society on 22 March 1990, quoted in Cope, D., *op. cit.*, p. 8.
59. Thompson, M., Commentary. In Clark, W. C. & Munn, R. E., *op. cit.*, p. 454.
60. 'Mainstream' is used here to suggest the bulk of domestic political issues as debated by the British parliament—for example, questions of social security, employment, public expenditure. Of course, it is recognised that the 'mainstream/environmental' distinction may conceal occasional areas of overlap or similarity.
61. *Acid Magazine*, **6** (1988) 37.
62. Haigh refers to some 278 multilateral treaties of interest to the protection of the environment signed between 1940 and 1983. (Haigh, N., In Clark, W. C. & Munn, R. E., *op. cit.*, p. 358.)
63. Majone, G., Sustainable institutions and the environment. In Clark, W. C. & Munn, R. E., *op. cit.*, pp. 351–8.
64. *Ibid.*, p. 352.
65. Pearce, D., Markandya, A. & Barbier, E. B., *Blueprint for a Green Economy.* Earthscan, London, UK, 1989.
66. '. . . attempting to control a technology is difficult, and not rarely impossible, because during its early stages, when it can be controlled, not enough can be known about its harmful social consequences to warrant controlling its development; but by the time these consequences are apparent, control has become costly and slow.' (Collingridge, D., *The Social Control of Technology.* Frances Pinter, London, UK, 1980, p. 19.)
67. Cope, D., *op. cit.*, p. 10.
68. Thus, a series of sociologists and anthropologists have examined the strong overlap between pollution beliefs and wider social and moral values. See, for example, Cotgrove, S. (Risk, value conflict and political legitimacy. In *Dealing with Risk,* ed. R. F. Griffiths. Manchester University Press, Manchester, UK, 1981, pp. 122–40.) and Douglas, M. & Wildavsky, A. (*Risk and Culture: an essay on the selection of technological and environmental dangers.* University of California Press, Berkeley, CA, USA, 1982.).
69. Sagoff, M., *The Economy of the Earth.* CUP, Cambridge, UK, 1988, p. 98.
70. Thus, an editorial in *Nature* observed (under the headline of 'Green politics make more trouble') that 'The widespread conversion of democratic governments to environmental causes is understandable: it keeps, or even catches, votes. But the consequences are increased costs and the tolerance of irrationality.' (*Nature*, **340** (1989) 489.)
71. Graham *et al.*, *op. cit.*, p. 198.
72. For a discussion of this point, see; James, P. and Thompson, M., The plural rationality approach. In Brown, J., *op. cit.*, pp. 87–94.

Index

GPSR Compliance

The European Union's (EU) General Product Safety Regulation (GPSR) is a set of rules that requires consumer products to be safe and our obligations to ensure this.

If you have any concerns about our products, you can contact us on ProductSafety@springernature.com

In case Publisher is established outside the EU, the EU authorized representative is:

Springer Nature Customer Service Center GmbH
Europaplatz 3
69115 Heidelberg, Germany

Batch number: 10809436

Printed by Printforce, the Netherlands